LONDON MATHEMATICAL SOCIETY LECTURE NOTE SERIES

Managing Editor: Professor Endre Süli, Mathematical Institute, Ur
6GG, United Kingdom

The titles below are available from booksellers, or from Cambridge
www.cambridge.org/mathematics

380 Real and complex singularities, M. MANOEL, M.C. ROME
381 Symmetries and integrability of difference equations, D. LE
P. WINTERNITZ (eds)
382 Forcing with random variables and proof complexity, J. KRAJÍČEK
383 Motivic integration and its interactions with model theory and non-Archimedean geometry I, R. CLUCKERS, J. NICAISE & J. SEBAG (eds)
384 Motivic integration and its interactions with model theory and non-Archimedean geometry II, R. CLUCKERS, J. NICAISE & J. SEBAG (eds)
385 Entropy of hidden Markov processes and connections to dynamical systems, B. MARCUS, K. PETERSEN & T. WEISSMAN (eds)
386 Independence-friendly logic, A.L. MANN, G. SANDU & M. SEVENSTER
387 Groups St Andrews 2009 in Bath I, C.M. CAMPBELL *et al* (eds)
388 Groups St Andrews 2009 in Bath II, C.M. CAMPBELL *et al* (eds)
389 Random fields on the sphere, D. MARINUCCI & G. PECCATI
390 Localization in periodic potentials, D.E. PELINOVSKY
391 Fusion systems in algebra and topology, M. ASCHBACHER, R. KESSAR & B. OLIVER
392 Surveys in combinatorics 2011, R. CHAPMAN (ed)
393 Non-abelian fundamental groups and Iwasawa theory, J. COATES *et al* (eds)
394 Variational problems in differential geometry, R. BIELAWSKI, K. HOUSTON & M. SPEIGHT (eds)
395 How groups grow, A. MANN
396 Arithmetic differential operators over the *p*-adic integers, C.C. RALPH & S.R. SIMANCA
397 Hyperbolic geometry and applications in quantum chaos and cosmology, J. BOLTE & F. STEINER (eds)
398 Mathematical models in contact mechanics, M. SOFONEA & A. MATEI
399 Circuit double cover of graphs, C.-Q. ZHANG
400 Dense sphere packings: a blueprint for formal proofs, T. HALES
401 A double Hall algebra approach to affine quantum Schur–Weyl theory, B. DENG, J. DU & Q. FU
402 Mathematical aspects of fluid mechanics, J.C. ROBINSON, J.L. RODRIGO & W. SADOWSKI (eds)
403 Foundations of computational mathematics, Budapest 2011, F. CUCKER, T. KRICK, A. PINKUS & A. SZANTO (eds)
404 Operator methods for boundary value problems, S. HASSI, H.S.V. DE SNOO & F.H. SZAFRANIEC (eds)
405 Torsors, étale homotopy and applications to rational points, A.N. SKOROBOGATOV (ed)
406 Appalachian set theory, J. CUMMINGS & E. SCHIMMERLING (eds)
407 The maximal subgroups of the low-dimensional finite classical groups, J.N. BRAY, D.F. HOLT & C.M. RONEY-DOUGAL
408 Complexity science: the Warwick master's course, R. BALL, V. KOLOKOLTSOV & R.S. MACKAY (eds)
409 Surveys in combinatorics 2013, S.R. BLACKBURN, S. GERKE & M. WILDON (eds)
410 Representation theory and harmonic analysis of wreath products of finite groups, T. CECCHERINI-SILBERSTEIN, F. SCARABOTTI & F. TOLLI
411 Moduli spaces, L. BRAMBILA-PAZ, O. GARCÍA-PRADA, P. NEWSTEAD & R.P. THOMAS (eds)
412 Automorphisms and equivalence relations in topological dynamics, D.B. ELLIS & R. ELLIS
413 Optimal transportation, Y. OLLIVIER, H. PAJOT & C. VILLANI (eds)
414 Automorphic forms and Galois representations I, F. DIAMOND, P.L. KASSAEI & M. KIM (eds)
415 Automorphic forms and Galois representations II, F. DIAMOND, P.L. KASSAEI & M. KIM (eds)
416 Reversibility in dynamics and group theory, A.G. O'FARRELL & I. SHORT
417 Recent advances in algebraic geometry, C.D. HACON, M. MUSTAŢĂ & M. POPA (eds)
418 The Bloch–Kato conjecture for the Riemann zeta function, J. COATES, A. RAGHURAM, A. SAIKIA & R. SUJATHA (eds)
419 The Cauchy problem for non-Lipschitz semi-linear parabolic partial differential equations, J.C. MEYER & D.J. NEEDHAM
420 Arithmetic and geometry, L. DIEULEFAIT *et al* (eds)
421 O-minimality and Diophantine geometry, G.O. JONES & A.J. WILKIE (eds)
422 Groups St Andrews 2013, C.M. CAMPBELL *et al* (eds)
423 Inequalities for graph eigenvalues, Z. STANIĆ
424 Surveys in combinatorics 2015, A. CZUMAJ *et al* (eds)
425 Geometry, topology and dynamics in negative curvature, C.S. ARAVINDA, F.T. FARRELL & J.-F. LAFONT (eds)
426 Lectures on the theory of water waves, T. BRIDGES, M. GROVES & D. NICHOLLS (eds)
427 Recent advances in Hodge theory, M. KERR & G. PEARLSTEIN (eds)
428 Geometry in a Fréchet context, C.T.J. DODSON, G. GALANIS & E. VASSILIOU
429 Sheaves and functions modulo *p*, L. TAELMAN
430 Recent progress in the theory of the Euler and Navier–Stokes equations, J.C. ROBINSON, J.L. RODRIGO, W. SADOWSKI & A. VIDAL-LÓPEZ (eds)
431 Harmonic and subharmonic function theory on the real hyperbolic ball, M. STOLL

432 Topics in graph automorphisms and reconstruction (2nd Edition), J. LAURI & R. SCAPELLATO
433 Regular and irregular holonomic D-modules, M. KASHIWARA & P. SCHAPIRA
434 Analytic semigroups and semilinear initial boundary value problems (2nd Edition), K. TAIRA
435 Graded rings and graded Grothendieck groups, R. HAZRAT
436 Groups, graphs and random walks, T. CECCHERINI-SILBERSTEIN, M. SALVATORI & E. SAVA-HUSS (eds)
437 Dynamics and analytic number theory, D. BADZIAHIN, A. GORODNIK & N. PEYERIMHOFF (eds)
438 Random walks and heat kernels on graphs, M.T. BARLOW
439 Evolution equations, K. AMMARI & S. GERBI (eds)
440 Surveys in combinatorics 2017, A. CLAESSON *et al* (eds)
441 Polynomials and the mod 2 Steenrod algebra I, G. WALKER & R.M.W. WOOD
442 Polynomials and the mod 2 Steenrod algebra II, G. WALKER & R.M.W. WOOD
443 Asymptotic analysis in general relativity, T. DAUDÉ, D. HÄFNER & J.-P. NICOLAS (eds)
444 Geometric and cohomological group theory, P.H. KROPHOLLER, I.J. LEARY, C. MARTÍNEZ-PÉREZ & B.E.A. NUCINKIS (eds)
445 Introduction to hidden semi-Markov models, J. VAN DER HOEK & R.J. ELLIOTT
446 Advances in two-dimensional homotopy and combinatorial group theory, W. METZLER & S. ROSEBROCK (eds)
447 New directions in locally compact groups, P.-E. CAPRACE & N. MONOD (eds)
448 Synthetic differential topology, M.C. BUNGE, F. GAGO & A.M. SAN LUIS
449 Permutation groups and cartesian decompositions, C.E. PRAEGER & C. SCHNEIDER
450 Partial differential equations arising from physics and geometry, M. BEN AYED *et al* (eds)
451 Topological methods in group theory, N. BROADDUS, M. DAVIS, J.-F. LAFONT & I. ORTIZ (eds)
452 Partial differential equations in fluid mechanics, C.L. FEFFERMAN, J.C. ROBINSON & J.L. RODRIGO (eds)
453 Stochastic stability of differential equations in abstract spaces, K. LIU
454 Beyond hyperbolicity, M. HAGEN, R. WEBB & H. WILTON (eds)
455 Groups St Andrews 2017 in Birmingham, C.M. CAMPBELL *et al* (eds)
456 Surveys in combinatorics 2019, A. LO, R. MYCROFT, G. PERARNAU & A. TREGLOWN (eds)
457 Shimura varieties, T. HAINES & M. HARRIS (eds)
458 Integrable systems and algebraic geometry I, R. DONAGI & T. SHASKA (eds)
459 Integrable systems and algebraic geometry II, R. DONAGI & T. SHASKA (eds)
460 Wigner-type theorems for Hilbert Grassmannians, M. PANKOV
461 Analysis and geometry on graphs and manifolds, M. KELLER, D. LENZ & R.K. WOJCIECHOWSKI
462 Zeta and L-functions of varieties and motives, B. KAHN
463 Differential geometry in the large, O. DEARRICOTT *et al* (eds)
464 Lectures on orthogonal polynomials and special functions, H.S. COHL & M.E.H. ISMAIL (eds)
465 Constrained Willmore surfaces, Á.C. QUINTINO
466 Invariance of modules under automorphisms of their envelopes and covers, A.K. SRIVASTAVA, A. TUGANBAEV & P.A.GUIL ASENSIO
467 The genesis of the Langlands program, J. MUELLER & F. SHAHIDI
468 (Co)end calculus, F. LOREGIAN
469 Computational cryptography, J.W. BOS & M. STAM (eds)
470 Surveys in combinatorics 2021, K.K. DABROWSKI *et al* (eds)
471 Matrix analysis and entrywise positivity preservers, A. KHARE
472 Facets of algebraic geometry I, P. ALUFFI *et al* (eds)
473 Facets of algebraic geometry II, P. ALUFFI *et al* (eds)
474 Equivariant topology and derived algebra, S. BALCHIN, D. BARNES, M. KĘDZIOREK & M. SZYMIK (eds)
475 Effective results and methods for Diophantine equations over finitely generated domains, J.-H. EVERTSE & K. GYŐRY
476 An indefinite excursion in operator theory, A. GHEONDEA
477 Elliptic regularity theory by approximation methods, E.A. PIMENTEL
478 Recent developments in algebraic geometry, H. ABBAN, G. BROWN, A. KASPRZYK & S. MORI (eds)
479 Bounded cohomology and simplicial volume, C. CAMPAGNOLO, F. FOURNIER-FACIO, N. HEUER & M. MORASCHINI (eds)
480 Stacks Project Expository Collection (SPEC), P. BELMANS, W. HO & A.J. DE JONG (eds)
481 Surveys in combinatorics 2022, A. NIXON & S. PRENDIVILLE (eds)
482 The logical approach to automatic sequences, J. SHALLIT
483 Rectifiability: a survey, P. MATTILA
484 Discrete quantum walks on graphs and digraphs, C. GODSIL & H. ZHAN
485 The Calabi problem for Fano threefolds, C. ARAUJO *et al*
486 Modern trends in algebra and representation theory, D. JORDAN, N. MAZZA & S. SCHROLL (eds)
487 Algebraic combinatorics and the Monster group, A.A. IVANOV (ed)
488 Maurer–Cartan methods in deformation theory, V. DOTSENKO, S. SHADRIN & B. VALLETTE
489 Künneth geometry, M.J.D. HAMILTON & D. KOTSCHICK

London Mathematical Society Lecture Note Series: 490

C^∞-Algebraic Geometry with Corners

KELLI FRANCIS-STAITE
University of Adelaide

DOMINIC JOYCE
University of Oxford

Shaftesbury Road, Cambridge CB2 8EA, United Kingdom

One Liberty Plaza, 20th Floor, New York, NY 10006, USA

477 Williamstown Road, Port Melbourne, VIC 3207, Australia

314–321, 3rd Floor, Plot 3, Splendor Forum, Jasola District Centre, New Delhi – 110025, India

103 Penang Road, #05-06/07, Visioncrest Commercial, Singapore 238467

Cambridge University Press is part of Cambridge University Press & Assessment, a department of the University of Cambridge.

We share the University's mission to contribute to society through the pursuit of education, learning and research at the highest international levels of excellence.

www.cambridge.org
Information on this title: www.cambridge.org/9781009400169

DOI: 10.1017/9781009400190

First published 2024

Printed in the United Kingdom by TJ Books Limited, Padstow, Cornwall

A catalogue record for this publication is available from the British Library

A Cataloging-in-Publication data record for this book is available from the Library of Congress

ISBN 978-1-009-40016-9 Paperback

Contents

1

Introduction

This book introduces and studies C^∞*-algebraic geometry with corners*. It extends the existing theory of C^∞-algebraic geometry [23, 42, 49, 52, 75], a version of Algebraic Geometry that generalizes smooth manifolds to a huge class of singular spaces. C^∞-algebraic geometry with corners is a generalization based on manifolds with corners rather than on manifolds. It is related to log geometry [78], as log structures are a kind of boundary in Algebraic Geometry.

The motivation for this book is two-fold. Firstly, it presents an introduction to C^∞-algebraic geometry with corners, along with many foundational results, in one self-contained volume. Secondly, it provides the foundations needed to extend current work on Derived Differential Geometry [6, 7, 8, 10, 11, 12, 13, 43, 44, 62, 88, 89, 87] to include derived C^∞-schemes and C^∞-stacks with corners, and hence derived manifolds and derived orbifolds with corners.

Derived orbifolds with corners have important applications in Symplectic Geometry, for example, in the most general versions of Lagrangian Floer cohomology [28, 29], Fukaya categories [3, 83], and Symplectic Field Theory [24]. In these areas one studies moduli spaces $\overline{\mathcal{M}}$ of J-holomorphic curves, which should be derived orbifolds with corners. (Related structures are the 'Kuranishi spaces with corners' of Fukaya–Oh–Ohta–Ono [29, 30] and 'polyfolds with corners' of Hofer–Wysocki–Zehnder [37, 38], but derived orbifolds with corners have better functoriality and are, we think, more beautiful.)

C^∞*-algebraic geometry* was originally suggested by William Lawvere in the late 1960's [58], and our primary reference is the second author's monograph [49]. In C^∞-algebraic geometry, rings are replaced with C^∞*-rings*, such as the space $C^\infty(X)$ of smooth functions $c : X \to \mathbb{R}$ of a manifold X. C^∞-rings are $\mathbb{R}$-algebras but with a far richer structure – for each smooth function $f : \mathbb{R}^n \to \mathbb{R}$ there is an n-fold operation $\Phi_f : C^\infty(X)^n \to C^\infty(X)$ acting by $\Phi_f : c_1, \ldots, c_n \mapsto f(c_1, \ldots, c_n)$, satisfying many natural identities.

The basic objects of C^∞-algebraic geometry are C^∞*-schemes*, which form a category $\mathbf{C^\infty Sch}$. They are special examples of (*local*) C^∞*-ringed spaces* $\underline{X} = (X, \mathcal{O}_X)$, a topological space X with a sheaf of C^∞-rings $\mathcal{O}_X$. These form categories $\mathbf{LC^\infty RS} \subset \mathbf{C^\infty RS}$. As in Algebraic Geometry, there is a *spectrum functor* $\mathrm{Spec} : \mathbf{C^\infty Rings}^{\mathrm{op}} \to \mathbf{LC^\infty RS}$, and a C^∞-scheme is a local C^∞-ringed space $\underline{X}$ covered by open subspaces $\underline{U} \subseteq \underline{X}$ with $\underline{U} \cong \mathrm{Spec}\,\mathfrak{C}$ for some $\mathfrak{C} \in \mathbf{C^\infty Rings}$. Any smooth manifold X determines a C^∞-scheme $\underline{X} = \mathrm{Spec}\, C^\infty(X)$, which is X with its sheaf of smooth functions $\mathcal{O}_X$, giving an embedding $\mathbf{Man} \subset \mathbf{C^\infty Sch}$ as a full subcategory.

C^∞-schemes are far more general than manifolds, and $\mathbf{C^\infty Sch}$ contains many singular or infinite-dimensional objects. In addition, the category of C^∞-schemes addresses several shortcomings of the category of smooth manifolds: it is Cartesian closed, and has all finite limits and directed colimits, unlike the category of manifolds. Thus all fibre products (not just transverse ones) exist in $\mathbf{C^\infty Sch}$, and many spaces constructed from fibre products of manifolds can be studied. These good properties are some of the reasons C^∞-algebraic geometry has been applied in the foundations of Synthetic Differential Geometry [23, 52, 73, 74, 75], and Derived Differential Geometry.

Manifolds with corners X are a generalization of manifolds locally modelled on $[0,\infty)^k \times \mathbb{R}^{m-k}$ rather than on $\mathbb{R}^m$. They occur in many places in Differential Geometry. There are several candidates for morphisms of manifolds with corners, but for reasons explained later we choose the 'b-maps' of Melrose [69, 70, 71, 72], which we just call 'smooth maps', to be the morphisms in the category $\mathbf{Man^c}$ of manifolds with corners.

A manifold with corners X of dimension n has a *boundary* ∂X, a manifold with corners of dimension $n-1$, with a (not necessarily injective) inclusion map $\Pi_X : \partial X \to X$. The *interior* is $X^\circ = X \setminus \Pi_X(\partial X)$, an ordinary n-manifold. A smooth map $f : X \to Y$ is called *interior* if $f(X^\circ) \subset Y^\circ$. We write $\mathbf{Man^c_{in}} \subset \mathbf{Man^c}$ for the subcategory with morphisms interior maps.

The symmetric group S_k acts freely on $\partial^k X$, and the quotient $\partial^k X/S_k$ is the k*-corners* $C_k(X)$, a manifold with corners of dimension $n-k$, with $C_0(X) = X$ and $C_1(X) = \partial X$. For example, the square $[0,1]^2$ has

$$\begin{aligned} &C_0\big([0,1]^2\big) = [0,1]^2, \qquad C_2\big([0,1]^2\big) = \{0,1\}^2, \\ &C_1\big([0,1]^2\big) = \partial\big([0,1]^2\big) = \big(\{0,1\} \times [0,1]\big) \amalg \big([0,1] \times \{0,1\}\big). \end{aligned}$$

The *corners* of X is $C(X) = \coprod_{k=0}^n C_k(X)$, a *manifold with corners of mixed dimension*, which form a category $\mathbf{\check{M}an^c}$. If $f : X \to Y$ is a smooth map of manifolds with corners (possibly of mixed dimension), we can define a natural interior map $C(f) : C(X) \to C(Y)$, which need not map $C_k(X) \to$

$C_k(Y)$. This defines a *corner functor* $C : \check{\mathbf{Man}}^{\mathbf{c}} \to \check{\mathbf{Man}}^{\mathbf{c}}_{\mathbf{in}}$, which is right adjoint to the inclusion inc : $\check{\mathbf{Man}}^{\mathbf{c}}_{\mathbf{in}} \hookrightarrow \check{\mathbf{Man}}^{\mathbf{c}}$. This means that not only do boundaries ∂X and corners $C_k(X)$ combine into a functorial object $C(X)$, but they are canonically determined just by the two notions of smooth map and interior map of manifolds with corners.

Manifolds with g-corners $\mathbf{Man^{gc}}$, as in [46] and §3.3, are a generalization of manifolds with corners whose local models X_P are more general than $[0,\infty)^k \times \mathbb{R}^{m-k}$, and depend on a weakly toric monoid P. They have nicer properties than manifolds with corners in some ways – for example, in the existence of b-transverse fibre products – and come up in some applications; for example, moduli spaces of 'quilts' in Symplectic Geometry [67, 68] may be manifolds with g-corners, but not manifolds with corners.

To extend C^∞-algebraic geometry to include manifolds with (g-)corners, we introduce the notion of a C^∞*-ring with corners* $\boldsymbol{\mathfrak{C}} = (\mathfrak{C}, \mathfrak{C}_{\rm ex})$, which is a pair consisting of a C^∞-ring $\mathfrak{C}$ and a commutative monoid $\mathfrak{C}_{\rm ex}$ with many intertwining relationships. If X is a manifold with corners, the corresponding C^∞-ring with corners is $\boldsymbol{C^\infty}(X) = (C^\infty(X), \mathrm{Ex}(X))$, where $C^\infty(X)$ is the C^∞-ring of smooth maps $X \to \mathbb{R}$, and $\mathrm{Ex}(X)$ is the monoid of smooth ('exterior') maps $X \to [0,\infty)$, with the monoid operation being multiplication. The monoid holds information about the corners structure of X.

The definition of C^∞-scheme with corners then follows those of schemes in Algebraic Geometry [36, §II], or C^∞-schemes. We introduce categories $\mathbf{LC^\infty RS^c} \subset \mathbf{C^\infty RS^c}$ of *(local)* C^∞*-ringed spaces with corners* $\boldsymbol{X} = (X, \boldsymbol{\mathcal{O}}_X)$, which are topological spaces X with sheaves of C^∞-ringed spaces with corners $\boldsymbol{\mathcal{O}}_X$. We construct a *spectrum functor* $\mathrm{Spec^c} : (\mathbf{C^\infty Rings^c})^{\rm op} \to \mathbf{LC^\infty RS^c}$, which is left adjoint to the global sections functor $\Gamma^{\rm c} : \mathbf{LC^\infty RS^c} \to (\mathbf{C^\infty Rings^c})^{\rm op}$. A C^∞*-scheme with corners* $\boldsymbol{X}$ is then a local C^∞-ringed space with corners which can be covered by open $\boldsymbol{U} \subseteq \boldsymbol{X}$ with $\boldsymbol{U} \cong \mathrm{Spec^c}\,\boldsymbol{\mathfrak{C}}$ for some C^∞-ring with corners $\boldsymbol{\mathfrak{C}}$.

We define many interesting subcategories of $\mathbf{C^\infty Sch^c}$, as in Figure 1.1, which reproduces a diagram (Figure 5.1) of subcategories of the category of C^∞-schemes with corners $\mathbf{C^\infty Sch^c}$ from §5.6. For example, *firm* C^∞-schemes with corners $\mathbf{C^\infty Sch^c_{fi}} \subset \mathbf{C^\infty Sch^c}$ have only finitely many boundary faces at each point. We define a subcategory $\mathbf{C^\infty Sch^c_{in}} \subset \mathbf{C^\infty Sch^c}$ of *interior* C^∞-schemes with corners, with *interior morphisms* corresponding to interior maps of manifolds with (g-)corners. We study categorical properties of $\mathbf{C^\infty Sch^c}$ and its subcategories. For example, fibre products and all finite limits exist in $\mathbf{C^\infty Sch^c_{fi}}$, and in most of the other subcategories in Figure 1.1.

We construct a *corner functor* $C : \mathbf{C^\infty Sch^c} \to \mathbf{C^\infty Sch^c_{in}}$, which is right adjoint to inc : $\mathbf{C^\infty Sch^c_{in}} \hookrightarrow \mathbf{C^\infty Sch^c}$. If $\boldsymbol{X}$ is a firm C^∞-scheme with

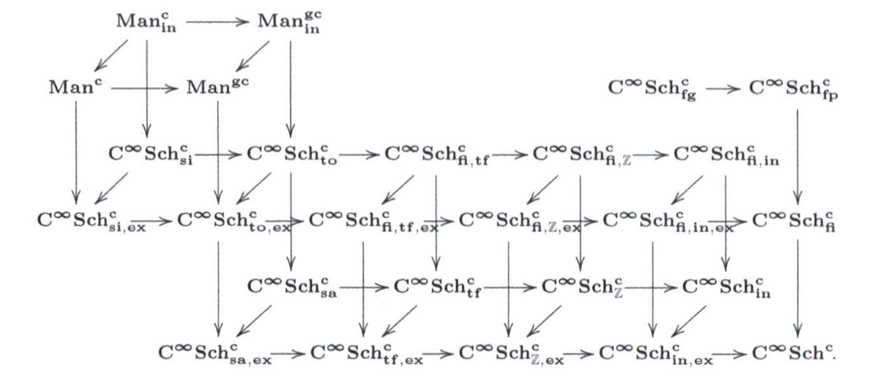

Figure 1.1 Subcategories of C^∞-schemes with corners $\mathbf{C^\infty Sch^c}$

corners we may write $C(\boldsymbol{X}) = \coprod_{k=0}^{\infty} C_k(\boldsymbol{X})$, where $C_k(\boldsymbol{X})$ is the *k-corners* of $\boldsymbol{X}$, with $C_0(\boldsymbol{X}) \cong \boldsymbol{X}$, and we define the *boundary* $\partial\boldsymbol{X} = C_1(\boldsymbol{X})$.

There are full and faithful inclusion functors $\mathbf{Man^c}, \mathbf{Man^{gc}} \hookrightarrow \mathbf{C^\infty Sch^c_{fi}}$ and $\mathbf{Man^c_{in}}, \mathbf{Man^{gc}_{in}} \hookrightarrow \mathbf{C^\infty Sch^c_{fi,in}}$, which commute with corner functors, boundaries ∂X, and k-corners $C_k(X)$, and preserve (b-)transverse fibre products. Thus, manifolds with (g-)corners can be regarded as special examples of C^∞-schemes with corners, and C^∞-schemes with corners generalize manifolds with (g-)corners.

The geometry of 'things with corners' turns out to be surprisingly interesting and complicated, and has received relatively little attention – for example, our observation that corners arise as right adjoints to inclusions like $\mathbf{Man^c_{in}} \hookrightarrow \mathbf{Man^c}$ appears to be new. We hope our book will inspire further research.

How this book came to be written

Beginning from foundational work of Lurie [62, §4.5] and Spivak [87], the second author has since 2009 been working on a theory of Derived Differential Geometry [43, 44], which is the study of *derived manifolds* and *derived orbifolds*. Here 'derived' is in the sense of Derived Algebraic Geometry [61, 62, 90, 91, 92]. This was aimed at applications in Symplectic Geometry, and is further explained in §2.9 and §8.4, though we summarize below.

Fukaya, Oh, Ohta and Ono [28, 29, 30, 31] were developing theories of Gromov–Witten invariants, Lagrangian Floer cohomology, and Fukaya categories, involving 'Kuranishi spaces', a geometric structure they put on moduli spaces of J-holomorphic curves. At the time there were problems with the definition and theory of Kuranishi spaces. The second author's view was that

Kuranishi spaces are actually derived orbifolds, and ideas from derived geometry were needed to understand them. Following [62, 87], he defined derived manifolds and derived orbifolds as special kinds of derived C^∞-schemes and derived C^∞-stacks.

For more advanced applications in Symplectic Geometry [28, 29], it was essential to have a theory of derived orbifolds *with corners*. To do this properly in the world of C^∞-algebraic geometry, it was necessary to go right back to the beginning, and introduce notions of C^∞-ring with corners and C^∞-scheme with corners, before defining derived C^∞-schemes and C^∞-stacks with corners.

The first attempt at this was the 2014 MSc thesis of Elana Kalashnikov [51], supervised by the second author, which defined notions of C^∞-rings with corners and C^∞-schemes with corners. Building on [51], the first author's 2019 PhD thesis [25] went into the subject in much more detail, again supervised by the second author.

One of the goals of the PhD was to find the best foundations for C^∞-algebraic geometry with corners, and its possible future applications, and the authors explored many options before settling on the definitions given here. This book is a rewritten and expanded version of [25]. It is designed in particular to provide the foundations of theories of 'derived C^∞-schemes with corners' and 'derived manifolds and orbifolds with corners' in the final version of the second author's book [44].

Why we chose these definitions

In Chapters 3 and 4 we have chosen particular definitions of the notions of smooth map of manifolds with corners, and of C^∞-ring with corners, which determine the course of the rest of the theory. The reader may wonder whether these choices could have been otherwise, leading to a different way of algebraizing 'things with corners', and a different theory of C^∞-algebraic geometry with corners. Our answer to this comes in two parts.

For the first part, knowing C^∞-algebraic geometry, the rough starting point for a theory of C^∞-algebraic geometry with corners is obvious. As in §2.2, a *categorical C^∞-ring* is a product-preserving functor $F : \mathbf{Euc} \to \mathbf{Sets}$, where $\mathbf{Euc} \subset \mathbf{Man}$ is the full subcategory of Euclidean spaces $\mathbb{R}^n$, and a C^∞-ring $\mathfrak{C}$ is an equivalent way of repackaging the data in F, with $\mathfrak{C} = F(\mathbb{R})$, and many operations on $\mathfrak{C}$. Here $\mathbf{Euc}$ is an Algebraic Theory [1], and a (categorical) C^∞-ring is an algebra over this Algebraic Theory.

Thus it is clear that a *categorical C^∞-ring with corners* should be a product-preserving functor $F : \mathbf{Euc^c} \to \mathbf{Sets}$, where $\mathbf{Euc^c} \subset \mathbf{Man^c}$ is the full subcategory of Euclidean corner spaces $\mathbb{R}^m_k = [0,\infty)^k \times \mathbb{R}^{m-k}$. Then a C^∞-*ring with corners* $\mathfrak{C} = (\mathfrak{C}, \mathfrak{C}_{\rm ex})$ should be an equivalent way of repackaging F, with $\mathfrak{C} = F(\mathbb{R})$ and $\mathfrak{C}_{\rm ex} = F([0,\infty))$, and many operations on $\mathfrak{C}, \mathfrak{C}_{\rm ex}$.

There are possible choices in two places in this definition. Firstly, how should we define morphisms in $\mathbf{Man^c}$ and $\mathbf{Euc^c}$? (As we will see in §3.1, there are many different classes of 'smooth maps' between manifolds with corners.) And secondly, do we want to impose any additional conditions on the functors F?

For the first, in order for 'product-preserving' to make sense, the class of morphisms of manifolds with corners X must be closed under products and direct products. Also, for good properties of boundaries ∂X we want the inclusion $\Pi_X : \partial X \to X$ to be a morphism in $\mathbf{Man^c}$ (this excludes interior morphisms, for example). With these restrictions, there are only really three sensible choices for the morphisms in $\mathbf{Man^c}$, which are called 'weakly smooth', 'smooth' (or 'b-maps'), and 'strongly smooth' in §3.1, with

$$\{\text{weakly smooth}\} \subset \{\text{smooth}\} \subset \{\text{strongly smooth}\}.$$

We rejected 'weakly smooth' maps, as they would cause C^∞-schemes with corners not to have well-behaved notions of boundary and corners. We found that smooth maps led to a richer and more interesting theory than strongly smooth maps, and the latter did not work for all the applications we had in mind (e.g., it does not correctly model manifolds with g-corners in §3.3). Also the strongly smooth theory is strictly contained in the smooth theory. So we decided on smooth maps as the morphisms in $\mathbf{Man^c}$ and $\mathbf{Euc^c}$.

For the second, we impose an extra condition for $\mathfrak{C} = (\mathfrak{C}, \mathfrak{C}_{\rm ex})$ coming from a product-preserving $F : \mathbf{Euc^c} \to \mathbf{Sets}$ to be a C^∞-ring with corners, that invertible functions in $\mathfrak{C}_{\rm ex}$ should have logs in $\mathfrak{C}$. To motivate this, note that if X is a manifold with corners and $f : X \to [0,\infty)$ is smooth with inverse $1/f$ then f maps $X \to (0,\infty)$, and so $\log f : X \to \mathbb{R}$ is well defined and smooth.

The second part of our answer is that we have defined not one theory of C^∞-rings and schemes with corners, but many, as in Figure 1.1; the overarching definitions of C^∞-rings and schemes with corners are the most general, which contain all the rest. General C^∞-schemes with corners $\boldsymbol{X} \in \mathbf{C^\infty Sch^c}$ may have 'corner behaviour' which is complicated and pathological, for example, $\boldsymbol{X} = [0,\infty)^{\mathbb{N}}$ has infinitely many boundary faces at a point. The subcategories of $\mathbf{C^\infty Sch^c}$ in Figure 1.1 have 'corner behaviour' which is progressively nicer, and more like ordinary manifolds with corners, as we impose more conditions.

The smallest categories $\mathbf{C^\infty Sch^c_{si}}, \mathbf{C^\infty Sch^c_{si,ex}}$, with 'simple corners', behave exactly like manifolds with corners, and $\mathbf{C^\infty Sch^c_{to}}, \mathbf{C^\infty Sch^c_{to,ex}}$, with 'toric corners', behave exactly like manifolds with g-corners. Different subcategories may be appropriate for different applications.

In the rest of this introduction we summarize the contents of the chapters.

Chapter 2: C^∞-rings and C^∞-schemes

We begin with background on category theory, and on C^∞-rings and C^∞-schemes, mostly following the second author [49]. A *categorical C^∞-ring* is a product-preserving functor $F : \mathbf{Euc} \to \mathbf{Sets}$, where $\mathbf{Euc} \subset \mathbf{Man}$ is the full subcategory of the category of smooth manifolds $\mathbf{Man}$ with objects the Euclidean spaces $\mathbb{R}^n$ for $n \geqslant 0$. These form a category $\mathbf{CC^\infty Rings}$, with morphisms natural transformations, which is equivalent to the category $\mathbf{C^\infty Rings}$ of C^∞-rings by mapping $F \mapsto \mathfrak{C} = F(\mathbb{R})$.

Here a *C^∞-ring* is data $\big(\mathfrak{C}, (\Phi_f)_{f:\mathbb{R}^n\to\mathbb{R}\ C^\infty}\big)$, where $\mathfrak{C}$ is a set and $\Phi_f : \mathfrak{C}^n \to \mathfrak{C}$ a map for all smooth maps $f : \mathbb{R}^n \to \mathbb{R}$, satisfying some axioms. Usually we write this as $\mathfrak{C}$, leaving the C^∞-operations Φ_f implicit. The motivating examples for X a smooth manifold are $\mathfrak{C} = C^\infty(X)$, the set of smooth functions $c : X \to \mathbb{R}$, with operations $\Phi_f(c_1, \ldots, c_n)(x) = f(c_1(x), \ldots, c_n(x))$. We discuss *$\mathfrak{C}$-modules*, which are just modules over $\mathfrak{C}$ considered as an $\mathbb{R}$-algebra, and the *cotangent module* $\Omega_{\mathfrak{C}}$, generalizing $\Gamma^\infty(T^*X)$, the vector space of smooth sections of T^*X, as a module over $C^\infty(X)$.

We then define the subcategory $\mathbf{C^\infty RS}$ of *C^∞-ringed spaces* $\underline{X} = (X, \mathcal{O}_X)$, where X is a topological space and $\mathcal{O}_X$ is a sheaf of C^∞-rings on X, and the subcategory $\mathbf{LC^\infty RS} \subset \mathbf{C^\infty RS}$ of *local C^∞-ringed spaces* for which the stalks $\mathcal{O}_{X,x}$ are local C^∞-rings. The global sections functor $\Gamma : \mathbf{LC^\infty RS} \to \mathbf{C^\infty Rings}^{\mathrm{op}}$ has a right adjoint $\mathrm{Spec} : \mathbf{C^\infty Rings}^{\mathrm{op}} \to \mathbf{LC^\infty RS}$, the *spectrum functor*. An object $\underline{X} \in \mathbf{LC^\infty RS}$ is called a *C^∞-scheme* if we may cover X with open $U \subseteq X$ with $(U, \mathcal{O}_X|_U) \cong \mathrm{Spec}\,\mathfrak{C}$ for $\mathfrak{C} \in \mathbf{C^\infty Rings}$. These form a full subcategory $\mathbf{C^\infty Sch} \subset \mathbf{LC^\infty RS}$.

There is a full embedding $\mathbf{Man} \hookrightarrow \mathbf{C^\infty Sch}$ mapping $X \mapsto \underline{X} = (X, \mathcal{O}_X)$, where $\mathcal{O}_X$ is the sheaf of local smooth functions $c : X \to \mathbb{R}$.

In classical Algebraic Geometry $\Gamma \circ \mathrm{Spec} \cong \mathrm{Id} : \mathbf{Rings}^{\mathrm{op}} \to \mathbf{Rings}^{\mathrm{op}}$, so $\mathbf{Rings}^{\mathrm{op}}$ is equivalent to the category $\mathbf{ASch}$ of affine schemes. In C^∞-algebraic geometry it is not true that $\Gamma \circ \mathrm{Spec} \cong \mathrm{Id}$, but we do have $\mathrm{Spec} \circ \Gamma \circ \mathrm{Spec} \cong \mathrm{Spec}$. Define a C^∞-ring $\mathfrak{C}$ to be *complete* if $\mathfrak{C} \cong \Gamma \circ \mathrm{Spec}\,\mathfrak{D}$ for some $\mathfrak{D}$. Write $\mathbf{C^\infty Rings_{co}} \subset \mathbf{C^\infty Rings}$ for the full subcategory of complete C^∞-rings. Then $(\mathbf{C^\infty Rings_{co}})^{\mathrm{op}}$ is equivalent to the category $\mathbf{AC^\infty Sch}$ of affine C^∞-schemes.

We define *$\mathcal{O}_X$-modules* on a C^∞-scheme $\underline{X}$, and the *cotangent sheaf* $T^*\underline{X}$, which generalizes cotangent bundles of manifolds. We review applications of C^∞-algebraic geometry in Synthetic Differential Geometry and Derived Differential Geometry.

Chapter 3: Manifolds with (g-)corners

We discuss *manifolds with corners*, locally modelled on $[0,\infty)^k \times \mathbb{R}^{m-k}$. The definition of manifold with corners X, Y is an obvious generalization of the definition of manifold. However, the definition of smooth map $f : X \to Y$ is *not* obvious: what conditions should we impose on f near the boundary and corners of X, Y? There are several non-equivalent definitions of morphisms of manifolds with corners in the literature. The one we choose is due to Melrose [71, §1.12], [55, §1], who calls them *b-maps*. This gives a category $\mathbf{Man^c}$.

A smooth map $f : X \to Y$ is called *interior* if $f(X^\circ) \subseteq Y^\circ$, where X° is the *interior* of X (that is, if X is locally modelled on $[0,\infty)^k \times \mathbb{R}^{m-k}$, then X° is locally modelled on $(0,\infty)^k \times \mathbb{R}^{m-k}$). Write $\mathbf{Man^c_{in}} \subset \mathbf{Man^c}$ for the subcategory with only interior morphisms.

A manifold with corners X has a *boundary* ∂X, a manifold with corners of dimension $\dim X - 1$, with a (non-injective) morphism $\Pi_X : \partial X \to X$. For example, if $X = [0,\infty)^2$, then ∂X is the disjoint union $(\{0\} \times [0,\infty)) \amalg ([0,\infty) \times \{0\})$ with the obvious map Π_X, so two points in ∂X lie over $(0,0) \in X$. There is a natural, free action of the symmetric group S_k on the kth boundary $\partial^k X$, permuting the boundary strata. The *k-corners* is $C_k(X) = \partial^k X / S_k$, of dimension $\dim X - k$. The *corners* of X is $C(X) = \coprod_{k=0}^{\dim X} C_k(X)$, an object in the category $\check{\mathbf{M}}\mathbf{an^c}$ of *manifolds with corners of mixed dimension.*

We call X a *manifold with faces* if $\Pi_X|_F : F \to X$ is injective for each connected component F of ∂X (called a *face*).

Now boundaries are not functorial: if $f : X \to Y$ is smooth, there is generally no natural morphism $\partial f : \partial X \to \partial Y$. However, there is a natural, interior morphism $C(f) : C(X) \to C(Y)$, giving the *corner functor* $C : \mathbf{Man^c} \to \check{\mathbf{M}}\mathbf{an^c_{in}}$ or $C : \check{\mathbf{M}}\mathbf{an^c} \to \check{\mathbf{M}}\mathbf{an^c_{in}}$. We explain that C is right adjoint to the inclusion $\mathrm{inc} : \check{\mathbf{M}}\mathbf{an^c_{in}} \hookrightarrow \check{\mathbf{M}}\mathbf{an^c}$. Thus, from a categorical point of view, boundaries and corners in $\mathbf{Man^c}$ are determined uniquely by the inclusion $\mathbf{Man^c_{in}} \hookrightarrow \mathbf{Man^c}$, that is, by the comparison between interior and smooth maps.

A manifold with corners X has two notions of (co)tangent bundle: the *tangent bundle* TX and its dual the *cotangent bundle* T^*X, and the *b-tangent bundle* ${}^bTX$ and its dual the *b-cotangent bundle* ${}^bT^*X$. Here TX is the obvious notion of tangent bundle. There is a morphism $I_X : {}^bTX \to TX$ which identifies $\Gamma^\infty({}^bTX)$ with the subspace of vector fields $v \in \Gamma^\infty(TX)$ which are tangent to every boundary stratum of X. Tangent bundles are functorial over smooth maps $f : X \to Y$ (that is, f lifts to $Tf : TX \to TY$ linear on the fibres). B-tangent bundles are functorial over interior maps $f : X \to Y$.

We also discuss the categories $\mathbf{Man^{gc}_{in}} \subset \mathbf{Man^{gc}}$ of *manifolds with gener-*

alized corners, or *manifolds with g-corners*, introduced by the second author in [46]. These allow more-complicated local models than $[0,\infty)^k \times \mathbb{R}^{m-k}$. The local models X_P are parametrized by *weakly toric monoids* P, with $X_P = [0,\infty)^k \times \mathbb{R}^{m-k}$ when $P = \mathbb{N}^k \times \mathbb{Z}^{m-k}$. So we provide an introduction to the theory of monoids. (All monoids in this book are commutative.)

The simplest manifold with g-corners which is not a manifold with corners is $X = \{(w,x,y,z) \in [0,\infty)^4 : wx = yz\}$. Most of the theory above extends to manifolds with g-corners, except that tangent bundles TX are not well-behaved, and we no longer have $C_k(X) \cong \partial^k X/S_k$.

We call morphisms $g : X \to Z$, $h : Y \to Z$ in $\mathbf{Man^c_{in}}$ or $\mathbf{Man^{gc}_{in}}$ *b-transverse* if ${}^bT_xg \oplus {}^bT_yh : {}^bT_xX \oplus {}^bT_yY \to {}^bT_zZ$ is surjective for all $x \in X$, $y \in Y$ with $g(x) = h(y) = z$. Manifolds with g-corners have the nice property that all b-transverse fibre products exist in $\mathbf{Man^{gc}_{in}}$, whereas fibre products only exist in $\mathbf{Man^c_{in}}, \mathbf{Man^c}$ under rather restrictive conditions. We can think of $\mathbf{Man^{gc}_{in}}$ as a kind of closure of $\mathbf{Man^c_{in}}$ under b-transverse fibre products.

We discuss applications of manifolds with (g-)corners in the literature, including in Topological Quantum Field Theories, analysis of partial differential equations, and moduli problems in Morse homology and Floer theories whose moduli spaces are manifolds with (g-)corners.

Chapter 4: (Pre) C^∞-rings with corners

We introduce *C^∞-rings with corners*. To decide on the correct definition, the obvious starting point is categorical C^∞-rings, that is, product-preserving functors $F : \mathbf{Euc} \to \mathbf{Sets}$. We define a *categorical pre C^∞-ring with corners* to be a product-preserving functor $F : \mathbf{Euc^c} \to \mathbf{Sets}$, where $\mathbf{Euc^c} \subset \mathbf{Man^c}$ is the full subcategory with objects $[0,\infty)^m \times \mathbb{R}^n$ for $m, n \geqslant 0$. These form a category $\mathbf{CPC^\infty Rings^c}$, with morphisms natural transformations.

Then we define the category $\mathbf{PC^\infty Rings^c}$ of *pre C^∞-rings with corners*, with objects $\big(\mathfrak{C}, \mathfrak{C}_{\rm ex}, (\Phi_f)_{f:\mathbb{R}^m\times[0,\infty)^n\to\mathbb{R}\ C^\infty}, (\Psi_g)_{g:\mathbb{R}^m\times[0,\infty)^n\to[0,\infty)\ C^\infty}\big)$ for $\mathfrak{C}, \mathfrak{C}_{\rm ex}$ sets and $\Phi_f : \mathfrak{C}^m \times \mathfrak{C}_{\rm ex}^n \to \mathfrak{C}$, $\Psi_g : \mathfrak{C}^m \times \mathfrak{C}_{\rm ex}^n \to \mathfrak{C}_{\rm ex}$ maps, satisfying some axioms. Usually we write this as $\boldsymbol{\mathfrak{C}} = (\mathfrak{C}, \mathfrak{C}_{\rm ex})$, leaving the C^∞-operations Φ_f, Ψ_g implicit. There is an equivalence $\mathbf{CPC^\infty Rings^c} \to \mathbf{PC^\infty Rings^c}$ mapping $F \mapsto (\mathfrak{C}, \mathfrak{C}_{\rm ex}) = (F(\mathbb{R}), F([0,\infty)))$. The motivating example, for X a manifold with (g-)corners, is $(\mathfrak{C}, \mathfrak{C}_{\rm ex}) = (C^\infty(X), \mathrm{Ex}(X))$, where $C^\infty(X)$ is the set of smooth maps $c : X \to \mathbb{R}$, and $\mathrm{Ex}(X)$ the set of *exterior* (i.e. smooth, but not necessarily interior) maps $c' : X \to [0,\infty)$.

We define the category of *C^∞-rings with corners* $\mathbf{C^\infty Rings^c}$ to be the full subcategory of $(\mathfrak{C}, \mathfrak{C}_{\rm ex})$ in $\mathbf{PC^\infty Rings^c}$ satisfying an extra condition, that if $c' \in \mathfrak{C}_{\rm ex}$ is invertible in $\mathfrak{C}_{\rm ex}$ then $c' = \Psi_{\exp}(c)$ for some $c \in \mathfrak{C}$. In terms of spaces X, this says that if $c' : X \to [0,\infty)$ is smooth and invertible (hence positive) then $c' = \exp c$ for some smooth $c = \log c' : X \to \mathbb{R}$.

Now we could also have started with $\mathbf{Man^c_{in}}$. So we write $\mathbf{Euc^c_{in}} \subset \mathbf{Man^c_{in}}$ for the full subcategory with objects $[0,\infty)^m \times \mathbb{R}^n$, and define the category $\mathbf{CPC^\infty Rings^c_{in}}$ of *categorical interior pre C^∞-rings with corners* to be product-preserving functors $F : \mathbf{Euc^c_{in}} \to \mathbf{Sets}$, with morphisms natural transformations. We define a subcategory $\mathbf{PC^\infty Rings^c_{in}} \subset \mathbf{PC^\infty Rings^c}$ of *interior pre C^∞-rings with corners*, and an equivalence $\mathbf{CPC^\infty Rings^c_{in}} \to \mathbf{PC^\infty Rings^c_{in}}$ mapping $F \mapsto (\mathfrak{C}, \mathfrak{C}_{\rm ex}) = (F(\mathbb{R}), F([0,\infty)) \amalg \{0\})$. The category $\mathbf{C^\infty Rings^c_{in}}$ of *interior C^∞-rings with corners* is the intersection $\mathbf{PC^\infty Rings^c_{in}} \cap \mathbf{C^\infty Rings^c}$ in $\mathbf{PC^\infty Rings^c}$. The motivating example, for X a manifold with (g-)corners, is $(\mathfrak{C}, \mathfrak{C}_{\rm ex}) = (C^\infty(X), \mathrm{In}(X) \amalg \{0\})$, for $\mathrm{In}(X)$ the interior maps $c' : X \to [0,\infty)$.

Although a C^∞-ring with corners $\boldsymbol{\mathfrak{C}} = (\mathfrak{C}, \mathfrak{C}_{\rm ex})$ has a huge number of C^∞-operations Φ_f, Ψ_g, it is helpful much of the time to focus on a small subset of these, giving $\mathfrak{C}, \mathfrak{C}_{\rm ex}$ smaller, more-manageable structures. In particular, we often think of $\mathfrak{C}$ as an $\mathbb{R}$-*algebra*, with addition and multiplication given by $\Phi_{f_+}, \Phi_{f_\cdot} : \mathfrak{C} \times \mathfrak{C} \to \mathfrak{C}$ for $f_+, f_\cdot : \mathbb{R}^2 \to \mathbb{R}$ given by $f_+(x,y) = x+y$, $f_\cdot(x,y) = xy$. And we think of $\mathfrak{C}_{\rm ex}$ as a *monoid*, with multiplication given by $\Psi_{g_\cdot} : \mathfrak{C}_{\rm ex} \times \mathfrak{C}_{\rm ex} \to \mathfrak{C}_{\rm ex}$ for $g_\cdot : [0,\infty)^2 \to [0,\infty)$ given by $g_\cdot(x,y) = xy$. Note that monoids also control the corner structure of manifolds with g-corners, in the same way.

We define various important subcategories of C^∞-rings with corners by imposing conditions on $\mathfrak{C}_{\rm ex}$ as a monoid. For example, a C^∞-ring with corners $\boldsymbol{\mathfrak{C}} = (\mathfrak{C}, \mathfrak{C}_{\rm ex})$ is *interior* if $\mathfrak{C}_{\rm ex} = \mathfrak{C}_{\rm in} \amalg \{0\}$ where $\mathfrak{C}_{\rm in}$ is a submonoid of $\mathfrak{C}_{\rm ex}$, that is, $\mathfrak{C}_{\rm ex}$ has no zero divisors. In terms of spaces X, we interpret $\mathfrak{C}_{\rm in}$ as the monoid $\mathrm{In}(X)$ of interior maps $c' : X \to [0,\infty)$. If $\boldsymbol{\mathfrak{C}}, \boldsymbol{\mathfrak{D}}$ are interior, a morphism $\boldsymbol{\phi} = (\phi, \phi_{\rm ex}) : \boldsymbol{\mathfrak{C}} \to \boldsymbol{\mathfrak{D}}$ is *interior* if $\phi_{\rm ex} : \mathfrak{C}_{\rm ex} \to \mathfrak{D}_{\rm ex}$ maps $\mathfrak{C}_{\rm in} \to \mathfrak{D}_{\rm in}$. The subcategory $\mathbf{C^\infty Sch^c_{fi}} \subset \mathbf{C^\infty Sch^c}$ of *firm* C^∞-rings with corners are those whose sharpening $\mathfrak{C}^\sharp_{\rm ex}$ is finitely generated.

Chapter 5: C^∞-schemes with corners

We define the category $\mathbf{C^\infty RS^c}$ of *C^∞-ringed spaces with corners* $\boldsymbol{X} = (X, \boldsymbol{\mathcal{O}}_X)$, where X is a topological space and $\boldsymbol{\mathcal{O}}_X = (\mathcal{O}_X, \mathcal{O}_X^{\rm ex})$ is a sheaf of C^∞-rings with corners on X, and the subcategory $\mathbf{LC^\infty RS^c} \subset \mathbf{C^\infty RS^c}$ of *local C^∞-ringed spaces with corners* for which the stalks $\boldsymbol{\mathcal{O}}_{X,x}$ for $x \in X$ are local C^∞-rings with corners. The global sections functor $\Gamma^c : \mathbf{LC^\infty RS^c} \to (\mathbf{C^\infty Rings^c})^{\rm op}$ has a right adjoint $\mathrm{Spec^c} : (\mathbf{C^\infty Rings^c})^{\rm op} \to \mathbf{LC^\infty RS^c}$, the *spectrum functor*. An object $\boldsymbol{X}$ in $\mathbf{LC^\infty RS^c}$ is called a *C^∞-scheme with* *corners* if we may cover X with open $U \subseteq X$ with $(U, \mathcal{O}_X|_U) \cong \mathrm{Spec^c}\,\mathfrak{C}$ for $\mathfrak{C}$ in $\mathbf{C^\infty Rings^c}$. These form a full subcategory $\mathbf{C^\infty Sch^c} \subset \mathbf{LC^\infty RS^c}$.

One might expect that to define corresponding subcategories $\mathbf{C^\infty RS^c_{in}} \subset \mathbf{C^\infty RS^c}, \mathbf{LC^\infty RS^c_{in}} \subset \mathbf{LC^\infty RS^c}$ of *interior* (local) C^∞-ringed spaces with corners, we should just replace $\mathbf{C^\infty Rings^c}$ by $\mathbf{C^\infty Rings^c_{in}} \subset \mathbf{C^\infty Rings^c}$. However, as $\mathrm{inc} : \mathbf{C^\infty Rings^c_{in}} \hookrightarrow \mathbf{C^\infty Rings^c}$ does not preserve limits, a sheaf of interior C^∞-rings with corners is only a presheaf of C^∞-rings with corners. So we define $\boldsymbol{X} = (X, \mathcal{O}_X) \in \mathbf{C^\infty RS^c}$ to be *interior* if $\mathcal{O}_X$ is the sheafification, as a sheaf valued in $\mathbf{C^\infty Rings^c}$, of a sheaf valued in $\mathbf{C^\infty Rings^c_{in}}$. Then the stalks $\mathcal{O}_{X,x}$ of $\mathcal{O}_X$ lie in $\mathbf{C^\infty Rings^c_{in}}$, so $\mathcal{O}^{\rm ex}_{X,x} = \mathcal{O}^{\rm in}_{X,x} \amalg \{0\}$.

We define $\Gamma^{\rm c}_{\rm in} : \mathbf{LC^\infty RS^c_{in}} \to (\mathbf{C^\infty Rings^c_{in}})^{\rm op}$ by $\Gamma^{\rm c}_{\rm in}(\boldsymbol{X}) = (\Gamma(\mathcal{O}_X), \Gamma(\mathcal{O}^{\rm in}_X) \amalg \{0\})$, where $\Gamma(\mathcal{O}^{\rm in}_X) \subset \Gamma(\mathcal{O}^{\rm ex}_X)$ is the subset of sections s with $s|_x \in \mathcal{O}^{\rm in}_{X,x} \subset \mathcal{O}^{\rm ex}_{X,x}$ for each $x \in X$. Then $\Gamma^{\rm c}_{\rm in}$ has a right adjoint $\mathrm{Spec^c_{in}} : (\mathbf{C^\infty Rings^c_{in}})^{\rm op} \to \mathbf{LC^\infty RS^c_{in}}$, where $\mathrm{Spec^c_{in}} = \mathrm{Spec^c}|_{(\mathbf{C^\infty Rings^c_{in}})^{\rm op}}$. An object $\boldsymbol{X} \in \mathbf{LC^\infty RS^c_{in}}$ is called an *interior C^∞-scheme with corners* if we may cover X with open $U \subseteq X$ with $(U, \mathcal{O}_X|_U) \cong \mathrm{Spec^c_{in}}\,\mathfrak{C}$ for $\mathfrak{C} \in \mathbf{C^\infty Rings^c_{in}}$. These form a full subcategory $\mathbf{C^\infty Sch^c_{in}} \subset \mathbf{LC^\infty RS^c_{in}}$, with $\mathbf{C^\infty Sch^c_{in}} \subset \mathbf{C^\infty Sch^c}$.

We define full embeddings $\mathbf{Man^{gc}} \hookrightarrow \mathbf{C^\infty Sch^c}$ or $\mathbf{Man^{gc}_{in}} \hookrightarrow \mathbf{C^\infty Sch^c_{in}}$ mapping $X \mapsto (X, (\mathcal{O}_X, \mathcal{O}^{\rm ex}_X))$ or $X \mapsto (X, (\mathcal{O}_X, \mathcal{O}^{\rm in}_X \amalg \{0\}))$, where $\mathcal{O}_X$, $\mathcal{O}^{\rm ex}_X, \mathcal{O}^{\rm in}_X$ are the sheaves of local smooth functions $X \to \mathbb{R}$, exterior functions $X \to [0,\infty)$, and interior functions $X \to [0,\infty)$, respectively. Thus, manifolds with (g-)corners may be regarded as special examples of C^∞-schemes with corners. Manifolds with (g-)faces map to *affine* C^∞-schemes with corners.

For C^∞-schemes, $\Gamma \circ \mathrm{Spec} \not\cong \mathrm{Id}$ but $\mathrm{Spec} \circ \Gamma \circ \mathrm{Spec} \cong \mathrm{Spec}$. Thus we can define a full subcategory $\mathbf{C^\infty Rings_{co}} \subset \mathbf{C^\infty Rings}$ of *complete C^∞-rings*, with $(\mathbf{C^\infty Rings_{co}})^{\rm op}$ equivalent to the category $\mathbf{AC^\infty Sch}$ of affine C^∞-schemes.

For C^∞-schemes with corners, the situation is worse: we have both $\Gamma^{\rm c} \circ \mathrm{Spec^c} \not\cong \mathrm{Id}$ and $\mathrm{Spec^c} \circ \Gamma^{\rm c} \circ \mathrm{Spec^c} \not\cong \mathrm{Spec^c}$. To see why, note that if X is a manifold with corners then smooth functions $c : X \to \mathbb{R}$ are essentially local objects – they can be glued using partitions of unity.

However, smooth functions $c' : X \to [0,\infty)$ have some *strictly global* behaviour: there is a locally constant function $\mu_{c'} : \partial X \to \mathbb{N} \amalg \{\infty\}$ giving the order of vanishing of c' along the boundary ∂X. So the behaviour of c' near distant points x, y in X is linked, if x, y lie in the image of the same connected component of ∂X. This means that smooth functions $c' : X \to [0,\infty)$ cannot always be glued using partitions of unity, and localizing a C^∞-ring with corners

$\mathfrak{C} = (\mathfrak{C}, \mathfrak{C}_{\rm ex})$ at an $\mathbb{R}$-point $x : \mathfrak{C} \to \mathbb{R}$, as one does to define $\mathrm{Spec^c}$, does not see only local behaviour around x.

Since $\mathrm{Spec^c} \circ \Gamma^{\rm c} \circ \mathrm{Spec^c} \not\cong \mathrm{Spec^c}$, we cannot define a subcategory of 'complete' C^∞-rings with corners equivalent to affine C^∞-schemes with corners. As a partial substitute, we define *semi-complete* C^∞-rings with corners $\mathfrak{C} = (\mathfrak{C}, \mathfrak{C}_{\rm ex})$, such that $\Gamma^{\rm c} \circ \mathrm{Spec^c}$ is an isomorphism on $\mathfrak{C}$ and injective on $\mathfrak{C}_{\rm ex}$.

If $\boldsymbol{X}, \boldsymbol{Y} \in \mathbf{C^\infty Sch^c}$, a morphism $\boldsymbol{f} : \boldsymbol{X} \to \boldsymbol{Y}$ in $\mathbf{C^\infty Sch^c}$ is a morphism in $\mathbf{LC^\infty RS^c}$. Although locally we can write $\boldsymbol{X} \cong \mathrm{Spec^c}\, \boldsymbol{\mathfrak{C}}$, $\boldsymbol{X} \cong \mathrm{Spec^c}\, \boldsymbol{\mathfrak{D}}$, because of the lack of a good notion of completeness, we do *not* know that locally we can write $\boldsymbol{f} = \mathrm{Spec^c}\, \boldsymbol{\phi}$ for some $\boldsymbol{\phi} : \boldsymbol{\mathfrak{D}} \to \boldsymbol{\mathfrak{C}}$ in $\mathbf{C^\infty Rings^c}$. One problem this causes is that $\boldsymbol{g} : \boldsymbol{X} \to \boldsymbol{Z}$, $\boldsymbol{h} : \boldsymbol{Y} \to \boldsymbol{Z}$ are morphisms in $\mathbf{C^\infty Sch^c}$, we do not know that the fibre product $\boldsymbol{X} \times_{\boldsymbol{g},\boldsymbol{Z},\boldsymbol{h}} \boldsymbol{Y}$ in $\mathbf{LC^\infty RS^c}$ (which always exists) lies in $\mathbf{C^\infty Sch^c}$, if $\boldsymbol{g}, \boldsymbol{h}$ are not locally of the form $\mathrm{Spec^c}\, \boldsymbol{\phi}$. So we do not know that all fibre products exist in $\mathbf{C^\infty Sch^c}$.

To get around this, we introduce the full subcategory $\mathbf{C^\infty Sch^c_{fi}} \subset \mathbf{C^\infty Sch^c}$ of *firm* C^∞-schemes with corners $\boldsymbol{X}$, which are locally of the form $\mathrm{Spec^c}\, \boldsymbol{\mathfrak{C}}$ for $\boldsymbol{\mathfrak{C}}$ a firm C^∞-ring with corners. Morphisms $\boldsymbol{f} : \boldsymbol{X} \to \boldsymbol{Y}$ in $\mathbf{C^\infty Sch^c_{fi}}$ are always locally of the form $\mathrm{Spec^c}\, \boldsymbol{\phi}$, so we can prove that $\mathbf{C^\infty Sch^c_{fi}}$ is closed under fibre products in $\mathbf{LC^\infty RS^c}$, and thus all fibre products exist in $\mathbf{C^\infty Sch^c_{fi}}$.

In general, $\mathbf{C^\infty Sch^c}$ contains a huge variety of objects, many of which are very singular and pathological, and do not fit with our intuitions about manifolds with corners. So it can be helpful to restrict to smaller subcategories of better-behaved objects in $\mathbf{C^\infty Sch^c}$, such as $\mathbf{C^\infty Sch^c_{fi}}$. For example, for $\boldsymbol{X}$ to be firm means that locally $\boldsymbol{X}$ has only finitely many boundary strata, which seems likely to hold in almost every interesting application.

We define many subcategories of $\mathbf{C^\infty Sch^c_{in}}, \mathbf{C^\infty Sch^c}$, and prove results such as existence of reflection functors between them, and existence of fibre products and finite limits in them. Two particularly interesting and well-behaved examples are the full subcategories $\mathbf{C^\infty Sch^c_{to}} \subset \mathbf{C^\infty Sch^c_{in}}$, $\mathbf{C^\infty Sch^c_{to,ex}} \subset \mathbf{C^\infty Sch^c}$ of *toric* C^∞-schemes with corners, whose corner structure is controlled by toric monoids in the same way that manifolds with g-corners are.

Chapter 6: Boundaries, corners, and the corner functor

One of the most important properties of manifolds with corners X is the existence of boundaries ∂X, and clearly we want to generalize this to C^∞-schemes with corners. Our starting point, as in Chapter 3, is that the boundary $\partial X =$

$C_1(X)$ is part of the corners $C(X) = \coprod_{k=0}^{\dim X} C_k(X)$, and the corner functor $C : \check{\mathbf{Man}}^{\mathbf{c}} \to \check{\mathbf{Man}}^{\mathbf{c}}_{\mathbf{in}}$ is right adjoint to the inclusion inc : $\check{\mathbf{Man}}^{\mathbf{c}}_{\mathbf{in}} \hookrightarrow \check{\mathbf{Man}}^{\mathbf{c}}$.

We construct a right adjoint *corner functor* $C : \mathbf{LC^\infty RS^c} \to \mathbf{LC^\infty RS^c_{in}}$ to the inclusion inc : $\mathbf{LC^\infty RS^c_{in}} \hookrightarrow \mathbf{LC^\infty RS^c}$. We prove that the restriction to $\mathbf{C^\infty Sch^c}$ maps to $\mathbf{C^\infty Sch^c_{in}}$, giving $C : \mathbf{C^\infty Sch^c} \to \mathbf{C^\infty Sch^c_{in}}$ right adjoint to inc : $\mathbf{C^\infty Sch^c_{in}} \hookrightarrow \mathbf{C^\infty Sch^c}$. For $\boldsymbol{X}$ in $\mathbf{LC^\infty RS^c}$, points in $C(X)$ are pairs (x, P) for $x \in X$ and $P \subset \mathcal{O}^{\rm ex}_{X,x}$ a prime ideal in the monoid $\mathcal{O}^{\rm ex}_{X,x}$ of the stalk $\boldsymbol{\mathcal{O}}_{X,x}$ of $\boldsymbol{\mathcal{O}}_X$ at x. This is an analogue, for $X \in \mathbf{Man^c}$, of a point in $C(X)$ being (x, γ) for $x \in X$ and γ a local corner component of X at x.

The corner functors for $\mathbf{Man^c}, \mathbf{Man^{gc}}$ and $\mathbf{C^\infty Sch^c}$ commute with the embeddings $\mathbf{Man^c}, \mathbf{Man^{gc}} \hookrightarrow \mathbf{C^\infty Sch^c}$.

To get an analogue of the decomposition $C(X) = \coprod_{k\geqslant 0} C_k(X)$ for C^∞-schemes with corners, we restrict to the subcategories $\mathbf{C^\infty Sch^c_{fi}} \subset \mathbf{C^\infty Sch^c}$ and $\mathbf{C^\infty Sch^c_{fi,in}} \subset \mathbf{C^\infty Sch^c_{in}}$ of *firm* C^∞-schemes with corners, where $C : \mathbf{C^\infty Sch^c_{fi}} \to \mathbf{C^\infty Sch^c_{fi,in}}$. For $\boldsymbol{X}$ firm there is a locally constant sheaf $\check{M}^{\rm ex}_{C(X)}$ of finitely generated monoids on $C(\boldsymbol{X})$, with $\check{M}^{\rm ex}_{C(X)}|_{(x,P)} = \mathcal{O}^{\rm ex}_{X,x}/[c' = 1$ if $c' \in \mathcal{O}^{\rm ex}_{X,x} \setminus P]$. We define a decomposition $C(\boldsymbol{X}) = \coprod_{k\geqslant 0} C_k(\boldsymbol{X})$ with $C_k(\boldsymbol{X})$ open and closed in $C(\boldsymbol{X})$, by saying that $(x, P) \in C_k(\boldsymbol{X})$ if the maximum length of a chain of prime ideals in $\check{M}^{\rm ex}_{C(X)}|_{(x,P)}$ is $k + 1$. This recovers the usual decomposition $C(X) = \coprod_{k\geqslant 0} C_k(X)$ if X is a manifold with (g-)corners. We define the *boundary* $\partial\boldsymbol{X} = C_1(\boldsymbol{X})$.

For toric C^∞-schemes with corners C maps $\mathbf{C^\infty Sch^c_{to,ex}} \to \mathbf{C^\infty Sch^c_{to}}$, and C preserves fibre products, and all fibre products exist in $\mathbf{C^\infty Sch^c_{to}}$. We use this to give criteria for when fibre products exist in $\mathbf{C^\infty Sch^c_{to,ex}}$. This is an analogue of results in [46] giving criteria for when fibre products exist in $\mathbf{Man^{gc}}$, given that b-transverse fibre products exist in $\mathbf{Man^{gc}_{in}}$.

Chapter 7: Modules, and sheaves of modules

If $\boldsymbol{\mathfrak{C}} = (\mathfrak{C}, \mathfrak{C}_{\rm ex})$ is a C^∞-ring with corners, we define a $\boldsymbol{\mathfrak{C}}$-*module* to be a module over $\mathfrak{C}$ as an $\mathbb{R}$-algebra. Similarly, if $\boldsymbol{X} = (X, (\mathcal{O}_X, \mathcal{O}^{\rm ex}_X))$ is a C^∞-scheme with corners, we consider $\mathcal{O}_X$-*modules* on X, which are just modules on the underlying C^∞-scheme $\underline{X} = (X, \mathcal{O}_X)$. So the theory of modules over C^∞-rings with corners and C^∞-schemes with corners lifts immediately from modules over C^∞-rings and C^∞-schemes in [49, §5], with no additional theory required.

If X is a manifold with corners then, as in Chapter 3, we have two notions of cotangent bundle $T^*X, {}^bT^*X$, where ${}^bT^*X$ is functorial only under interior morphisms. Similarly, if $\boldsymbol{\mathfrak{C}} = (\mathfrak{C}, \mathfrak{C}_{\rm ex})$ is a C^∞-ring with corners, we have the *cotangent module* $\Omega_{\mathfrak{C}}$ of $\mathfrak{C}$ from [49, §5]. If $\boldsymbol{\mathfrak{C}}$ is interior we also define the *b-cotangent module* ${}^b\Omega_{\boldsymbol{\mathfrak{C}}}$, which uses the corner structure. If X is a manifold with (g-)faces and $\boldsymbol{\mathfrak{C}} = \boldsymbol{C}^\infty_{\rm in}(X)$ then $\Omega_{\mathfrak{C}} = \Gamma^\infty(T^*X)$ and ${}^b\Omega_{\boldsymbol{\mathfrak{C}}} = \Gamma^\infty({}^bT^*X)$. We show that b-cotangent modules are functorial under interior morphisms and have exact sequences for pushouts.

If $\boldsymbol{X}$ is a C^∞-scheme with corners we define the *cotangent sheaf* $T^*\boldsymbol{X}$, and if $\boldsymbol{X}$ is interior the *b-cotangent sheaf* ${}^bT^*\boldsymbol{X}$, by sheafifying the (b-)cotangent modules of $\boldsymbol{\mathcal{O}}_X(U)$ for open $U \subset X$. If $\boldsymbol{X} = F_{\mathbf{Man^c}}(X)$ for $X \in \mathbf{Man^c}$ these are the sheaves of sections of T^*X and ${}^bT^*X$. We show that Cartesian squares in subcategories such as $\mathbf{C^\infty Sch^c_{to}}, \mathbf{C^\infty Sch^c_{fi,in}} \subset \mathbf{C^\infty Sch^c_{in}}$ yield exact sequences of b-cotangent sheaves. On the corners $C(\boldsymbol{X})$ we construct an exact sequence relating ${}^bT^*C(\boldsymbol{X})$, $\boldsymbol{\Pi}^*_{\boldsymbol{X}}({}^bT^*\boldsymbol{X})$ and $\check{M}^{\rm ex}_{C(X)} \otimes_{\mathbb{N}} \mathcal{O}_{C(X)}$.

Chapter 8: Further generalizations and applications

Finally we propose four directions in which this book could be generalized and applied.

Synthetic Differential Geometry with corners Synthetic Differential Geometry is a subject in which one proves theorems about manifolds in Differential Geometry by reasoning using 'infinitesimals', as in Kock [52, 53]. C^∞-schemes are used to provide a 'model' for Synthetic Differential Geometry, and so show that the axioms of Synthetic Differential Geometry are consistent. In a similar way, one could develop a theory of 'Synthetic Differential Geometry with corners', for proving theorems about manifolds with corners using infinitesimals, and C^∞-schemes with corners could be used to show that it is consistent.

C^∞-stacks with corners In classical Algebraic Geometry, schemes are generalized to (Deligne–Mumford or Artin) stacks. The second author [49] extended the theory of C^∞-schemes to *C^∞-stacks*, including *Deligne–Mumford C^∞-stacks*. This corresponds to generalizing manifolds to orbifolds.

We discuss a theory of *C^∞-stacks with corners*, including *Deligne–Mumford C^∞-stacks with corners*. These generalize *orbifolds with (g-)corners*. Much of the theory follows from [49] with only cosmetic changes.

C^∞-rings and C^∞-schemes with a-corners Our theory starts with the categories $\mathbf{Man^c_{in}} \subset \mathbf{Man^c}$ of manifolds with corners defined in Chapter 3. The second author [47] also defined categories $\mathbf{Man^{ac}_{in}} \subset \mathbf{Man^{ac}}$ of *manifolds with analytic corners*, or *manifolds with a-corners*. Even the simplest objects $[\![0,\infty)$ in $\mathbf{Man^{ac}}$ and $[0,\infty)$ in $\mathbf{Man^c}$ have different smooth structures. There is also a category $\mathbf{Man^{c,ac}}$ of *manifolds with corners and a-corners* containing both $\mathbf{Man^c}$ and $\mathbf{Man^{ac}}$.

Manifolds with a-corners have applications to partial differential equations with boundary conditions of asymptotic type, and to moduli spaces with boundary and corners, such as moduli spaces of Morse flow lines, in which (we argue) manifolds with a-corners give the correct smooth structure.

This entire book could be rewritten over $\mathbf{Man^{ac}_{in}} \subset \mathbf{Man^{ac}}$ or $\mathbf{Man^{c,ac}_{in}} \subset \mathbf{Man^{c,ac}}$ rather than $\mathbf{Man^c_{in}} \subset \mathbf{Man^c}$. We explain the first steps in this.

Derived manifolds and derived orbifolds with corners Classical Algebraic Geometry has been generalized to Derived Algebraic Geometry, which is now a major area of mathematics. As in §2.9, classical Differential Geometry can be generalized to *Derived Differential Geometry*, the study of *derived manifolds* and *derived orbifolds*, regarded as special examples of *derived C^∞-schemes* and *derived C^∞-stacks*. Some references are [6, 7, 8, 10, 11, 12, 13, 43, 44, 62, 87, 88, 89].

It is desirable to extend the subject to *derived manifolds with corners* and *derived orbifolds with corners*, regarded as special examples of *derived C^∞-schemes with corners* and *derived C^∞-stacks with corners*. This will be done by the second author in [44], with this book as its foundations (see also Steffens [88, 89]). Derived orbifolds with corners will have important applications in Symplectic Geometry, as (we argue) they are the correct way to make Fukaya–Ohta–Oh–Ono's 'Kuranishi spaces with corners' [28, 29, 30, 31] into well-behaved geometric spaces.

Acknowledgements The first author would like to thank The University of Oxford, St John's College, the Rhodes Trust, the Centre for Quantum Geometry of Moduli Spaces in Aarhus, and her family, for their support while writing her PhD thesis [25], on which this book is based.

2
Background on C^∞-schemes

One can think of C^∞-rings as a specific type of commutative $\mathbb{R}$-algebra, where there are not only the addition and multiplication operations, but operations corresponding to every smooth function $\mathbb{R}^n \to \mathbb{R}$. Alternatively, C^∞-rings can also be considered as certain product-preserving functors, and we will introduce both definitions in this chapter.

C^∞-rings, along with a spectrum functor, form the building blocks of C^∞-schemes. As in ordinary Algebraic Geometry, the image of a C^∞-ring under the spectrum functor gives an affine C^∞-scheme. However, while there will be many similarities to ordinary Algebraic Geometry, C^∞-algebraic geometry has several differences that may challenge the reader's intuition. These include the following.

- C^∞-rings are non-noetherian, so finitely presented C^∞-rings are not necessarily finitely generated.
- The spectrum functor uses maximal ideals with residue field $\mathbb{R}$, not prime ideals. This makes affine C^∞-schemes Hausdorff and regular.
- Affine C^∞-schemes are very general, enough so that all manifolds can be represented as affine C^∞-schemes, and study can be restricted to affine C^∞-schemes in many cases. However, manifolds with corners will not always be affine C^∞-schemes with corners in Chapter 5.
- C^∞-rings are not in 1-1 correspondence with affine C^∞-schemes, and the spectrum functor is neither full nor faithful. However, the full subcategory of *complete* C^∞-rings is in 1-1 correspondence with affine C^∞-schemes, and the spectrum functor is full and faithful on this subcategory.

Our motivating example throughout this chapter will be a manifold X. We will see that its set of smooth functions $C^\infty(X)$ is a complete C^∞-ring, and applying the spectrum functor returns the affine C^∞-scheme with underlying topological space X along with its sheaf of smooth functions. Transverse fibre products of manifolds map to fibre products of (affine) C^∞-schemes. Unlike the category of manifolds, all fibre products of (affine) C^∞-schemes exist, which is one of the motivating reasons for considering this category.

Our main reference for this chapter is the second author [49, §2–§4]. See also [42], Dubuc [23], Moerdijk and Reyes [75], and Kock [52].

2.1 Introduction to category theory

We begin with some well-known definitions from category theory. All the material of this section can be found in MacLane [64].

2.1.1 Categories and functors

Definition 2.1 A *category* $\mathcal{C}$ consists of a proper class of *objects* $\mathrm{Obj}(\mathcal{C})$, and for all $X, Y \in \mathrm{Obj}(\mathcal{C})$ a set $\mathrm{Hom}(X, Y)$ of *morphisms* f from X to Y, written $f : X \to Y$, and for all $X, Y, Z \in \mathrm{Obj}(\mathcal{C})$ a *composition map* $\circ : \mathrm{Hom}(X,Y) \times \mathrm{Hom}(Y,Z) \to \mathrm{Hom}(X,Z)$, written $(f,g) \mapsto g \circ f$. Composition must be *associative*, that is, if $f : W \to X$, $g : X \to Y$ and $h : Y \to Z$ are morphisms in $\mathcal{C}$ then $(h \circ g) \circ f = h \circ (g \circ f)$. For each $X \in \mathrm{Obj}(\mathcal{C})$ there must exist an *identity morphism* $\mathrm{id}_X : X \to X$ such that $f \circ \mathrm{id}_X = f = \mathrm{id}_Y \circ f$ for all $f : X \to Y$ in $\mathcal{C}$. A morphism $f : X \to Y$ is an *isomorphism* if there exists $f^{-1} : Y \to X$ with $f^{-1} \circ f = \mathrm{id}_X$ and $f \circ f^{-1} = \mathrm{id}_Y$.

If $\mathcal{C}$ is a category, the *opposite category* $\mathcal{C}^{\mathrm{op}}$ is $\mathcal{C}$ with the directions of all morphisms reversed. That is, we define $\mathrm{Obj}(\mathcal{C}^{\mathrm{op}}) = \mathrm{Obj}(\mathcal{C})$, and for all $X, Y, Z \in \mathrm{Obj}(\mathcal{C})$ we define $\mathrm{Hom}_{\mathcal{C}^{\mathrm{op}}}(X,Y) = \mathrm{Hom}_{\mathcal{C}}(Y,X)$, and for $f : X \to Y$, $g : Y \to Z$ in $\mathcal{C}$ we define $f \circ_{\mathcal{C}^{\mathrm{op}}} g = g \circ_{\mathcal{C}} f$, and $\mathrm{id}_{\mathcal{C}^{\mathrm{op}}} X = \mathrm{id}_{\mathcal{C}} X$.

We call $\mathcal{D}$ a *subcategory* of $\mathcal{C}$ if $\mathrm{Obj}(\mathcal{D}) \subseteq \mathrm{Obj}(\mathcal{C})$, and $\mathrm{Hom}_{\mathcal{D}}(X,Y) \subseteq \mathrm{Hom}_{\mathcal{C}}(X,Y)$ for all $X, Y \in \mathrm{Obj}(\mathcal{D})$, and compositions and identities in $\mathcal{D}$ agree with those in $\mathcal{C}$. We call $\mathcal{D}$ a *full* subcategory if also $\mathrm{Hom}_{\mathcal{D}}(X,Y) = \mathrm{Hom}_{\mathcal{C}}(X,Y)$ for all X, Y in $\mathcal{D}$.

Definition 2.2 Let $\mathcal{C}, \mathcal{D}$ be categories. A (*covariant*) *functor* $F : \mathcal{C} \to \mathcal{D}$ gives for all objects X in $\mathcal{C}$ an object $F(X)$ in $\mathcal{D}$, and for all morphisms $f : X \to Y$ in $\mathcal{C}$ a morphism $F(f) : F(X) \to F(Y)$ in $\mathcal{D}$, such that $F(g \circ f) = F(g) \circ F(f)$ for all $f : X \to Y$, $g : Y \to Z$ in $\mathcal{C}$, and $F(\mathrm{id}_X) = \mathrm{id}_{F(X)}$ for all $X \in \mathrm{Obj}(\mathcal{C})$. A *contravariant functor* $F : \mathcal{C} \to \mathcal{D}$ is a covariant functor $F : \mathcal{C}^{\mathrm{op}} \to \mathcal{D}$.

Functors compose in the obvious way. Each category $\mathcal{C}$ has an obvious *identity functor* $\mathrm{id}_{\mathcal{C}} : \mathcal{C} \to \mathcal{C}$ with $\mathrm{id}_{\mathcal{C}}(X) = X$ and $\mathrm{id}_{\mathcal{C}}(f) = f$ for all

X, f. A functor $F : \mathcal{C} \to \mathcal{D}$ is called *full* if the maps $\mathrm{Hom}_{\mathcal{C}}(X,Y) \to \mathrm{Hom}_{\mathcal{D}}(F(X), F(Y))$, $f \mapsto F(f)$ are surjective for all $X, Y \in \mathrm{Obj}(\mathcal{C})$, and *faithful* if the maps $\mathrm{Hom}_{\mathcal{C}}(X,Y) \to \mathrm{Hom}_{\mathcal{D}}(F(X), F(Y))$ are injective for all $X, Y \in \mathrm{Obj}(\mathcal{C})$.

A functor $F : \mathcal{C} \to \mathcal{D}$ is called *essentially surjective* if for every object $Y \in \mathcal{D}$ there exists $X \in \mathcal{C}$ such that $Y \cong F(X)$ in $\mathcal{D}$.

Let $\mathcal{C}, \mathcal{D}$ be categories and $F, G : \mathcal{C} \to \mathcal{D}$ be functors. A *natural transformation* $\eta : F \Rightarrow G$ gives, for all objects X in $\mathcal{C}$, a morphism $\eta(X) : F(X) \to G(X)$ such that if $f : X \to Y$ is a morphism in $\mathcal{C}$ then $\eta(Y) \circ F(f) = G(f) \circ \eta(X)$ as a morphism $F(X) \to G(Y)$ in $\mathcal{D}$. We call η a *natural isomorphism* if $\eta(X)$ is an isomorphism for all $X \in \mathrm{Obj}(\mathcal{C})$.

An *equivalence* between categories $\mathcal{C}, \mathcal{D}$ is a functor $F : \mathcal{C} \to \mathcal{D}$ such that there exists a functor $G : \mathcal{D} \to \mathcal{C}$ and natural isomorphisms $\eta : G \circ F \Rightarrow \mathrm{id}_{\mathcal{C}}$ and $\zeta : F \circ G \Rightarrow \mathrm{id}_{\mathcal{D}}$. That is, F is invertible up to natural isomorphism. Then we call $\mathcal{C}, \mathcal{D}$ *equivalent categories*. A functor $F : \mathcal{C} \to \mathcal{D}$ is an equivalence if and only if it is full, faithful, and essentially surjective.

It is a fundamental principle of category theory that equivalent categories $\mathcal{C}, \mathcal{D}$ should be thought of as being 'the same', and naturally isomorphic functors $F, G : \mathcal{C} \to \mathcal{D}$ should be thought of as being 'the same'. Note that equivalence of categories $\mathcal{C}, \mathcal{D}$ is much weaker than strict isomorphism: isomorphism classes of objects in $\mathcal{C}$ are naturally in bijection with isomorphism classes of objects in $\mathcal{D}$, but there is no relation between the sizes of the isomorphism classes, so that $\mathcal{C}$ could have many more objects than $\mathcal{D}$, for instance.

2.1.2 Limits, colimits and fibre products in categories

We shall be interested in various kinds of *limits* and *colimits* in our categories of spaces. These are objects in the category with a universal property with respect to some class of diagrams.

Definition 2.3 Let $\mathcal{C}$ be a category. A *diagram* Δ in $\mathcal{C}$ is a class of objects S_i in $\mathcal{C}$ for $i \in I$, and a class of morphisms $\rho_j : S_{b(j)} \to S_{e(j)}$ in $\mathcal{C}$ for $j \in J$, where $b, e : J \to I$. The diagram is called *small* if I, J are sets (rather than something too large to be a set), and *finite* if I, J are finite sets.

A *limit* of the diagram Δ is an object L in $\mathcal{C}$ and morphisms $\pi_i : L \to S_i$ for $i \in I$ such that $\rho_j \circ \pi_{b(j)} = \pi_{e(j)}$ for all $j \in J$, with the universal property that given $L' \in \mathcal{C}$ and $\pi'_i : L' \to S_i$ for $i \in I$ with $\rho_j \circ \pi'_{b(j)} = \pi'_{e(j)}$ for all $j \in J$, there is a unique morphism $\lambda : L' \to L$ with $\pi'_i = \pi_i \circ \lambda$ for all $i \in I$. The limit is called *small*, or *finite*, if Δ is small, or finite.

Here are some important kinds of limit.

(i) A *terminal object* is a limit of the empty diagram.

(ii) Let X, Y be objects in $\mathcal{C}$. A *product* $X \times Y$ is a limit of the diagram with two objects X, Y and no morphisms.

(iii) Let $g : X \to Z$, $h : Y \to Z$ be morphisms in $\mathcal{C}$. A *fibre product* is a limit of a diagram $X \xrightarrow{g} Z \xleftarrow{h} Y$. The limit object W is often written $X \times_{g,Z,h} Y$ or $X \times_Z Y$. Explicitly, a fibre product is an object W and morphisms $e : W \to X$ and $f : W \to Y$ in $\mathcal{C}$, such that $g \circ e = h \circ f$, with the universal property that if $e' : W' \to X$ and $f' : W' \to Y$ are morphisms in $\mathcal{C}$ with $g \circ e' = h \circ f'$ then there is a unique morphism $b : W' \to W$ with $e' = e \circ b$ and $f' = f \circ b$. The commutative diagram

$$\begin{array}{ccc} W & \xrightarrow{\quad f \quad} & Y \\ \downarrow{\scriptstyle e} & & {\scriptstyle h}\downarrow \\ X & \xrightarrow{\quad g \quad} & Z \end{array} \tag{2.1}$$

is called a *Cartesian square.*

If Z is a terminal object then $X \times_Z Y$ is a product $X \times Y$.

A *colimit* of the diagram Δ is an object L in $\mathcal{C}$ and morphisms $\lambda_i : S_i \to L$ for $i \in I$ such that $\lambda_{b(j)} = \lambda_{e(j)} \circ \rho_j$ for all $j \in J$, which has the universal property that given $L' \in \mathcal{C}$ and $\lambda'_i : S_i \to L'$ for $i \in I$ with $\lambda'_{b(j)} = \lambda'_{e(j)} \circ \rho_j$ for all $j \in J$, there is a unique morphism $\pi : L \to L'$ with $\lambda'_i = \pi \circ \lambda_i$ for all $i \in I$.

Here are some important kinds of colimit.

(iv) An *initial object* is a colimit of the empty diagram.

(v) Let X, Y be objects in $\mathcal{C}$. A *coproduct* $X \amalg Y$ is a colimit of the diagram with two objects X, Y and no morphisms.

(vi) Let $e : W \to X$, $f : W \to Y$ be morphisms in $\mathcal{C}$. A *pushout* is a colimit of a diagram $X \xleftarrow{e} W \xrightarrow{f} Y$. The colimit object Z is often written $X \amalg_{e,W,f} Y$ or $X \amalg_W Y$. Explicitly, a pushout is an object Z and morphisms $g : X \to Z$ and $h : Y \to Z$ in $\mathcal{C}$, such that $g \circ e = h \circ f$, with the universal property that if $g' : X \to Z'$ and $h' : Y \to Z'$ are morphisms in $\mathcal{C}$ with $g' \circ e = h' \circ f$ then there is a unique morphism $b : Z \to Z'$ with $g' = b \circ g$ and $h' = b \circ h$. The diagram (2.1) is then called a *co-Cartesian square.*

If W is an initial object then $X \amalg_W Y$ is a coproduct $X \amalg Y$.

Limits and colimits may not may not exist. If a limit or colimit exists, it is unique up to canonical isomorphism in $\mathcal{C}$. We say that *all limits*, or *all small limits*, or *all finite limits exist in* $\mathcal{C}$, if limits exist for all diagrams, or all small diagrams, or all finite diagrams respectively; and similarly for colimits.

A category $\mathcal{C}$ is called *complete* if all small limits exist in $\mathcal{C}$, and *cocomplete* if all small colimits exist in $\mathcal{C}$.

Limits in $\mathcal{C}$ are equivalent to colimits (of the opposite diagram) in the opposite category $\mathcal{C}^{\mathrm{op}}$. So, for example, fibre products in $\mathcal{C}$ are pushouts in $\mathcal{C}^{\mathrm{op}}$.

A *directed colimit* is a colimit for which the diagram Δ is an *upward-directed set*, that is, Δ is a *preorder* (a category in which there is at most one morphism $S_i \to S_j$ between any two objects) in which every finite subset has an upper bound. Confusingly, directed colimits are also called *inductive limits* or *direct limits*, although they are actually colimits. So we can say that *all directed colimits exist in* $\mathcal{C}$.

The dual concept, called *codirected limit*, will not be used in this book.

By category theory general nonsense, one can prove the following.

Proposition 2.4 *Suppose a category* $\mathcal{C}$ *has a terminal object, and all fibre products exist in* $\mathcal{C}$*. Then all finite limits exist in* $\mathcal{C}$*.*

Example 2.5 Let $\mathcal{C}$ be a category of spaces, for instance, topological spaces $\mathbf{Top}$, manifolds $\mathbf{Man}$, schemes $\mathbf{Sch}$, or C^∞-schemes $\mathbf{C^\infty Sch}$ in §2.5. Then we have the following.

(i) The *terminal object* is the point $*$, and exists for all sensible $\mathcal{C}$.
(ii) *Products* $X \times Y$ are the usual products of manifolds, topological spaces, ...
(iii) *Fibre products* may or may not exist, depending on $\mathcal{C}$. All fibre products $W = X \times_{g,Z,h} Y$ exist in $\mathbf{Top}$, with $W = \{(x,y) \in X \times Y : g(x) = h(y)\}$, with the subspace topology as a subset of $X \times Y$. All fibre products also exist in $\mathbf{Sch}$, $\mathbf{C^\infty Sch}$. Fibre products $X \times_{g,Z,h} Y$ exist in $\mathbf{Man}$ if g, h are *transverse*, but not in general.
(iv) The *initial object* is the empty set $\emptyset$, and exists for all sensible $\mathcal{C}$.
(v) *Coproducts* $X \amalg Y$ are disjoint unions of the spaces X, Y. In $\mathbf{Man}$ this exists if $\dim X = \dim Y$.
(vi) General *pushouts* in categories such as $\mathbf{Man}$, $\mathbf{Sch}$, $\mathbf{C^\infty Sch}, \ldots$ tend not to exist, and have not been a focus of research.

Many important constructions in categories of spaces can be expressed as finite limits. For example, the intersection $X \cap Y$ of submanifolds $X, Y \subset Z$ is a fibre product $X \times_Z Y$. By Proposition 2.4, the existence of finite limits reduces to that of fibre products. So in our study of C^∞-schemes with corners, we will be particularly interested in existence and properties of fibre products.

Example 2.6 Let $\mathcal{C}$ be a category of (generalized) commutative algebras over a field $\mathbb{K}$, for example, commutative $\mathbb{C}$-algebras $\mathbf{Alg}_{\mathbb{C}}$, or C^∞-rings $\mathbf{C^\infty Rings}$ in §2.2 with $\mathbb{K} = \mathbb{R}$. Then we have the following.

(i) The *terminal object* is the zero algebra 0.
(ii) *Products* $B \times C$ are direct sums $B \oplus C$.
(iii) For morphisms $\beta : B \to D$, $\gamma : C \to D$, the *fibre product* $B \times_{\beta,D,\gamma} C$ is the subalgebra $\{(b,c) \in B \oplus C : \beta(b) = \gamma(c)\}$ in $B \oplus C$.
(iv) The *initial object* is the field $\mathbb{K}$.
(v) *Coproducts* $B \amalg C$ are (possibly completed) tensor products $B \otimes_{\mathbb{K}} C$.
(vi) For morphisms $\alpha : A \to B$, $\beta : A \to C$, a *pushout* is a (possibly completed) tensor product $B \otimes_{\alpha,A,\beta} C$.

2.1.3 Adjoint functors

Definition 2.7 Let $\mathcal{C}, \mathcal{D}$ be categories. An *adjunction* (F, G, φ) between $\mathcal{C}$ and $\mathcal{D}$ consists of functors $F : \mathcal{C} \to \mathcal{D}$ and $G : \mathcal{D} \to \mathcal{C}$ and bijections

$$\varphi(X,Y) : \mathrm{Hom}_{\mathcal{D}}(F(X), Y) \xrightarrow{\cong} \mathrm{Hom}_{\mathcal{C}}(X, G(Y))$$

for all objects $X \in \mathcal{C}$ and $Y \in \mathcal{D}$, which are natural in X, Y, that is, if $f : X_1 \to X_2$ and $g : Y_1 \to Y_2$ are morphisms in $\mathcal{C}, \mathcal{D}$ then the following commutes:

$$\begin{array}{ccccc}
\mathrm{Hom}_{\mathcal{D}}(F(X_2), Y_1) & \xrightarrow[g\circ -]{} & \mathrm{Hom}_{\mathcal{D}}(F(X_2), Y_2) & \xrightarrow[-\circ F(f)]{} & \mathrm{Hom}_{\mathcal{D}}(F(X_1), Y_2) \\
\downarrow{\scriptstyle \varphi(X_2,Y_1)} & & & & {\scriptstyle \varphi(X_1,Y_2)}\downarrow \\
\mathrm{Hom}_{\mathcal{C}}(X_2, G(Y_1)) & \xrightarrow{-\circ f} & \mathrm{Hom}_{\mathcal{C}}(X_1, G(Y_1)) & \xrightarrow{G(g)\circ -} & \mathrm{Hom}_{\mathcal{C}}(X_1, G(Y_2)).
\end{array}$$

Adjunctions are often written like this:

$$\mathcal{C} \underset{G}{\overset{F}{\rightleftarrows}} \mathcal{D}.$$

Then we say that F *is left adjoint to* G, and G *is right adjoint to* F. We say that $F : \mathcal{C} \to \mathcal{D}$ *has a right adjoint* if it can be completed to an adjunction (F, G, φ). We say that $G : \mathcal{D} \to \mathcal{C}$ *has a left adjoint* if it can be completed to an adjunction (F, G, φ).

Suppose $\mathcal{C}$ is a category, and $\mathcal{D} \subset \mathcal{C}$ is a full subcategory of $\mathcal{C}$. We say that $\mathcal{D}$ is a *reflective subcategory* of $\mathcal{C}$ if the inclusion $\mathrm{inc} : \mathcal{D} \hookrightarrow \mathcal{C}$ has a left adjoint. This left adjoint $R : \mathcal{C} \to \mathcal{D}$ is called a *reflection functor*.

Dually, if $\mathcal{C} \subset \mathcal{D}$ is a full subcategory, we say $\mathcal{C}$ is a *coreflective subcategory* of $\mathcal{D}$ if the inclusion $\mathrm{inc} : \mathcal{C} \hookrightarrow \mathcal{D}$ has a right adjoint. This right adjoint $C : \mathcal{D} \to \mathcal{C}$ is called a *coreflection functor*.

Here are some properties of adjoint functors.

Theorem 2.8 **(a)** *In Definition 2.7 there are natural transformations $\eta : \mathrm{Id}_{\mathcal{C}} \Rightarrow G \circ F$, called the* **unit of the adjunction**, *and $\epsilon : F \circ G \Rightarrow \mathrm{Id}_{\mathcal{D}}$, called the* **counit of the adjunction**, *such that for $X \in \mathcal{C}$ and $Y \in \mathcal{D}$ we have*

$$\eta(X) = \varphi(X, F(X))(\mathrm{id}_{F(X)}) : X \longrightarrow G(F(X)),$$
$$\epsilon(Y) = \varphi(G(Y), Y)^{-1}(\mathrm{id}_{G(Y)}) : F(G(Y)) \longrightarrow Y.$$

(b) *F,G are both equivalences of categories if and only if F,G are both full and faithful, if and only if η,ϵ are both natural isomorphisms.*

(c) *If $F : \mathcal{C} \to \mathcal{D}$ has a right adjoint, then the right adjoint $G : \mathcal{D} \to \mathcal{C}$ is determined up to natural isomorphism by F.*

(d) *If $G : \mathcal{D} \to \mathcal{C}$ has a left adjoint, then the left adjoint $F : \mathcal{C} \to \mathcal{D}$ is determined up to natural isomorphism by G.*

(e) *If $F : \mathcal{C} \to \mathcal{D}$ has a right adjoint then it* **preserves colimits**, *that is, F maps a colimit in $\mathcal{C}$ to the corresponding colimit in $\mathcal{D}$ (which is guaranteed to exist in $\mathcal{D}$, if the initial colimit exists in $\mathcal{C}$).*

(f) *If $G : \mathcal{D} \to \mathcal{C}$ has a left adjoint then it* **preserves limits**, *that is, G maps a limit in $\mathcal{D}$ to the corresponding limit in $\mathcal{C}$ (which is guaranteed to exist in $\mathcal{C}$, if the initial limit exists in $\mathcal{D}$).*

(g) *Let $\mathcal{D} \subset \mathcal{C}$ be a reflective subcategory, with reflection functor $R : \mathcal{C} \to \mathcal{D}$. Suppose some class of colimits (e.g. all small colimits, or all pushouts) exists in $\mathcal{C}$. Then the same class of colimits exists in $\mathcal{D}$. We can obtain the colimit of a diagram in $\mathcal{D}$ by taking the colimit in $\mathcal{C}$ and then applying R.*

(h) *Let $\mathcal{C} \subset \mathcal{D}$ be a coreflective subcategory, with coreflection functor $C : \mathcal{D} \to \mathcal{C}$. Suppose some class of limits (e.g. all small limits, or all fibre products) exists in $\mathcal{D}$. Then the same class of limits exists in $\mathcal{C}$. We can obtain the limit of a diagram in $\mathcal{C}$ by taking the limit in $\mathcal{D}$ and then applying C.*

Remark 2.9 We will use adjoint functors in two main ways below. Firstly, by Theorem 2.8(e)–(h), we can use them to prove results on existence of (co)limits. The second is more philosophical, and we illustrate it by two examples.

(a) In §2.5 we define an adjunction

$$\mathbf{LC^\infty RS} \underset{\mathrm{Spec}}{\overset{\Gamma}{\rightleftarrows}} \mathbf{C^\infty Rings}$$

between the categories $\mathbf{C^\infty Rings}$ of C^∞-rings and $\mathbf{LC^\infty RS}$ of locally C^∞-ringed spaces. Here the definition of the global sections functor Γ is simple and obvious, but that of the spectrum functor Spec is complicated and apparently arbitrary. However, Theorem 2.8(c) implies that Spec is determined up to natural isomorphism by Γ and the adjoint property. This justifies the definition of Spec, showing it could not have been otherwise.

(b) In §6.2 we define an adjunction

$$\mathbf{C^\infty Sch^c_{in}} \underset{C}{\overset{\text{inc}}{\rightleftarrows}} \mathbf{C^\infty Sch^c}.$$

Here $\mathbf{C^\infty Sch^c}$ is the category of C^∞-schemes with corners, $\mathbf{C^\infty Sch^c_{in}}$ is the non-full category with only interior morphisms, with inclusion inc : $\mathbf{C^\infty Sch^c_{in}} \hookrightarrow \mathbf{C^\infty Sch^c}$, and $C : \mathbf{C^\infty Sch^c} \to \mathbf{C^\infty Sch^c_{in}}$ is the 'corner functor', which encodes the notions of boundary $\partial\boldsymbol{X}$ and k-corners $C_k(\boldsymbol{X})$ of a (firm) C^∞-scheme with corners $\boldsymbol{X}$ in a functorial way.

Again, the definition of inc is simple and obvious, and that of C is complicated and contrived. But the adjoint property shows that C is determined up to natural isomorphism by inc, and so justifies the definition.

2.2 C^∞-rings

Here are two equivalent definitions of C^∞-ring. The first definition describes C^∞-rings as functors while the second definition describes C^∞-rings as sets in the style of classical algebra.

Definition 2.10 Write $\mathbf{Man}$ for the category of manifolds, and $\mathbf{Euc}$ for the full subcategory of $\mathbf{Man}$ with objects the Euclidean spaces $\mathbb{R}^n$. That is, the objects of $\mathbf{Euc}$ are $\mathbb{R}^n$ for $n = 0, 1, 2, \ldots$, and the morphisms in $\mathbf{Euc}$ are smooth maps $f : \mathbb{R}^m \to \mathbb{R}^n$. Write $\mathbf{Sets}$ for the category of sets. In both $\mathbf{Euc}$ and $\mathbf{Sets}$ we have notions of (finite) products of objects (that is, $\mathbb{R}^{m+n} = \mathbb{R}^m \times \mathbb{R}^n$, and products $S \times T$ of sets S, T), and products of morphisms.

Define a *categorical C^∞-ring* to be a product-preserving functor $F : \mathbf{Euc} \to \mathbf{Sets}$. Here F should also preserve the empty product, that is, it maps $\mathbb{R}^0$ in $\mathbf{Euc}$ to the terminal object in $\mathbf{Sets}$, the point $*$. If $F, G : \mathbf{Euc} \to \mathbf{Sets}$ are categorical C^∞-rings, a *morphism* $\eta : F \to G$ is a natural transformation $\eta : F \Rightarrow G$. We write $\mathbf{CC^\infty Rings}$ for the category of categorical C^∞-rings.

Categorical C^∞-rings are an example of an *Algebraic Theory* in the sense of Adámek, Rosický, and Vitale [1], and many of the basic categorical properties of C^∞-rings follow from this.

Definition 2.11 A *C^∞-ring* is a set $\mathfrak{C}$ together with operations

$$\Phi_f : \mathfrak{C}^n = \overbrace{\mathfrak{C} \times \cdots \times \mathfrak{C}}^{n \text{ copies}} \longrightarrow \mathfrak{C}$$

for all $n \geqslant 0$ and smooth maps $f : \mathbb{R}^n \to \mathbb{R}$, where by convention when $n = 0$ we define $\mathfrak{C}^0$ to be the single point $\{\emptyset\}$. These operations must satisfy the following relations: suppose $m, n \geqslant 0$, and $f_i : \mathbb{R}^n \to \mathbb{R}$ for $i = 1, \dots, m$ and $g : \mathbb{R}^m \to \mathbb{R}$ are smooth functions. Define a smooth function $h : \mathbb{R}^n \to \mathbb{R}$ by

$$h(x_1, \dots, x_n) = g\big(f_1(x_1, \dots, x_n), \dots, f_m(x_1 \dots, x_n)\big),$$

for all $(x_1, \dots, x_n) \in \mathbb{R}^n$. Then for all $(c_1, \dots, c_n) \in \mathfrak{C}^n$ we have

$$\Phi_h(c_1, \dots, c_n) = \Phi_g\big(\Phi_{f_1}(c_1, \dots, c_n), \dots, \Phi_{f_m}(c_1, \dots, c_n)\big).$$

We also require that for all $1 \leqslant j \leqslant n$, defining $\pi_j : \mathbb{R}^n \to \mathbb{R}$ by $\pi_j : (x_1, \dots, x_n) \mapsto x_j$, we have $\Phi_{\pi_j}(c_1, \dots, c_n) = c_j$ for all $(c_1, \dots, c_n) \in \mathfrak{C}^n$. Usually we refer to $\mathfrak{C}$ as the C^∞-ring, leaving the C^∞-operations Φ_f implicit.

A *morphism* of C^∞-rings $\big(\mathfrak{C}, (\Phi_f)_{f:\mathbb{R}^n \to \mathbb{R}\ C^\infty}\big), \big(\mathfrak{D}, (\Psi_f)_{f:\mathbb{R}^n \to \mathbb{R}\ C^\infty}\big)$ is a map $\phi : \mathfrak{C} \to \mathfrak{D}$ such that $\Psi_f\big(\phi(c_1), \dots, \phi(c_n)\big) = \phi \circ \Phi_f(c_1, \dots, c_n)$ for all smooth $f : \mathbb{R}^n \to \mathbb{R}$ and $c_1, \dots, c_n \in \mathfrak{C}$. We will write $\mathbf{C^\infty Rings}$ for the category of C^∞-rings.

Each C^∞-ring has an underlying commutative $\mathbb{R}$-algebra structure. The addition map $f : \mathbb{R}^2 \to \mathbb{R}$, $f : (x, y) \mapsto x + y$ gives addition '+' on $\mathfrak{C}$ as an $\mathbb{R}$-algebra by $c + d = \Phi_f(c, d)$ for $c, d \in \mathfrak{C}$. The multiplication map $g : \mathbb{R}^2 \to \mathbb{R}$, $g : (x, y) \mapsto xy$ gives multiplication '$\cdot$' on $\mathfrak{C}$ by $c \cdot d = \Phi_g(c, d)$. For each $\lambda \in \mathbb{R}$ write $\lambda' : \mathbb{R} \to \mathbb{R}$, $\lambda' : x \mapsto \lambda x$, and then scalar multiplication is $\lambda c = \Phi_{\lambda'}(c)$. Let $0', 1' : \mathbb{R}^0 \to \mathbb{R}$ map $*$ to $0, 1$. Then $0 = \Phi_{0'}$ and $1 = \Phi_{1'}$ are the zero element and identity element for $\mathfrak{C}$. The projection and composition relations show this gives $\mathfrak{C}$ the structure of a commutative $\mathbb{R}$-algebra. However, an $\mathbb{R}$-algebra allows only for operations corresponding to polynomials, whereas a C^∞-ring allows for operations corresponding to all smooth functions and so has a richer structure.

Proposition 2.12 *There is an equivalence* $\mathbf{C^\infty Rings} \cong \mathbf{CC^\infty Rings}$. *This identifies* $\mathfrak{C}$ *in* $\mathbf{C^\infty Rings}$ *with* $F : \mathbf{Euc} \to \mathbf{Sets}$ *in* $\mathbf{CC^\infty Rings}$ *such that* $F(\mathbb{R}^n) = \mathfrak{C}^n$ *for* $n \geqslant 0$, *and for smooth* $f : \mathbb{R}^n \to \mathbb{R}$ *then* $F(f)$ *is identified with* Φ_f.

Proving Proposition 2.12 is straightforward and relies on F being product-preserving. We leave it as an exercise to guide the reader's intuition.

The following example motivates these definitions.

Example 2.13 **(a)** Let X be a manifold. Define a functor $F_X : \mathbf{Euc} \to \mathbf{Sets}$ by $F_X(\mathbb{R}^n) = \mathrm{Hom}_{\mathbf{Man}}(X, \mathbb{R}^n)$, and $F_X(g) = g\circ : \mathrm{Hom}_{\mathbf{Man}}(X, \mathbb{R}^m) \to \mathrm{Hom}_{\mathbf{Man}}(X, \mathbb{R}^n)$ for each morphism $g : \mathbb{R}^m \to \mathbb{R}^n$ in $\mathbf{Euc}$. Then F_X is a categorical C^∞-ring. If $f : X \to Y$ is a smooth map of manifolds, define a natural transformation $F_f : F_Y \Rightarrow F_X$ by $F_f(\mathbb{R}^n) = \circ f : \mathrm{Hom}_{\mathbf{Man}}(Y, \mathbb{R}^n) \to \mathrm{Hom}_{\mathbf{Man}}(X, \mathbb{R}^n)$. Then F_f is a morphism in $\mathbf{CC^\infty Rings}$. Define a functor $F_{\mathbf{Man}}^{\mathbf{CC^\infty Rings}} : \mathbf{Man} \to \mathbf{CC^\infty Rings}^{\mathrm{op}}$ to map $X \mapsto F_X$ and $f \mapsto F_f$.

(b) Let X be a manifold. Write $C^\infty(X)$ for the set of smooth functions $c : X \to \mathbb{R}$. For non-negative integers n and smooth $f : \mathbb{R}^n \to \mathbb{R}$, define C^∞-operations $\Phi_f : C^\infty(X)^n \to C^\infty(X)$ by composition

$$\big(\Phi_f(c_1, \ldots, c_n)\big)(x) = f\big(c_1(x), \ldots, c_n(x)\big), \tag{2.2}$$

for all $c_1, \ldots, c_n \in C^\infty(X)$ and $x \in X$. The composition and projection relations follow directly from the definition of Φ_f, so that $C^\infty(X)$ forms a C^∞-ring. If we consider the $\mathbb{R}$-algebra structure of $C^\infty(X)$ as a C^∞-ring, this is the canonical $\mathbb{R}$-algebra structure on $C^\infty(X)$. If $f : X \to Y$ is a smooth map of manifolds, then $f^* : C^\infty(Y) \to C^\infty(X)$ mapping $c \mapsto c \circ f$ is a morphism of C^∞-rings.

Define $F_{\mathbf{Man}}^{\mathbf{C^\infty Rings}} : \mathbf{Man} \to \mathbf{C^\infty Rings}^{\mathrm{op}}$ to map $X \mapsto C^\infty(X)$ and $f \mapsto f^*$. Moerdijk and Reyes [75, Th. I.2.8] show that $F_{\mathbf{Man}}^{\mathbf{C^\infty Rings}}$ is full and faithful, and takes transverse fibre products in $\mathbf{Man}$ to fibre products in $\mathbf{C^\infty Rings}^{\mathrm{op}}$. This fact is non-trivial, as it relies on knowing that all manifolds X can be embedded as a closed subspace of $\mathbb{R}^n$ for some large n, that $C^\infty(X)$ is a *finitely generated* C^∞-ring (as defined later in Proposition 2.17), and that manifolds admit partitions of unity that behave well with respect to smooth maps.

There are many more C^∞-rings than those that come from manifolds. For example, if X is a smooth manifold of positive dimension, then the set $C^k(X)$ of k-differentiable maps $f : X \to \mathbb{R}$ is a C^∞-ring with operations Φ_f defined as in (2.2), and each of these C^∞-rings is different for all $k = 0, 1, \ldots$.

Example 2.14 Consider $X = *$ the point, so $\dim X = 0$, then $C^\infty(*) = \mathbb{R}$ and Example 2.13 shows the C^∞-operations $\Phi_f : \mathbb{R}^n \to \mathbb{R}$ given by $\Phi_f(x_1, \ldots, x_n) = f(x_1, \ldots, x_n)$ make $\mathbb{R}$ into a C^∞-ring. It is the initial object in $\mathbf{C^\infty Rings}$, and the simplest non-zero example of a C^∞-ring. The zero C^∞-ring is the set $\{0\}$, where all C^∞-operations $\Phi_f : \{0\} \to \{0\}$ send $0 \mapsto 0$, and this is the final object in $\mathbf{C^\infty Rings}$.

Definition 2.15 An *ideal* I in $\mathfrak{C}$ is an ideal in $\mathfrak{C}$ when $\mathfrak{C}$ is considered as a commutative $\mathbb{R}$-algebra. We do not require it to be closed under all C^∞-operations, as this would force $I = \mathfrak{C}$.

We can make the $\mathbb{R}$-algebra quotient $\mathfrak{C}/I$ into a C^∞-ring using Hadamard's Lemma. That is, if $f : \mathbb{R}^n \to \mathbb{R}$ is smooth, define $\Phi_f^I : (\mathfrak{C}/I)^n \to \mathfrak{C}/I$ by

$$\big(\Phi_f^I(c_1 + I, \ldots, c_n + I)\big)(x) = \Phi_f\big(c_1(x), \ldots, c_n(x)\big) + I.$$

Then Hadamard's Lemma says for any smooth function $f : \mathbb{R}^n \to \mathbb{R}$, there exists $g_i : \mathbb{R}^{2n} \to \mathbb{R}$ for $i = 1, \ldots, n$, such that

$$f(x_1, \ldots, x_n) - f(y_1, \ldots, y_n) = \sum_{i=1}^{n} (x_i - y_i) g_i(x_1, \ldots, x_n, y_1, \ldots, y_n).$$

If $d_1, \ldots, d_n$ are alternative choices for $c_1, \ldots, c_n$, then $c_i - d_i \in I$ for each $i = 1, \ldots, n$ and

$$\Phi_f(c_1, \ldots, c_n) - \Phi_f(d_1, \ldots, d_n) = \sum_{i=1}^{n} (c_i - d_i) \Phi_f(c_1, \ldots, c_n, d_1, \ldots, d_n)$$

lies in I, so Φ_f^I is independent of the choice of representatives $c_1, \ldots, c_n$ in $\mathfrak{C}$ and is well defined.

The next definition and proposition come from Adámek et al. [1, Rem. 11.21, Props. 11.26, 11.28, 11.30 and Cor. 11.33].

Definition 2.16 If A is a set then by [1, Rem. 11.21] we can define the *free C^∞-ring $\mathfrak{F}^A$ generated by A*. We may think of $\mathfrak{F}^A$ as $C^\infty(\mathbb{R}^A)$, where $\mathbb{R}^A = \{(x_a)_{a \in A} : x_a \in \mathbb{R}\}$. Explicitly, we define $\mathfrak{F}^A$ to be the set of maps $c : \mathbb{R}^A \to \mathbb{R}$ which depend smoothly on only finitely many variables x_a, and operations Φ_f are defined as in (2.2). Regarding $x_a : \mathbb{R}^A \to \mathbb{R}$ as functions for $a \in A$, we have $x_a \in \mathfrak{F}^A$, and we call x_a the *generators* of $\mathfrak{F}^A$.

Then $\mathfrak{F}^A$ has the universal property that if $\mathfrak{C}$ is any C^∞-ring then a choice of map $\alpha : A \to \mathfrak{C}$ uniquely determines a morphism $\phi : \mathfrak{F}^A \to \mathfrak{C}$ with $\phi(x_a) = \alpha(a)$ for $a \in A$. When $A = \{1, \ldots, n\}$ we have $\mathfrak{F}^A \cong C^\infty(\mathbb{R}^n)$, as in [52, Prop. III.5.1].

Proposition 2.17 **(a)** *Every object $\mathfrak{C}$ in* $\mathbf{C^\infty Rings}$ *admits a surjective morphism $\phi : \mathfrak{F}^A \to \mathfrak{C}$ from some free C^∞-ring $\mathfrak{F}^A$. We call $\mathfrak{C}$* ***finitely generated*** *if this holds with A finite. The kernel of ϕ, $\ker(\phi)$, is an ideal in $\mathfrak{F}^A$ and the quotient $\mathfrak{F}^A / \ker(\phi)$ is isomorphic to $\mathfrak{C}$.*

(b) *Every object $\mathfrak{C}$ in* $\mathbf{C^\infty Rings}$ *fits into a* ***coequalizer diagram***

$$\mathfrak{F}^B \underset{\beta}{\overset{\alpha}{\rightrightarrows}} \mathfrak{F}^A \xrightarrow{\phi} \mathfrak{C}, \tag{2.3}$$

that is, $\mathfrak{C}$ is the colimit of $\mathfrak{F}^B \rightrightarrows \mathfrak{F}^A$ in $\mathbf{C^\infty Rings}$, *where ϕ is automatically surjective. We call $\mathfrak{C}$* ***finitely presented*** *if this holds with A, B finite.*

Actually as any relation $f = g$ in $\mathfrak{C}$ is equivalent to the relation $f - g = 0$, we can simplify (2.3) by taking β to map $x_b \mapsto 0$ for all $b \in B$. This means that finitely presented is equivalent to requiring $\ker(\phi)$ from Proposition 2.17(a) to be finitely generated as an ideal. But the analogue of this for C^∞-rings with corners in §4.5 will not hold. In addition, $C^\infty(\mathbb{R}^n)$ is not noetherian, so ideals in a finitely generated C^∞-ring may not be finitely generated. This implies that finitely presented C^∞-rings are a proper subcategory of finitely generated C^∞-rings, in contrast to ordinary Algebraic Geometry, where they are equal.

We now study limits and colimits in $\mathbf{C^\infty Rings}$. For the pushout of morphisms $\phi : \mathfrak{C} \to \mathfrak{D}$, $\psi : \mathfrak{C} \to \mathfrak{E}$ in $\mathbf{C^\infty Rings}$, we write $\mathfrak{D} \amalg_{\phi,\mathfrak{C},\psi} \mathfrak{E}$ or $\mathfrak{D} \amalg_{\mathfrak{C}} \mathfrak{E}$. In the special case $\mathfrak{C} = \mathbb{R}$ the coproduct $\mathfrak{D} \amalg_{\mathbb{R}} \mathfrak{E}$ will be written as $\mathfrak{D} \otimes_\infty \mathfrak{E}$. Recall that the coproduct of $\mathbb{R}$-algebras A, B is the tensor product $A \otimes B$; however, $\mathfrak{D} \otimes_\infty \mathfrak{E}$ is usually different from their tensor product $\mathfrak{D} \otimes \mathfrak{E}$, as discussed for $C^\infty(\mathbb{R}^n)$ and $C^\infty(\mathbb{R}^m)$ above. For example, for $m, n > 0$, then $C^\infty(\mathbb{R}^m) \otimes_\infty C^\infty(\mathbb{R}^n) \cong C^\infty(\mathbb{R}^{m+n})$ as in [75, p. 22], which contains $C^\infty(\mathbb{R}^m) \otimes C^\infty(\mathbb{R}^n)$ but is larger than this, as it includes elements such as $\exp(fg)$ for $f \in C^\infty(\mathbb{R}^m)$ and $g \in C^\infty(\mathbb{R}^n)$.

By Moerdijk and Reyes [75, pp. 21–22] and Adámek et al. [1, Props. 1.21, 2.5 and Th. 4.5] we have the following.

Proposition 2.18 *The category* $\mathbf{C^\infty Rings}$ *of* C^∞*-rings has all small limits and all small colimits. The forgetful functor* $\Pi : \mathbf{C^\infty Rings} \to \mathbf{Sets}$ *preserves limits and directed colimits, and can be used to compute such (co)limits pointwise; however, it does not preserve general colimits such as pushouts.*

The proof of Proposition 2.18 is straightforward, firstly by proving that the small limits and directed colimits in the category of sets inherit their universal properties and a C^∞-ring structure. Proving separately that coproducts exist follows from the universal property and considering simple cases. This includes that for $m, n > 0$ then the coproduct of $C^\infty(\mathbb{R}^n)$ and $C^\infty(\mathbb{R}^m)$ is $C^\infty(\mathbb{R}^{n+m})$, and that for ideals I, J we have the coproduct

$$\left(C^\infty(\mathbb{R}^n)/I\right) \otimes_\infty \left(C^\infty(\mathbb{R}^m)/J\right) \cong \left(C^\infty(\mathbb{R}^n) \otimes_\infty C^\infty(\mathbb{R}^m)\right)/(I, J).$$

A similar result holds for all finitely generated C^∞-rings. The proof then uses that any C^∞-ring is a directed colimit of finitely generated C^∞-rings to deduce the result.

We will need *local* C^∞-rings and *localizations* of C^∞-rings to define local C^∞-ringed spaces and C^∞-schemes in §2.5.

Definition 2.19 Recall that a *local* $\mathbb{R}$*-algebra* is an $\mathbb{R}$-algebra R with a unique maximal ideal $\mathfrak{m}$. The *residue field* of R is the field (isomorphic to) $R/\mathfrak{m}$. A C^∞-ring $\mathfrak{C}$ is called *local* if, regarded as an $\mathbb{R}$-algebra, $\mathfrak{C}$ is a local $\mathbb{R}$-algebra with residue field $\mathbb{R}$. The quotient morphism gives a (necessarily unique) morphism of C^∞-rings $\pi : \mathfrak{C} \to \mathbb{R}$ with the property that $c \in \mathfrak{C}$ is invertible if and only if $\pi(c) \neq 0$. Equivalently, if such a morphism $\pi : \mathfrak{C} \to \mathbb{R}$ exists with this property, then $\mathfrak{C}$ is local with maximal ideal $\mathfrak{m}_{\mathfrak{C}} \cong \operatorname{Ker} \pi$. Write $\mathbf{C^\infty Rings_{lo}} \subset \mathbf{C^\infty Rings}$ for the full subcategory of local C^∞-rings.

Usually morphisms of local rings are required to send maximal ideals into maximal ideals. However, if $\phi : \mathfrak{C} \to \mathfrak{D}$ is any morphism of local C^∞-rings, we see that $\phi^{-1}(\mathfrak{m}_{\mathfrak{D}}) = \mathfrak{m}_{\mathfrak{C}}$ as the residue fields in both cases are $\mathbb{R}$, so there is no difference between local morphisms and morphisms for C^∞-rings.

Remark 2.20 We use the term 'local C^∞-ring' following Dubuc [23, Def. 4] and the second author [42]. They are known by different names in other references, such as *Archimedean local C^∞-rings* in [73, §3], *C^∞-local rings* in Dubuc [23, Def. 2.13], and *pointed local C^∞-rings* in [75, §I.3]. Moerdijk and Reyes [73, 74, 75] use 'local C^∞-ring' to mean a C^∞-ring which is a local $\mathbb{R}$-algebra, and require no restriction on its residue field.

The next proposition may be found in Moerdijk and Reyes [75, §I.3] and Dubuc [23, Prop. 5].

Proposition 2.21 *All finite colimits exist in $\mathbf{C^\infty Rings_{lo}}$, and agree with the corresponding colimits in $\mathbf{C^\infty Rings}$.*

In [25], the first author shows how to extend this theorem to small colimits, and how small limits also exist in $\mathbf{C^\infty Rings_{lo}}$ but usually do not agree with small limits in $\mathbf{C^\infty Rings}$. A key fact used in the proof is the existence of bump functions for $\mathbb{R}^n$.

Localizations of C^∞-rings were studied in [22, 23, 49, 73, 74, 75].

Definition 2.22 A *localization* $\mathfrak{C}(s^{-1} : s \in S) = \mathfrak{D}$ of a C^∞-ring $\mathfrak{C}$ at a subset $S \subset \mathfrak{C}$ is a C^∞-ring $\mathfrak{D}$ and a morphism $\pi : \mathfrak{C} \to \mathfrak{D}$ such that $\pi(s)$ is invertible in $\mathfrak{D}$ (as an $\mathbb{R}$-algebra) for all $s \in S$, which has the universal property that for any morphism of C^∞-rings $\phi : \mathfrak{C} \to \mathfrak{E}$ such that $\phi(s)$ is invertible in $\mathfrak{E}$ for all $s \in S$, there is a unique morphism $\psi : \mathfrak{D} \to \mathfrak{E}$ with $\phi = \psi \circ \pi$. We call $\pi : \mathfrak{C} \to \mathfrak{D}$ the *localization morphism* for $\mathfrak{D}$.

By adding an extra generator s^{-1} and extra relation $s \cdot s^{-1} - 1 = 0$ for each $s \in S$ to $\mathfrak{C}$, it can be shown that localizations $\mathfrak{C}(s^{-1} : s \in S)$ always exist and are unique up to unique isomorphism. When $S = \{c\}$ then $\mathfrak{C}(c^{-1}) \cong (\mathfrak{C} \otimes_\infty C^\infty(\mathbb{R}))/I$, where I is the ideal generated by $\iota_1(c) \cdot \iota_2(x) - 1$, and x is the generator of $C^\infty(\mathbb{R})$, and ι_1, ι_2 are the coproduct morphisms $\iota_1 : \mathfrak{C} \to \mathfrak{C} \otimes_\infty C^\infty(\mathbb{R})$ and $\iota_2 : C^\infty(\mathbb{R}) \to \mathfrak{C} \otimes_\infty C^\infty(\mathbb{R})$.

An example of this is that if $f \in C^\infty(\mathbb{R}^n)$ is a smooth function, and $U = f^{-1}(\mathbb{R} \setminus \{0\}) \subseteq \mathbb{R}^n$, then using partitions of unity one can show that $C^\infty(U) \cong C^\infty(\mathbb{R}^n)(f^{-1})$, as in [75, Prop. I.1.6].

Definition 2.23 A C^∞-ring morphism $x : \mathfrak{C} \to \mathbb{R}$, where $\mathbb{R}$ is regarded as a C^∞-ring as in Example 2.14, is called an $\mathbb{R}$-*point*. Note that a map $x : \mathfrak{C} \to \mathbb{R}$ is a morphism of C^∞-rings whenever it is a morphism of the underlying $\mathbb{R}$-algebras, as in [75, Prop. I.3.6]. We define $\mathfrak{C}_x$ as the localization $\mathfrak{C}_x = \mathfrak{C}(s^{-1} : s \in \mathfrak{C},\ x(s) \neq 0)$, and denote the projection morphism by $\pi_x : \mathfrak{C} \to \mathfrak{C}_x$. Importantly, [74, Lem. 1.1] shows $\mathfrak{C}_x$ is a local C^∞-ring.

We will use $\mathbb{R}$-points $x : \mathfrak{C} \to \mathbb{R}$ to define our spectrum functor in §2.5. We can describe $\mathfrak{C}_x$ explicitly as in [49, Prop. 2.14].

Proposition 2.24 *Let $x : \mathfrak{C} \to \mathbb{R}$ be an $\mathbb{R}$-point of a C^∞-ring $\mathfrak{C}$, and consider the projection morphism $\pi_x : \mathfrak{C} \to \mathfrak{C}_x$. Then $\mathfrak{C}_x \cong \mathfrak{C}/\operatorname{Ker}\pi_x$. This kernel is* $\operatorname{Ker}\pi_x = I$, *where*

$$I = \{c \in \mathfrak{C} : \text{there exists } d \in \mathfrak{C} \text{ with } x(d) \neq 0 \text{ in } \mathbb{R} \text{ and } c \cdot d = 0 \text{ in } \mathfrak{C}\}. \quad (2.4)$$

While this localization morphism $\pi_x : \mathfrak{C} \to \mathfrak{C}_x$ is surjective, general localizations of C^∞-rings need not have surjective localization morphisms. Here is an important example of localization of C^∞-rings.

Example 2.25 Let $C^\infty_p(\mathbb{R}^n)$ be the set of germs of smooth functions $c : \mathbb{R}^n \to \mathbb{R}$ at $p \in \mathbb{R}^n$ for $n \geqslant 0$ and $p \in \mathbb{R}^n$. We give $C^\infty_p(\mathbb{R}^n)$ a C^∞-ring structure by using (2.2) on germs of functions. There are several equivalent definitions.

(i) $C^\infty_p(\mathbb{R}^n)$ is the set of $\sim$-equivalence classes $[U, c]$ of pairs (U, c), where $p \in U \subseteq \mathbb{R}^n$ is open and $c : U \to \mathbb{R}$ is smooth, and $(U, c) \sim (U', c')$ if there exists open $p \in U'' \subseteq U \cap U'$ with $c|_{U''} = c'|_{U''}$.
(ii) $C^\infty_p(\mathbb{R}^n) \cong C^\infty(\mathbb{R}^n)/I_p$, where $I_p \subset C^\infty(\mathbb{R}^n)$ is the ideal of functions vanishing near p.
(iii) $C^\infty_p(\mathbb{R}^n) \cong C^\infty(\mathbb{R}^n)(f^{-1} : f \in C^\infty(\mathbb{R}^n),\ f(p) \neq 0)$.

Then $C^\infty_p(\mathbb{R}^n)$ is local, with maximal ideal $\mathfrak{m}_p = \{[U, c] \in C^\infty_p(\mathbb{R}^n) : c(x) = 0\}$.

Finally, we prove some facts about exponentials and logs in (local) C^∞-rings. These will be used in defining C^∞-rings with corners.

Proposition 2.26 **(a)** *Let $\mathfrak{C}$ be a C^∞-ring. Then the C^∞-operation $\Phi_{\exp} : \mathfrak{C} \to \mathfrak{C}$ induced by* $\exp : \mathbb{R} \to \mathbb{R}$ *is injective.*

(b) *Let $\mathfrak{C}$ be a local C^∞-ring, with morphism $\pi : \mathfrak{C} \to \mathbb{R}$. If $a \in \mathfrak{C}$ with $\pi(a) > 0$ then there exists $b \in \mathfrak{C}$ with $\Phi_{\exp}(b) = a$. This b is unique by* **(a)**.

Proof For (a), let $a \in \mathfrak{C}$ with $b = \Phi_{\exp}(a) \in \mathfrak{C}$. Then $\Phi_{\exp}(-a)$ is the inverse b^{-1} of b. The map $t \mapsto \exp(t) - \exp(-t)$ is a diffeomorphism $\mathbb{R} \to \mathbb{R}$. Let $e : \mathbb{R} \to \mathbb{R}$ be its inverse. Define smooth $f : \mathbb{R}^2 \to \mathbb{R}$ by $f(x,y) = e(x-y)$. Then $f(\exp t, \exp(-t)) = t$. Hence in the C^∞-ring $\mathfrak{C}$ we have

$$\Phi_f(b, b^{-1}) = \Phi_f(\Phi_{\exp}(a), \Phi_{\exp\circ -}(a)) = \Phi_{f\circ(\exp,\exp\circ -)}(a) = \Phi_{\mathrm{id}}(a) = a.$$

But b determines b^{-1} uniquely, so $\Phi_f(b, b^{-1}) = a$ implies that $b = \Phi_{\exp}(a)$ determines a uniquely, and $\Phi_{\exp} : \mathfrak{C} \to \mathfrak{C}$ is injective.

For (b), choose smooth $g, h : \mathbb{R} \to \mathbb{R}$ with $g(x) = \log x$ for $x \geqslant \frac{1}{2}\pi(a) > 0$, $h(\pi(a)) > 0$, and $h(x) = 0$ for $x \leqslant \frac{1}{2}\pi(a)$. Set $b = \Phi_g(a)$ and $c = \Phi_h(a)$. Then

$$c \cdot (\Phi_{\exp}(b) - a) = \Phi_h(a) \cdot (\Phi_{\exp\circ g}(a) - a) = \Phi_{h(x)\cdot(\exp\circ g(x) - x)}(a) = 0,$$

since $h(x) \cdot (x - \exp\circ g(x)) = 0$. Also $\pi(c) = h(\pi(a)) > 0$, so c is invertible. Thus $\Phi_{\exp}(b) = a$. □

2.3 Modules and cotangent modules of C^∞-rings

We discuss modules and cotangent modules for C^∞-rings, following [49, §5].

Definition 2.27 A *module* M over a C^∞-ring $\mathfrak{C}$ is a module over $\mathfrak{C}$ as a commutative $\mathbb{R}$-algebra, and morphisms of $\mathfrak{C}$-modules are the usual morphisms of $\mathbb{R}$-algebra modules. Denote $\mu_M : \mathfrak{C} \times M \to M$ the multiplication map, and write $\mu_M(c, m) = c \cdot m$ for $c \in \mathfrak{C}$ and $m \in M$. The category $\mathfrak{C}$-mod of $\mathfrak{C}$-modules is an abelian category.

If a $\mathfrak{C}$-module M fits into an exact sequence $\mathfrak{C} \otimes \mathbb{R}^n \to M \to 0$ in $\mathfrak{C}$-mod then it is *finitely generated*; if it further fits into an exact sequence $\mathfrak{C} \otimes \mathbb{R}^m \to \mathfrak{C} \otimes \mathbb{R}^n \to M \to 0$ it is *finitely presented*. This second condition is not automatic from the first as C^∞-rings are generally not noetherian.

For a morphism $\phi : \mathfrak{C} \to \mathfrak{D}$ of C^∞-rings and M in $\mathfrak{C}$-mod we have $\phi_*(M) = M \otimes_{\mathfrak{C}} \mathfrak{D}$ in $\mathfrak{D}$-mod, giving a functor $\phi_* : \mathfrak{C}\text{-mod} \to \mathfrak{D}\text{-mod}$. For N in $\mathfrak{D}$-mod there is a $\mathfrak{C}$-module $\phi^*(N) = N$ with $\mathfrak{C}$-action $\mu_{\phi^*(N)}(c, n) = \mu_N(\phi(c), n)$. This gives a functor $\phi^* : \mathfrak{D}\text{-mod} \to \mathfrak{C}\text{-mod}$.

Example 2.28 Let $\Gamma^\infty(E)$ be the collection of smooth sections e of a vector bundle $E \to X$ of a manifold X, so $\Gamma^\infty(E)$ is a vector space and a module over $C^\infty(X)$. If $\lambda : E \to F$ is a morphism of vector bundles over X, then there is a morphism of $C^\infty(X)$-modules $\lambda_* : \Gamma^\infty(E) \to \Gamma^\infty(F)$, where $\lambda_* : e \mapsto \lambda \circ e$.

For each smooth map of manifolds $f : X \to Y$ there is a morphism of

C^∞-rings $f^* : C^\infty(Y) \to C^\infty(X)$. Each vector bundle $E \to Y$ gives a vector bundle $f^*(E) \to X$. Using $(f^*)_* : C^\infty(Y)\text{-mod} \to C^\infty(X)\text{-mod}$ from Definition 2.27, then $(f^*)_*(\Gamma^\infty(E)) = \Gamma^\infty(E) \otimes_{C^\infty(Y)} C^\infty(X)$ is isomorphic to $\Gamma^\infty(f^*(E))$ in $C^\infty(X)$-mod.

The definition of $\mathfrak{C}$-module used only the commutative $\mathbb{R}$-algebra structure of $\mathfrak{C}$; however, the *cotangent module* $\Omega_{\mathfrak{C}}$ of $\mathfrak{C}$ uses the full C^∞-ring structure, making it smaller in some sense than the corresponding classical Algebraic Geometry version of *Kähler differentials* from [36, II 8].

Definition 2.29 Take a C^∞-ring $\mathfrak{C}$ and $M \in \mathfrak{C}$-mod, then a *C^∞-derivation* is a map $\mathrm{d} : \mathfrak{C} \to M$ that satisfies the following: for any smooth $f : \mathbb{R}^n \to \mathbb{R}$ and elements $c_1, \ldots, c_n \in \mathfrak{C}$, then

$$\mathrm{d}\Phi_f(c_1, \ldots, c_n) - \sum_{i=1}^{n} \Phi_{\frac{\partial f}{\partial x_i}}(c_1, \ldots, c_n) \cdot \mathrm{d}c_i = 0. \tag{2.5}$$

This implies that d is $\mathbb{R}$-linear and is a derivation of $\mathfrak{C}$ as a commutative $\mathbb{R}$-algebra, that is, $\mathrm{d}(c_1c_2) = c_1 \cdot \mathrm{d}c_2 + c_2 \cdot \mathrm{d}c_1$ for all $c_1, c_2 \in \mathfrak{C}$.

The pair (M, d) is called a *cotangent module* for $\mathfrak{C}$ if it is universal in the sense that for any $M' \in \mathfrak{C}$-mod with C^∞-derivation $\mathrm{d}' : \mathfrak{C} \to M'$, there exists a unique morphism of $\mathfrak{C}$-modules $\lambda : M \to M'$ with $\mathrm{d}' = \lambda \circ \mathrm{d}$. Then a cotangent module is unique up to unique isomorphism. We can explicitly construct a cotangent module for $\mathfrak{C}$ by considering the free $\mathfrak{C}$-module over the symbols $\mathrm{d}c$ for all $c \in \mathfrak{C}$, and quotienting by all relations (2.5) for smooth $f : \mathbb{R}^n \to \mathbb{R}$ and elements $c_1, \ldots, c_n \in \mathfrak{C}$. We call this construction 'the' cotangent module of $\mathfrak{C}$, and write it as $\mathrm{d}_{\mathfrak{C}} : \mathfrak{C} \to \Omega_{\mathfrak{C}}$.

If we have a morphism of C^∞-rings $\mathfrak{C} \to \mathfrak{D}$ then $\Omega_{\mathfrak{D}} = \phi^*(\Omega_{\mathfrak{D}})$ can be considered as a $\mathfrak{C}$-module with C^∞-derivation $\mathrm{d}_{\mathfrak{D}} \circ \phi : \mathfrak{C} \to \Omega_{\mathfrak{D}}$. The universal property of $\Omega_{\mathfrak{C}}$ gives a unique morphism $\Omega_\phi : \Omega_{\mathfrak{C}} \to \Omega_{\mathfrak{D}}$ of $\mathfrak{C}$-modules such that $\mathrm{d}_{\mathfrak{D}} \circ \phi = \Omega_\phi \circ \mathrm{d}_{\mathfrak{C}}$. From this we have a morphism of $\mathfrak{D}$-modules $(\Omega_\phi)_* : \Omega_{\mathfrak{C}} \otimes_{\mathfrak{C}} \mathfrak{D} \to \Omega_{\mathfrak{D}}$. If we have two morphisms of C^∞-rings $\phi : \mathfrak{C} \to \mathfrak{D}$, $\psi : \mathfrak{D} \to \mathfrak{E}$ then uniqueness implies that $\Omega_{\psi \circ \phi} = \Omega_\psi \circ \Omega_\phi : \Omega_{\mathfrak{C}} \to \Omega_{\mathfrak{E}}$.

Here is our motivating example.

Example 2.30 As in Example 2.28, if X is a manifold, its cotangent bundle T^*X is a vector bundle over X, and its global sections $\Gamma^\infty(T^*X)$ form a $C^\infty(X)$-module, with C^∞-derivation $\mathrm{d} : C^\infty(X) \to \Gamma^\infty(T^*X)$, $\mathrm{d} : c \mapsto \mathrm{d}c$ the usual exterior derivative, and equation (2.5) following from the chain rule.

One can show that $(\Gamma^\infty(T^*X), \mathrm{d})$ has the universal property in Definition 2.29, and so forms a cotangent module for $C^\infty(X)$. This is stated in [49, Ex. 5.4], and proved in greater generality in Theorem 7.7(a) below.

If we have a smooth map of manifolds $f : X \to Y$, then $f^*(T^*Y)$, T^*X are vector bundles over X, and the derivative $\mathrm{d}f : f^*(T^*Y) \to T^*X$ is a vector bundle morphism. This induces a morphism of $C^\infty(X)$-modules $(\mathrm{d}f)_* : \Gamma^\infty(f^*(T^*Y)) \to \Gamma^\infty(T^*X)$, which is identified with $(\Omega_{f^*})_*$ from Definition 2.29 using that $\Gamma^\infty(f^*(T^*Y)) \cong \Gamma^\infty(T^*Y) \otimes_{C^\infty(Y)} C^\infty(X)$.

This example shows that Definition 2.29 abstracts the notion of sections of a cotangent bundle of a manifold to a concept that is well defined for any C^∞-ring. Here are further helpful examples that we later generalize for C^∞-rings with corners.

Example 2.31 **(a)** Let A be a set and $\mathfrak{F}^A$ the free C^∞-ring from Definition 2.16, with generators $x_a \in \mathfrak{F}^A$ for $a \in A$. Then there is a natural isomorphism

$$\Omega_{\mathfrak{F}^A} \cong \langle \mathrm{d}x_a : a \in A \rangle_{\mathbb{R}} \otimes_{\mathbb{R}} \mathfrak{F}^A.$$

(b) Suppose $\mathfrak{C}$ is defined by a coequalizer diagram (2.3) in $\mathbf{C^\infty Rings}$. Then writing $(x_a)_{a\in A}$, $(\tilde{x}_b)_{b\in B}$ for the generators of $\mathfrak{F}^A, \mathfrak{F}^B$, we have an exact sequence

$$\langle \mathrm{d}\tilde{x}_b : b \in B \rangle_{\mathbb{R}} \otimes_{\mathbb{R}} \mathfrak{C} \xrightarrow{\gamma} \langle \mathrm{d}x_a : a \in A \rangle_{\mathbb{R}} \otimes_{\mathbb{R}} \mathfrak{C} \xrightarrow{\delta} \Omega_{\mathfrak{C}} \longrightarrow 0$$

in $\mathfrak{C}$-mod, where if α, β in (2.3) map $\alpha : \tilde{x}_b \mapsto f_b\big((x_a)_{a\in A}\big)$, $\beta : \tilde{x}_b \mapsto g_b\big((x_a)_{a\in A}\big)$, for f_b, g_b depending only on finitely many x_a, then γ, δ are given by

$$\gamma(\mathrm{d}\tilde{x}_b) = \sum_{a\in A} \phi\Big(\frac{\partial f_b}{\partial x_a}\big((x_a)_{a\in A}\big) - \frac{\partial g_b}{\partial x_a}\big((x_a)_{a\in A}\big)\Big)\mathrm{d}x_a, \; \delta(\mathrm{d}x_a) = \mathrm{d}_{\mathfrak{C}} \circ \phi(x_a).$$

Hence as in [49, Prop. 5.6], if $\mathfrak{C}$ is finitely generated (or finitely presented) in the sense of Proposition 2.17, then $\Omega_{\mathfrak{C}}$ is finitely generated (or finitely presented).

Cotangent modules behave well under localization, as in the following proposition from [49, Prop. 5.7]. This proposition is used to prove the theorem that follows it, and we use it to generalize these results to C^∞-rings with corners.

Proposition 2.32 *Let $\mathfrak{C}$ be a C^∞-ring, $S \subseteq \mathfrak{C}$, and let $\mathfrak{D} = \mathfrak{C}(s^{-1} : s \in S)$ be the localization of $\mathfrak{C}$ at S with projection $\pi : \mathfrak{C} \to \mathfrak{D}$, as in Definition 2.22. Then $(\Omega_\pi)_* : \Omega_{\mathfrak{C}} \otimes_{\mathfrak{C}} \mathfrak{D} \to \Omega_{\mathfrak{D}}$ is an isomorphism of $\mathfrak{D}$-modules.*

Finally, here is [49, Th. 5.8] which shows how pushouts of C^∞-rings give exact sequences of cotangent modules.

Theorem 2.33 *Suppose we are given a pushout diagram of C^∞-rings:*

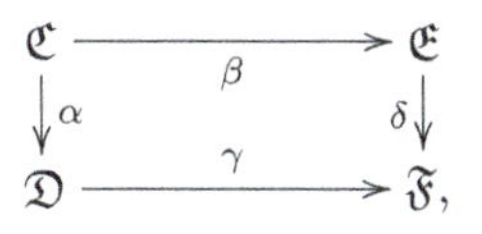

so that $\mathfrak{F} = \mathfrak{D} \amalg_{\mathfrak{C}} \mathfrak{E}$. Then the following sequence of $\mathfrak{F}$-modules is exact:

$$\Omega_{\mathfrak{C}} \otimes_{\mathfrak{C},\gamma\circ\alpha} \mathfrak{F} \xrightarrow{(\Omega_\alpha)_* \oplus -(\Omega_\beta)_*} \begin{array}{c} \Omega_{\mathfrak{D}} \otimes_{\mathfrak{D},\gamma} \mathfrak{F} \\ \oplus\, \Omega_{\mathfrak{E}} \otimes_{\mathfrak{E},\delta} \mathfrak{F} \end{array} \xrightarrow{(\Omega_\gamma)_* \oplus (\Omega_\delta)_*} \Omega_{\mathfrak{F}} \longrightarrow 0. \quad (2.6)$$

Here $(\Omega_\alpha)_* : \Omega_{\mathfrak{C}} \otimes_{\mathfrak{C},\gamma\circ\alpha} \mathfrak{F} \to \Omega_{\mathfrak{D}} \otimes_{\mathfrak{D},\gamma} \mathfrak{F}$ is induced by $\Omega_\alpha : \Omega_{\mathfrak{C}} \to \Omega_{\mathfrak{D}}$, and so on. Note the sign of $-(\Omega_\beta)_*$ in (2.6).

2.4 Sheaves

In this section we explain presheaves and sheaves with values in a (nice) category $\mathcal{A}$, following Godement [33] and MacLane and Moerdijk [65]. Throughout we suppose $\mathcal{A}$ is *complete*, that is, all small limits exist in $\mathcal{A}$, and *cocomplete*, that is, all small colimits exist in $\mathcal{A}$. The categories of sets, abelian groups, rings, C^∞-rings, monoids, etc., all satisfy this, as will (interior) C^∞-rings with corners. Sometimes it is helpful to suppose that objects of $\mathcal{A}$ are sets with extra structure, so there is a faithful functor $\mathcal{A} \to$ **Sets** taking each object to its underlying set. We use presheaves and sheaves and the facts that follow to define and study local C^∞-ringed spaces and C^∞-schemes.

Definition 2.34 A *presheaf* $\mathcal{E}$ on a topological space X valued in $\mathcal{A}$ gives an object $\mathcal{E}(U) \in \mathcal{A}$ for every open set $U \subseteq X$, and a morphism $\rho_{UV} : \mathcal{E}(U) \to \mathcal{E}(V)$ in $\mathcal{A}$ called the *restriction map* for every inclusion $V \subseteq U \subseteq X$ of open sets, satisfying the conditions that

(i) $\rho_{UU} = \mathrm{id}_{\mathcal{E}(U)} : \mathcal{E}(U) \to \mathcal{E}(U)$ for all open $U \subseteq X$; and
(ii) $\rho_{UW} = \rho_{VW} \circ \rho_{UV} : \mathcal{E}(U) \to \mathcal{E}(W)$ for all open $W \subseteq V \subseteq U \subseteq X$.

A presheaf $\mathcal{E}$ is called a *sheaf* if for all open covers $\{U_i\}_{i\in I}$ of U, then

$$\mathcal{E}(U) \to \prod_{i\in I} \mathcal{E}(U_i) \rightrightarrows \prod_{i,j\in I} \mathcal{E}(U_i \cap U_j)$$

forms an equalizer diagram in $\mathcal{A}$. This implies the following.

(iii) $\mathcal{E}(\emptyset) = 0$, where 0 is the final object in $\mathcal{A}$.

If there is a faithful functor $F : \mathcal{A} \to \mathbf{Sets}$ taking an object of $\mathcal{A}$ to its underlying set that preserves limits, then a presheaf $\mathcal{E}$ valued in $\mathcal{A}$ on X is a sheaf if it equivalently satisfies the following.

(iv) (Uniqueness.) If $U \subseteq X$ is open, $\{V_i : i \in I\}$ is an open cover of U, and $s, t \in F(\mathcal{E}(U))$ with $F(\rho_{UV_i})(s) = F(\rho_{UV_i})(t)$ in $F(\mathcal{E}(V_i))$ for all $i \in I$, then $s = t$ in $F(\mathcal{E}(U))$.

(v) (Gluing.) If $U \subseteq X$ is open, $\{V_i : i \in I\}$ is an open cover of U, and we are given elements $s_i \in F(\mathcal{E}(V_i))$ for all $i \in I$ such that $F(\rho_{V_i(V_i \cap V_j)})(s_i) = F(\rho_{V_j(V_i \cap V_j)})(s_j)$ in $F(\mathcal{E}(V_i \cap V_j))$ for all $i, j \in I$, then there exists $s \in F(\mathcal{E}(U))$ with $F(\rho_{UV_i})(s) = s_i$ for all $i \in I$.

If $s \in F(\mathcal{E}(U))$ and $V \subseteq U$ is open we write $s|_V = F(\rho_{UV})(s)$.

If $\mathcal{E}, \mathcal{F}$ are presheaves or sheaves valued in $\mathcal{A}$ on X, then a *morphism* $\phi : \mathcal{E} \to \mathcal{F}$ is a morphism $\phi(U) : \mathcal{E}(U) \to \mathcal{F}(U)$ in $\mathcal{A}$ for all open $U \subseteq X$ such that the following diagram commutes for all open $V \subseteq U \subseteq X$

$$\begin{array}{ccc} \mathcal{E}(U) & \xrightarrow{\phi(U)} & \mathcal{F}(U) \\ \downarrow{\rho_{UV}} & & \downarrow{\rho'_{UV}} \\ \mathcal{E}(V) & \xrightarrow{\phi(V)} & \mathcal{F}(V), \end{array}$$

where ρ_{UV} is the restriction map for $\mathcal{E}$, and ρ'_{UV} the restriction map for $\mathcal{F}$. We write $\mathrm{PreSh}(X, \mathcal{A})$ and $\mathrm{Sh}(X, \mathcal{A})$ for the categories of presheaves and sheaves on a topological space X valued in $\mathcal{A}$.

Definition 2.35 For $\mathcal{E}$ a presheaf valued in $\mathcal{A}$ on a topological space X, then we can define the *stalk* $\mathcal{E}_x \in \mathcal{A}$ at a point $x \in X$ to be the direct limit of the $\mathcal{E}(U)$ in $\mathcal{A}$ for all $U \subseteq X$ with $x \in U$, using the restriction maps ρ_{UV}.

If there is a faithful functor $F : \mathcal{A} \to \mathbf{Sets}$ taking an object of $\mathcal{A}$ to its underlying set that preserves colimits, then explicitly it can be written as a set of equivalence classes of sections $s \in F(\mathcal{E}(U))$ for any open U which contains x, where the equivalence relation is such that $s_1 \sim s_2$ for $s_1 \in F(\mathcal{E}(U))$ and $s_2 \in F(\mathcal{E}(V))$ with $x \in U, V$ if there is an open set $W \subset V \cap U$ with $x \in W$ and $s_1|_W = s_2|_W$ in $F(\mathcal{E}(W))$.

The stalk is an object of $\mathcal{A}$, and the restriction morphisms give rise to morphisms $\rho_{U,x} : \mathcal{E}(U) \to \mathcal{E}_x$. A morphism of presheaves $\phi : \mathcal{E} \to \mathcal{F}$ induces morphisms $\phi_x : \mathcal{E}_x \to \mathcal{F}_x$ for all $x \in X$. If $\mathcal{E}, \mathcal{F}$ are sheaves then ϕ is an isomorphism if and only if ϕ_x is an isomorphism for all $x \in X$.

Definition 2.36 There is a *sheafification* functor $\mathrm{PreSh}(X, \mathcal{A}) \to \mathrm{Sh}(X, \mathcal{A})$, which is left adjoint to the inclusion $\mathrm{Sh}(X, \mathcal{A}) \hookrightarrow \mathrm{PreSh}(X, \mathcal{A})$. We write $\hat{\mathcal{E}}$ for the sheafification of a presheaf $\mathcal{E}$. The adjoint property gives a morphism

$\pi : \mathcal{E} \to \hat{\mathcal{E}}$ and a universal property: whenever we have a morphism $\phi : \mathcal{E} \to \mathcal{F}$ of presheaves on X and $\mathcal{F}$ is a sheaf, then there is a unique morphism $\hat{\phi} : \hat{\mathcal{E}} \to \mathcal{F}$ with $\phi = \hat{\phi} \circ \pi$. Thus sheafification is unique up to canonical isomorphism.

Sheafifications always exist for our categories $\mathcal{A}$, and there are isomorphisms of stalks $\mathcal{E}_x \cong \hat{\mathcal{E}}_x$ for all $x \in X$. If there is a faithful functor $F : \mathcal{A} \to \mathbf{Sets}$ taking an object of $\mathcal{A}$ to its underlying set that preserves colimits and limits, it can be constructed (as in [36, Prop. II.1.2]) by defining $\hat{\mathcal{E}}(U)$ as the subset of all functions $t : U \to \amalg_{x\in U}\mathcal{E}_x$ such that for all $x \in U$, then $t(x) = F(\rho_{V,x})(s) \in \mathcal{E}_x$ for some $s \in F(\mathcal{E}(V))$ for open $V \subset U, x \in V$.

If $f : X \to Y$ is a continuous map of topological spaces, we can consider *pushforwards* and *pullbacks* of sheaves by f. We will use both of these definitions when defining C^∞-schemes (with corners).

Definition 2.37 If $f : X \to Y$ is a continuous map of topological spaces, and $\mathcal{E}$ is a sheaf valued in $\mathcal{A}$ on X, then the *direct image* (or *pushforward*) sheaf $f_*(\mathcal{E})$ on Y is defined by $\big(f_*(\mathcal{E})\big)(U) = \mathcal{E}\big(f^{-1}(U)\big)$ for all open $U \subseteq V$. Here, we have restriction maps $\rho'_{UV} = \rho_{f^{-1}(U)f^{-1}(V)} : \big(f_*(\mathcal{E})\big)(U) \to \big(f_*(\mathcal{E})\big)(V)$ for all open $V \subseteq U \subseteq Y$ so that $f_*(\mathcal{E})$ is a sheaf valued in $\mathcal{A}$ on Y.

For a morphism $\phi : \mathcal{E} \to \mathcal{F}$ in $\mathrm{Sh}(X, \mathcal{A})$ we can define $f_*(\phi) : f_*(\mathcal{E}) \to f_*(\mathcal{F})$ by $\big(f_*(\phi)\big)(U) = \phi\big(f^{-1}(U)\big)$ for all open $U \subseteq Y$. This gives a morphism $f_*(\phi)$ in $\mathrm{Sh}(Y, \mathcal{A})$, and a functor $f_* : \mathrm{Sh}(X, \mathcal{A}) \to \mathrm{Sh}(Y, \mathcal{A})$. For two continuous maps of topological spaces, $f : X \to Y$, $g : Y \to Z$, then $(g \circ f)_* = g_* \circ f_*$.

Definition 2.38 For a continuous map $f : X \to Y$ and a sheaf $\mathcal{E}$ valued in $\mathcal{A}$ on Y, we define the *pullback* (*inverse image*) of $\mathcal{E}$ under f to be the sheafification of the presheaf $U \mapsto \lim_{A\supseteq f(U)} \mathcal{E}(A)$ for open $U \subseteq X$, where the direct limit is taken over all open $A \subseteq Y$ containing $f(U)$, using the restriction maps ρ_{AB} in $\mathcal{E}$. We write this sheaf as $f^{-1}(\mathcal{E})$. If $\phi : \mathcal{E} \to \mathcal{F}$ is a morphism in $\mathrm{Sh}(Y, \mathcal{A})$, there is a *pullback morphism* $f^{-1}(\phi) : f^{-1}(\mathcal{E}) \to f^{-1}(\mathcal{F})$.

Remark 2.39 For a continuous map $f : X \to Y$ of topological spaces we have functors $f_* : \mathrm{Sh}(X, \mathcal{A}) \to \mathrm{Sh}(Y, \mathcal{A})$, and $f^{-1} : \mathrm{Sh}(Y, \mathcal{A}) \to \mathrm{Sh}(X, \mathcal{A})$. Hartshorne [36, Ex. II.1.18] gives a natural bijection

$$\mathrm{Hom}_X\big(f^{-1}(\mathcal{E}), \mathcal{F}\big) \cong \mathrm{Hom}_Y\big(\mathcal{E}, f_*(\mathcal{F})\big) \tag{2.7}$$

for all $\mathcal{E} \in \mathrm{Sh}(Y, \mathcal{A})$ and $\mathcal{F} \in \mathrm{Sh}(X, \mathcal{A})$, so that f_* is right adjoint to f^{-1}, as in §2.1.3. This will be important in several proofs later.

2.5 C^∞-schemes

We recall the definition of (local) C^∞-ringed spaces, following [49]. These allow us to define a spectrum functor and C^∞-schemes.

Definition 2.40 A C^∞*-ringed space* $\underline{X} = (X, \mathcal{O}_X)$ is a topological space X with a sheaf $\mathcal{O}_X$ of C^∞-rings on X.

A *morphism* $\underline{f} = (f, f^\sharp) : (X, \mathcal{O}_X) \to (Y, \mathcal{O}_Y)$ of C^∞ ringed spaces consists of a continuous map $f : X \to Y$ and a morphism $f^\sharp : f^{-1}(\mathcal{O}_Y) \to \mathcal{O}_X$ of sheaves of C^∞-rings on X, for $f^{-1}(\mathcal{O}_Y)$ the inverse image sheaf as in Definition 2.38. From (2.7), we know f_* is right adjoint to f^{-1}, so there is a natural bijection

$$\mathrm{Hom}_X\big(f^{-1}(\mathcal{O}_Y), \mathcal{O}_X\big) \cong \mathrm{Hom}_Y\big(\mathcal{O}_Y, f_*(\mathcal{O}_X)\big). \tag{2.8}$$

We will write $f_\sharp : \mathcal{O}_Y \to f_*(\mathcal{O}_X)$ for the morphism of sheaves of C^∞-rings on Y corresponding to the morphism $f^\sharp$ under (2.8), so that

$$f^\sharp : f^{-1}(\mathcal{O}_Y) \longrightarrow \mathcal{O}_X \quad \rightsquigarrow \quad f_\sharp : \mathcal{O}_Y \longrightarrow f_*(\mathcal{O}_X). \tag{2.9}$$

Given two C^∞-ringed space morphisms $\underline{f} : \underline{X} \to \underline{Y}$ and $\underline{g} : \underline{Y} \to \underline{Z}$ we can compose them to form

$$\underline{g} \circ \underline{f} = \big(g \circ f, (g \circ f)^\sharp\big) = \big(g \circ f, f^\sharp \circ f^{-1}(g^\sharp)\big).$$

If we consider $f_\sharp : \mathcal{O}_Y \to f_*(\mathcal{O}_X)$, then the composition is

$$(g \circ f)_\sharp = g_*(f_\sharp) \circ g_\sharp : \mathcal{O}_Z \longrightarrow (g \circ f)_*(\mathcal{O}_X) = g_* \circ f_*(\mathcal{O}_X).$$

We call $\underline{X} = (X, \mathcal{O}_X)$ a *local* C^∞*-ringed space* if it is C^∞-ringed space for which the stalks $\mathcal{O}_{X,x}$ of $\mathcal{O}_X$ at x are local C^∞-rings for all $x \in X$. As in Definition 2.19, since morphisms of local C^∞-rings are automatically local morphisms, morphisms of local C^∞-ringed spaces $(X, \mathcal{O}_X), (Y, \mathcal{O}_Y)$ are just morphisms of C^∞-ringed spaces without any additional locality condition. Local C^∞-ringed spaces are called *Archimedean* C^∞*-spaces* in Moerdijk, van Quê, and Reyes [73, §3].

We will follow the notation of [49] and write $\mathbf{C^\infty RS}$ for the category of C^∞-ringed spaces, and $\mathbf{LC^\infty RS}$ for the full subcategory of local C^∞-ringed spaces. We write underlined upper case letters such as $\underline{X}, \underline{Y}, \underline{Z}, \ldots$ to represent C^∞-ringed spaces $(X, \mathcal{O}_X), (Y, \mathcal{O}_Y), (Z, \mathcal{O}_Z), \ldots$, and underlined lower case letters $\underline{f}, \underline{g}, \ldots$ to represent morphisms of C^∞-ringed spaces $(f, f^\sharp), (g, g^\sharp), \ldots$. When we write '$x \in \underline{X}$' we mean that $\underline{X} = (X, \mathcal{O}_X)$ and $x \in X$. If we write '$\underline{U}$ *is open in* $\underline{X}$' we will mean that $\underline{U} = (U, \mathcal{O}_U)$ and $\underline{X} = (X, \mathcal{O}_X)$ with $U \subseteq X$ an open set and $\mathcal{O}_U = \mathcal{O}_X|_U$.

Here is our motivating example.

Example 2.41 For a manifold X, we have a C^∞-ringed space $\underline{X} = (X, \mathcal{O}_X)$ with topological space X and its sheaf of smooth functions $\mathcal{O}_X(U) = C^\infty(U)$ for each open subset $U \subseteq X$, with $C^\infty(U)$ defined in Example 2.13. If $V \subseteq U \subseteq X$ then the restriction morphisms $\rho_{UV} : C^\infty(U) \to C^\infty(V)$ are the usual restriction of a function to an open subset $\rho_{UV} : c \mapsto c|_V$.

As the stalks $\mathcal{O}_{X,x}$ at $x \in X$ are local C^∞-rings, isomorphic to the ring of germs as in Example 2.25, then using partitions of unity we can show that $\underline{X}$ is a local C^∞-ringed space.

For a smooth map of manifolds $f : X \to Y$ with corresponding local C^∞-ringed spaces $(X, \mathcal{O}_X), (Y, \mathcal{O}_Y)$ as previously we define $f_\sharp(U) : \mathcal{O}_Y(U) = C^\infty(U) \to \mathcal{O}_X(f^{-1}(U)) = C^\infty(f^{-1}(U))$ for each open $U \subseteq Y$ by $f_\sharp(U) : c \mapsto c \circ f$ for all $c \in C^\infty(U)$. This gives a morphism $f_\sharp : \mathcal{O}_Y \to f_*(\mathcal{O}_X)$ of sheaves of C^∞-rings on Y. Then $\underline{f} = (f, f^\sharp) : (X, \mathcal{O}_X) \to (Y, \mathcal{O}_Y)$ is a morphism of (local) C^∞-ringed spaces with $f^\sharp : f^{-1}(\mathcal{O}_Y) \to \mathcal{O}_X$ corresponding to $f_\sharp$ under (2.9).

To define a spectrum functor taking a C^∞-ring to an element of $\mathbf{LC^\infty RS}$ we require the following definition.

Definition 2.42 Let $\mathfrak{C}$ be a C^∞-ring, and write $X_\mathfrak{C}$ for the set of all $\mathbb{R}$-points x of $\mathfrak{C}$, as in Definition 2.22. Write $\mathcal{T}_\mathfrak{C}$ for the topology on $X_\mathfrak{C}$ that has basis of open sets $U_c = \{x \in X_\mathfrak{C} : x(c) \neq 0\}$ for all $c \in \mathfrak{C}$. For each $c \in \mathfrak{C}$ define a map $c_* : X_\mathfrak{C} \to \mathbb{R}$ such that $c_* : x \mapsto x(c)$.

For a morphism $\phi : \mathfrak{C} \to \mathfrak{D}$ of C^∞-rings, we can define $f_\phi : X_\mathfrak{D} \to X_\mathfrak{C}$ by $f_\phi(x) = x \circ \phi$, which is continuous.

From [49, Lem. 4.15], this definition implies that $\mathcal{T}_\mathfrak{C}$ is the weakest topology on $X_\mathfrak{C}$ such that the $c_* : X_\mathfrak{C} \to \mathbb{R}$ are continuous for all $c \in \mathfrak{C}$. Also $(X_\mathfrak{C}, \mathcal{T}_\mathfrak{C})$ is a regular, Hausdorff topological space. We now define the spectrum functor.

Definition 2.43 For a C^∞-ring $\mathfrak{C}$, we will define the *spectrum* of $\mathfrak{C}$, written $\operatorname{Spec}\mathfrak{C}$. Here, $\operatorname{Spec}\mathfrak{C}$ is a local C^∞-ringed space $(X, \mathcal{O}_X)$, with X the topological space $X_\mathfrak{C}$ from Definition 2.42. If $U \subseteq X$ is open then $\mathcal{O}_X(U)$ is the set of functions $s : U \to \coprod_{x \in U} \mathfrak{C}_x$, where we write s_x for the image of x under s, such that around each point $x \in U$ there is an open subset $x \in W \subseteq U$ and element $c \in \mathfrak{C}$ with $s_y = \pi_y(c) \in \mathfrak{C}_y$ for all $y \in W$. This is a C^∞-ring with the operations Φ_f on $\mathcal{O}_X(U)$ defined using the operations Φ_f on $\mathfrak{C}_x$ for $x \in U$.

For $s \in \mathcal{O}_X(U)$, the restriction map of functions $s \mapsto s|_V$ for open $V \subseteq U \subseteq X$ is a morphism of C^∞-rings, giving the restriction map $\rho_{UV} : \mathcal{O}_X(U)$

$\to \mathcal{O}_X(V)$. The stalk $\mathcal{O}_{X,x}$ at $x \in X$ is isomorphic to $\mathfrak{C}_x$, which is a local C^∞-ring. Hence $(X, \mathcal{O}_X)$ is a local C^∞-ringed space.

For a morphism $\phi : \mathfrak{C} \to \mathfrak{D}$ of C^∞-rings, we have an induced morphism of local C^∞-rings, $\phi_x : \mathfrak{C}_{f_\phi(x)} \to \mathfrak{D}_x$. If we let $(X, \mathcal{O}_X) = \operatorname{Spec}\mathfrak{C}$, $(Y, \mathcal{O}_Y) = \operatorname{Spec}\mathfrak{D}$, then for open $U \subseteq X$ define $(f_\phi)_\sharp(U) : \mathcal{O}_X(U) \to \mathcal{O}_Y(f_\phi^{-1}(U))$ by $(f_\phi)_\sharp(U)s : x \mapsto \phi_x(s_{f_\phi(x)})$. This gives a morphism $(f_\phi)_\sharp : \mathcal{O}_X \to (f_\phi)_*(\mathcal{O}_Y)$ of sheaves of C^∞-rings on X. Then $\underline{f}_\phi = (f_\phi, f_\phi^\sharp) : (Y, \mathcal{O}_Y) \to (X, \mathcal{O}_X)$ is a morphism of local C^∞-ringed spaces, where $f_\phi^\sharp$ corresponds to $(f_\phi)_\sharp$ under (2.9). Then Spec is a functor $\mathbf{C^\infty Rings}^{\mathrm{op}} \to \mathbf{LC^\infty RS}$, called the *spectrum functor*, where $\operatorname{Spec}\phi : \operatorname{Spec}\mathfrak{D} \to \operatorname{Spec}\mathfrak{C}$ is defined by $\operatorname{Spec}\phi = \underline{f}_\phi$.

This definition of spectrum functor is different to the classical Algebraic Geometry version [36, §II.2], as the topological space corresponds only to maximal ideals of the C^∞-ring, instead of also including all prime ideals. In this sense, it is coarser; however, this corresponds to the topology of a manifold as in the following example.

Example 2.44 For a manifold X then $\operatorname{Spec} C^\infty(X)$ is isomorphic to the local C^∞-ringed space $\underline{X}$ constructed in Example 2.41.

Here is [49, Lem 4.28] which shows how the spectrum functor behaves with respect to localizations and open sets. A proof is contained in [25, Lem 2.4.6], which relies on the existence of bump functions for $\mathbb{R}^n$.

Lemma 2.45 *Let $\mathfrak{C}$ be a C^∞-ring, set $\underline{X} = \operatorname{Spec}\mathfrak{C} = (X, \mathcal{O}_X)$, and let $c \in \mathfrak{C}$. If we write $U_c = \{x \in X : x(c) \neq 0\}$ as in Definition 2.42, then $U_c \subseteq X$ is open and $\underline{U}_c = (U_c, \mathcal{O}_X|_{U_c}) \cong \operatorname{Spec}\mathfrak{C}(c^{-1})$.*

We now define the global sections functor and describe its relationship to the spectrum functor.

Definition 2.46 The *global sections functor* $\Gamma : \mathbf{LC^\infty RS} \to \mathbf{C^\infty Rings}^{\mathrm{op}}$ takes $(X, \mathcal{O}_X)$ to $\mathcal{O}_X(X)$ and morphisms $(f, f^\sharp) : (X, \mathcal{O}_X) \to (Y, \mathcal{O}_Y)$ to $\Gamma : (f, f^\sharp) \mapsto f_\sharp(Y)$, for $f_\sharp$ relating $f^\sharp$ as in (2.9).

For each C^∞-ring $\mathfrak{C}$ we can define a morphism $\Xi_\mathfrak{C} : \mathfrak{C} \to \Gamma \circ \operatorname{Spec}\mathfrak{C}$. Here, for $c \in \mathfrak{C}$ then $\Xi_\mathfrak{C}(c) : X_\mathfrak{C} \to \coprod_{x \in X_\mathfrak{C}} \mathfrak{C}_x$ is defined by $\Xi_\mathfrak{C}(c)_x = \pi_x(c) \in \mathfrak{C}_x$, so $\Xi_\mathfrak{C}(c) \in \mathcal{O}_{X_\mathfrak{C}}(X_\mathfrak{C}) = \Gamma \circ \operatorname{Spec}\mathfrak{C}$. This $\Xi_\mathfrak{C}$ is a C^∞-ring morphism as it is built from C^∞-ring morphisms $\pi_x : \mathfrak{C} \to \mathfrak{C}_x$, and the C^∞-operations on $\mathcal{O}_{X_\mathfrak{C}}(X_\mathfrak{C})$ are defined pointwise in the $\mathfrak{C}_x$. This defines a natural transformation $\Xi : \mathrm{Id}_{\mathbf{C^\infty Rings}} \Rightarrow \Gamma \circ \operatorname{Spec}$ of functors $\mathbf{C^\infty Rings} \to \mathbf{C^\infty Rings}$.

Theorem 2.47 *The functor* Spec : $\mathbf{C^\infty Rings}^{\rm op} \to \mathbf{LC^\infty RS}$ *is **right adjoint** to* $\Gamma : \mathbf{LC^\infty RS} \to \mathbf{C^\infty Rings}^{\rm op}$, *and* Ξ *is the unit of the adjunction. This implies that* Spec *preserves limits as in* §2.1.3. *Hence if we have* C^∞*-ring morphisms* $\phi : \mathfrak{F} \to \mathfrak{D}$, $\psi : \mathfrak{F} \to \mathfrak{E}$ *in* $\mathbf{C^\infty Rings}$ *then their pushout* $\mathfrak{C} = \mathfrak{D} \amalg_{\mathfrak{F}} \mathfrak{E}$ *has image that is isomorphic to the fibre product* $\operatorname{Spec}\mathfrak{C} \cong \operatorname{Spec}\mathfrak{D} \times_{\operatorname{Spec}\mathfrak{F}} \operatorname{Spec}\mathfrak{E}$.

We extend this theorem to C^∞-schemes with corners in §5.3.

Remark 2.48 Our definition of spectrum functor follows [49] and Dubuc [23], and is called the *Archimedean spectrum* in Moerdijk et al. [73, §3]. They also show it is a right adjoint to the global sections functor as above.

Definition 2.49 Objects $\underline{X} \in \mathbf{LC^\infty RS}$ that are isomorphic to $\operatorname{Spec}\mathfrak{C}$ for some $\mathfrak{C} \in \mathbf{C^\infty Rings}$ are called *affine* C^∞*-schemes*. Elements $\underline{X} \in \mathbf{LC^\infty RS}$ that are locally isomorphic to $\operatorname{Spec}\mathfrak{C}$ for some $\mathfrak{C} \in \mathbf{C^\infty Rings}$ (depending upon the open sets) are called C^∞*-schemes*.

We define $\mathbf{C^\infty Sch}$ and $\mathbf{AC^\infty Sch}$ to be the full subcategories of C^∞-schemes and affine C^∞-schemes in $\mathbf{LC^\infty RS}$ respectively.

Remark 2.50 **(a)** Unlike ordinary Algebraic Geometry, affine C^∞-schemes are very general objects. All manifolds are affine, and all their fibre products are affine. But not all manifolds with corners are affine C^∞-schemes with corners.

(b) **(Alternatives to C^∞-schemes.)** We briefly review other generalizations of manifolds similar to C^∞-schemes. Such generalizations usually fall into a 'maps out' (based on maps $X \to \mathbb{R}$) or a 'maps in' (based on maps $\mathbb{R}^n \to X$) approach. C^∞-algebraic geometry uses 'maps out', as do the C^∞*-differentiable spaces* of Navarro González and Sancho de Salas [77], which form a subcategory of C^∞-schemes. Sikorski [84], Spallek [86], Buchner et al. [9] and González and Sancho de Salas [77] describe other 'maps out' approaches.

'Maps in' approaches include the *diffeological spaces* of Souriau [85] and Iglesias-Zemmour [39], and the various *Chen spaces* from Chen [15, 16, 17, 18]. Usually 'maps out' approaches deal well with finite limits, and 'maps in' approaches work well for infinite-dimensional spaces and quotient spaces.

2.6 Complete C^∞-rings

In ordinary Algebraic Geometry, if A is a commutative ring then $\Gamma \circ \operatorname{Spec} A \cong A$, and $\operatorname{Spec} : \mathbf{Rings}^{\rm op} \to \mathbf{ASch}$ is an equivalence of categories, with inverse Γ. For C^∞-rings $\mathfrak{C}$, in general $\Gamma \circ \operatorname{Spec}\mathfrak{C} \not\cong \mathfrak{C}$, and $\operatorname{Spec} : \mathbf{C^\infty Rings}^{\rm op} \to$

$\mathbf{AC^\infty Sch}$ is neither full nor faithful. But as in [49, Prop. 4.34], we have the following.

Proposition 2.51 *For each C^∞-ring $\mathfrak{C}$, $\mathrm{Spec}\,\Xi_{\mathfrak{C}} : \mathrm{Spec}\circ\Gamma\circ\mathrm{Spec}\,\mathfrak{C} \to \mathrm{Spec}\,\mathfrak{C}$ is an isomorphism in $\mathbf{LC^\infty RS}$.*

This motivates the following definition [49, Def. 4.35].

Definition 2.52 A C^∞-ring $\mathfrak{C}$ is called *complete* if $\Xi_{\mathfrak{C}} : \mathfrak{C} \to \Gamma\circ\mathrm{Spec}\,\mathfrak{C}$ is an isomorphism. We define $\mathbf{C^\infty Rings_{co}}$ to be the full subcategory in $\mathbf{C^\infty Rings}$ of complete C^∞-rings. By Proposition 2.51 we see that $\mathbf{C^\infty Rings_{co}}$ is equivalent to the image of the functor $\Gamma\circ\mathrm{Spec} : \mathbf{C^\infty Rings} \to \mathbf{C^\infty Rings}$, which gives a left adjoint to the inclusion of $\mathbf{C^\infty Rings_{co}}$ into $\mathbf{C^\infty Rings}$. Write this left adjoint as the functor $\Pi_{\mathrm{all}}^{\mathrm{co}} = \Gamma\circ\mathrm{Spec} : \mathbf{C^\infty Rings} \to \mathbf{C^\infty Rings_{co}}$.

An example of a non-complete C^∞-ring is the quotient $\mathfrak{C} = C^\infty(\mathbb{R}^n)/I_{\mathrm{cs}}$ of $C^\infty(\mathbb{R}^n)$ for $n > 0$ by the ideal I_{cs} of compactly supported functions, and $\Pi_{\mathrm{all}}^{\mathrm{co}}(\mathfrak{C}) = 0 \ncong \mathfrak{C}$. The next theorem comes from [49, Prop. 4.11 and Th. 4.25].

Theorem 2.53 **(a)** $\mathrm{Spec}\,|_{(\mathbf{C^\infty Rings_{co}})^{\mathrm{op}}} : (\mathbf{C^\infty Rings_{co}})^{\mathrm{op}} \to \mathbf{LC^\infty RS}$ *is full and faithful, and an equivalence* $(\mathbf{C^\infty Rings_{co}})^{\mathrm{op}} \to \mathbf{AC^\infty Sch}$.

(b) *Let $\underline{X}$ be an affine C^∞-scheme. Then $\underline{X} \cong \mathrm{Spec}\,\mathcal{O}_X(X)$, where $\mathcal{O}_X(X)$ is a complete C^∞-ring.*

(c) *The functor* $\Pi_{\mathrm{all}}^{\mathrm{co}} : \mathbf{C^\infty Rings} \to \mathbf{C^\infty Rings_{co}}$ *is left adjoint to the inclusion functor* $\mathrm{inc} : \mathbf{C^\infty Rings_{co}} \hookrightarrow \mathbf{C^\infty Rings}$*. That is,* $\Pi_{\mathrm{all}}^{\mathrm{co}}$ *is a **reflection functor**.*

(d) *All small colimits exist in* $\mathbf{C^\infty Rings_{co}}$*, although they may not coincide with the corresponding small colimits in* $\mathbf{C^\infty Rings}$.

(e) $\mathrm{Spec}\,|_{(\mathbf{C^\infty Rings_{co}})^{\mathrm{op}}} = \mathrm{Spec}\circ\mathrm{inc} : (\mathbf{C^\infty Rings_{co}})^{\mathrm{op}} \to \mathbf{LC^\infty RS}$ *is right adjoint to* $\Pi_{\mathrm{all}}^{\mathrm{co}}\circ\Gamma : \mathbf{LC^\infty RS} \to (\mathbf{C^\infty Rings_{co}})^{\mathrm{op}}$*. Thus* $\mathrm{Spec}\,|_{\dots}$ *takes limits in* $(\mathbf{C^\infty Rings_{co}})^{\mathrm{op}}$ *(equivalently, colimits in* $\mathbf{C^\infty Rings_{co}}$*) to limits in* $\mathbf{LC^\infty RS}$.

Using (a), that small limits exist in the category of $\mathbf{C^\infty Rings}$, and that $\Gamma : \mathbf{LC^\infty RS} \to \mathbf{C^\infty Rings}$ is a left adjoint with image in $(\mathbf{C^\infty Rings_{co}})^{\mathrm{op}}$ when restricted to $\mathbf{AC^\infty Sch}$, then small limits in $\mathbf{C^\infty Rings_{co}}$ exist and coincide with small limits in $\mathbf{C^\infty Rings}$. As $(\mathbf{C^\infty Rings_{co}})^{\mathrm{op}} \to \mathbf{AC^\infty Sch}$ is an equivalence of categories, then $\mathbf{AC^\infty Sch}$ also has all small colimits and small limits. As Spec is a right adjoint, then limits in $\mathbf{AC^\infty Sch}$ coincide with limits in $\mathbf{C^\infty Sch}$ and $\mathbf{LC^\infty RS}$; however, it is not necessarily true that colimits in $\mathbf{AC^\infty Sch}$ coincide with colimits in $\mathbf{C^\infty Sch}$ and $\mathbf{LC^\infty RS}$.

In the following theorem we summarize results found in Dubuc [23, Th. 16], Moerdijk and Reyes [74, § II. Prop. 1.2], and the second author [49, Cor. 4.27].

Theorem 2.54 *There is a full and faithful functor* $F_{\mathbf{Man}}^{\mathbf{AC^\infty Sch}} : \mathbf{Man} \to \mathbf{AC^\infty Sch}$ *that takes a manifold* X *to the affine* C^∞*-scheme* $\underline{X} = (X, \mathcal{O}_X)$*, where* $\mathcal{O}_X(U) = C^\infty(U)$ *is the usual smooth functions on* U*. Here* $(X, \mathcal{O}_X) \cong \operatorname{Spec}(C^\infty(X))$ *and hence* $\underline{X}$ *is affine. The functor* $F_{\mathbf{Man}}^{\mathbf{AC^\infty Sch}}$ *sends transverse fibre products of manifolds to fibre products of* C^∞*-schemes.*

2.7 Sheaves of $\mathcal{O}_X$-modules on C^∞-ringed spaces

This section follows [49, §5.3], where we give the basics of sheaves of $\mathcal{O}_X$-modules for C^∞-ringed spaces. This includes the pullback of a sheaf of modules and the cotangent sheaf. Our definition of $\mathcal{O}_X$-module is the usual definition of sheaf of modules on a ringed space as in Hartshorne [36, §II.5] and Grothendieck [35, §0.4.1], using the $\mathbb{R}$-algebra structure on our C^∞-rings. The cotangent sheaf uses the cotangent modules of §2.3.

Definition 2.55 For each C^∞-ringed space $\underline{X} = (X, \mathcal{O}_X)$ we define a category $\mathcal{O}_X$-mod. The objects are *sheaves of* $\mathcal{O}_X$*-modules* (or simply $\mathcal{O}_X$*-modules*) $\mathcal{E}$ on X. Here, $\mathcal{E}$ is a functor on open sets $U \subseteq X$ such that $\mathcal{E} : U \mapsto \mathcal{E}(U)$ in $\mathcal{O}_X(U)$-mod is a sheaf as in Definition 2.34. This means we have linear restriction maps $\mathcal{E}_{UV} : \mathcal{E}(U) \to \mathcal{E}(V)$ for each inclusion of open sets $V \subseteq U \subseteq X$, such that the following commutes:

$$\begin{array}{ccc} \mathcal{O}_X(U) \times \mathcal{E}(U) & \longrightarrow & \mathcal{E}(U) \\ \downarrow{\scriptstyle \rho_{UV} \times \mathcal{E}_{UV}} & & {\scriptstyle \mathcal{E}_{UV}}\downarrow \\ \mathcal{O}_X(V) \times \mathcal{E}(V) & \longrightarrow & \mathcal{E}(V), \end{array}$$

where the horizontal arrows are module multiplication. Morphisms in $\mathcal{O}_X$-mod are sheaf morphisms $\phi : \mathcal{E} \to \mathcal{F}$ commuting with the $\mathcal{O}_X$-actions. An $\mathcal{O}_X$-module $\mathcal{E}$ is called a *vector bundle* if it is locally free, that is, around every point there is an open set $U \subseteq X$ with $\mathcal{E}|_U \cong \mathcal{O}_X|_U \otimes_{\mathbb{R}} \mathbb{R}^n$.

Definition 2.56 We define the *pullback* $\underline{f}^*(\mathcal{E})$ of a sheaf of modules $\mathcal{E}$ on $\underline{Y}$ by a morphism $\underline{f} = (f, f^\sharp) : \underline{X} \to \underline{Y}$ of C^∞-ringed spaces as $\underline{f}^*(\mathcal{E}) = f^{-1}(\mathcal{E}) \otimes_{f^{-1}(\mathcal{O}_Y)} \mathcal{O}_X$. Here $f^{-1}(\mathcal{E})$ is as in Definition 2.38, so that $\underline{f}^*(\mathcal{E})$ is a sheaf of modules on $\underline{X}$. Morphisms of $\mathcal{O}_Y$-modules $\phi : \mathcal{E} \to \mathcal{F}$ give morphisms of $\mathcal{O}_X$-modules $\underline{f}^*(\phi) = f^{-1}(\phi) \otimes \mathrm{id}_{\mathcal{O}_X} : \underline{f}^*(\mathcal{E}) \to \underline{f}^*(\mathcal{F})$.

Definition 2.57 Let $\underline{X} = (X, \mathcal{O}_X)$ be a C^∞-ringed space. Define a presheaf $\mathcal{P}T^*\underline{X}$ of $\mathcal{O}_X$-modules on X such that $\mathcal{P}T^*\underline{X}(U)$ is the cotangent module $\Omega_{\mathcal{O}_X(U)}$ of Definition 2.29, regarded as a module over the C^∞-ring $\mathcal{O}_X(U)$. For open sets $V \subseteq U \subseteq X$ we have restriction morphisms $\Omega_{\rho_{UV}} : \Omega_{\mathcal{O}_X(U)} \to \Omega_{\mathcal{O}_X(V)}$ associated to the morphisms of C^∞-rings $\rho_{UV} : \mathcal{O}_X(U) \to \mathcal{O}_X(V)$ so that the following commutes:

$$\begin{array}{ccc} \mathcal{O}_X(U) \times \Omega_{\mathcal{O}_X(U)} & \xrightarrow{\mu_{\mathcal{O}_X(U)}} & \Omega_{\mathcal{O}_X(U)} \\ \downarrow{\scriptstyle \rho_{UV} \times \Omega_{\rho_{UV}}} & & {\scriptstyle \Omega_{\rho_{UV}}}\downarrow \\ \mathcal{O}_X(V) \times \Omega_{\mathcal{O}_X(V)} & \xrightarrow{\mu_{\mathcal{O}_X(V)}} & \Omega_{\mathcal{O}_X(V)}. \end{array}$$

Definition 2.29 implies $\Omega_{\psi\circ\phi} = \Omega_\psi \circ \Omega_\phi$, so this is a well defined presheaf of $\mathcal{O}_X$-modules. The *cotangent sheaf* $T^*\underline{X}$ *of* X is the sheafification of $\mathcal{P}T^*\underline{X}$.

The universal property of sheafification shows that for open $U \subseteq X$ we have an isomorphism of $\mathcal{O}_X|_U$-modules

$$T^*\underline{U} = T^*(U, \mathcal{O}_X|_U) \cong T^*\underline{X}|_U.$$

For $\underline{f} : \underline{X} \to \underline{Y}$ in $\mathbf{C^\infty RS}$ we have $\underline{f}^*(T^*\underline{Y}) = f^{-1}(T^*\underline{Y}) \otimes_{f^{-1}(\mathcal{O}_Y)} \mathcal{O}_X$. The universal properties of sheafification imply that $\underline{f}^*(T^*\underline{Y})$ is the sheafification of the presheaf $\mathcal{P}(\underline{f}^*(T^*\underline{Y}))$, where

$$U \longmapsto \mathcal{P}(\underline{f}^*(T^*\underline{Y}))(U) = \lim_{V \supseteq f(U)} \Omega_{\mathcal{O}_Y(V)} \otimes_{\mathcal{O}_Y(V)} \mathcal{O}_X(U).$$

This gives a presheaf morphism $\mathcal{P}\Omega_{\underline{f}} : \mathcal{P}(\underline{f}^*(T^*\underline{Y})) \to \mathcal{P}T^*\underline{X}$ on X, where

$$(\mathcal{P}\Omega_{\underline{f}})(U) = \lim_{V \supseteq f(U)} (\Omega_{\rho_{f^{-1}(V)U} \circ f_\sharp(V)})_*.$$

Here, we have morphisms $f_\sharp(V) : \mathcal{O}_Y(V) \to \mathcal{O}_X(f^{-1}(V))$ from $f_\sharp : \mathcal{O}_Y \to f_*(\mathcal{O}_X)$ corresponding to $f^\sharp$ in $\underline{f}$ as in (2.9), and $\rho_{f^{-1}(V)U} : \mathcal{O}_X(f^{-1}(V)) \to \mathcal{O}_X(U)$ in $\mathcal{O}_X$ so that $(\Omega_{\rho_{f^{-1}(V)U} \circ f_\sharp(V)})_* : \Omega_{\mathcal{O}_Y(V)} \otimes_{\mathcal{O}_Y(V)} \mathcal{O}_X(U) \to \Omega_{\mathcal{O}_X(U)} = (\mathcal{P}T^*\underline{X})(U)$ is constructed as in Definition 2.29. Then write $\Omega_{\underline{f}} : \underline{f}^*(T^*\underline{Y}) \to T^*\underline{X}$ for the induced morphism of the associated sheaves. This corresponds to the morphism $\mathrm{d}f : f^*(T^*Y) \to T^*X$ of vector bundles over a manifold X and smooth map of manifolds $f : X \to Y$ as in Example 2.30.

2.8 Sheaves of $\mathcal{O}_X$-modules on C^∞-schemes

We define the module spectrum functor MSpec as in [49, Defs. 5.16, 5.17 and 5.25], and its corresponding global sections functor, and recall their properties.

Definition 2.58 Let $\mathfrak{C}$ be a C^∞-ring and set $\underline{X} = (X, \mathcal{O}_X) = \operatorname{Spec} \mathfrak{C}$. Let $M \in \mathfrak{C}$-mod be a $\mathfrak{C}$-module. For each open subset $U \subseteq X$ there is a natural morphism $\mathfrak{C} \to \mathcal{O}_X(U)$ in $\mathbf{C^\infty Rings}$. Using this we make $M \otimes_{\mathfrak{C}} \mathcal{O}_X(U)$ into an $\mathcal{O}_X(U)$-module. This assignment $U \mapsto M \otimes_{\mathfrak{C}} \mathcal{O}_X(U)$ is naturally a presheaf $\mathcal{P}\,\mathrm{MSpec}\, M$ of $\mathcal{O}_X$-modules. Define $\mathrm{MSpec}\, M \in \mathcal{O}_X$-mod to be its sheafification.

A morphism $\mu : M \to N$ in $\mathfrak{C}$-mod induces $\mathcal{O}_X(U)$-module morphisms $M \otimes_{\mathfrak{C}} \mathcal{O}_X(U) \to N \otimes_{\mathfrak{C}} \mathcal{O}_X(U)$ for all open $U \subseteq X$, and hence a presheaf morphism, which descends to a morphism $\mathrm{MSpec}\, \mu : \mathrm{MSpec}\, M \to \mathrm{MSpec}\, N$ in $\mathcal{O}_X$-mod. This defines a functor $\mathrm{MSpec} : \mathfrak{C}\text{-mod} \to \mathcal{O}_X\text{-mod}$. It is an exact functor of abelian categories.

There is also a global sections functor $\Gamma : \mathcal{O}_X\text{-mod} \to \mathfrak{C}\text{-mod}$ mapping $\Gamma : \mathcal{E} \mapsto \mathcal{E}(X)$, where the $\mathcal{O}_X(X)$-module $\mathcal{E}(X)$ is viewed as a $\mathfrak{C}$-module via the natural morphism $\mathfrak{C} \to \mathcal{O}_X(X)$.

For any $M \in \mathfrak{C}$-mod there is a natural morphism $\Xi_M : M \to \Gamma \circ \mathrm{MSpec}\, M$ in $\mathfrak{C}$-mod, by composing $M \to M \otimes_{\mathfrak{C}} \mathcal{O}_X(X) = \mathcal{P}\,\mathrm{MSpec}\, M(X)$ with the sheafification morphism $\mathcal{P}\,\mathrm{MSpec}\, M(X) \to \mathrm{MSpec}\, M(X) = \Gamma \circ \mathrm{MSpec}\, M$. Generalizing Definition 2.52, we call M *complete* if Ξ_M is an isomorphism. Write $\mathfrak{C}\text{-mod}_{\mathrm{co}} \subseteq \mathfrak{C}$-mod for the full subcategory of complete $\mathfrak{C}$-modules.

Here is [49, Th. 5.19, Prop. 5.20, Th. 5.26, and Prop. 5.31].

Theorem 2.59 **(a)** *In Definition* 2.58, $\mathrm{MSpec} : \mathfrak{C}\text{-mod} \to \mathcal{O}_X\text{-mod}$ *is left adjoint to* $\Gamma : \mathcal{O}_X\text{-mod} \to \mathfrak{C}\text{-mod}$, *generalizing Theorem* 2.47.

(b) *There is a natural isomorphism* $\mathrm{MSpec} \circ \Gamma \Rightarrow \mathrm{Id}_{\mathcal{O}_X\text{-mod}}$. *This gives a natural isomorphism* $\mathrm{MSpec} \circ \Gamma \circ \mathrm{MSpec} \Rightarrow \mathrm{MSpec}$, *generalizing Proposition* 2.51.

(c) $\mathrm{MSpec}\,|_{\mathfrak{C}\text{-mod}_{\mathrm{co}}} : \mathfrak{C}\text{-mod}_{\mathrm{co}} \to \mathcal{O}_X\text{-mod}$ *is an equivalence of categories, generalizing Theorem* 2.53(a).

(d) *The functor* $\Pi^{\mathrm{co}}_{\mathrm{all}} = \Gamma \circ \mathrm{MSpec} : \mathfrak{C}\text{-mod} \to \mathfrak{C}\text{-mod}_{\mathrm{co}}$ *is left adjoint to the inclusion functor* $\mathrm{inc} : \mathfrak{C}\text{-mod}_{\mathrm{co}} \hookrightarrow \mathfrak{C}$-mod, *generalizing Theorem* 2.53(c). *That is,* $\Pi^{\mathrm{co}}_{\mathrm{all}}$ *is a* ***reflection functor***.

(e) *There is a natural isomorphism* $T^*\underline{X} \cong \mathrm{MSpec}\, \Omega_{\mathfrak{C}}$ *in* $\mathcal{O}_X$-mod.

Remark 2.60 **(a)** In [49, §5.4], following conventional Algebraic Geometry as in Hartshorne [36, §II.5], the first author defined a notion of *quasi-coherent sheaf* $\mathcal{E}$ on a C^∞-scheme $\underline{X}$, which is that we may cover $\underline{X}$ by open $\underline{U} \subseteq \underline{X}$ with $\underline{U} \cong \mathrm{Spec}\, \mathfrak{C}$ and $\mathcal{E}|_{\underline{U}} \cong \mathrm{MSpec}\, M$ for $\mathfrak{C} \in \mathbf{C^\infty Rings}$ and $M \in \mathfrak{C}$-mod. But then [49, Cor. 5.22] uses Theorem 2.59(c) to show that every $\mathcal{O}_X$-module is quasi-coherent, that is, $\mathrm{qcoh}(\underline{X}) = \mathcal{O}_X$-mod, which is not true in conventional Algebraic Geometry. So here we will not bother with the language of quasi-coherent sheaves.

(b) In conventional Algebraic Geometry one also defines *coherent sheaves* [36, §II.5] to be $\mathcal{O}_X$-modules $\mathcal{E}$ locally modelled on MSpec M for M a finitely generated $\mathfrak{C}$-module. However, as in [49, Rem. 5.23(b)], coherent sheaves are only well-behaved on *noetherian* C^∞-schemes, and most interesting C^∞-rings, such as $C^\infty(\mathbb{R}^n)$ for $n > 0$, are not noetherian. So coherent sheaves do not seem to be a useful idea in C^∞-algebraic geometry. For example, $\mathrm{coh}(\underline{X})$ is not closed under kernels in $\mathcal{O}_X$-mod, and is not an abelian category.

Here is [49, Th. 5.32], where part (b) is deduced from Theorem 2.33.

Theorem 2.61 **(a)** *Let $\underline{f} : \underline{X} \to \underline{Y}$ and $\underline{g} : \underline{Y} \to \underline{Z}$ be morphisms of C^∞-schemes. Then in $\mathcal{O}_X$-*mod *we have*

$$\Omega_{\underline{g}\circ\underline{f}} = \Omega_{\underline{f}} \circ \underline{f}^*(\Omega_{\underline{g}}) : (\underline{g}\circ\underline{f})^*(T^*\underline{Z}) \longrightarrow T^*\underline{X}.$$

(b) *Suppose we are given a Cartesian square in* $\mathbf{C^\infty Sch}$*:*

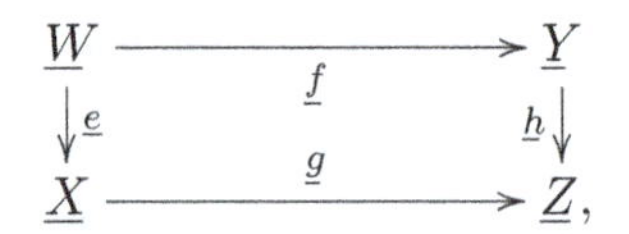

*so that $\underline{W} = \underline{X} \times_{\underline{Z}} \underline{Y}$. Then the following is exact in $\mathcal{O}_W$-*mod*:*

$$(\underline{g}\circ\underline{e})^*(T^*\underline{Z}) \xrightarrow{\underline{e}^*(\Omega_{\underline{g}})\oplus -\underline{f}^*(\Omega_{\underline{h}})} \underline{e}^*(T^*\underline{X})\oplus\underline{f}^*(T^*\underline{Y}) \xrightarrow{\Omega_{\underline{e}}\oplus\Omega_{\underline{f}}} T^*\underline{W} \longrightarrow 0.$$

2.9 Applications of C^∞-rings and C^∞-schemes

Since the work of Grothendieck, the theory of schemes in Algebraic Geometry has become an enormously powerful tool, and the language in which most modern Algebraic Geometry is written. As a result, Algebraic Geometers are far better at dealing with singular spaces than Differential Geometers are.

It seems desirable to have a theory of schemes in Differential Geometry – C^∞-schemes, or something similar – that could in future be used for the same purposes as schemes in Algebraic Geometry. For example, it seems very likely that many moduli spaces $\mathcal{M}$ of differential-geometric objects are naturally C^∞-schemes, as well as topological spaces.

As another example, suppose (X, g) is a Riemannian manifold, and we are interested in the moduli space $\mathcal{M}$ of some class of special embedded submanifolds $Y \subset X$, for example, minimal, or calibrated. We could imagine trying to define a compactification $\overline{\mathcal{M}}$ of $\mathcal{M}$ by regarding submanifolds $Y \subset X$

as C^∞-subschemes of X, and taking the closure of $\mathcal{M}$ in the space of C^∞-subschemes.

For the present, to the authors' knowledge, applications of C^∞-algebraic geometry in the literature are confined to two areas: Synthetic Differential Geometry and Derived Differential Geometry, which we now discuss.

2.9.1 Synthetic Differential Geometry

Synthetic Differential Geometry is a subject in which one proves theorems about manifolds in Differential Geometry using 'infinitesimals'. It was used non-rigorously in the nineteenth century by authors such as Sophus Lie. In the 1960s William Lawvere [58] suggested a way to make it rigorous, and the subject has since been developed in detail by Anders Kock [52, 53] and others.

One supposes that the real numbers $\mathbb{R}$ can be enlarged to a 'number line' $R \supset \mathbb{R}$, a ring containing non-zero 'infinitesimal' elements $x \in R$ with $x^n = 0$ for some $n > 1$. An important rôle in the theory is played by the 'double point'

$$D = \{x \in R : x^2 = 0\}. \tag{2.10}$$

One assumes smooth functions $f : \mathbb{R} \to \mathbb{R}$, and manifolds X, can all be enlarged by infinitesimals in this way. Here are examples of how these are used.

(a) If $f : \mathbb{R} \to \mathbb{R}$ is smooth, we can define the *derivative* $\frac{\mathrm{d}f}{\mathrm{d}x}$ by $f(x + y) = f(x) + y\frac{\mathrm{d}f}{\mathrm{d}x}$ for $y \in R$ with $y^2 = 0$.
(b) If X is a manifold, the *tangent bundle* TX is the mapping space $X^D = \mathrm{Map}_{C^\infty}(D, X)$.
(c) A *vector field* on a manifold X can be defined to be a smooth map $v : X \times D \to X$ with $v|_{X\times\{0\}} = \mathrm{id}_X$.

The theory is developed axiomatically, with axioms on the properties of infinitesimals, and theorems about manifolds (including classical results not involving infinitesimals) are proved from them. The logic is unusual: since one treats infinitesimals as ordinary points of the 'set' R, though it turns out that R is not really an honest set, then only constructive logic is allowed, and the law of the excluded middle may not be used.

For the enterprise to be at all credible, we need to know that the axioms of Synthetic Differential Geometry are consistent, as otherwise one could prove any statement from them, true or false. Consistency is proved by constructing a 'model' for Synthetic Differential Geometry, that is, a category (in fact, a topos) of spaces $\mathcal{C}$ which includes $\mathbf{Man}$ as a full subcategory, and contains other 'infinitesimal' objects such as the double point D, such that the axioms can be interpreted as true statements in $\mathcal{C}$, and thus proofs in Synthetic Differential Geometry can be reinterpreted as reasoning (with ordinary logic) in $\mathcal{C}$.

The connection to C^∞-algebraic geometry is that, as in Dubuc [23] and Moerdijk and Reyes [75], this category $\mathcal{C}$ may be taken to be the category $\mathbf{C^\infty Sch}$ of C^∞-schemes, with $D = \mathrm{Spec}\bigl(\mathbb{R}[x]/(x^2)\bigr)$, as a C^∞-subscheme of $\mathbb{R}$. Most early work on C^∞-schemes was directed towards proving properties of $\mathbf{C^\infty Sch}$ needed to verify consistency of various sets of axioms in Synthetic Differential Geometry.

Knowing their axioms were consistent, Synthetic Differential Geometers had little reason to study C^∞-schemes further, so the subject became inactive.

2.9.2 Derived Differential Geometry

Derived Algebraic Geometry is a generalization of classical Algebraic Geometry, in which schemes (and stacks) X are replaced by *derived schemes* (and *derived stacks*) $\boldsymbol{X}$, which have a richer geometric structure. A derived scheme $\boldsymbol{X}$ has a classical truncation $X = t_0(\boldsymbol{X})$, an ordinary scheme. The foundations were developed in the 2000s by Bertrand Toën and Gabriele Vezzosi [90, 91, 92] and Jacob Lurie [61, 62], and it has now become a major area in Algebraic Geometry. Toën [90, 91] gives accessible surveys.

Quasi-smooth derived schemes are a class of derived schemes $\boldsymbol{X}$ that behave in some ways like smooth schemes, although their classical truncations $X = t_0(\boldsymbol{X})$ may be very singular. This is known as the 'hidden smoothness' philosophy of Kontsevich [54]. A proper quasi-smooth derived $\mathbb{C}$-scheme $\boldsymbol{X}$ has a *dimension* $\dim_{\mathbb{C}} \boldsymbol{X}$, and a *virtual class* $[\boldsymbol{X}]_{\mathrm{virt}}$ in homology $H_{2\dim_{\mathbb{C}}\boldsymbol{X}}(\boldsymbol{X},\mathbb{Z})$, which is the analogue of the fundamental class of a compact complex manifold.

Quasi-smooth derived schemes have important applications in *enumerative geometry* (as does the older notion of *scheme with obstruction theory*, which turns out to be a semi-classical truncation of a quasi-smooth derived scheme). Various moduli problems, such as moduli of stable coherent sheaves on a projective surface, have derived moduli schemes which are quasi-smooth, and the virtual class is used to define invariants 'counting' such moduli spaces.

Two characteristic features of Derived Algebraic Geometry are as follows.

(i) Derived Algebraic Geometry is always done in ∞-*categories*, not ordinary categories, as truncating to ordinary categories loses too much information. For example, the structure sheaf $\boldsymbol{\mathcal{O}}_X$ of a derived scheme is an ∞-sheaf (homotopy sheaf), but truncating to ordinary categories loses the sheaf property.

(ii) To pass from smooth algebraic geometry to Derived Algebraic Geometry, we replace vector bundles by perfect complexes of coherent sheaves. So a smooth scheme X has tangent and cotangent bundles TX, T^*X, but a derived scheme $\boldsymbol{X}$ has a *tangent complex* $\mathbb{T}_{\boldsymbol{X}}$ and *cotangent complex* $\mathbb{L}_{\boldsymbol{X}}$, which are concentrated in degrees $[0,1], [-1,0]$ if $\boldsymbol{X}$ is quasi-smooth.

We can now ask whether there is an analogous 'derived' version of Differential Geometry. In particular, can one define 'derived manifolds' and 'derived orbifolds' $\boldsymbol{X}$ which would be C^∞ analogues of quasi-smooth derived schemes? We might hope that a compact, oriented derived manifold or orbifold $\boldsymbol{X}$ would have a well-defined dimension and virtual class in homology, and could be applied to enumerative invariant problems in Differential Geometry.

In the last paragraph of [62, §4.5], Jacob Lurie outlined how to use his huge framework to define an ∞-category of derived C^∞-schemes, including derived manifolds. In 2008 Lurie's student David Spivak [87] worked out the details of this, defining an ∞-category of *derived C^∞-schemes* $\boldsymbol{X} = (X, \boldsymbol{\mathcal{O}}_X)$ which are topological spaces X with an ∞-sheaf $\boldsymbol{\mathcal{O}}_X$ of simplicial C^∞-rings, and a full ∞-subcategory of *derived manifolds*, which are derived C^∞-schemes locally modelled on fibre products $X \times_Z Y$ for X, Y, Z manifolds. Spivak also gave a list of axioms for an ∞-category of 'derived manifolds' to satisfy, and showed they hold for his ∞-category.

Some years before the invention of Derived Algebraic Geometry, Fukaya–Oh–Ohta–Ono [28, 29, 30, 31] were working on theories of Gromov–Witten invariants and Lagrangian Floer theory in Symplectic Geometry. Their theories involved giving moduli spaces $\bar{\mathcal{M}}$ of J-holomorphic curves in a symplectic manifold (X, ω) the structure of a *Kuranishi space* $\bar{\boldsymbol{\mathcal{M}}}$, and defining a virtual class/chain $[\bar{\boldsymbol{\mathcal{M}}}]_{\mathrm{virt}}$ in homology. In the 2000's there were still significant problems with the definition and theory of Kuranishi spaces, and the subject was under dispute.

When the second author read Spivak's thesis [87], he realized that *Kuranishi spaces are really derived orbifolds*. This explained the problems in the theory: vital ideas from Derived Algebraic Geometry were missing, especially the need for higher categories, as these were unknown when Kuranishi spaces were invented. The second author then developed theories of derived manifolds and derived orbifolds [42, 43, 44, 45, 48, 49, 50] with a view to applications in Symplectic Geometry, as a substitute for Fukaya–Oh–Ohta–Ono's Kuranishi spaces.

The second author found Spivak's ∞-category far too complicated to work with. So he defined a simplified version, 'd-manifolds' and 'd-orbifolds' [43, 44], which form 2-categories $\mathbf{dMan}, \mathbf{dOrb}$ rather than ∞-categories. (Although this would not work in Derived Algebraic Geometry, it turns out that

2-categories are sufficient in C^∞ geometry because of the existence of partitions of unity.) As part of the foundations of this, he developed C^∞-algebraic geometry in new directions [42, 49], in particular $\mathcal{O}_X$-modules and C^∞-stacks.

Later, the second author [45, 48, 50] found a definition of Kuranishi spaces using an atlas of charts in the style of Fukaya–Oh–Ohta–Ono, which yielded a 2-category **Kur** equivalent to **dOrb**, fixing the problems with the original definition. This gives two different models for Derived Differential Geometry, one starting from derived C^∞-schemes, and one from Kuranishi spaces. To understand the relationship, observe that there are two ways to define manifolds.

(A) A manifold is a Hausdorff, second countable topological space X equipped with a sheaf $\mathcal{O}_X$ of $\mathbb{R}$-algebras (or C^∞-rings) such that $(X, \mathcal{O}_X)$ is locally isomorphic to $\mathbb{R}^n$ with its sheaf of smooth functions $\mathcal{O}_{\mathbb{R}^n}$.

(B) A manifold is a Hausdorff, second countable topological space X equipped with a maximal atlas of charts $\{(U_i, \phi_i) : i \in I\}$.

If we try to define derived manifolds by generalizing approach (A), we get some kind of derived C^∞-scheme, as in [6, 7, 8, 10, 11, 12, 13, 43, 44, 62, 87, 88, 89]; if we try to generalize (B), we get something like Kuranishi spaces in [28, 29, 30, 31, 45, 48, 50].

Derived manifolds and orbifolds are interesting for many reasons, including the following.

(a) Much of classical Differential Geometry extends nicely to the derived case.

(b) Many mathematical objects are naturally derived manifolds, for example

  (i) The solution set of $f_1(x_1, \ldots, x_n) = \cdots = f_k(x_1, \ldots, x_n) = 0$, where $x_1, \ldots, x_n$ are real variables and $f_1, \ldots, f_k$ are smooth functions.

  (ii) (Non-transverse) intersections $X \cap Y$ of submanifolds $X, Y \subset Z$.

  (iii) Moduli spaces $\mathcal{M}$ of solutions of nonlinear elliptic equations on compact manifolds. Also, if we consider moduli spaces $\mathcal{M}$ for nonlinear equations which are elliptic modulo symmetries, and restrict to objects with finite automorphism groups, then $\mathcal{M}$ is a derived orbifold.

(c) A compact, oriented derived manifold (or orbifold) $\boldsymbol{X}$ has a *virtual class* $[\boldsymbol{X}]_{\mathrm{virt}}$ in (Steenrod/Čech) homology $H_{\mathrm{vdim}\,\boldsymbol{X}}(X, \mathbb{Z})$ (or $H_{\mathrm{vdim}\,\boldsymbol{X}}(X, \mathbb{Q})$), with deformation invariance properties. Combining this with (b)(iii), we can use derived orbifolds as tools in enumerative invariant theories such as Gromov–Witten invariants in Symplectic Geometry.

Now for applications in Symplectic Geometry, especially Lagrangian Floer theory [28, 29] and Fukaya categories [3, 83], it is important to have a theory of derived orbifolds *with corners*. (Some applications involving 'quilts' also require derived orbifolds *with g-corners*.) To get satisfactory notions of derived manifold or orbifold with corners in the derived C^∞-scheme approach, it is necessary to go right back to the beginning, and introduce C^∞-rings and C^∞-schemes with corners. The second author pursued these ideas with his students Elana Kalashnikov [51] and the first author [25], which led to this book.

For further references on Derived Differential Geometry see Behrend–Liao–Xu [6], Borisov [7], Borisov–Noel [8], Carchedi [10], Carchedi–Roytenberg [11, 12], Carchedi–Steffens [13], and Steffens [88, 89].

3

Background on manifolds with (g-)corners

Next we discuss *manifolds with corners*, following Melrose [69, 70, 71, 72] and the second author [41], [46, §2], as well as a generalization of them, *manifolds with g-corners*, introduced in [46].

While the spaces underlying a manifold with corners are generally agreed as based on $\mathbb{R}^m_k \cong [0,\infty)^k \times \mathbb{R}^{m-k}$, there is more than one notion of smooth map between manifolds with corners. The one we choose was introduced by Melrose [69, 70, 71, 72], who calls them *b-maps*. It is stricter than requiring that all partial derivatives exist and are continuous. We call this latter definition of smooth *weakly smooth*, while b-maps are called *smooth* in this text. Smooth maps $f : X \to Y$ are compatible with the corners stratifications of X, Y, whereas weakly smooth maps need not be. We call a smooth map $f : X \to Y$ *interior* if $f(X^\circ) \subseteq Y^\circ$.

The theory of manifolds with corners is generally similar to that of manifolds, including the existence of partitions of unity. However, there are subtle differences. For example, while manifolds with corners admit partitions of unity [72, Lem. 1.6.1], constructions involving smooth maps and partitions of unity may return weakly smooth but not smooth maps. In this sense, the geometry of manifolds with corners is somewhat more global. A key definition in this chapter is a *manifold with faces*, which is a manifold with corners where the local and the global behaviours coincide.

Manifolds with corners can be considered as C^∞-schemes, and there is a faithful embedding of manifolds with corners into the category of affine C^∞-schemes. However, this embedding is not full, as morphisms of C^∞-schemes with corners correspond to weakly smooth maps of manifolds with corners.

Manifolds with g-corners are a generalization of manifolds with corners which allow more general local models than $\mathbb{R}^m_k$. These local models X_P depend on a *weakly toric monoid* P, and we give an introduction to monoids in §3.2, before proceeding with the definition of manifolds with g-corners in §3.3.

We discuss boundaries and corners for manifolds with (g-)corners in §3.4. We will see that the corners $C(X)$ of a manifold with (g-)corners X behave functorially, so that smooth maps $f : X \to Y$ of manifolds with (g-)corners lift to interior maps $C(f) : C(X) \to C(Y)$ of the corners. In §3.5 we discuss (co)tangent bundles and b-(co)tangent bundles of manifolds with (g-)corners.

3.1 Manifolds with corners

We define manifolds with corners, following the second author [46, §2].

Definition 3.1 Use the notation $\mathbb{R}^m_k = [0,\infty)^k \times \mathbb{R}^{m-k}$ for $0 \leqslant k \leqslant m$, and write points of $\mathbb{R}^m_k$ as $u = (u_1, \ldots, u_m)$ for $u_1, \ldots, u_k \in [0,\infty)$, $u_{k+1}, \ldots, u_m \in \mathbb{R}$. Let $U \subseteq \mathbb{R}^m_k$ and $V \subseteq \mathbb{R}^n_l$ be open, and let $f = (f_1, \ldots, f_n) : U \to V$ be a continuous map, so that $f_j = f_j(u_1, \ldots, u_m)$ maps $U \to [0,\infty)$ for $j = 1, \ldots, l$ and $U \to \mathbb{R}$ for $j = l+1, \ldots, n$. Then we say the following.

(a) f is *weakly smooth* if all derivatives $\frac{\partial^{a_1+\cdots+a_m}}{\partial u_1^{a_1}\cdots\partial u_m^{a_m}} f_j(u_1, \ldots, u_m) : U \to \mathbb{R}$ exist and are continuous for all $j = 1, \ldots, n$ and $a_1, \ldots, a_m \geqslant 0$, including one-sided derivatives where $u_i = 0$ for $i = 1, \ldots, k$. By Seeley's Extension Theorem, this is equivalent to requiring f_j to extend to a smooth function $f'_j : U' \to \mathbb{R}$ on open neighbourhood U' of U in $\mathbb{R}^m$.

(b) f is *smooth* if it is weakly smooth and every $u = (u_1, \ldots, u_m) \in U$ has an open neighbourhood $\tilde{U}$ in U such that for each $j = 1, \ldots, l$, either:

  (i) we may write $f_j(\tilde{u}_1, \ldots, \tilde{u}_m) = F_j(\tilde{u}_1, \ldots, \tilde{u}_m) \cdot \tilde{u}_1^{a_{1,j}} \cdots \tilde{u}_k^{a_{k,j}}$ for all $(\tilde{u}_1, \ldots, \tilde{u}_m) \in \tilde{U}$, where $F_j : \tilde{U} \to (0,\infty)$ is weakly smooth and $a_{1,j}, \ldots, a_{k,j} \in \mathbb{N} = \{0, 1, 2, \ldots\}$, with $a_{i,j} = 0$ if $u_i \neq 0$; or

  (ii) $f_j|_{\tilde{U}} = 0$.

(c) f is *interior* if it is smooth, and case (b)(ii) does not occur.

(d) f is *strongly smooth* if it is smooth, and in case (b)(i), for each $j = 1, \ldots, l$ we have $a_{i,j} = 1$ for at most one $i = 1, \ldots, k$, and $a_{i,j} = 0$ otherwise.

(e) f is a *diffeomorphism* if it is a smooth bijection with smooth inverse.

Definition 3.2 Let X be a second countable Hausdorff topological space. An *m-dimensional chart on X* is a pair (U, ϕ), where $U \subseteq \mathbb{R}^m_k$ is open for some $0 \leqslant k \leqslant m$, and $\phi : U \to X$ is a homeomorphism with an open set $\phi(U) \subseteq X$.

Let $(U, \phi), (V, \psi)$ be m-dimensional charts on X. We call (U, ϕ) and (V, ψ) *compatible* if $\psi^{-1} \circ \phi : \phi^{-1}(\phi(U) \cap \psi(V)) \to \psi^{-1}(\phi(U) \cap \psi(V))$ is a diffeomorphism between open subsets of $\mathbb{R}^m_k, \mathbb{R}^m_l$, in the sense of Definition 3.1(e).

An *m-dimensional atlas* for X is a system $\{(U_a, \phi_a) : a \in A\}$ of pairwise compatible m-dimensional charts on X with $X = \bigcup_{a \in A} \phi_a(U_a)$. We call such an atlas *maximal* if it is not a proper subset of any other atlas. Any atlas $\{(U_a, \phi_a) : a \in A\}$ is contained in a unique maximal atlas, the set of all charts (U, ϕ) of this type on X which are compatible with (U_a, ϕ_a) for all $a \in A$.

An *m-dimensional manifold with corners* is a second countable Hausdorff topological space X equipped with a maximal m-dimensional atlas. Usually we refer to X as the manifold, leaving the atlas implicit, and by a *chart (U, ϕ) on X*, we mean an element of the maximal atlas.

Now let X, Y be manifolds with corners of dimensions m, n, and $f : X \to Y$ a continuous map. We call f *weakly smooth*, or *smooth*, or *interior*, or *strongly smooth*, if whenever $(U, \phi), (V, \psi)$ are charts on X, Y with $U \subseteq \mathbb{R}^m_k$, $V \subseteq \mathbb{R}^n_l$ open, then $\psi^{-1} \circ f \circ \phi : (f \circ \phi)^{-1}(\psi(V)) \to V$ is weakly smooth, or smooth, or interior, or strongly smooth, respectively, as maps between open subsets of $\mathbb{R}^m_k, \mathbb{R}^n_l$ in the sense of Definition 3.1. We call $f : X \to Y$ a *diffeomorphism* if it is a bijection and f, f^{-1} are smooth.

Write $\mathbf{Man^c_{in}}, \mathbf{Man^c_{st}} \subset \mathbf{Man^c} \subset \mathbf{Man^c_{we}}$ for the categories with objects manifolds with corners, and morphisms interior maps, and strongly smooth maps, and smooth maps, and weakly smooth maps, respectively. We will be interested almost exclusively in the categories $\mathbf{Man^c_{in}} \subset \mathbf{Man^c}$.

We write $\check{\mathbf{M}}\mathbf{an^c}$ for the category with objects disjoint unions $\coprod_{m=0}^{\infty} X_m$, where X_m is a manifold with corners of dimension m (we call these *manifolds with corners of mixed dimension*), allowing $X_m = \emptyset$, and morphisms continuous maps $f : \coprod_{m=0}^{\infty} X_m \to \coprod_{n=0}^{\infty} Y_n$, such that $f_{mn} := f|_{X_m \cap f^{-1}(Y_n)} : X_m \cap f^{-1}(Y_n) \to Y_n$ is a smooth map of manifolds with corners for all $m, n \geqslant 0$. We write $\check{\mathbf{M}}\mathbf{an^c_{in}} \subset \check{\mathbf{M}}\mathbf{an^c}$ for the category with the same objects, and with morphisms f such that f_{mn} is interior for all $m, n \geqslant 0$. There are obvious full and faithful embeddings $\mathbf{Man^c} \subset \check{\mathbf{M}}\mathbf{an^c}$, $\mathbf{Man^c_{in}} \subset \check{\mathbf{M}}\mathbf{an^c_{in}}$.

Remark 3.3 Some references on manifolds with corners are Cerf [14], Douady [21], Gillam and Molcho [32, §6.7], Kottke and Melrose [55], Margalef-Roig and Outerelo Dominguez [66], Melrose [69, 70, 71, 72], Monthubert [76], and the second author [41, 46]. Just as objects, without considering morphisms, most authors define manifolds with corners X as in Definition 3.2. However, Melrose [69, 70, 71, 72] and authors who follow him impose an extra condition, that X should be a *manifold with faces* in the sense of Definition 3.13 below.

There is no general agreement in the literature on how to define smooth maps, or morphisms, of manifolds with corners.

(i) Our 'smooth maps' in Definitions 3.1–3.2 are due to Melrose [71, §1.12], [55, §1], who calls them *b-maps*. 'Interior maps' are also due to Melrose.
(ii) The author [41] defined and studied 'strongly smooth maps' above (which were just called 'smooth maps' in [41]).

(iii) Gillam and Molcho's *morphisms of manifolds with corners* [32, §6.7] coincide with our 'interior maps'.
(iv) Most other authors, such as Cerf [14, §I.1.2], define smooth maps of manifolds with corners to be weakly smooth maps, in our notation.

We will base our theory of (interior) C^∞-rings and C^∞-schemes with corners on the categories $\mathbf{Man^c_{in}} \subset \mathbf{Man^c}$. Section 8.3 will discuss theories based on other categories of 'manifolds with corners' including the category $\mathbf{Man^{ac}}$ of *manifolds with analytic corners*, or *manifolds with a-corners*, defined by the second author [47], which are different to the categories above.

Definition 3.4 Let X be a manifold with corners (or a manifold with g-corners in §3.3). Smooth maps $g : X \to [0,\infty)$ will be called *exterior maps*, to contrast them with interior maps. We write $C^\infty(X)$ for the set of smooth maps $f : X \to \mathbb{R}$, and $\mathrm{In}(X)$ for the set of interior maps $g : X \to [0,\infty)$, and $\mathrm{Ex}(X)$ for the set of exterior maps $g : X \to [0,\infty)$. Thus, we have three sets:

(a) $C^\infty(X)$ of smooth maps $f : X \to \mathbb{R}$;
(b) $\mathrm{In}(X)$ of interior maps $g : X \to [0,\infty)$; and
(c) $\mathrm{Ex}(X)$ of exterior (smooth) maps $g : X \to [0,\infty)$, with $\mathrm{In}(X) \subseteq \mathrm{Ex}(X)$.

Much of the book will be concerned with algebraic structures on these three sets, and generalizations to other spaces X. In Chapter 4 we give $(C^\infty(X), \mathrm{In}(X) \amalg \{0\})$ and $(C^\infty(X), \mathrm{Ex}(X))$ the (large and complicated) structure of '(pre) C^∞-rings with corners'. But a lot of the time, it will be enough that $C^\infty(X)$ is an $\mathbb{R}$-algebra, and $\mathrm{In}(X), \mathrm{Ex}(X)$ are monoids under multiplication, as in §3.2.

3.2 Monoids

We will use monoids to define manifolds with g-corners in §3.3, and in the study of C^∞-rings with corners in Chapter 4. Here we recall some facts about monoids, in the style of log geometry. A good reference is Ogus [78, §I].

Definition 3.5 A (commutative) monoid is a set P equipped with an associative commutative binary operation $+ : P \times P \to P$ that has an identity element 0. All monoids in this book will be commutative. A morphism of monoids $P \to Q$ is a map of sets that respects the binary operations and sends the identity to the identity. Write $\mathbf{Mon}$ for the category of monoids. For any $n \in \mathbb{N}$ and $p \in P$ we will write $np = n \cdot p = \overbrace{p + \cdots + p}^{n \text{ copies}}$, and set $0 \cdot p = 0$.

The above is *additive notation* for monoids. We will also very often use *multiplicative notation*, in which the binary operation is written $\cdot : P \times P \to P$, thought of as multiplication, and the identity element is written 1, and we write p^n rather than np, with $p^0 = 1$.

If P is a monoid written in multiplicative notation, a *zero element* is $0 \in P$ with $0 \cdot p = 0$ for all $p \in P$. Zero elements need not exist, but are unique if they do. One should not confuse zero elements with identities.

The rest of this definition will use additive notation.

A *submonoid* Q of a monoid P is a subset that is closed under the binary operation and contains the identity element. We can form the *quotient monoid* P/Q which is the set of all $\sim$-equivalence classes $[p]$ of $p \in P$ such that $p \sim p'$ if there are $q, q' \in Q$ with $p + q = p' + q' \in P$. It has an induced monoid structure from the monoid P. There is a morphism $\pi : P \to P/Q$. This quotient satisfies the following universal property: it is a monoid P/Q with a morphism $\pi : P \to P/Q$ such that $\pi(Q) = \{0\}$ and if $\mu : P \to R$ is a monoid morphism with $\mu(Q) = \{0\}$ then $\mu = \nu \circ \pi$ for a unique morphism $\nu : P/Q \to R$.

A *unit* in P is an element $p \in P$ that has a (necessarily unique) inverse under the binary operation, p', so that $p' + p = 0$. Write $P^\times$ for the set of all units of P. It is a submonoid of P, and an abelian group. A monoid P is an abelian group if and only if $P = P^\times$.

An *ideal* I in a monoid P is a non-empty proper subset $\emptyset \neq I \subsetneq P$ such that if $p \in P$ and $i \in I$ then $i + p \in I$, so it is necessarily closed under P's binary operation. It must not contain any units. An ideal I is called *prime* if whenever $a + b \in I$ for $a, b \in P$ then either a or b is in I. We say the complement $P \setminus I$ of a prime ideal I is a *face* which is automatically a submonoid of P. If we have elements $p_j \in P$ for j in some indexing set J then we can consider the *ideal generated* by the p_j, which we write as $\langle p_j \rangle_{j \in J}$. It consists of all elements in P of the form $a + p_j$ for any $a \in P$ and any $j \in J$. Note that if any of the p_j are units then the 'ideal' generated by these p_j is a misnomer, as $\langle p_j \rangle_{j \in J}$ is not an ideal and instead equal to P.

For any monoid P there is an associated abelian group $P^{\rm gp}$ called the *groupification* of P, with a monoid morphism $\pi^{\rm gp} : P \to P^{\rm gp}$. This has the universal property that any morphism from P to an abelian group factors through $\pi^{\rm gp}$, so $P^{\rm gp}$ is unique up to canonical isomorphism. It can be shown to be isomorphic to the quotient monoid $(P \times P)/\Delta_P$, where $\Delta_P = \{(p, p) : p \in P\}$ is the diagonal submonoid of $P \times P$, and $\pi^{\rm gp} : p \mapsto [p, 0]$.

For a monoid P we make the following definitions.

(i) If there is a surjective morphism $\mathbb{N}^k \to P$ for some $k \geqslant 0$, we call P *finitely generated*. This morphism can be uniquely written as $(n_1, \ldots, n_k)$

$\mapsto n_1p_1 + \cdots + n_kp_k$ for some $p_1, \ldots, p_k \in P$, which we call the *generators* of P. This implies that P^{gp} is finitely generated. If there is an isomorphism $P \cong \mathbb{N}^A$ for some set A (e.g. if $P \cong \mathbb{N}^k$ for $k \geqslant 0$) then P is called *free*.

(ii) If $P^\times = \{0\}$ we call P *sharp*. Any monoid has an associated *sharpening* $P^\sharp$ which is the sharp quotient monoid $P/P^\times$ with surjection $\pi^\sharp : P \to P^\sharp$.

(iii) If $\pi^{\mathrm{gp}} : P \to P^{\mathrm{gp}}$ is injective we call P *integral* or *cancellative*. This occurs if and only if $p+p' = p+p''$ implies $p' = p''$ for all $p, p', p'' \in P$. Then P is isomorphic to its image under π^{gp}, so we consider it a subset of P^{gp}.

(iv) If P is integral and whenever $p \in P^{\mathrm{gp}}$ with $np \in P \subset P^{\mathrm{gp}}$ for some $n \geqslant 1$ implies $p \in P$ then we call P *saturated*.

(v) If P^{gp} is a torsion-free group, then we call P *torsion-free*. That is, if there is $n \geqslant 0$ and $p \in P^{\mathrm{gp}}$ such that $np = 0$ then $p = 0$.

(vi) If P is finitely generated, integral, saturated, and torsion-free then it is called *weakly toric*. It has *rank* $\operatorname{rank} P = \dim_{\mathbb{R}}(P \otimes_{\mathbb{N}} \mathbb{R})$. For a weakly toric P there is an isomorphism $P^\times \cong \mathbb{Z}^l$ and $P^\sharp$ is a toric monoid (see (vii)). The exact sequence $0 \to P^\times \to P \to P^\sharp \to 0$ splits, so that $P \cong P^\sharp \times \mathbb{Z}^l$. Then the rank of P is equal to $\operatorname{rank} P = \operatorname{rank} P^{\mathrm{gp}} = \operatorname{rank} P^\sharp + l$.

(vii) If P is a weakly toric monoid and is also sharp we call P *toric* (note that saturated and sharp together imply torsion-free). For a toric monoid P its associated group P^{gp} is a finitely generated, torsion-free abelian group, so $P^{\mathrm{gp}} \cong \mathbb{Z}^k$ for $k \geqslant 0$. Then the rank of P is $\operatorname{rank} P = k$.

Parts (vi)–(vii) are not universally agreed in the literature. For example, Ogus [78, p. 13], calls our weakly toric monoids *toric monoids*, and our toric monoids *sharp toric monoids*.

Example 3.6 **(a)** The most basic toric monoid is $\mathbb{N}^k$ under addition for $k = 0, 1, \ldots$, with $(\mathbb{N}^k)^{\mathrm{gp}} \cong \mathbb{Z}^k$.

(b) $\mathbb{Z}^k$ under addition for $k > 0$ is weakly toric, but not toric.

(c) $\big([0,\infty), \cdot, 1\big)$ under multiplication is a monoid that is not finitely generated. It has identity 1 and zero element 0. We have $[0,\infty)^{\mathrm{gp}} = \{1\}$, so $[0,\infty)$ is not integral, and $[0,\infty)^\times = (0,\infty)$, so $[0,\infty)$ is not sharp.

3.3 Manifolds with g-corners

The second author [46] defined the category $\mathbf{Man^{gc}}$ of *manifolds with generalized corners*, or *manifolds with g-corners*. These extend manifolds with corners X in §3.1, but rather than being locally modelled on $\mathbb{R}^m_k = [0,\infty)^k \times \mathbb{R}^{m-k}$, they are modelled on spaces X_P for P a weakly toric monoid in §3.2.

Definition 3.7 Let P be a weakly toric monoid. As in [46, §3.2] we define $X_P = \mathrm{Hom}(P,[0,\infty))$ to be the set of monoid morphisms $x : P \to [0,\infty)$, where the target is considered as a monoid under multiplication as in Example 3.6(c). The *interior* of X_P is defined to be $X_P^\circ = \mathrm{Hom}(P,(0,\infty))$, where $(0,\infty)$ is a submonoid of $[0,\infty)$, so that $X_P^\circ \subset X_P$.

For $p \in P$ there is a corresponding function $\lambda_p : X_P \to [0,\infty)$ such that $\lambda_p(x) = x(p)$. If $p,q \in P$ then $\lambda_{p+q} = \lambda_p \cdot \lambda_q$, and $\lambda_0 = 1$. Define a topology on X_P to be the weakest topology such that each λ_p is continuous. Then X_P is locally compact and Hausdorff and X_P° is an open subset of X_P. The *interior* U° of an open set $U \subset X_P$ is defined to be $U \cap X_P^\circ$.

As P is weakly toric we can take a presentation for P with generators $p_1,\ldots,p_m$ and relations

$$a_1^j p_1 + \cdots + a_m^j p_m = b_1^j p_1 + \cdots + b_m^j p_m \quad \text{in } P \text{ for } j = 1,\ldots,k,$$

for $a_i^j, b_i^j \in \mathbb{N}$, $i = 1,\ldots,m$, $j = 1,\ldots,k$. Then we have a continuous function $\lambda_{p_1} \times \cdots \times \lambda_{p_m} : X_P \to [0,\infty)^m$ that is a homeomorphism onto its image

$$X'_P = \{(x_1,\ldots,x_m) \in [0,\infty)^m : x_1^{a_1^j} \cdots x_m^{a_m^j} = x_1^{b_1^j} \cdots x_m^{b_m^j},\ j = 1,\ldots,k\},$$

which is a closed subset of $[0,\infty)^m$.

Let U be an open subset of X_P, and $U' = \lambda_{p_1} \times \cdots \times \lambda_{p_m}(U) \subset X'_P$. Then we say a continuous function $f : U \to \mathbb{R}$ or $f : U \to [0,\infty)$ is *smooth* if there exists an open neighbourhood W' of U' in $[0,\infty)^m$ and a smooth function $g : W' \to \mathbb{R}$ or $g : W' \to [0,\infty)$ that is smooth in the sense above, such that $f = g \circ \lambda_{p_1} \times \cdots \times \lambda_{p_m}$. As in [46, Prop. 3.14], this is independent of the choice of generators $p_1,\ldots,p_m$ for P.

Suppose Q is another weakly toric monoid, and consider open $V \subseteq X_Q$ and a continuous function $f : U \to V$. Then f is *smooth* if $\lambda_q \circ f : U \to [0,\infty)$ is smooth for all $q \in Q$ in the sense above. We call smooth f *interior* if $f(U^\circ) \subseteq V^\circ$, and a *diffeomorphism* if it is bijective with smooth inverse.

Example 3.8 If $P = \mathbb{N}^k \times \mathbb{Z}^{m-k}$ then P is weakly toric. We can take generators $p_1 = (1,0,\ldots,0), p_2 = (0,1,0,\ldots,0),\ldots,p_m = (0,\ldots,0,1), p_{m+1} =$

$(0, \ldots, 0, -1, \ldots, -1)$ with p_{m+1} having -1 in the $k+1$ to $m+1$ entries, so the only relation is $p_{k+1} + \cdots + p_{m+1} = 0$. Then X_P is homeomorphic to

$$X'_P = \{(x_1, \ldots, x_{m+1}) \in [0, \infty)^{m+1} : x_{k+1} \cdots x_{m+1} = 1\}.$$

This means that for $(x_1, \ldots, x_{m+1}) \in X'_P$ we have $x_{k+1}, \ldots, x_{m+1} > 0$ with $x_{m+1} = x_{k+1}^{-1} \cdots x_m^{-1}$. So there is a homeomorphism from $X_P \to \mathbb{R}^m_k$ mapping $(x_1, \ldots, x_{m+1}) \mapsto (x_1, \ldots, x_k, \log(x_{k+1}), \ldots, \log(x_m))$. In [46, Ex. 3.15] we show that this identification $X_P \cong \mathbb{R}^m_k$ identifies the topology, and the notion of smooth maps to $\mathbb{R}$, $[0, \infty)$ and between open subsets of $X_P \cong \mathbb{R}^m_k$ and $X_Q \cong \mathbb{R}^n_l$, in Definition 3.1 for $\mathbb{R}^m_k$ and above for X_P. Thus, the spaces X_P generalize the spaces $\mathbb{R}^m_k$ used as local models for manifolds with corners.

Following [46, §3.3] we define manifolds with g-corners.

Definition 3.9 Let X be a topological space. Define a *g-chart* on X to be a triple (P, U, ϕ), where P is a weakly toric monoid, $U \subset X_P$ is open, and $\phi : U \to X$ is a homeomorphism with an open subset $\phi(U) \subset X$. If $\operatorname{rank} P = m$ we call (P, U, ϕ) *m-dimensional*. For set-theory reasons (to ensure a maximal atlas is a set not a class) we suppose P is a submonoid of $\mathbb{Z}^k$ for some $k \geqslant 0$.

We call m-dimensional g-charts (P, U, ϕ) and (Q, V, ψ) on X *compatible* if $\psi^{-1} \circ \phi : \phi^{-1}(\phi(U) \cap \psi(V)) \to \psi^{-1}(\phi(U) \cap \psi(V))$ is a diffeomorphism between open subsets of X_P and X_Q. A *g-atlas* on X is a family $\mathcal{A} = \{(P_i, U_i, \phi_i) : i \in I\}$ of pairwise compatible g-charts (P_i, U_i, ϕ_i) on X with the same dimension m, with $X = \bigcup_{i \in I} \phi_i(U_i)$. We call $\mathcal{A}$ *maximal* if it is not a proper subset of any other g-atlas. We define a *manifold with g-corners* $(X, \mathcal{A})$ to be a Hausdorff, second countable topological space X with a maximal g-atlas $\mathcal{A}$.

We define smooth maps, and interior maps, $f : X \to Y$ between manifolds with g-corners X, Y as in Definition 3.2. We write $\mathbf{Man^{gc}}$ for the category of manifolds with g-corners and smooth maps, and $\mathbf{Man^{gc}_{in}} \subset \mathbf{Man^{gc}}$ for the subcategory of manifolds with g-corners and interior maps.

As in Example 3.8, if $P \cong \mathbb{N}^k \times \mathbb{Z}^{m-k}$ we may identify $X_P \cong \mathbb{R}^m_k$. So as in [46, Def. 3.22], we may identify $\mathbf{Man^c} \subset \mathbf{Man^{gc}}$ and $\mathbf{Man^c_{in}} \subset \mathbf{Man^{gc}_{in}}$ as full subcategories, where a manifold with g-corners $(X, \mathcal{A})$ is a manifold with corners if $\mathcal{A}$ has a g-subatlas of g-charts (P_i, U_i, ϕ_i) with $P_i \cong \mathbb{N}^k \times \mathbb{Z}^{m-k}$.

We write $\check{\mathbf{M}}\mathbf{an^{gc}}$ for the category with objects disjoint unions $\coprod_{m=0}^{\infty} X_m$, where X_m is a manifold with g-corners of dimension m (we call these *manifolds with g-corners of mixed dimension*), allowing $X_m = \emptyset$, and morphisms continuous $f : \coprod_{m=0}^{\infty} X_m \to \coprod_{n=0}^{\infty} Y_n$, such that $f_{mn} := f|_{X_m \cap f^{-1}(Y_n)} : X_m \cap f^{-1}(Y_n) \to Y_n$ is a smooth map of manifolds with g-corners for all $m, n \geqslant 0$. We write $\check{\mathbf{M}}\mathbf{an^c_{in}} \subset \check{\mathbf{M}}\mathbf{an^c}$ for the category with the same objects, and with morphisms f such that f_{mn} is interior for all $m, n \geqslant 0$. There are obvious full embeddings $\mathbf{Man^{gc}} \subset \check{\mathbf{M}}\mathbf{an^{gc}}$ and $\mathbf{Man^{gc}_{in}} \subset \check{\mathbf{M}}\mathbf{an^{gc}_{in}}$.

Remark 3.10 Any weakly toric monoid P is isomorphic to $P^\sharp \times \mathbb{Z}^l$, where $P^\sharp$ is toric and $l \geqslant 0$. Then $X_P \cong X_{P^\sharp} \times X_{\mathbb{Z}^l} \cong X_{P^\sharp} \times \mathbb{R}^l$. Hence manifolds with g-corners have local models $X_Q \times \mathbb{R}^l$ for toric monoids Q and $l \geqslant 0$, where $X_{\mathbb{N}^k} \cong [0, \infty)^k$. Each toric monoid Q has a natural point $\delta_0 \in X_Q$ called the *vertex* of X_Q, which acts by taking $0 \in Q$ to $1 \in [0, \infty)$ and all non-zero $q \in Q$ to zero. Given a manifold with g-corners X and a point $x \in X$, there is a toric monoid Q such that X near x is modelled on $X_Q \times \mathbb{R}^l$ near $(\delta_0, 0) \in X_Q \times \mathbb{R}^l$, where $\operatorname{rank} Q + l = \dim X$.

From [46, Ex. 3.23] we have the simplest example of a manifold with g-corners that is not a manifold with corners.

Example 3.11 Let P be the weakly toric monoid of rank 3 with

$$P = \{(a, b, c) \in \mathbb{Z}^3 : a \geqslant 0,\ b \geqslant 0,\ a + b \geqslant c \geqslant 0\}.$$

This has generators $p_1 = (1, 0, 0)$, $p_2 = (0, 1, 1)$, $p_3 = (0, 1, 0)$, and $p_4 = (1, 0, 1)$, and one relation $p_1 + p_2 = p_3 + p_4$. The local model it induces is

$$X_P \cong X'_P = \{(x_1, x_2, x_3, x_4) \in [0, \infty)^4 : x_1 x_2 = x_3 x_4\}. \tag{3.1}$$

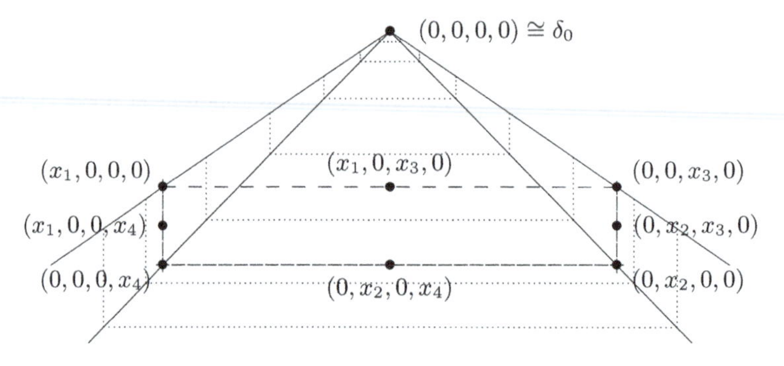

Figure 3.1 A 3-manifold with g-corners $X'_P \cong X_P$ in (3.1)

Figure 3.1 is a sketch of X'_P as a square-based three-dimensional infinite pyramid. As in Remark 3.10, X'_P has vertex $(0, 0, 0, 0)$ corresponding to $\delta_0 \in X_P$. It has one-dimensional edges of points $(x_1, 0, 0, 0)$, $(0, x_2, 0, 0)$, $(0, 0, x_3, 0)$, $(0, 0, 0, x_4)$, and two-dimensional faces of points $(x_1, 0, x_3, 0)$, $(x_1, 0, 0, x_4)$, $(0, x_2, x_3, 0)$, $(0, x_2, 0, x_4)$. Its interior $X'^\circ_P \cong \mathbb{R}^3$ consists of points (x_1, x_2, x_3, x_4) with $x_1, \dots, x_4$ non-zero and $x_1 x_2 = x_3 x_4$. Then

$X_P \setminus \{\delta_0\}$ is a 3-manifold with corners, but X_P is not a manifold with corners near δ_0, as it is not locally isomorphic to $\mathbb{R}^m_k$.

We give more material on manifolds with g-corners in §3.4, §3.5, and §6.7.1.

3.4 Boundaries, corners, and the corner functor

This section follows the second author [41] and [46, §2.2 and §3.4].

Definition 3.12 Let X be a manifold with corners, or g-corners, of dimension m. For each $x \in X$ we define the *depth* $\mathrm{depth}_X(x) \in \{0, 1, \ldots, m\}$ as follows: if X is a manifold with corners then $\mathrm{depth}_X(x) = k$ if X near x is locally modelled on $\mathbb{R}^m_k$ near 0. And if X is a manifold with g-corners then $\mathrm{depth}_X(x) = k$ if X near x is locally modelled on $X_P \times \mathbb{R}^{m-k}$ near $(\delta_0, 0)$, using Remark 3.10, where P is a toric monoid of rank k.

Write $S^k(X) = \{x \in X : \mathrm{depth}_X(x) = k\}$ for $k = 0, \ldots, m$, the *codimension k boundary stratum*. This defines a stratification $X = \coprod_{k=0}^m S^k(X)$ called the *depth stratification*. Then $S^k(X)$ has a natural structure of a manifold without boundary of dimension $m - k$. Its closure is $\overline{S^k(X)} = \coprod_{l=k}^m S^l(X)$. The *interior* of X is $X^\circ := S^0(X)$.

Define a *local k-corner component γ of X at $x \in X$* to be a local choice of connected component of $S^k(X)$ near x. That is, for any sufficiently small open neighbourhood V of x in X, then γ assigns a choice of connected component W of $V \cap S^k(X)$ with $x \in \overline{W}$, and if V', W' are alternative choices then $x \in \overline{W \cap W'}$. The local 1-corner components are called *local boundary components* of X.

As sets, we define the *boundary* and *k-corners* of X for $k = 0, \ldots, m$ by

$$\begin{aligned} \partial X &= \big\{(x,\beta) : x \in X,\ \beta \text{ is a local boundary component of } X \text{ at } x\big\}, \\ C_k(X) &= \big\{(x,\gamma) : x \in X,\ \gamma \text{ is a local } k\text{-corner component of } X \text{ at } x\big\}, \end{aligned} \tag{3.2}$$

for $k = 0, 1, \ldots, m$. This implies that $\partial X = C_1(X)$ and $C_0(X) \cong X$. In [46, §2.2 and §3.4] we define natural structures of a manifold with corners, or g-corners (depending on which X is) on ∂X and $C_k(X)$, so that $\dim \partial X = m-1$ and $\dim C_k(X) = m - k$. The interiors are $(\partial X)^\circ \cong S^1(X)$ and $C_k(X)^\circ \cong S^k(X)$. We define smooth maps $\Pi_X : \partial X \to X$, $\Pi_X : C_k(X) \to X$ by $\Pi_X : (x, \beta) \mapsto x$ and $\Pi_X : (x, \gamma) \mapsto x$. These are generally neither interior, not injective.

The next definition will be important in Chapter 5, as a manifold with corners corresponds to an *affine* C^∞-scheme with corners if it is a manifold with faces. We extend Definition 3.13 to manifolds with g-corners in Definition 4.37.

Definition 3.13 A manifold with corners X is called a *manifold with faces* if $\Pi_X|_F : F \to X$ is injective for each connected component F of ∂X. The *faces* of X are the components of ∂X, regarded as subsets of X. Melrose [69, 70, 71, 72] assumes his manifolds with corners are manifolds with faces. Write $\mathbf{Man^f_{in}} \subset \mathbf{Man^c_{in}}$, $\mathbf{Man^f} \subset \mathbf{Man^c}$ for the full subcategories of manifolds with faces.

Example 3.14 Define the *teardrop* to be the subset $T = \{(x,y) \in \mathbb{R}^2 : x \geqslant 0,\ y^2 \leqslant x^2 - x^4\}$, as in [46, Ex. 2.8]. As shown in Figure 3.2, T is a manifold with corners of dimension 2. The teardrop is not a manifold with faces as in Definition 3.13, since the boundary is diffeomorphic to $[0,1]$, and so is connected, but the map $\Pi_T : \partial T \to T$ is not injective.

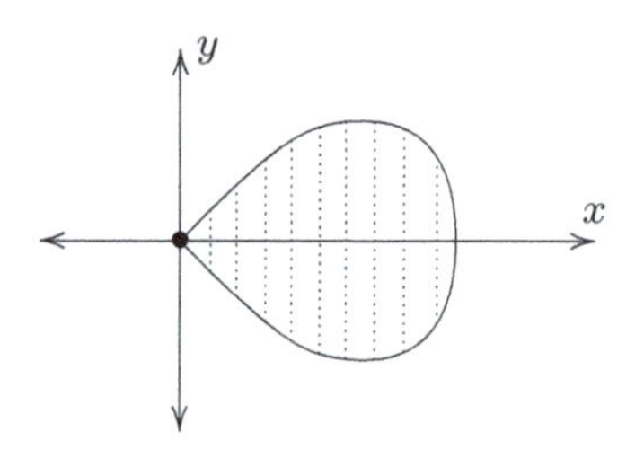

Figure 3.2 The teardrop T

Remark 3.15 Let X be a manifold with (g-)corners. Then smooth functions $f : X \to \mathbb{R}$ are local objects: we can combine them using partitions of unity. In contrast, exterior functions $g : X \to [0,\infty)$ as in Definition 3.4 cannot be combined using partitions of unity, and have some non-local, global features.

To see this, observe that if $g : X \to [0,\infty)$ is an exterior function, we may define a *multiplicity function* $\mu_g : \partial X \to \mathbb{N} \cup \{\infty\}$ giving the order of vanishing of g on the boundary faces of X. If $x \in X$ and $(x_1,\ldots,x_m) \in \mathbb{R}^m_k$ are local coordinates on X near x with $x = (0,\ldots,0)$ then either $g = x_1^{a_1} \cdots x_k^{a_k} F(x_1,\ldots,x_m)$ near x for $F > 0$ smooth, and we set $\mu_g(x, \{x_i = 0\}) = a_i$ for $i = 1,\ldots,k$, or $g = 0$ near x, when we write $\mu_g(x, \{x_i = 0\}) = \infty$.

Then $\mu_g : \partial X \to \mathbb{N} \cup \{\infty\}$ is locally constant, and hence constant on each connected component of ∂X. Also g interior implies μ_g maps to $\mathbb{N} \subset \mathbb{N} \cup \{\infty\}$. If two points $x_1, x_2 \in X$ lie in the image of the same connected component of ∂X then the behaviour of g near x_1, x_2 is linked even if x_1, x_2 are far away in X. Interior functions $\{g_i : i \in I\}$ can be combined with a partition of unity $\{\eta_i : i \in I\}$ to give interior $\sum_{i \in I} \eta_i g_i$ if the μ_{g_i} for $i \in I$ are all equal.

The boundary ∂X and corners $C_k(X)$ are not functorial under smooth or interior maps: smooth $f : X \to Y$ generally do not lift to smooth $\partial f : \partial X \to \partial Y$. However, as in the second author [41, §4] and [46, §2.2 and §3.4], the corners $C(X) = \coprod_{k=0}^{\dim X} C_k(X)$ do behave functorially.

Definition 3.16 Let X be a manifold with corners, or with g-corners. The *corners* of X is the manifold with (g-)corners of mixed dimension

$$C(X) = \coprod_{k=0}^{\dim X} C_k(X),$$

as an object of $\check{\mathbf{Man}}^{\mathbf{c}}$ or $\check{\mathbf{Man}}^{\mathbf{gc}}$ in Definitions 3.2 and 3.9. By (3.2) we have

$$C(X) = \{(x, \gamma) : x \in X, \gamma \text{ is a local } k\text{-corner component of } X \text{ at } x, \ k \geqslant 0\}.$$

Define morphisms $\Pi_X : C(X) \to X$ and $\iota_X : X \to C(X)$ in $\check{\mathbf{Man}}^{\mathbf{c}}$ or $\check{\mathbf{Man}}^{\mathbf{gc}}$ by $\Pi_X : (x, \gamma) \mapsto x$ and $\iota_X : x \mapsto (x, X^\circ)$, that is, $\iota_X : X \xrightarrow{\cong} C^0(X) \subset C(X)$. Here ι_X is interior, but Π_X generally is not.

Now let $f : X \to Y$ be a morphism in $\mathbf{Man^c}$ or $\mathbf{Man^{gc}}$. Suppose γ is a local k-corner component of X at $x \in X$. For each small open neighbourhood V of x in X, γ gives a connected component W of $V \cap S^k(X)$ with $x \in \overline{W}$. As f preserves the stratifications $X = \coprod_{k \geqslant 0} S^k(X)$, $Y = \coprod_{l \geqslant 0} S^l(Y)$, we have $f(W) \subseteq S^l(Y)$ for some $l \geqslant 0$. Since f is continuous, $f(W)$ is connected, and $f(x) \in \overline{f(W)}$. Thus there is a unique l-corner component $f_*(\gamma)$ of Y at $f(x)$, such that if $\tilde{V}$ is a sufficiently small open neighbourhood of $f(x)$ in Y, then the connected component $\tilde{W}$ of $\tilde{V} \cap S^l(Y)$ given by $f_*(\gamma)$ has $\tilde{W} \cap f(W) \neq \emptyset$. This $f_*(\gamma)$ is independent of the choice of sufficiently small $V, \tilde{V}$, so is well defined.

Define a map $C(f) : C(X) \to C(Y)$ by $C(f) : (x, \gamma) \mapsto (f(x), f_*(\gamma))$. As in [46, §2.2 and §3.4], this $C(f)$ is smooth, and interior, and so is a morphism in $\check{\mathbf{Man}}^{\mathbf{gc}}_{\mathbf{in}}$. Also $C(g \circ f) = C(g) \circ C(f)$ and $C(\mathrm{id}_X) = \mathrm{id}_{C(X)}$. Thus we have defined functors $C : \mathbf{Man^c} \to \check{\mathbf{Man}}^{\mathbf{c}}_{\mathbf{in}}$ and $C : \mathbf{Man^{gc}} \to \check{\mathbf{Man}}^{\mathbf{gc}}_{\mathbf{in}}$, which we call the *corner functors*. We extend them to $C : \check{\mathbf{Man}}^{\mathbf{c}} \to \check{\mathbf{Man}}^{\mathbf{c}}_{\mathbf{in}}$ and $C : \check{\mathbf{Man}}^{\mathbf{gc}} \to \check{\mathbf{Man}}^{\mathbf{gc}}_{\mathbf{in}}$ by $C(\coprod_{m \geqslant 0} X_m) = \coprod_{m \geqslant 0} C(X_m)$.

Consider the inclusions of subcategories $\mathrm{inc} : \check{\mathbf{Man}}^{\mathbf{c}}_{\mathbf{in}} \hookrightarrow \check{\mathbf{Man}}^{\mathbf{c}}$ and $\mathrm{inc} : \check{\mathbf{Man}}^{\mathbf{gc}}_{\mathbf{in}} \hookrightarrow \check{\mathbf{Man}}^{\mathbf{gc}}$. The morphisms $\Pi_X : C(X) \to X$ give a natural transformation $\Pi : \mathrm{inc} \circ C \Rightarrow \mathrm{Id}$ in $\check{\mathbf{Man}}^{\mathbf{c}}$ or $\check{\mathbf{Man}}^{\mathbf{gc}}$. The morphisms $\iota_X : X \to C(X)$ give a natural transformation $\iota : \mathrm{Id} \Rightarrow C \circ \mathrm{inc}$ in $\check{\mathbf{Man}}^{\mathbf{c}}_{\mathbf{in}}$ or $\check{\mathbf{Man}}^{\mathbf{gc}}_{\mathbf{in}}$.

Remark 3.17 If X is an object in $\check{\mathbf{Man}}^{\mathbf{c}}$ or $\check{\mathbf{Man}}^{\mathbf{gc}}$, and Y an object in

$\check{\mathbf{Man}}^{\mathbf{c}}_{\mathbf{in}}$ or $\check{\mathbf{Man}}^{\mathbf{gc}}_{\mathbf{in}}$, it is easy to check that we have inverse 1-1 correspondences

$$\mathrm{Hom}_{\check{\mathbf{Man}}^{\mathbf{c}} \text{ or } \check{\mathbf{Man}}^{\mathbf{gc}}}(\mathrm{inc}\, X, Y) \underset{g \longmapsto \Pi_Y \circ g}{\overset{f \longmapsto C(f)\circ \iota_X}{\rightleftarrows}} \mathrm{Hom}_{\check{\mathbf{Man}}^{\mathbf{c}}_{\mathbf{in}} \text{ or } \check{\mathbf{Man}}^{\mathbf{gc}}_{\mathbf{in}}}(X, C(Y)).$$

Therefore $C : \check{\mathbf{Man}}^{\mathbf{c}} \to \check{\mathbf{Man}}^{\mathbf{c}}_{\mathbf{in}}$ and $C : \check{\mathbf{Man}}^{\mathbf{gc}} \to \check{\mathbf{Man}}^{\mathbf{gc}}_{\mathbf{in}}$ are *right adjoint* to $\mathrm{inc} : \check{\mathbf{Man}}^{\mathbf{c}}_{\mathbf{in}} \hookrightarrow \check{\mathbf{Man}}^{\mathbf{c}}$ and $\mathrm{inc} : \check{\mathbf{Man}}^{\mathbf{gc}}_{\mathbf{in}} \hookrightarrow \check{\mathbf{Man}}^{\mathbf{gc}}$, and Π, ι are the units of the adjunction. (This is a new observation, not in [41, 46].)

This implies that the corner functors C, and hence the boundary ∂X and corners $C_k(X)$ of a manifold with (g-)corners X, are not arbitrary constructions, but are determined up to natural isomorphism by the inclusions of subcategories $\mathbf{Man}^{\mathbf{c}}_{\mathbf{in}} \subset \mathbf{Man}^{\mathbf{c}}$ and $\mathbf{Man}^{\mathbf{gc}}_{\mathbf{in}} \subset \mathbf{Man}^{\mathbf{gc}}$.

3.5 Tangent bundles and b-tangent bundles

Finally we briefly discuss (co)tangent bundles of manifolds with (g-)corners. For a manifold with corners X there are two kinds: the (*ordinary*) *tangent bundle* TX, the obvious generalization of tangent bundles of manifolds, and the *b-tangent bundle* ${}^bTX$ introduced by Melrose [70, §2], [71, §2.2], [72, §I.10]. The duals of the tangent bundle and b-tangent bundle are the *cotangent bundle* T^*X and *b-cotangent bundle* ${}^bT^*X$. If X is a manifold with g-corners, then ${}^bTX$ is well defined, but in general TX is not. We follow [46, §2.3 and §3.5].

Remark 3.18 **(a)** Let X be a manifold with corners of dimension m. The *tangent bundle* $TX \to X$, *cotangent bundle* $T^*X \to X$, *b-tangent bundle* ${}^bTX \to X$, and *b-cotangent bundle* ${}^bT^*X \to X$, are natural rank m vector bundles on X, defined in detail in [46, §2.3]. In local coordinates, if $(x_1, \ldots, x_k, x_{k+1}, \ldots, x_m) \in \mathbb{R}^m_k$ are local coordinates on an open set $\phi(U) \subset X$, from a chart (U, ϕ) with $U \subset \mathbb{R}^m_k$ open, then $TX, \ldots, {}^bT^*X$ are given on $\phi(U)$ by the bases of sections:

$$\begin{aligned}
TX|_{\phi(U)} &= \Bigl\langle \frac{\partial}{\partial x_1}, \ldots, \frac{\partial}{\partial x_k}, \frac{\partial}{\partial x_{k+1}}, \ldots, \frac{\partial}{\partial x_m} \Bigr\rangle_{\mathbb{R}}, \\
T^*X|_{\phi(U)} &= \bigl\langle \mathrm{d}x_1, \ldots, \mathrm{d}x_k, \mathrm{d}x_{k+1}, \ldots, \mathrm{d}x_m \bigr\rangle_{\mathbb{R}}, \\
{}^bTX|_{\phi(U)} &= \Bigl\langle x_1 \frac{\partial}{\partial x_1}, \ldots, x_k \frac{\partial}{\partial x_k}, \frac{\partial}{\partial x_{k+1}}, \ldots, \frac{\partial}{\partial x_m} \Bigr\rangle_{\mathbb{R}}, \\
{}^bT^*X|_{\phi(U)} &= \bigl\langle x_1^{-1}\mathrm{d}x_1, \ldots, x_k^{-1}\mathrm{d}x_k, \mathrm{d}x_{k+1}, \ldots, \mathrm{d}x_m \bigr\rangle_{\mathbb{R}}.
\end{aligned}$$

Note that "$x_i \frac{\partial}{\partial x_i}$" is a non-zero section of ${}^bTX$ even where $x_i = 0$ for $i = 1, \ldots, k$, and "$x_i^{-1}\mathrm{d}x_i$" is a well-defined section of ${}^bT^*X$ even where

$x_i = 0$. These are formal symbols, but are useful as they determine the transition functions for ${}^bTX, {}^bT^*X$ under change of coordinates $(x_1, \dots, x_m) \rightsquigarrow (\tilde{x}_1, \dots, \tilde{x}_m)$, etc.

There is a natural morphism of vector bundles $I_X : {}^bTX \to TX$ mapping $x_i \frac{\partial}{\partial x_i} \mapsto x_i \cdot \frac{\partial}{\partial x_i}$ for $i = 1, \dots, k$ and $\frac{\partial}{\partial x_i} \mapsto \frac{\partial}{\partial x_i}$ for $i = k+1, \dots, m$. It is an isomorphism over X°, and has kernel of rank k over $S^k(X)$.

(b) Tangent bundles TX in $\mathbf{Man^c}$ are functorial under smooth maps. That is, for any morphism $f : X \to Y$ in $\mathbf{Man^c}$ there is a corresponding morphism $Tf : TX \to TY$ defined in [46, Def. 2.14], functorial in f and linear on the fibres of TX, TY, in a commuting diagram

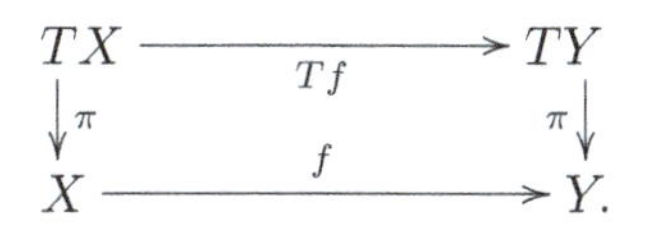

Equivalently, we may write Tf as a vector bundle morphism $\mathrm{d}f : TX \to f^*(TY)$ on X. It has a dual morphism $\mathrm{d}f^* : f^*(T^*Y) \to T^*X$ on cotangent bundles.

Similarly, b-tangent bundles ${}^bTX$ in $\mathbf{Man^c}$ are functorial under *interior* maps. That is, for any morphism $f : X \to Y$ in $\mathbf{Man^c_{in}}$ there is a corresponding morphism ${}^bTf : {}^bTX \to {}^bTY$ defined in [46, Def. 2.15], functorial in f and linear on the fibres of ${}^bTX, {}^bTY$, in a commuting diagram

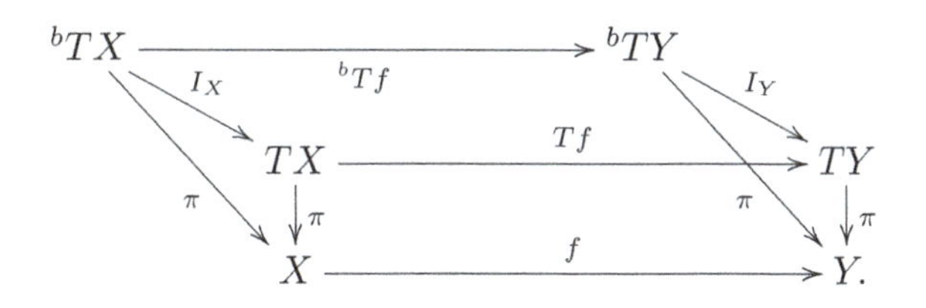

Equivalently, we have a vector bundle morphism ${}^b\mathrm{d}f : {}^bTX \to f^*({}^bTY)$ on X, with a dual morphism ${}^b\mathrm{d}f^* : f^*({}^bT^*Y) \to {}^bT^*X$ on b-cotangent bundles.

(c) As in [46, §3.5], tangent bundles $TX \to X$ do not extend to manifolds with g-corners X. That is, for each $x \in X$ we can define a tangent space T_xX with the usual functorial properties of tangent spaces, but $\dim T_xX$ need not be locally constant on X, so T_xX, $x \in X$ are not the fibres of a vector bundle.

However, b-(co)tangent bundles ${}^bTX \to X$, ${}^bT^*X \to X$ do extend nicely to manifolds with g-corners. If X is locally modelled on X_P for a weakly toric monoid P, as in §3.3, then ${}^bTX$ is locally trivial with fibre $\mathrm{Hom}_{\mathbf{Mon}}(P, \mathbb{R})$, and ${}^bT^*X$ is locally trivial with fibre $P \otimes_{\mathbb{N}} \mathbb{R}$. B-tangent bundles are functorial under interior maps, as in **(b)**. So if $f : X \to Y$ is a morphism in $\mathbf{Man^{gc}_{in}}$, as in [46, Def. 3.43] we have a morphism ${}^b Tf : {}^b TX \to {}^b TY$, and vector bundle morphisms ${}^b\mathrm{d}f : {}^b TX \to f^*({}^b TY)$ and ${}^b\mathrm{d}f^* : f^*({}^b T^*Y) \to {}^b T^*X$ on X.

(d) Let X be a manifold with (g-)corners. Then [46, §3.6] constructs a natural exact sequence of vector bundles of mixed rank on $C(X)$:

$$0 \longrightarrow {}^b N_{C(X)} \xrightarrow{\;{}^b i_T\;} \Pi_X^*({}^b TX) \xrightarrow{\;{}^b \pi_T\;} {}^b T(C(X)) \longrightarrow 0. \tag{3.3}$$

Here ${}^b N_{C(X)}$ is called the *b-normal bundle* of $C(X)$ in X. It has rank k on $C_k(X)$ for $k \geqslant 0$. Note that ${}^b\pi_T$ is *not* ${}^b\mathrm{d}\Pi_X$ from **(c)** for $\Pi_X : C(X) \to X$; ${}^b\mathrm{d}\Pi_X$ is undefined as Π_X is not interior, and would go the opposite way to ${}^b\pi_T$.

Here is the dual complex to (3.3), with ${}^b N^*_{C(X)}$ the *b-conormal bundle:*

$$0 \longrightarrow {}^b T^*(C(X)) \xrightarrow{\;{}^b \pi_T^*\;} \Pi_X^*({}^b T^*X) \xrightarrow{\;{}^b i_T^*\;} {}^b N^*_{C(X)} \longrightarrow 0. \tag{3.4}$$

In [46, §3.6] we define the *monoid bundle* $M_{C(X)} \to C(X)$, which is a locally constant bundle of toric monoids with an embedding $M_{C(X)} \subset {}^b N_{C(X)}$ such that ${}^b N_{C(X)} \cong M_{C(X)} \otimes_{\mathbb{N}} \mathbb{R}$. The monoid fibres have rank k on $C_k(X)$. If X is a manifold with corners then $M_{C(X)}$ has fibre $\mathbb{N}^k$ on $C_k(X)$, and the embedding $M_{C(X)} \subset {}^b N_{C(X)}$ is locally modelled on $\mathbb{N}^k \subset \mathbb{R}^k$. We also have a dual *comonoid bundle* $M^\vee_{C(X)} \to C(X)$, with an embedding $M^\vee_{C(X)} \subset {}^b N^*_{C(X)}$.

If $(x,\gamma) \in C(X)^\circ$, then X near x is locally diffeomorphic to $X_P \times \mathbb{R}^l$ for P a toric monoid, such that γ is identified with $\{\delta_0\} \times \mathbb{R}^l$ for δ_0 the vertex in X_P. Then near (x,γ) in $C(X)$, the fibres of ${}^b N_{C(X)}, {}^b N^*_{C(X)}, M_{C(X)}, M^\vee_{C(X)}$ are $\mathrm{Hom}_{\mathbf{Mon}}(P,\mathbb{R}), P \otimes_{\mathbb{N}} \mathbb{R}, P^\vee, P$, respectively.

Example 3.19 Let X be a manifold with corners, and $(x,\gamma) \in C_k(X)^\circ \subset C(X)$. We will explain the ideas of Remark 3.18(d) near (x,γ). We can choose local coordinates $(x_1,\ldots,x_m) \in \mathbb{R}^m_k$ near x in X, such that the local boundary component γ of X at x is $x_1 = \cdots = x_k = 0$ in coordinates. Then $(x_{k+1},\ldots,x_m)$ in $\mathbb{R}^{m-k}$ are local coordinates on $C_k(X)^\circ$ near (x,γ), and (3.3)–(3.4) become

$$\begin{array}{ccccccc}
0 \longrightarrow {}^b N_{C(X)} & \xrightarrow[{}^b i_T]{} & \Pi_X^*({}^b TX) & \xrightarrow[{}^b \pi_T]{} & {}^b T(C(X)) \longrightarrow 0 \\
\| & & \| & & \| \\
\langle x_1 \frac{\partial}{\partial x_1},\ldots,x_k\frac{\partial}{\partial x_k}\rangle_{\mathbb{R}} & \longrightarrow & \begin{array}{c}\langle x_1 \frac{\partial}{\partial x_1},\ldots,x_k\frac{\partial}{\partial x_k}, \\ \frac{\partial}{\partial x_{k+1}},\ldots,\frac{\partial}{\partial x_m}\rangle_{\mathbb{R}}\end{array} & \longrightarrow & \langle \frac{\partial}{\partial x_{k+1}},\ldots,\frac{\partial}{\partial x_m}\rangle_{\mathbb{R}},
\end{array}$$

$$\begin{array}{ccccccc}
0 \longrightarrow {}^b T^*(C(X)) & \xrightarrow[{}^b \pi_T^*]{} & \Pi_X^*({}^b T^*X) & \xrightarrow[{}^b i_T^*]{} & {}^b N^*_{C(X)} \longrightarrow 0 \\
\| & & \| & & \| \\
\langle \mathrm{d}x_{k+1},\ldots,\mathrm{d}x_m\rangle_{\mathbb{R}} & \longrightarrow & \begin{array}{c}\langle x_1^{-1}\mathrm{d}x_1,\ldots,x_k^{-1}\mathrm{d}x_k, \\ \mathrm{d}x_{k+1},\ldots,\mathrm{d}x_m\rangle_{\mathbb{R}}\end{array} & \longrightarrow & \langle x_1^{-1}\mathrm{d}x_1,\ldots,x_k^{-1}\mathrm{d}x_k\rangle_{\mathbb{R}}.
\end{array}$$

Under these identifications we have

$$M_{C(X)} = \left\langle x_1\tfrac{\partial}{\partial x_1},\ldots,x_k\tfrac{\partial}{\partial x_k}\right\rangle_{\mathbb{N}}, \quad M^\vee_{C(X)} = \left\langle x_1^{-1}\mathrm{d}x_1,\ldots,x_k^{-1}\mathrm{d}x_k\right\rangle_{\mathbb{N}}.$$

3.6 Applications of manifolds with (g-)corners

Manifolds with corners appear in many places in the literature. Here, with no attempt at completeness, we review a selection of areas involving manifolds with corners. Those in §3.6.4–§3.6.6 are also relevant to C^∞-schemes with corners and derived manifolds with corners, as in §8.4. We also point out where our particular definition of smooth map in §3.1 is important, and a few applications of manifolds with g-corners.

3.6.1 Extended Topological Quantum Field Theories

As in Atiyah [2] and Freed [27], an *(unextended, oriented, unitary) Topological Quantum Field Theory* (*TQFT*) Z of dimension n assigns the following data.

(a) To each compact, oriented $(n-1)$-manifold N, a complex Hilbert space $Z(N)$. We require that $Z(\emptyset)=\mathbb{C}$, and $Z(-N)=\overline{Z(N)}$, where $-N$ is N with the opposite orientation, and $Z(N_1\amalg N_2)\cong Z(N_1)\hat{\otimes}_{\mathbb{C}}Z(N_2)$.

(b) To each compact, oriented n-manifold with boundary M, an element $Z(M)$ in $Z(\partial M)$. If $\partial M_{12}=-N_1\amalg N_2$ we interpret $Z(M_{12})\in |!\overline{Z(N_1)}\hat{\otimes}_{\mathbb{C}}Z(N_2)$ as a bounded linear map $Z(M_{12}):Z(N_1)\to Z(N_2)$ using the Hilbert inner product on N_1. We require that if $\partial M_{12}=-N_1\amalg N_2$ and $\partial M_{23}=-N_2\amalg N_3$, and $M_{13}=M_{12}\amalg_{N_2}M_{23}$ is obtained by gluing M_{12},M_{23} at their common boundary component N_2, then $Z(M_{13})=Z(M_{23})\circ Z(M_{12})$.

This can be written as a functor $\mathbf{Bord}^+_{\langle n-1,n\rangle}\to\mathbf{Hilb}_{\mathbb{C}}$ from an 'oriented bordism category' $\mathbf{Bord}^+_{\langle n-1,n\rangle}$ to the category of complex Hilbert spaces $\mathbf{Hilb}_{\mathbb{C}}$. TQFTs arose from Physics, but are also important in Mathematics, for example Donaldson invariants of 4-manifolds [20] and instanton Floer homology of 3-manifolds [19] can be encoded as a kind of four-dimensional TQFT.

As in Freed [26], *extended TQFTs* generalize this idea to manifolds with corners. The details are complicated. For $0<d\leqslant n$ one can define a symmetric monoidal d-category $\mathbf{Bord}^+_{\langle n-d,n\rangle}$ in which the k-morphisms for $0\leqslant k\leqslant d$ are compact, oriented $(n+k-d)$-manifolds with corners N_k, such that $\partial^l N_k=\emptyset$ for $l>k$. A *d-extended TQFT of dimension* n is then a d-functor Z :

$\mathbf{Bord}^{+}_{\langle n-d,n\rangle} \to \mathcal{C}$, where $\mathcal{C}$ is a suitable $\mathbb{C}$-linear symmetric monoidal d-category generalizing $\mathbf{Hilb}_{\mathbb{C}}$. We call Z *fully extended* if $d = n$.

The *Baez–Dolan cobordism hypothesis* [5, 26] says fully extended TQFTs Z should be classified by the object $Z(*) \in \mathcal{C}$, which should be 'fully dualizable', where $*$ is the point, as a 0-manifold. A proof was outlined by Jacob Lurie [63]. This is an active area.

3.6.2 Manifolds with corners in asymptotic geometry and analysis

Manifolds with corners are used extensively in geometric analysis by Richard Melrose [69, 70, 71, 72], in his 'b-calculus', and by his many mathematical followers. Grieser [34] gives a good introduction.

A key idea in Melrose's philosophy is this: suppose you have a non-compact manifold X and a smooth function $f : X \to \mathbb{R}$ (or a Riemannian metric g on X, etc.) and you want to describe the asymptotic behaviour of f (or g) at infinity in X. Then you should compactify X to a compact manifold with faces $\bar{X}$ with interior $\bar{X}^\circ = X$, such that $\partial\bar{X} = Y_1 \amalg \cdots \amalg Y_k$ with Y_i connected.

Choose a smooth function $h_i : \bar{X} \to [0, \infty)$ for $i = 1, \ldots, k$ which vanishes to order 1 along Y_i and is positive on $\bar{X} \setminus Y_i$. Then we should describe the asymptotic behaviour of f (or g) near infinity in X (that is, near $\partial\bar{X}$ in $\bar{X}$) by bounding it by functions of $h_1, \ldots, h_k$, such as $\prod_{i=1}^{k} h_i^{\alpha_i}$ for $\alpha_1, \ldots, \alpha_k \in \mathbb{R}$. For this to work the compactification $\bar{X}$ should depend on f (or g).

As a typical application, suppose we are given a (possibly singular) elliptic operator $P : \Gamma^\infty(E) \to \Gamma^\infty(F)$ on a manifold X, for vector bundles $E, F \to X$, and we wish to solve the equation $Pu = v$. In good situations we may write

$$u(x) = \int_{y \in X} Q(x, y) v(y) \mathrm{d}y,$$

where $Q \in \Gamma(E \boxtimes F^*) = \Gamma(\Pi_X^*(E) \otimes \Pi_Y^*(F))$ is a *Schwartz kernel* (*Green's function*) for P. Here Q is defined on a dense open subset $U \subset X \times X$, for example, $U = (X \times X) \setminus \Delta_X$, and may have a pole along the diagonal Δ_X. Many important properties of P depend only on the asymptotic behaviour of Q, which can be described by extending U to a compact manifold with faces $\bar{U}$ as above. The construction of $\bar{U}$ could involve taking a *real blow up* of $X \times X$ along the diagonal Δ_X.

The 'b-calculus' provides a tool-kit for understanding the asymptotic behaviour of Schwartz kernels of pseudo-differential operators. An important rôle is played by smooth maps $f : X \to Y$ of manifolds with corners (e.g., real blow ups), in the sense of §3.1 (Melrose calls these 'b-maps'), and by special classes of these (e.g. 'b-fibrations'). There are theorems [70] which describe how the asymptotic behaviour of Schwartz kernels transform under pullbacks f^* and pushforwards f_* along such maps f. The b-tangent bundle ${}^bTX \to X$ is also important in the theory.

There is also a place in the Melrose theory for *manifolds with g-corners*, as in §3.3: these are basically the 'interior binomial varieties' of Kottke–Melrose [55], on which the second author's definition of manifolds with g-corners [46] is based. For example, given a Schwartz kernel Q defined on $U \subset X \times X$, the *minimal* compactification $\bar{U}$ of U on which Q has a good asymptotic description might be a manifold with g-corners. We can turn $\bar{U}$ into a manifold with ordinary corners by taking further real blow ups of corner strata of $\bar{U}$.

3.6.3 Partial differential equations on manifolds with corners

A second way in which manifolds with corners appear in geometry and analysis is in the study of partial differential equations $Pu = v$ whose domain is a manifold with corners X, with some boundary conditions on u at ∂X. There are two different types of such problems.

(i) The boundary ∂X is at 'infinite distance'. In this case, the partial differential equation is really on the interior X°, and the boundary conditions at ∂X prescribe the asymptotic behaviour of u at infinity in X°.
(ii) The boundary ∂X is at 'finite distance', for example Dirichlet or Neumann boundary conditions for Laplace's equation.

Type (i) is closely related to the ideas of §3.6.2, and has been studied by Richard Melrose and his school, and others. See, for example, Melrose [71] on type (i) analysis for X a manifold with boundary, and Melrose's student Loya [59, 60] on Fredholm properties of elliptic operators on manifolds X with corners in codimension 2.

Problems of type (ii) occur, for example, in Lagrangian Floer theory in Symplectic Geometry [28, 29]. Let (X, ω) be a symplectic manifold, J an almost complex structure on X compatible with ω, and $L_1, L_2 \subset X$ be embedded, transversely intersecting Lagrangians. To define the Lagrangian Floer cohomology $HF^*(L_1, L_2)$, one must consider moduli spaces of J-holomorphic maps $u : \Sigma \to X$, where Σ is a complex disc, with the boundary condition that $u(\partial\Sigma) \subset L_1 \cup L_2$. A typical such disc is shown in Figure 3.3.

Really one should regard Σ as a *compact 2-manifold with corners*, where components of $\partial\Sigma$ are mapped to L_1 or L_2, and codimension 2 corners of Σ are mapped to $L_1 \cap L_2$. There are similar, analytically well-behaved problems on manifolds with corners of higher dimension, although they are not much studied. For example, if X is a Calabi–Yau m-fold and $D \subset X$ is a divisor with

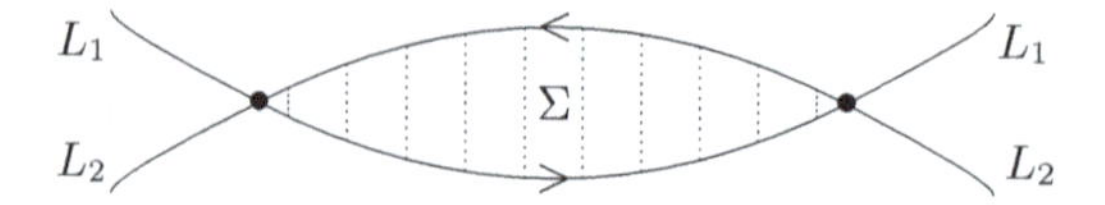

Figure 3.3 J-holomorphic disc Σ with boundary in $L_1 \cup L_2$

simple normal crossings, one could consider special Lagrangian submanifolds $L \subset X$ with corners, with $\partial L \subset D$.

In [47] the second author argues that for problems of type (i), the correct smooth structure on X is actually that of a *manifold with a-corners*, to be discussed in §8.3.

3.6.4 Moduli spaces of flow lines in Morse theory

In §3.6.4–§3.6.6 we discuss moduli problems in Differential Geometry in which the moduli space $\bar{\mathcal{M}}$ is a manifold with corners. Often this happens when we start with a moduli space $\mathcal{M}$ of non-singular objects, which is a manifold, and compactify it by adding a boundary $\bar{\mathcal{M}} \setminus \mathcal{M}$ of singular objects, such that $\bar{\mathcal{M}}$ is a compact manifold with corners, with interior $\mathcal{M} = \bar{\mathcal{M}}^\circ \subset \bar{\mathcal{M}}$.

A well-known example of this is moduli spaces of flow lines in Morse theory, as in Schwarz [82] or Austin and Braam [4]. Let X be a compact manifold, and let $f : X \to \mathbb{R}$ be smooth. We call f *Morse* if the critical locus $\mathrm{Crit}(f)$ consists of isolated points $x \in X$ with Hessian $\mathrm{Hess}\, f = \nabla^2 f \in S^2 T_x^* X$ non-degenerate. For $x \in \mathrm{Crit}(f)$ we define $\mu(x)$ to be the number of negative eigenvalues of $\mathrm{Hess}\, f|_x$. Fix a Riemannian metric g on X, and let $\nabla f \in \Gamma^\infty(TX)$ be the associated gradient vector field of f.

For $x, y \in \mathrm{Crit}(f)$, consider the moduli space $\mathcal{M}(x, y)$ of $\sim$-equivalence classes $[\gamma]$ of smooth flow-lines $\gamma : \mathbb{R} \to X$ of $-\nabla f$ (that is, $\frac{\mathrm{d}}{\mathrm{d}t}\gamma(t) = -\nabla f|_{\gamma(t)}$ for $t \in \mathbb{R}$) with $\lim_{t \to -\infty} \gamma(t) = x$ and $\lim_{t \to \infty} \gamma(t) = y$, with equivalence relation $\gamma \sim \gamma'$ if $\gamma(t) = \gamma'(t + c)$ for some $c \in \mathbb{R}$ and all t.

Next, enlarge $\mathcal{M}(x, y)$ to a compactification $\bar{\mathcal{M}}(x, y)$ by including 'broken flow-lines', which are sequences $([\gamma_1], \ldots, [\gamma_k])$ for $[\gamma_i] \in \mathcal{M}(z_{i-1}, z_i)$ with $z_0 = x$ and $z_k = y$. As discussed by Wehrheim [93, §1], there is a widely believed folklore result, which we state (*without* claiming it is true).

'Folklore Theorem'. *Let X be a compact manifold, $f : X \to \mathbb{R}$ a Morse function, and g a generic Riemannian metric on X. Then $\bar{\mathcal{M}}(x, y)$ has the canonical structure of a compact manifold with corners of dimension $\mu(x) - \mu(y) - 1$ for all $x, y \in \mathrm{Crit}(f)$, with interior $\bar{\mathcal{M}}(x, y)^\circ = \mathcal{M}(x, y)$, where $\bar{\mathcal{M}}(x, y) = \emptyset$ if $\mu(x) \leqslant \mu(y)$. There is a canonical diffeomorphism*

$$\partial \bar{\mathcal{M}}(x, y) \cong \textstyle\coprod_{z \in \mathrm{Crit}(f)} \bar{\mathcal{M}}(x, z) \times \bar{\mathcal{M}}(z, y).$$

Moduli spaces $\bar{\mathcal{M}}(x,y)$ of dimensions 0 and 1 are used to construct the *Morse homology groups* $H_*^{\mathrm{Mo}}(X,\mathbb{Z})$ of X, as in [4, 82]. The higher-dimensional moduli spaces contain information on the homotopy type of X.

Remark 3.20 **(a)** Actually proving the 'Folklore Theorem' and, in particular, describing the smooth structure on $\bar{\mathcal{M}}(x,y)$ near $\partial\bar{\mathcal{M}}(x,y)$, seems to be surprisingly difficult. References such as [4, Lem. 2.5], [82, Th. 3, p. 69] show only that $\mathcal{M}(x,y)$ is a manifold, and then indicate how to glue the strata of $\bar{\mathcal{M}}(x,y)$ together topologically, but without smooth structures. In [47, §6.2.2] the second author argues that the correct smooth structure on $\bar{\mathcal{M}}(x,y)$ is that of a *manifold with a-corners*, to be discussed in §8.3.

(b) For $\bar{\mathcal{M}}(x,y)$ to be a manifold with corners, it is essential that g should be generic, so that the deformation theory of flow-lines of ∇f is unobstructed. For non-generic g, $\bar{\mathcal{M}}(x,y)$ should be singular, and we expect $\bar{\mathcal{M}}(x,y)$ to be a C^∞*-scheme with corners*, or a *derived manifold with corners* in the sense of §8.4. This is our first example of the idea that C^∞-algebraic geometry with corners is relevant to the study of moduli spaces, see also §3.6.5–§3.6.6.

3.6.5 Moduli spaces of stable Riemann surfaces with boundary

This section follows Fukaya et al. [29, §2]. A *Riemann surface with boundary* Σ is a complex 1-manifold with real boundary $\partial\Sigma$. A *prestable* or *nodal Riemann surface with boundary* Σ is a singular complex 1-manifold with boundary $\partial\Sigma$, whose only singularities are nodes, of two kinds.

(a) Σ has an *interior node* p in Σ° locally modelled on $(0,0)$ in $\{(x,y)\in\mathbb{C}^2 : xy=0\}$. Such a node can occur as the limit of a family of non-singular Riemann surfaces with boundary Σ_ϵ modelled on $\{(x,y)\in\mathbb{C}^2 : xy=\epsilon\}$, for $\epsilon\in\mathbb{C}\setminus\{0\}$.

(b) Σ has a *boundary node* q locally modelled on $(0,0)$ in $\{(x,y)\in\mathbb{C}^2 : xy= 0,\ \mathrm{Im}\,x\geqslant 0,\ \mathrm{Im}\,y\geqslant 0\}$. Such a node can occur as the limit of a family of non-singular Σ_ϵ modelled on $\{(x,y)\in\mathbb{C}^2 : xy=-\epsilon,\ \mathrm{Im}\,x\geqslant 0,\ \mathrm{Im}\,y\geqslant 0\}$, for $\epsilon>0$.

If Σ is a prestable Riemann surface with boundary, the *smoothing* Σ' of Σ is the oriented 2-manifold with boundary obtained by smoothing the nodes of Σ, using the families Σ_ϵ above. The *genus* and *number of boundary components* of Σ are the genus and number of boundary components of the smoothing Σ'.

Fix $g,h,m,n\geqslant 0$. Consider triples $(\Sigma,\boldsymbol{y},\boldsymbol{z})$, where:

(i) Σ is a compact, connected, prestable Riemann surface with boundary, of genus g, with h boundary components;
(ii) $\boldsymbol{y} = (y_1, \ldots, y_m)$ for $y_1, \ldots, y_m$ distinct non-singular points of Σ°, called *interior marked points*;
(iii) $\boldsymbol{z} = (z_1, \ldots, z_n)$ for $z_1, \ldots, z_n$ distinct non-singular points of $\partial\Sigma$, called *boundary marked points*. We could also require the z_i to lie on prescribed boundary components, or to have a prescribed cyclic order around a boundary component, but we will not bother.

Such a triple $(\Sigma, \boldsymbol{y}, \boldsymbol{z})$ is called *stable* if $\mathrm{Aut}(\Sigma, \boldsymbol{y}, \boldsymbol{z})$ is finite. Write $\overline{\mathcal{M}}_{g,h,m,n}$ for the set of isomorphism classes $[\Sigma, \boldsymbol{y}, \boldsymbol{z}]$ of stable triples $(\Sigma, \boldsymbol{y}, \boldsymbol{z})$. Then $\overline{\mathcal{M}}_{g,h,m,n}$ has the natural structure of a *compact orbifold with corners*, with

$$\dim \overline{\mathcal{M}}_{g,h,m,n} = 6g + 3h + 2m + n - 6.$$

The orbifold group of a point $[\Sigma, \boldsymbol{y}, \boldsymbol{z}]$ is $\mathrm{Aut}(\Sigma, \boldsymbol{y}, \boldsymbol{z})$. The codimension k boundary stratum $S^k(\overline{\mathcal{M}}_{g,h,m,n})$, in the sense of Definition 3.12, is the subset of $[\Sigma, \boldsymbol{y}, \boldsymbol{z}]$ such that Σ has k boundary nodes. Points of the boundary $\partial\overline{\mathcal{M}}_{g,h,m,n}$ are isomorphism classes $[\Sigma, \boldsymbol{y}, \boldsymbol{z}, q]$ of a triple $(\Sigma, \boldsymbol{y}, \boldsymbol{z})$ together with a chosen boundary node $q \in \Sigma$. By regarding Σ with chosen boundary node q as being obtained by gluing two other Σ', Σ'' at boundary marked points z', z'' by identifying $z' = z''$ to make q, we can identify $\partial\overline{\mathcal{M}}_{g,h,m,n}$ with a disjoint union of products of other $\overline{\mathcal{M}}_{g',h',m',n'}$. If $h = 0$, which forces $n = 0$, then $\overline{\mathcal{M}}_{g,0,m,0}$ is a compact complex orbifold, without boundary.

One can also study maps between moduli spaces, for example the map

$$\overline{\mathcal{M}}_{g,h,m,n+1} \longrightarrow \overline{\mathcal{M}}_{g,h,m,n}$$

which acts on $(\Sigma, \boldsymbol{y}, \boldsymbol{z})$ by forgetting the last boundary marked point z_{n+1}, and then stabilizing $(\Sigma, \boldsymbol{y}, (z_1, \ldots, z_n))$ if necessary. Such maps tend to be smooth (Melrose's 'b-maps') in the sense of §3.1, but not strongly smooth; for example, they may be locally modelled on $[0,\infty)^2 \times \mathbb{R}^k \to [0,\infty) \times \mathbb{R}^k$, $(x, y, \boldsymbol{z}) \mapsto (xy, \boldsymbol{z})$. This is one reason for our choice of definition of smooth maps in §3.1.

Quilted discs are generalizations of prestable holomorphic discs Σ (the case $g = 0$, $h = 1$ above), which have a real circle C in Σ tangent to a point of $\partial\Sigma$. They are used in Symplectic Geometry, for example, to construct functors between Fukaya categories $F_L : \mathscr{F}(X_1, \omega_1) \to \mathscr{F}(X_2, \omega_2)$ from a Lagrangian correspondence $L \subset (X_1 \times X_2, -\omega_1 \boxplus \omega_2)$. Ma'u and Woodward [67, 68] define moduli spaces $\overline{\mathcal{M}}_{n,1}$ of 'stable n-marked quilted discs'. As in [68, §6], for $n \geqslant 4$ these are not manifolds with corners, but have an exotic corner structure; in the language of §3.3, the $\overline{\mathcal{M}}_{n,1}$ are *manifolds with g-corners*.

3.6.6 Moduli spaces with corners in Floer theories

Floer theories are infinite-dimensional generalizations of Morse theory, as in §3.6.4, which are used to construct exotic cohomology theories in Symplectic Geometry and low-dimensional differential topology. They involve moduli spaces $\bar{\mathcal{M}}$ of geometric objects, such as J-holomorphic curves or instantons. In good cases $\bar{\mathcal{M}}$ is a manifold with corners.

Often the boundaries $\partial\bar{\mathcal{M}}$ may be written as (fibre) products of other moduli spaces $\bar{\mathcal{M}}', \bar{\mathcal{M}}''$, and the relations this implies between fundamental or virtual chains $[\bar{\mathcal{M}}]_{\rm virt}$ are used to prove identities in the theory. For example, often one constructs a complex (CF^*, d) in which d involves virtual counts $\#[\bar{\mathcal{M}}]_{\rm virt}$ for $\mathrm{vdim}\,\bar{\mathcal{M}} = 0$, and boundary relations for moduli spaces $\bar{\mathcal{M}}$ with $\mathrm{vdim}\,\bar{\mathcal{M}} = 1$ are used to prove that $\mathrm{d}^2 = 0$, so that the cohomology HF^* of (CF^*, d) is well defined. Thus, the corner structure is essential to the theory.

Some examples of Floer theories are as follows.

(a) *Instanton Floer homology* $IH_*(Y)$ of a compact, oriented 3-manifold Y, as in Donaldson [19]. This involves studying moduli spaces $\bar{\mathcal{M}}$ of instantons (anti-self-dual connections on a principal $\mathrm{SU}(2)$-bundle) on $Y \times \mathbb{R}$.

(b) *Seiberg–Witten Floer homology* $HM_*(Y)$ of a compact, oriented 3-manifold Y, as in Kronheimer–Mrowka [56]. This involves studying moduli spaces $\bar{\mathcal{M}}$ of Seiberg–Witten monopoles on $Y \times \mathbb{R}$.

(c) *Lagrangian Floer cohomology* $HF^*(L_1, L_2)$ of transversely intersecting Lagrangians L_1, L_2 in a symplectic manifold (X, ω), as in Fukaya–Oh–Ohta–Ono [28, 29]. This involves choosing an almost complex structure J on X compatible with ω, and studying moduli spaces $\bar{\mathcal{M}}$ of J-holomorphic maps $u : \Sigma \to X$ with $u(\Sigma) \subset L_1 \cup L_2$ for Σ a prestable holomorphic disc, as in §3.6.3 and §3.6.5. Then $HF^0(L_1, L_2)$ is the morphisms $\mathrm{Hom}(L_1, L_2)$ in the *Fukaya category* $\mathscr{F}(X, \omega)$ of X.

In each of (a)–(c), in good cases the moduli spaces $\bar{\mathcal{M}}$ have the structure of manifolds with corners, but otherwise can be singular, and could be modelled as C^∞-schemes with corners or C^∞-stacks with corners, or as derived manifolds or orbifolds with corners in the sense of §8.4. Thus, the ideas of this book could have applications in Floer theories.

As in §3.6.5, natural morphisms between moduli spaces of J-holomorphic maps in Symplectic Geometry, for example, maps forgetting a boundary marked point, tend to be smooth in the sense of §3.1, but not strongly smooth. This is one reason for our choice of definition of smooth maps in §3.1.

As for quilted discs in §3.6.5, there are also moduli spaces $\bar{\mathcal{M}}$ used in Floer theories which in good cases have the structure of *manifolds with g-corners*, and in general should be *derived orbifolds with g-corners*, as in §8.4. Ma'u, Wehrheim and Woodward [67, 68, 94, 95, 96] study moduli spaces $\bar{\mathcal{M}}$ of *pseudoholomorphic quilts*, which are J-holomorphic maps whose domain is a quilted disc, and these should have g-corners rather than ordinary corners. Pardon [80, Rem. 2.16] uses moduli spaces of J-holomorphic curves with g-corners to define contact homology of Legendrian submanifolds. Work of Chris Kottke (private communication) suggests that natural compactifications of $\mathrm{SU}(2)$ magnetic monopole spaces may have the structure of manifolds with g-corners.

4

(Pre) C^∞-rings with corners

This chapter defines and studies *(pre) C^∞-rings with corners*. As in §2.2, a *categorical C^∞-ring* is a product-preserving functor $F : \mathbf{Euc} \to \mathbf{Sets}$, where $\mathbf{Euc} \subset \mathbf{Man}$ is the full subcategory of Euclidean spaces $\mathbb{R}^n$, and a C^∞-ring $\mathfrak{C}$ is an equivalent way of repackaging the data in F, with $\mathfrak{C} = F(\mathbb{R})$, and many operations on $\mathfrak{C}$. Here $\mathbf{Euc}$ is an Algebraic Theory [1], and a (categorical) C^∞-ring is an algebra over this Algebraic Theory.

Similarly, we define a *categorical pre C^∞-ring with corners* to be a product-preserving functor $F : \mathbf{Euc^c} \to \mathbf{Sets}$, where $\mathbf{Euc^c} \subset \mathbf{Man^c}$ is the full subcategory of Euclidean corner spaces $\mathbb{R}^m_k = [0,\infty)^k \times \mathbb{R}^{m-k}$. A *pre C^∞-ring with corners* $\boldsymbol{\mathfrak{C}} = (\mathfrak{C}, \mathfrak{C}_{\rm ex})$ is an equivalent way of repackaging F, with $\mathfrak{C} = F(\mathbb{R})$ and $\mathfrak{C}_{\rm ex} = F([0,\infty))$, and many operations on $\mathfrak{C}, \mathfrak{C}_{\rm ex}$. These were introduced by Kalashnikov [51], who called them C^∞-rings with corners. Again, $\mathbf{Euc^c}$ is an Algebraic Theory, and a (categorical) pre C^∞-ring with corners is an algebra over this Algebraic Theory, so we can quote results from Algebraic Theories on existence of (co)limits of pre C^∞-rings with corners.

We define *C^∞-rings with corners* in §4.4 to be the full subcategory of pre C^∞-rings with corners satisfying an extra condition, that invertible functions in $\mathfrak{C}_{\rm ex}$ should have logs in $\mathfrak{C}$. To motivate this, note that if X is a manifold with corners and $f : X \to [0,\infty)$ is smooth with an inverse $1/f$ then f maps $X \to (0,\infty)$, and so $\log f : X \to \mathbb{R}$ is well defined and smooth. This condition makes C^∞-rings and C^∞-schemes with corners better behaved.

Sections 4.5–4.7 develop the theory of C^∞-rings with corners. This includes studying localizations and local C^∞-rings with corners, which are used to define a spectrum functor in §5.3. We also introduce special classes of C^∞-rings with corners that are useful in later chapters. Throughout the chapter, we will return to our motivating example of manifolds with (g-)corners, and show how they can be represented as (pre) C^∞-rings with corners.

4.1 Categorical pre C^∞-rings with corners

Here is the obvious generalization of Definition 2.10.

Definition 4.1 Write $\mathbf{Euc^c}, \mathbf{Euc^c_{in}}$ for the full subcategories of $\mathbf{Man^c}$ and $\mathbf{Man^c_{in}}$ with objects the Euclidean spaces with corners $\mathbb{R}^m_k = [0,\infty)^k \times \mathbb{R}^{m-k}$ for $0 \leqslant k \leqslant m$. We have inclusions of subcategories

$$\mathbf{Euc} \subset \mathbf{Euc^c_{in}} \subset \mathbf{Euc^c}. \tag{4.1}$$

Define a *categorical pre C^∞-ring with corners* to be a product-preserving functor $F : \mathbf{Euc^c} \to \mathbf{Sets}$. Here F should also preserve the trivial product, that is, it maps $\mathbb{R}^0 \times [0,\infty)^0 = \{\emptyset\}$ in $\mathbf{Euc^c}$ to the terminal object in $\mathbf{Sets}$, the point $*$.

If $F, G : \mathbf{Euc^c} \to \mathbf{Sets}$ are categorical pre C^∞-rings with corners, a *morphism* $\eta : F \to G$ is a natural transformation $\eta : F \Rightarrow G$. Such natural transformations are automatically product-preserving. We write $\mathbf{CPC^\infty Rings^c}$ for the category of categorical pre C^∞-rings with corners. Define a *categorical interior pre C^∞-ring with corners* to be a product-preserving functor $F : \mathbf{Euc^c_{in}} \to \mathbf{Sets}$. They form a category $\mathbf{CPC^\infty Rings^c_{in}}$, with morphisms natural transformations.

Define a commutative triangle of functors

$$\begin{array}{ccc} \mathbf{CPC^\infty Rings^c} & \xrightarrow{\Pi^{\mathbf{CC^\infty Rings}}_{\mathbf{CPC^\infty Rings^c}}} & \mathbf{CC^\infty Rings} \\ & \searrow_{\Pi^{\mathbf{CPC^\infty Rings^c_{in}}}_{\mathbf{CPC^\infty Rings^c}}} \quad \nearrow_{\Pi^{\mathbf{CC^\infty Rings}}_{\mathbf{CPC^\infty Rings^c_{in}}}} & \\ & \mathbf{CPC^\infty Rings^c_{in}} & \end{array} \tag{4.2}$$

by restriction to subcategories in (4.1), so that, for example, $\Pi^{\mathbf{CPC^\infty Rings^c_{in}}}_{\mathbf{CPC^\infty Rings^c}}$ maps $F : \mathbf{Euc^c} \to \mathbf{Sets}$ to $F|_{\mathbf{Euc^c_{in}}} : \mathbf{Euc^c_{in}} \to \mathbf{Sets}$.

Here is the motivating example.

Example 4.2 **(a)** Let X be a manifold with corners. Define a categorical pre C^∞-ring with corners $F : \mathbf{Euc^c} \to \mathbf{Sets}$ by $F = \mathrm{Hom}_{\mathbf{Man^c}}(X, -)$. That is, for objects $\mathbb{R}^m \times [0,\infty)^n$ in $\mathbf{Euc^c} \subset \mathbf{Man^c}$ we have

$$F(\mathbb{R}^m \times [0,\infty)^n) = \mathrm{Hom}_{\mathbf{Man^c}}(X, \mathbb{R}^m \times [0,\infty)^n),$$

and for morphisms $g : \mathbb{R}^m \times [0,\infty)^n \to \mathbb{R}^{m'} \times [0,\infty)^{n'}$ in $\mathbf{Euc^c}$ we have

$$F(g) = g\circ : \mathrm{Hom}_{\mathbf{Man^c}}(X, \mathbb{R}^m \times [0,\infty)^n) \longrightarrow \mathrm{Hom}_{\mathbf{Man^c}}(X, \mathbb{R}^{m'} \times [0,\infty)^{n'})$$

mapping $F(g) : h \mapsto g \circ h$. Let $f : X \to Y$ be a smooth map of manifolds with corners, and $F, G : \mathbf{Euc^c} \to \mathbf{Sets}$ the functors corresponding to X, Y. Define a natural transformation $\eta : G \Rightarrow F$ by

$$\eta(\mathbb{R}^m \times [0,\infty)^n) = \circ f : \mathrm{Hom}(Y, \mathbb{R}^m \times [0,\infty)^n) \longrightarrow \mathrm{Hom}(X, \mathbb{R}^m \times [0,\infty)^n)$$

mapping $\eta : h \mapsto h \circ f$.

Define a functor $F_{\mathbf{Man^c}}^{\mathbf{CPC^\infty Rings^c}} : \mathbf{Man^c} \to (\mathbf{CPC^\infty Rings^c})^{\mathrm{op}}$ to map $X \mapsto F$ on objects, and $f \mapsto \eta$ on morphisms, for X, Y, F, G, f, η as above.

All of this also works if X, Y are manifolds with g-corners, as in §3.3, giving a functor $F_{\mathbf{Man^{gc}}}^{\mathbf{CPC^\infty Rings^c}} : \mathbf{Man^{gc}} \to (\mathbf{CPC^\infty Rings^c})^{\mathrm{op}}$.

(b) Similarly, if X is a manifold with (g-)corners, define a categorical interior pre C^∞-ring with corners $F : \mathbf{Euc^c_{in}} \to \mathbf{Sets}$ by $F = \mathrm{Hom}_{\mathbf{Man^c_{in}}}(X, -)$. This gives functors $F_{\mathbf{Man^c_{in}}}^{\mathbf{CPC^\infty Rings^c_{in}}} : \mathbf{Man^c_{in}} \to (\mathbf{CPC^\infty Rings^c_{in}})^{\mathrm{op}}$ and $F_{\mathbf{Man^{gc}_{in}}}^{\mathbf{CPC^\infty Rings^c_{in}}} : \mathbf{Man^{gc}_{in}} \to (\mathbf{CPC^\infty Rings^c_{in}})^{\mathrm{op}}$.

In the language of Algebraic Theories, as in Adámek, Rosický, and Vitale [1], $\mathbf{Euc}, \mathbf{Euc^c}, \mathbf{Euc^c_{in}}$ are *algebraic theories* (i.e., small categories with finite products), and $\mathbf{CC^\infty Rings}, \mathbf{CPC^\infty Rings^c}, \mathbf{CPC^\infty Rings^c_{in}}$ are the corresponding *categories of algebras*. Also the inclusions of subcategories (4.1) are *morphisms of algebraic theories*, and the functors (4.2) the corresponding morphisms. So, as for Proposition 2.18, Adámek et al. [1, Props. 1.21, 2.5, 9.3, and Th. 4.5] give important results on their categorical properties.

Theorem 4.3 **(a)** *All small limits and directed colimits exist in the categories* $\mathbf{CPC^\infty Rings^c}, \mathbf{CPC^\infty Rings^c_{in}}$*, and they may be computed objectwise in* $\mathbf{Euc^c}, \mathbf{Euc^c_{in}}$ *by taking the corresponding small limits/directed colimits in* $\mathbf{Sets}$.

(b) *All small colimits exist in* $\mathbf{CPC^\infty Rings^c}, \mathbf{CPC^\infty Rings^c_{in}}$*, though in general they are not computed objectwise in* $\mathbf{Euc^c}, \mathbf{Euc^c_{in}}$ *as colimits in* $\mathbf{Sets}$.

(c) *The functors* $\Pi_{\mathbf{CPC^\infty Rings^c}}^{\mathbf{CC^\infty Rings}}, \ldots, \Pi_{\mathbf{CPC^\infty Rings^c}}^{\mathbf{CPC^\infty Rings^c_{in}}}$ *in* (4.2) *have left adjoints, so they preserve limits.*

4.2 Pre C^∞-rings with corners

The next definition relates to Definition 4.1 in the same way that Definition 2.11 relates to Definition 2.10.

Definition 4.4 A *pre C^∞-ring with corners* $\mathfrak{C}$ assigns the data:

(a) Two sets $\mathfrak{C}$ and $\mathfrak{C}_{\mathrm{ex}}$.

(b) Operations $\Phi_f : \mathfrak{C}^m \times \mathfrak{C}_{\rm ex}^n \to \mathfrak{C}$ for all smooth maps $f : \mathbb{R}^m \times [0,\infty)^n \to \mathbb{R}$.

(c) Operations $\Psi_g : \mathfrak{C}^m \times \mathfrak{C}_{\rm ex}^n \to \mathfrak{C}_{\rm ex}$ for all exterior $g : \mathbb{R}^m \times [0,\infty)^n \to [0,\infty)$.

Here we allow one or both of m, n to be zero, and consider S^0 to be the single point $\{\emptyset\}$ for any set S. These operations must satisfy the following relations.

(i) Suppose $k, l, m, n \geqslant 0$, and $e_i : \mathbb{R}^k \times [0,\infty)^l \to \mathbb{R}$ is smooth for $i = 1, \ldots, m$, and $f_j : \mathbb{R}^k \times [0,\infty)^l \to [0,\infty)$ is exterior for $j = 1, \ldots, n$, and $g : \mathbb{R}^m \times [0,\infty)^n \to \mathbb{R}$ is smooth. Define smooth $h : \mathbb{R}^k \times [0,\infty)^l \to \mathbb{R}$ by

$$\begin{aligned} h(x_1,\ldots,x_k,y_1,\ldots,y_l) = g\bigl(&e_1(x_1,\ldots,y_l),\ldots,e_m(x_1,\ldots,y_l),\\ &f_1(x_1,\ldots,y_l),\ldots,f_n(x_1,\ldots,y_l)\bigr). \end{aligned} \tag{4.3}$$

Then for all $(c_1,\ldots,c_k,c_1',\ldots,c_l') \in \mathfrak{C}^k \times \mathfrak{C}_{\rm ex}^l$ we have

$$\begin{aligned} \Phi_h(c_1,\ldots,c_k,c_1',\ldots,c_l') = \Phi_g\bigl(&\Phi_{e_1}(c_1,\ldots,c_l'),\ldots,\Phi_{e_m}(c_1,\ldots,c_l'),\\ &\Psi_{f_1}(c_1,\ldots,c_l'),\ldots,\Psi_{f_n}(c_1,\ldots,c_l')\bigr). \end{aligned}$$

(ii) Suppose $k, l, m, n \geqslant 0$, and $e_i : \mathbb{R}^k \times [0,\infty)^l \to \mathbb{R}$ is smooth for $i = 1, \ldots, m$, and $f_j : \mathbb{R}^k \times [0,\infty)^l \to [0,\infty)$ is exterior for $j = 1, \ldots, n$, and $g : \mathbb{R}^m \times [0,\infty)^n \to [0,\infty)$ is exterior. Define exterior $h : \mathbb{R}^k \times [0,\infty)^l \to [0,\infty)$ by (4.3). Then for all $(c_1,\ldots,c_k,c_1',\ldots,c_l') \in \mathfrak{C}^k \times \mathfrak{C}_{\rm ex}^l$ we have

$$\begin{aligned} \Psi_h(c_1,\ldots,c_k,c_1',\ldots,c_l') = \Psi_g\bigl(&\Phi_{e_1}(c_1,\ldots,c_l'),\ldots,\Phi_{e_m}(c_1,\ldots,c_l'),\\ &\Psi_{f_1}(c_1,\ldots,c_l'),\ldots,\Psi_{f_n}(c_1,\ldots,c_l')\bigr). \end{aligned}$$

(iii) Write $\pi_i : \mathbb{R}^m \times [0,\infty)^n \to \mathbb{R}$ for projection to the ith coordinate of $\mathbb{R}^m$ for $i = 1, \ldots, m$, and $\pi_j' : \mathbb{R}^m \times [0,\infty)^n \to [0,\infty)$ for projection to the jth coordinate of $[0,\infty)^n$ for $j = 1, \ldots, n$. Then for all $(c_1,\ldots,c_m,c_1',\ldots,c_n')$ in $\mathfrak{C}^m \times \mathfrak{C}_{\rm ex}^n$ and all $i = 1, \ldots, m$, $j = 1, \ldots, n$ we have

$$\Phi_{\pi_i}(c_1,\ldots,c_m,c_1',\ldots,c_n') = c_i, \quad \Psi_{\pi_j'}(c_1,\ldots,c_m,c_1',\ldots,c_n') = c_j'.$$

We will refer to the operations Φ_f, Ψ_g as the *C^∞-operations*, and we often write a pre C^∞-ring with corners as a pair $\boldsymbol{\mathfrak{C}} = (\mathfrak{C}, \mathfrak{C}_{\rm ex})$, leaving the C^∞ operations implicit.

Let $\boldsymbol{\mathfrak{C}} = (\mathfrak{C}, \mathfrak{C}_{\rm ex})$ and $\boldsymbol{\mathfrak{D}} = (\mathfrak{D}, \mathfrak{D}_{\rm ex})$ be pre C^∞-rings with corners. A *morphism* $\boldsymbol{\phi} : \boldsymbol{\mathfrak{C}} \to \boldsymbol{\mathfrak{D}}$ is a pair $\boldsymbol{\phi} = (\phi, \phi_{\rm ex})$ of maps $\phi : \mathfrak{C} \to \mathfrak{D}$ and

$\phi_{\rm ex}:\mathfrak{C}_{\rm ex}\rightarrow\mathfrak{D}_{\rm ex}$, which commute with all the operations Φ_f,Ψ_g on $\boldsymbol{\mathfrak{C}},\boldsymbol{\mathfrak{D}}$. Write $\mathbf{PC^\infty Rings^c}$ for the category of pre C^∞-rings with corners.

As for Proposition 2.12, we have the following.

Proposition 4.5 *There is an equivalence of categories from* $\mathbf{CPC^\infty Rings^c}$ *to* $\mathbf{PC^\infty Rings^c}$*, which identifies* $F:\mathbf{Euc^c}\rightarrow\mathbf{Sets}$ *in* $\mathbf{CPC^\infty Rings^c}$ *with* $\boldsymbol{\mathfrak{C}}=(\mathfrak{C},\mathfrak{C}_{\rm ex})$ *in* $\mathbf{PC^\infty Rings^c}$ *such that* $F\bigl(\mathbb{R}^m\times[0,\infty)^n\bigr)=\mathfrak{C}^m\times\mathfrak{C}_{\rm ex}^n$ *for* $m,n\geqslant 0$.

Under this equivalence, for a smooth function $f:\mathbb{R}^m_k\rightarrow\mathbb{R}$, we identify $F(f)$ with Φ_f, and for an exterior function $g:\mathbb{R}^m_k\rightarrow[0,\infty)$, we identify $F(g)$ with Ψ_g. The proof of the proposition then follows from F being a product-preserving functor, and the definition of pre C^∞-ring with corners.

It is often helpful to work with a small subset of C^∞-operations. The next definition explains this small subset. Monoids were discussed in §3.2.

Definition 4.6 Let $\boldsymbol{\mathfrak{C}}=(\mathfrak{C},\mathfrak{C}_{\rm ex})$ be a pre C^∞-ring with corners. Then (as we will see in Definition 4.11) $\mathfrak{C}$ is a C^∞-ring, and thus a commutative $\mathbb{R}$-algebra. The $\mathbb{R}$-algebra structure makes $\mathfrak{C}$ into a monoid in two ways: under multiplication '$\cdot$' with identity 1, and under addition '$+$' with identity 0.

Define $g:[0,\infty)^2\rightarrow[0,\infty)$ by $g(x,y)=xy$. Then g induces $\Psi_g:\mathfrak{C}_{\rm ex}\times\mathfrak{C}_{\rm ex}\rightarrow\mathfrak{C}_{\rm ex}$. Define multiplication $\cdot:\mathfrak{C}_{\rm ex}\times\mathfrak{C}_{\rm ex}\rightarrow\mathfrak{C}_{\rm ex}$ by $c'\cdot c''=\Psi_g(c',c'')$. The map $1:\mathbb{R}^0\rightarrow[0,\infty)$ gives an operation $\Psi_1:\{\emptyset\}\rightarrow\mathfrak{C}_{\rm ex}$. The *identity* in $\mathfrak{C}_{\rm ex}$ is $1_{\mathfrak{C}_{\rm ex}}=\Psi_1(\emptyset)$. Then $(\mathfrak{C}_{\rm ex},\cdot,1_{\mathfrak{C}_{\rm ex}})$ is a monoid.

The map $0:\mathbb{R}^0\rightarrow[0,\infty)$ gives an operation $\Psi_0:\{\emptyset\}\rightarrow\mathfrak{C}_{\rm ex}$. This gives a distinguished element $0_{\mathfrak{C}_{\rm ex}}=\Psi_0(\emptyset)$ in $\mathfrak{C}_{\rm ex}$, which satisfies $c'\cdot 0_{\mathfrak{C}_{\rm ex}}=0_{\mathfrak{C}_{\rm ex}}$ for all $c'\in\mathfrak{C}_{\rm ex}$, that is, $0_{\mathfrak{C}_{\rm ex}}$ is a *zero element* in the monoid $(\mathfrak{C}_{\rm ex},\cdot,1_{\mathfrak{C}_{\rm ex}})$. We have the following.

(a) $\mathfrak{C}$ is a commutative $\mathbb{R}$-algebra.

(b) $\mathfrak{C}_{\rm ex}$ is a monoid, in multiplicative notation, with zero element $0_{\mathfrak{C}_{\rm ex}}\in\mathfrak{C}_{\rm ex}$.
We write $\mathfrak{C}_{\rm ex}^\times$ for the group of invertible elements in $\mathfrak{C}_{\rm ex}$.

(c) $\Phi_i:\mathfrak{C}_{\rm ex}\rightarrow\mathfrak{C}$ is a monoid morphism, for $i:[0,\infty)\hookrightarrow\mathbb{R}$ the inclusion and $\mathfrak{C}$ a monoid under multiplication.

(d) $\Psi_{\rm exp}:\mathfrak{C}\rightarrow\mathfrak{C}_{\rm ex}$ is a monoid morphism, for $\exp:\mathbb{R}\rightarrow[0,\infty)$ and $\mathfrak{C}$ a monoid under addition.

(e) $\Phi_{\rm exp}=\Phi_i\circ\Psi_{\rm exp}:\mathfrak{C}\rightarrow\mathfrak{C}$ for $\exp:\mathbb{R}\rightarrow\mathbb{R}$ and $\Psi_{\rm exp}$ in (d).

Many of our definitions will use only the structures (a)–(e). When we write $\Phi_i,\Psi_{\rm exp},\Phi_{\rm exp}$ without further explanation, we mean those in (c)–(e).

Proposition 4.7 *Let $\boldsymbol{\mathfrak{C}} = (\mathfrak{C}, \mathfrak{C}_{\rm ex})$ be a pre C^∞-ring with corners, and suppose c' lies in the group $\mathfrak{C}_{\rm ex}^\times$ of invertible elements in the monoid $\mathfrak{C}_{\rm ex}$. Then there exists a unique $c \in \mathfrak{C}$ such that $\Phi_{\exp}(c) = \Phi_i(c')$ in $\mathfrak{C}$.*

Proof Since $c' \in \mathfrak{C}_{\rm ex}^\times$ we have a unique inverse $c'^{-1} \in \mathfrak{C}_{\rm ex}^\times$. As in the proof of Proposition 2.26(a), let $e : \mathbb{R} \to \mathbb{R}$ be the inverse of $t \mapsto \exp(t) - \exp(-t)$. Define smooth $g : [0,\infty)^2 \to \mathbb{R}$ by $g(x,y) = e(x-y)$. Observe that if $(x,y) \in [0,\infty)^2$ with $xy = 1$ then $x = \exp t$, $y = \exp(-t)$ for $t = \log x$, and so

$$\begin{aligned}\exp \circ g(x,y) &= \exp \circ g(\exp t, \exp(-t))\\ &= \exp \circ e(\exp t - \exp(-t)) = \exp(t) = x.\end{aligned}$$

Therefore there is a unique smooth function $h : [0,\infty)^2 \to \mathbb{R}$ with

$$\exp \circ g(x,y) - x = h(x,y)(xy-1). \tag{4.4}$$

We have operations $\Phi_g, \Phi_h : \mathfrak{C}_{\rm ex}^2 \to \mathfrak{C}$. Define $c = \Phi_g(c', c'^{-1})$. Then

$$\begin{aligned}\Phi_{\exp}(c) - \Phi_i(c') &= \Phi_{\exp \circ g(x,y)-x}(c', c'^{-1}) = \Phi_{h(x,y)(xy-1)}(c', c'^{-1})\\ &= \Phi_h(c', c'^{-1}) \cdot (\Phi_i(c' \cdot c'^{-1}) - 1_{\mathfrak{C}}) = 0,\end{aligned}$$

using Definition 4.1(i) in the first and third steps, and (4.4) in the second. Hence $\Phi_{\exp}(c) = \Phi_i(c')$. Uniqueness of c follows from Proposition 2.26(a). □

We could define 'interior pre C^∞-rings with corners' following Definition 4.4, but replacing exterior maps by interior maps throughout. Instead we will do something equivalent but more complicated, and define interior pre C^∞-rings with corners as special examples of pre C^∞-rings with corners, and interior morphisms as special morphisms between (interior) pre C^∞-rings with corners. The advantage of this is that we can work with both interior and non-interior pre C^∞-rings with corners and their morphisms in a single theory.

Definition 4.8 Let $\boldsymbol{\mathfrak{C}} = (\mathfrak{C}, \mathfrak{C}_{\rm ex})$ be a pre C^∞-ring with corners. Then $\mathfrak{C}_{\rm ex}$ is a monoid and $0_{\mathfrak{C}_{\rm ex}} \in \mathfrak{C}_{\rm ex}$ with $c' \cdot 0_{\mathfrak{C}_{\rm ex}} = 0_{\mathfrak{C}_{\rm ex}}$ for all $c' \in \mathfrak{C}_{\rm ex}$.

We call $\boldsymbol{\mathfrak{C}}$ an *interior* pre C^∞-ring with corners if $0_{\mathfrak{C}_{\rm ex}} \neq 1_{\mathfrak{C}_{\rm ex}}$, and there do not exist $c', c'' \in \mathfrak{C}_{\rm ex}$ with $c' \neq 0_{\mathfrak{C}_{\rm ex}} \neq c''$ and $c' \cdot c'' = 0_{\mathfrak{C}_{\rm ex}}$. That is, $\mathfrak{C}_{\rm ex}$ should have no zero divisors. Write $\mathfrak{C}_{\rm in} = \mathfrak{C}_{\rm ex} \setminus \{0_{\mathfrak{C}_{\rm ex}}\}$. Then $\mathfrak{C}_{\rm ex} = \mathfrak{C}_{\rm in} \amalg \{0_{\mathfrak{C}_{\rm ex}}\}$, where $\amalg$ is the disjoint union. Since $\mathfrak{C}_{\rm ex}$ has no zero divisors, $\mathfrak{C}_{\rm in}$ is closed under multiplication, and $1_{\mathfrak{C}_{\rm ex}} \in \mathfrak{C}_{\rm in}$ as $0_{\mathfrak{C}_{\rm ex}} \neq 1_{\mathfrak{C}_{\rm ex}}$. Thus $\mathfrak{C}_{\rm in}$ is a submonoid of $\mathfrak{C}_{\rm ex}$. We write $1_{\mathfrak{C}_{\rm in}} = 1_{\mathfrak{C}_{\rm ex}}$.

Let $\boldsymbol{\mathfrak{C}}, \boldsymbol{\mathfrak{D}}$ be interior pre C^∞-rings with corners, and $\boldsymbol{\phi} = (\phi, \phi_{\rm ex}) : \boldsymbol{\mathfrak{C}} \to \boldsymbol{\mathfrak{D}}$ be a morphism in $\mathbf{PC^\infty Rings^c}$. We call $\boldsymbol{\phi}$ *interior* if $\phi_{\rm ex}(\mathfrak{C}_{\rm in}) \subseteq \mathfrak{D}_{\rm in}$. Then we write $\phi_{\rm in} = \phi_{\rm ex}|_{\mathfrak{C}_{\rm in}} : \mathfrak{C}_{\rm in} \to \mathfrak{D}_{\rm in}$. Interior morphisms are closed under composition and include the identity morphisms.

Write $\mathbf{PC^\infty Rings^c_{in}}$ for the (non-full) subcategory of $\mathbf{PC^\infty Rings^c}$ with objects interior pre C^∞-rings with corners, and morphisms interior morphisms.

Lemma 4.9 *Let $\mathfrak{C} = (\mathfrak{C}, \mathfrak{C}_{\rm ex})$ be an interior pre C^∞-ring with corners. Then $\mathfrak{C}_{\rm ex}^\times \subseteq \mathfrak{C}_{\rm in}$. If $g : \mathbb{R}^m \times [0,\infty)^n \to [0,\infty)$ is interior, then $\Psi_g : \mathfrak{C}^m \times \mathfrak{C}_{\rm ex}^n \to \mathfrak{C}_{\rm ex}$ maps $\mathfrak{C}^m \times \mathfrak{C}_{\rm in}^n \to \mathfrak{C}_{\rm in}$.*

Proof Clearly $0_{\mathfrak{C}_{\rm ex}}$ is not invertible, so $0_{\mathfrak{C}_{\rm ex}} \notin \mathfrak{C}_{\rm ex}^\times$, and $\mathfrak{C}_{\rm ex}^\times \subseteq \mathfrak{C}_{\rm ex} \setminus \{0_{\mathfrak{C}_{\rm ex}}\} = \mathfrak{C}_{\rm in}$. As g is interior we may write

$$g(x_1,\ldots,x_m,y_1,\ldots,y_n) = y_1^{a_1}\cdots y_n^{a_n}\cdot \exp\circ h(x_1,\ldots,y_n), \tag{4.5}$$

for $a_1,\ldots,a_n \in \mathbb{N}$ and $h : \mathbb{R}^m \times [0,\infty)^n \to \mathbb{R}$ smooth. Then for $c_1,\ldots,c_m \in \mathfrak{C}$ and $c_1',\ldots,c_n' \in \mathfrak{C}_{\rm in}$ we have

$$\Psi_g\bigl(c_1,\ldots,c_m,c_1',\ldots,c_n'\bigr) = c_1'^{a_1}\cdots c_n'^{a_n}\cdot\Psi_{\exp}\bigl[\Phi_h\bigl(c_1,\ldots,c_m,c_1',\ldots,c_n'\bigr)\bigr].$$

Here $c_1'^{a_1}\cdots c_n'^{a_n} \in \mathfrak{C}_{\rm in}$ as $\mathfrak{C}_{\rm in}$ is a submonoid of $\mathfrak{C}_{\rm ex}$, and $\Psi_{\exp}[\cdots] \in \mathfrak{C}_{\rm in}$ as $\Psi_{\exp}$ maps to $\mathfrak{C}_{\rm ex}^\times \subseteq \mathfrak{C}_{\rm in} \subseteq \mathfrak{C}_{\rm ex}$. Thus $\Psi_g\bigl(c_1,\ldots,c_m,c_1',\ldots,c_n'\bigr) \in \mathfrak{C}_{\rm in}$. □

Here is the analogue of Proposition 4.5.

Proposition 4.10 *There is an equivalence*

$$\mathbf{CPC^\infty Rings^c_{in}} \cong \mathbf{PC^\infty Rings^c_{in}},$$

which identifies $F : \mathbf{Euc^c_{in}} \to \mathbf{Sets}$ in $\mathbf{CPC^\infty Rings^c_{in}}$ with $\mathfrak{C} = (\mathfrak{C}, \mathfrak{C}_{\rm ex})$ in $\mathbf{PC^\infty Rings^c_{in}}$ such that $F\bigl(\mathbb{R}^m \times [0,\infty)^n\bigr) = \mathfrak{C}^m \times \mathfrak{C}_{\rm in}^n$ for $m,n \geqslant 0$.

Proof Let $F : \mathbf{Euc^c_{in}} \to \mathbf{Sets}$ be a categorical interior pre C^∞-ring with corners. Define sets $\mathfrak{C} = F(\mathbb{R})$, $\mathfrak{C}_{\rm in} = F([0,\infty))$, and $\mathfrak{C}_{\rm ex} = \mathfrak{C}_{\rm in} \amalg \{0_{\mathfrak{C}_{\rm ex}}\}$, where $\amalg$ is the disjoint union. Then $F\bigl(\mathbb{R}^m \times [0,\infty)^n\bigr) = \mathfrak{C}^m \times \mathfrak{C}_{\rm in}^n$, as F is product-preserving. Let $f : \mathbb{R}^m \times [0,\infty)^n \to \mathbb{R}$ and $g : \mathbb{R}^m \times [0,\infty)^n \to [0,\infty)$ be smooth. We must define maps $\Phi_f : \mathfrak{C}^m \times \mathfrak{C}_{\rm ex}^n \to \mathfrak{C}$ and $\Psi_g : \mathfrak{C}^m \times \mathfrak{C}_{\rm ex}^n \to \mathfrak{C}_{\rm ex}$.

Let $c_1,\ldots,c_m \in \mathfrak{C}$ and $c_1',\ldots,c_n' \in \mathfrak{C}_{\rm ex}$. Then some of $c_1',\ldots,c_n'$ lie in $\mathfrak{C}_{\rm in}$ and the rest in $\{0_{\mathfrak{C}_{\rm ex}}\}$. For simplicity suppose that $c_1',\ldots,c_k' \in \mathfrak{C}_{\rm in}$ and $c_{k+1}' = \cdots = c_n' = 0_{\mathfrak{C}_{\rm ex}}$ for $0 \leqslant k \leqslant n$. Define smooth $d : \mathbb{R}^m \times [0,\infty)^k \to \mathbb{R}$, $e : \mathbb{R}^m \times [0,\infty)^k \to [0,\infty)$ by

$$\begin{aligned} d(x_1,\ldots,x_m,y_1,\ldots,y_k) &= f(x_1,\ldots,x_m,y_1,\ldots,y_k,0,\ldots,0),\\ e(x_1,\ldots,x_m,y_1,\ldots,y_k) &= g(x_1,\ldots,x_m,y_1,\ldots,y_k,0,\ldots,0). \end{aligned} \tag{4.6}$$

Then $F(d)$ maps $\mathfrak{C}^m \times \mathfrak{C}_{\rm in}^k \to \mathfrak{C}$. Set

$$\Phi_f(c_1,\ldots,c_m,c'_1,\ldots,c'_k,0_{\mathfrak{C}_{\rm ex}},\ldots,0_{\mathfrak{C}_{\rm ex}}) = F(d)(c_1,\ldots,c_m,c'_1,\ldots,c'_k).$$

Either $e:\mathbb{R}^m\times[0,\infty)^k\to[0,\infty)$ is interior, or $e=0$. If e is interior define

$$\Psi_g(c_1,\ldots,c_m,c'_1,\ldots,c'_k,0_{\mathfrak{C}_{\rm ex}},\ldots,0_{\mathfrak{C}_{\rm ex}}) = F(e)(c_1,\ldots,c_m,c'_1,\ldots,c'_k).$$

If $e=0$ set $\Psi_g(c_1,\ldots,c_m,c'_1,\ldots,c'_k,0_{\mathfrak{C}_{\rm ex}},\ldots,0_{\mathfrak{C}_{\rm ex}}) = 0_{\mathfrak{C}_{\rm ex}}$. This defines Φ_f,Ψ_g, and makes $\boldsymbol{\mathfrak{C}}=(\mathfrak{C},\mathfrak{C}_{\rm ex})$ into an interior pre C^∞-ring with corners.

Conversely, let $\boldsymbol{\mathfrak{C}}=(\mathfrak{C},\mathfrak{C}_{\rm ex})$ be an interior pre C^∞-ring with corners. Then $\mathfrak{C}_{\rm ex}=\mathfrak{C}_{\rm in}\amalg\{0\}$ and we define a product-preserving functor $F:\mathbf{Euc}^{\mathbf{c}}_{\mathbf{in}}\to\mathbf{Sets}$ with $F(\mathbb{R}^m\times[0,\infty)^n)=\mathfrak{C}^m\times\mathfrak{C}^n_{\rm in}$, using the fact from Lemma 4.9 that $\Psi_g:\mathfrak{C}^m\times\mathfrak{C}^n_{\rm ex}\to\mathfrak{C}_{\rm ex}$ maps $\mathfrak{C}^m\times\mathfrak{C}^n_{\rm in}\to\mathfrak{C}_{\rm in}$ for g interior. The rest of the proof follows that of Proposition 4.5. □

4.3 Adjoints and (co)limits for pre C^∞-rings with corners

The equivalences of categories in Propositions 2.12, 4.5, and 4.10 mean that facts about $\mathbf{CPC^\infty Rings^c}$, $\mathbf{CPC^\infty Rings^c_{in}}$, $\mathbf{CC^\infty Rings}$ in §4.1 correspond to facts about $\mathbf{PC^\infty Rings^c}$, $\mathbf{PC^\infty Rings^c_{in}}$, $\mathbf{C^\infty Rings}$. Hence (4.2) should correspond to a commutative triangle of functors

$$\begin{array}{ccc}\mathbf{PC^\infty Rings^c} & \xrightarrow{\Pi^{\mathbf{C^\infty Rings}}_{\mathbf{PC^\infty Rings^c}}} & \mathbf{C^\infty Rings}. \\ {\scriptstyle \Pi^{\mathbf{PC^\infty Rings^c_{in}}}_{\mathbf{PC^\infty Rings^c}}} \searrow & & \nearrow {\scriptstyle \Pi^{\mathbf{C^\infty Rings}}_{\mathbf{PC^\infty Rings^c_{in}}}} \\ & \mathbf{PC^\infty Rings^c_{in}} & \end{array} \qquad (4.7)$$

The explicit definitions, which we give next, follow from the correspondences.

Definition 4.11 The functor $\Pi^{\mathbf{C^\infty Rings}}_{\mathbf{PC^\infty Rings^c}}:\mathbf{PC^\infty Rings^c}\to\mathbf{C^\infty Rings}$ in (4.7) acts on objects by $\boldsymbol{\mathfrak{C}}=(\mathfrak{C},\mathfrak{C}_{\rm ex})\mapsto\mathfrak{C}$, where the C^∞-ring $\mathfrak{C}$ has C^∞-operations $\Phi_f:\mathfrak{C}^m\to\mathfrak{C}$ from smooth $f:\mathbb{R}^m\to\mathbb{R}$ in Definition 4.4(b) with $n=0$, and on morphisms by $\boldsymbol{\phi}=(\phi,\phi_{\rm ex})\mapsto\phi$.

Define $\Pi^{\mathbf{C^\infty Rings}}_{\mathbf{PC^\infty Rings^c_{in}}}:\mathbf{PC^\infty Rings^c_{in}}\to\mathbf{C^\infty Rings}$ in (4.7) to be the restriction of $\Pi^{\mathbf{C^\infty Rings}}_{\mathbf{PC^\infty Rings^c}}$ to $\mathbf{PC^\infty Rings^c_{in}}$.

To define $\Pi^{\mathbf{PC^\infty Rings^c_{in}}}_{\mathbf{PC^\infty Rings^c}}$ in (4.7), let $\boldsymbol{\mathfrak{C}}=(\mathfrak{C},\mathfrak{C}_{\rm ex})$ be a pre C^∞-ring with corners. We will define an interior pre C^∞-ring with corners $\tilde{\boldsymbol{\mathfrak{C}}}=(\mathfrak{C},\tilde{\mathfrak{C}}_{\rm ex})$ where $\tilde{\mathfrak{C}}_{\rm ex}=\mathfrak{C}_{\rm ex}\amalg\{0_{\tilde{\mathfrak{C}}_{\rm ex}}\}$, and set $\Pi^{\mathbf{PC^\infty Rings^c_{in}}}_{\mathbf{PC^\infty Rings^c}}(\boldsymbol{\mathfrak{C}})=\tilde{\boldsymbol{\mathfrak{C}}}$. Here $\mathfrak{C}_{\rm ex}$ already contains a zero element $0_{\mathfrak{C}_{\rm ex}}$, but we are adding an extra $0_{\tilde{\mathfrak{C}}_{\rm ex}}$ with $0_{\mathfrak{C}_{\rm ex}}\neq 0_{\tilde{\mathfrak{C}}_{\rm ex}}$.

Let $f:\mathbb{R}^m\times[0,\infty)^n\to\mathbb{R}$ and $g:\mathbb{R}^m\times[0,\infty)^n\to[0,\infty)$ be smooth, and write Φ_f,Ψ_g for the operations in $\boldsymbol{\mathfrak{C}}$. We must define maps $\tilde{\Phi}_f:\mathfrak{C}^m\times\tilde{\mathfrak{C}}^n_{\rm ex}\to\mathfrak{C}$

and $\tilde\Psi_g : \mathfrak{C}^m \times \tilde{\mathfrak{C}}_{\rm ex}^n \to \tilde{\mathfrak{C}}_{\rm ex}$. Let $c_1,\ldots,c_m \in \mathfrak{C}$ and $c'_1,\ldots,c'_n \in \tilde{\mathfrak{C}}_{\rm ex}$. Then some of $c'_1,\ldots,c'_n$ lie in $\mathfrak{C}_{\rm ex}$ and the rest in $\{0_{\tilde{\mathfrak{C}}_{\rm ex}}\}$. For simplicity suppose that $c'_1,\ldots,c'_k \in \mathfrak{C}_{\rm ex}$ and $c'_{k+1} = \cdots = c'_n = 0_{\tilde{\mathfrak{C}}_{\rm ex}}$ for $0 \leqslant k \leqslant n$. Define smooth $d : \mathbb{R}^m \times [0,\infty)^k \to \mathbb{R}$, $e : \mathbb{R}^m \times [0,\infty)^k \to [0,\infty)$ by (4.6). Set

$$\tilde\Phi_f(c_1,\ldots,c_m,c'_1,\ldots,c'_k,0_{\tilde{\mathfrak{C}}_{\rm ex}},\ldots,0_{\tilde{\mathfrak{C}}_{\rm ex}}) = \Phi_d(c_1,\ldots,c_m,c'_1,\ldots,c'_k).$$

Either $e : \mathbb{R}^m \times [0,\infty)^k \to [0,\infty)$ is interior, or $e = 0$. If e is interior define

$$\tilde\Psi_g(c_1,\ldots,c_m,c'_1,\ldots,c'_k,0_{\tilde{\mathfrak{C}}_{\rm ex}},\ldots,0_{\tilde{\mathfrak{C}}_{\rm ex}}) = \Psi_e(c_1,\ldots,c_m,c'_1,\ldots,c'_k).$$

If $e = 0$ define $\tilde\Psi_g(c_1,\ldots,c_m,c'_1,\ldots,c'_k,0_{\tilde{\mathfrak{C}}_{\rm ex}},\ldots,0_{\tilde{\mathfrak{C}}_{\rm ex}}) = 0_{\tilde{\mathfrak{C}}_{\rm ex}}$. This defines the maps $\tilde\Phi_f, \tilde\Psi_g$. It is easy to check that these make $\tilde{\boldsymbol{\mathfrak{C}}} = (\mathfrak{C}, \tilde{\mathfrak{C}}_{\rm ex})$ into an interior pre C^∞-ring with corners.

Now let $\boldsymbol\phi : \boldsymbol{\mathfrak{C}} \to \boldsymbol{\mathfrak{D}}$ be a morphism in $\mathbf{PC^\infty Rings^c}$, and define $\tilde{\boldsymbol{\mathfrak{C}}}, \tilde{\boldsymbol{\mathfrak{D}}}$ as above. Define $\tilde\phi_{\rm ex} : \tilde{\mathfrak{C}}_{\rm ex} \to \tilde{\mathfrak{D}}_{\rm ex}$ by $\tilde\phi_{\rm ex}|_{\mathfrak{C}_{\rm ex}} = \phi_{\rm ex}$ and $\tilde\phi_{\rm ex}(0_{\tilde{\mathfrak{C}}_{\rm ex}}) = 0_{\tilde{\mathfrak{D}}_{\rm ex}}$. Then $\tilde{\boldsymbol\phi} = (\phi, \tilde\phi_{\rm ex}) : \tilde{\boldsymbol{\mathfrak{C}}} \to \tilde{\boldsymbol{\mathfrak{D}}}$ is a morphism in $\mathbf{PC^\infty Rings^c_{in}}$. Define $\Pi_{\mathbf{PC^\infty Rings^c}}^{\mathbf{PC^\infty Rings^c_{in}}} : \mathbf{PC^\infty Rings^c} \hookrightarrow \mathbf{PC^\infty Rings^c_{in}}$ by $\Pi_{\mathbf{PC^\infty Rings^c}}^{\mathbf{PC^\infty Rings^c_{in}}} : \boldsymbol{\mathfrak{C}} \mapsto \tilde{\boldsymbol{\mathfrak{C}}}$ and $\Pi_{\mathbf{PC^\infty Rings^c}}^{\mathbf{PC^\infty Rings^c_{in}}} : \boldsymbol\phi \mapsto \tilde{\boldsymbol\phi}$, for $\boldsymbol{\mathfrak{C}},\ldots,\tilde{\boldsymbol\phi}$ as above.

We will show that $\Pi_{\mathbf{PC^\infty Rings^c}}^{\mathbf{PC^\infty Rings^c_{in}}}$ is right adjoint to $\mathrm{inc} : \mathbf{PC^\infty Rings^c_{in}} \hookrightarrow \mathbf{PC^\infty Rings^c}$. Suppose that $\boldsymbol{\mathfrak{C}}, \boldsymbol{\mathfrak{D}}$ are pre C^∞-rings with corners with $\boldsymbol{\mathfrak{C}}$ interior. Then we can define a 1-1 correspondence

$$\begin{aligned}\mathrm{Hom}_{\mathbf{PC^\infty Rings^c}}\bigl(\mathrm{inc}(\boldsymbol{\mathfrak{C}}), \boldsymbol{\mathfrak{D}}\bigr) &= \mathrm{Hom}_{\mathbf{PC^\infty Rings^c}}(\boldsymbol{\mathfrak{C}}, \boldsymbol{\mathfrak{D}}) \\ &\cong \mathrm{Hom}_{\mathbf{PC^\infty Rings^c_{in}}}\bigl(\boldsymbol{\mathfrak{C}}, \Pi_{\mathbf{PC^\infty Rings^c}}^{\mathbf{PC^\infty Rings^c_{in}}}(\boldsymbol{\mathfrak{D}})\bigr),\end{aligned}$$

identifying $\boldsymbol\phi : \mathrm{inc}(\boldsymbol{\mathfrak{C}}) \to \boldsymbol{\mathfrak{D}}$ with $\hat{\boldsymbol\phi} : \boldsymbol{\mathfrak{C}} \to \Pi_{\mathbf{PC^\infty Rings^c}}^{\mathbf{PC^\infty Rings^c_{in}}}(\boldsymbol{\mathfrak{D}})$, where $\boldsymbol\phi = (\phi, \phi_{\rm ex})$ and $\hat{\boldsymbol\phi} = (\phi, \hat\phi_{\rm ex})$ with $\hat\phi_{\rm ex}|_{\mathfrak{C}_{\rm in}} = \phi_{\rm ex}|_{\mathfrak{C}_{\rm in}}$ and $\hat\phi_{\rm ex}(0_{\mathfrak{C}_{\rm ex}}) = 0_{\tilde{\mathfrak{D}}_{\rm ex}}$. This is functorial in $\boldsymbol{\mathfrak{C}}, \boldsymbol{\mathfrak{D}}$, and so proves that $\Pi_{\mathbf{PC^\infty Rings^c}}^{\mathbf{PC^\infty Rings^c_{in}}}$ is right adjoint to inc.

Also define functors $\Pi_{\rm sm}, \Pi_{\rm ex} : \mathbf{PC^\infty Rings^c} \to \mathbf{Sets}$ by $\Pi_{\rm sm} : \boldsymbol{\mathfrak{C}} \mapsto \mathfrak{C}$, $\Pi_{\rm ex} : \boldsymbol{\mathfrak{C}} \mapsto \mathfrak{C}_{\rm ex}$ on objects, and $\Pi_{\rm sm} : \boldsymbol\phi \mapsto \phi$, $\Pi_{\rm ex} : \boldsymbol\phi \mapsto \phi_{\rm ex}$ on morphisms, where 'sm' and 'ex' are short for 'smooth' and 'exterior'. Define functors $\Pi_{\rm sm}, \Pi_{\rm in} : \mathbf{PC^\infty Rings^c_{in}} \to \mathbf{Sets}$ by $\Pi_{\rm sm} : \boldsymbol{\mathfrak{C}} \mapsto \mathfrak{C}$, $\Pi_{\rm in} : \boldsymbol{\mathfrak{C}} \mapsto \mathfrak{C}_{\rm in}$ on objects, and $\Pi_{\rm sm} : \boldsymbol\phi \mapsto \phi$, $\Pi_{\rm in} : \boldsymbol\phi \mapsto \phi_{\rm in}$ on morphisms, where 'in' is short for 'interior'.

Combining the equivalences in Propositions 2.12, 4.5, and 4.10 with Theorem 4.3 and Definition 4.11 yields the following.

Theorem 4.12 **(a)** *All small limits and directed colimits exist in the categories* $\mathbf{PC^\infty Rings^c}$,$\mathbf{PC^\infty Rings^c_{in}}$. *The functors* $\Pi_{\rm sm}$,$\Pi_{\rm ex}$: $\mathbf{PC^\infty Rings^c} \to$

$\mathbf{Sets}$ *and* $\Pi_{\rm sm},\Pi_{\rm in} : \mathbf{PC^\infty Rings^c_{in}} \to \mathbf{Sets}$ *preserve limits and directed colimits.*

(b) *All small colimits exist in* $\mathbf{PC^\infty Rings^c}, \mathbf{PC^\infty Rings^c_{in}}$*, though in general they are not preserved by* $\Pi_{\rm sm},\Pi_{\rm ex}$ *and* $\Pi_{\rm sm},\Pi_{\rm in}$.

(c) *The functors* $\Pi^{\mathbf{C^\infty Rings}}_{\mathbf{PC^\infty Rings^c}}, \Pi^{\mathbf{C^\infty Rings}}_{\mathbf{PC^\infty Rings^c_{in}}}, \Pi^{\mathbf{PC^\infty Rings^c_{in}}}_{\mathbf{PC^\infty Rings^c}}$ *in (4.7) have left adjoints, so they preserve limits. The left adjoint of* $\Pi^{\mathbf{PC^\infty Rings^c_{in}}}_{\mathbf{PC^\infty Rings^c}}$ *is the inclusion* inc : $\mathbf{PC^\infty Rings^c_{in}} \hookrightarrow \mathbf{PC^\infty Rings^c}$*, so* inc *preserves colimits.*

Example 4.13 The inclusion inc : $\mathbf{PC^\infty Rings^c_{in}} \hookrightarrow \mathbf{PC^\infty Rings^c}$ in general does not preserve limits, and therefore cannot have a left adjoint.

For example, suppose $\boldsymbol{\mathfrak{C}}, \boldsymbol{\mathfrak{D}}$ are interior pre C^∞-rings with corners, and write $\boldsymbol{\mathfrak{E}} = \boldsymbol{\mathfrak{C}} \times \boldsymbol{\mathfrak{D}}$ and $\boldsymbol{\mathfrak{F}} = \boldsymbol{\mathfrak{C}} \times_{\rm in} \boldsymbol{\mathfrak{D}}$ for the products in $\mathbf{PC^\infty Rings^c}$ and $\mathbf{PC^\infty Rings^c_{in}}$. Then Theorem 4.12(a) implies that $\mathfrak{E} = \mathfrak{C} \times \mathfrak{D}$, $\mathfrak{E}_{\rm ex} = \mathfrak{C}_{\rm ex} \times \mathfrak{D}_{\rm ex}$, $\mathfrak{F} = \mathfrak{C} \times \mathfrak{D}$, $\mathfrak{F}_{\rm in} = \mathfrak{C}_{\rm in} \times \mathfrak{D}_{\rm in}$. Since $\mathfrak{C}_{\rm ex} = \mathfrak{C}_{\rm in} \amalg \{0_{\mathfrak{C}_{\rm ex}}\}$, etc., this gives

$$
\begin{aligned}
\mathfrak{E}_{\rm ex} &= (\mathfrak{C}_{\rm in} \times \mathfrak{D}_{\rm in}) \amalg (\mathfrak{C}_{\rm in} \times \{0_{\mathfrak{D}_{\rm ex}}\}) \amalg (\{0_{\mathfrak{C}_{\rm ex}}\} \times \mathfrak{D}_{\rm in}) \amalg (\{0_{\mathfrak{C}_{\rm ex}}\} \times \{0_{\mathfrak{D}_{\rm ex}}\}),\\
\mathfrak{F}_{\rm ex} &= (\mathfrak{C}_{\rm in} \times \mathfrak{D}_{\rm in}) \amalg (\{0_{\mathfrak{C}_{\rm ex}}\} \times \{0_{\mathfrak{D}_{\rm ex}}\}).
\end{aligned}
$$

Thus $\boldsymbol{\mathfrak{E}} \not\cong \boldsymbol{\mathfrak{F}}$. Moreover in $\mathfrak{E}_{\rm ex}$ we have $(1_{\mathfrak{C}_{\rm ex}}, 0_{\mathfrak{D}_{\rm ex}}) \cdot (0_{\mathfrak{C}_{\rm ex}}, 1_{\mathfrak{D}_{\rm ex}}) = (0_{\mathfrak{C}_{\rm ex}}, 0_{\mathfrak{D}_{\rm ex}})$, so $\mathfrak{E}_{\rm ex}$ has zero divisors, and $\boldsymbol{\mathfrak{E}}$ is not an object in $\mathbf{PC^\infty Rings^c_{in}}$.

Proposition 4.14 *The functors* $\Pi^{\mathbf{C^\infty Rings}}_{\mathbf{PC^\infty Rings^c}}, \Pi^{\mathbf{C^\infty Rings}}_{\mathbf{PC^\infty Rings^c_{in}}}$ *in* (4.7) *have right adjoints, so they preserve colimits.*

Proof We will define a functor $F_{\geqslant 0} : \mathbf{C^\infty Rings} \to \mathbf{PC^\infty Rings^c}$, and show it is right adjoint to $\Pi^{\mathbf{C^\infty Rings}}_{\mathbf{PC^\infty Rings^c}}$. Let $\mathfrak{C}$ be a C^∞-ring, and define $\mathfrak{C}_{\geqslant 0}$ to be the subset of elements $c \in \mathfrak{C}$ that satisfy the following condition: for all smooth $f : \mathbb{R}^n \to \mathbb{R}$ such that $f|_{[0,\infty)\times\mathbb{R}^{n-1}} = 0$ and for $d_1, \ldots, d_{n-1} \in \mathfrak{C}$, then $\Phi_f(c, d_1, \ldots, d_{n-1}) = 0 \in \mathfrak{C}$. Note that $\mathfrak{C}_{\geqslant 0}$ is non-empty, as it contains $\Phi_{\exp}(\mathfrak{C}) \amalg \{0\}$. Also, any pre C^∞-ring with corners $(\mathfrak{D}, \mathfrak{D}_{\rm ex})$ has $\Phi_i(\mathfrak{D}_{\rm ex}) \subseteq \mathfrak{D}_{\geqslant 0}$.

We will make $(\mathfrak{C}, \mathfrak{C}_{\geqslant 0})$ into a pre C^∞-ring with corners. Take a smooth map $f : \mathbb{R}^m_k \to [0,\infty)$. Then there is a (non-unique) smooth extension $g : \mathbb{R}^m \to \mathbb{R}$ such that $g|_{\mathbb{R}^m_k} = i \circ f$ for $i : [0,\infty) \to \mathbb{R}$ the inclusion map. We define

$$
\Psi_f(c'_1, \ldots, c'_k, c_{k+1}, \ldots, c_m) = \Phi_g(c'_1, \ldots, c'_k, c_{k+1}, \ldots, c_m)
$$

for $c'_i \in \mathfrak{C}_{\geqslant 0}$ and $c_i \in \mathfrak{C}$. To show this is well defined, say h is another such extension, then $(g-h)|_{\mathbb{R}^m_k} = 0$. We must prove $\Phi_{g-h}(c'_1, \ldots, c'_k, c_{k+1}, \ldots, c_m) = 0$ for all $c'_i \in \mathfrak{C}_{\geqslant 0}$ and $c_i \in \mathfrak{C}$.

As $g - h$ satisfies the hypothesis of Lemma 4.15 below, we can assume $g - h = f_1 + \cdots + f_k$ such that $f_i|_{\mathbb{R}^{i-1}\times[0,\infty)\times\mathbb{R}^{m-i-1}} = 0$. Then

$$\begin{aligned}\Phi_{g-h}(c'_1,\ldots,c'_k,c_{k+1},\ldots,c_m) &= \Phi_{f_1+\cdots+f_k}(c'_1,\ldots,c'_k,c_{k+1},\ldots,c_m)\\ &= \Phi_{f_1}(c'_1,\ldots,c'_k,c_{k+1},\ldots,c_m)+\cdots+\Phi_{f_k}(c'_1,\ldots,c'_k,c_{k+1},\ldots,c_m)=0,\end{aligned}$$

as c'_i are in $\mathfrak{C}_{\geqslant 0}$. Therefore Ψ_f is independent of the choice of extension g.

To show that Ψ_f maps to $\mathfrak{C}_{\geqslant 0} \subseteq \mathfrak{C}$, suppose $e : \mathbb{R}^n \to \mathbb{R}$ is smooth with $e|_{[0,\infty)\times\mathbb{R}^{n-1}} = 0$, and let $d_1,\ldots,d_{n-1} \in \mathfrak{C}$. Then

$$\begin{aligned}&\Phi_e(\Psi_f(c'_1,\ldots,c'_k,c_{k+1},\ldots,c_m),d_1,\ldots,d_{n-1})\\ &\quad= \Phi_e(\Phi_g(c'_1,\ldots,c'_k,c_{k+1},\ldots,c_m),d_1,\ldots,d_{n-1}) \qquad (4.8)\\ &\quad= \Phi_{e\circ(g\times\mathrm{id}_{\mathbb{R}^{n-1}})}(c'_1,\ldots,c'_k,c_{k+1},\ldots,c_m,d_1,\ldots,d_{n-1}).\end{aligned}$$

Here $e\circ(g\times\mathrm{id}_{\mathbb{R}^{n-1}}) : \mathbb{R}^{m+n-1} \to \mathbb{R}$ restricts to zero on $\mathbb{R}^{m+n-1}_k \subset \mathbb{R}^{m+n-1}$, since $(g \times \mathrm{id}_{\mathbb{R}^{n-1}})|_{\mathbb{R}^{m+n-1}_k} = f \times \mathrm{id}_{\mathbb{R}^{n-1}}$, which maps to $[0,\infty) \times \mathbb{R}^{n-1}$, and $e|_{[0,\infty)\times\mathbb{R}^{n-1}} = 0$. Regarding $e\circ(g\times\mathrm{id}_{\mathbb{R}^{n-1}})$ as an extension of $0 : \mathbb{R}^{m+n-1}_k \to \mathbb{R}$, the previous argument shows (4.8) is independent of extension, so we can replace $e \circ (g \times \mathrm{id}_{\mathbb{R}^{n-1}})$ by $0 : \mathbb{R}^{m+n-1} \to \mathbb{R}$, and (4.8) is zero. As this holds for all e, n, we have $\Psi_f(c'_1,\ldots,c_m) \in \mathfrak{C}_{\geqslant 0}$, as we want.

Similarly, smooth functions $f : \mathbb{R}^m_k \to \mathbb{R}$ give well-defined C^∞-operations Φ_f. Hence $(\mathfrak{C},\mathfrak{C}_{\geqslant 0})$ is a pre C^∞-ring with corners. Set $F_{\geqslant 0}(\mathfrak{C}) = (\mathfrak{C},\mathfrak{C}_{\geqslant 0})$.

On morphisms, $F_{\geqslant 0}$ sends $\phi : \mathfrak{C} \to \mathfrak{D}$ to $(\phi,\phi_{\mathrm{ex}}) : (\mathfrak{C},\mathfrak{C}_{\geqslant 0}) \to (\mathfrak{D},\mathfrak{D}_{\geqslant 0})$, where $\phi_{\mathrm{ex}} = \phi|_{\mathfrak{C}_{\geqslant 0}}$. As ϕ respects the C^∞-operations, the image of $\phi(\mathfrak{C}_{\geqslant 0}) \subseteq \mathfrak{D}_{\geqslant 0} \subseteq \mathfrak{D}$, so ϕ_{ex} is well defined.

To show $F_{\geqslant 0}$ is right adjoint to $\Pi^{\mathbf{C^\infty Rings}}_{\mathbf{PC^\infty Rings^c}}$, we describe the unit and counit of the adjunction. The unit acts on $(\mathfrak{C},\mathfrak{C}_{\mathrm{ex}}) \in \mathbf{PC^\infty Rings^c}$ as $(\mathrm{id}_{\mathfrak{C}},\Phi_i) : (\mathfrak{C},\mathfrak{C}_{\mathrm{ex}}) \to (\mathfrak{C},\mathfrak{C}_{\geqslant 0})$. The counit acts on $\mathfrak{C} \in \mathbf{C^\infty Rings}$ as $\mathrm{id}_{\mathfrak{C}} : \mathfrak{C} \to \mathfrak{C}$. The compositions $F_{\geqslant 0} \Rightarrow F_{\geqslant 0} \circ \Pi^{\mathbf{C^\infty Rings}}_{\mathbf{PC^\infty Rings^c}} \circ F_{\geqslant 0} \Rightarrow F_{\geqslant 0}$ and $\Pi^{\mathbf{C^\infty Rings}}_{\mathbf{PC^\infty Rings^c}} \Rightarrow \Pi^{\mathbf{C^\infty Rings}}_{\mathbf{PC^\infty Rings^c}} \circ F_{\geqslant 0} \circ \Pi^{\mathbf{C^\infty Rings}}_{\mathbf{PC^\infty Rings^c}} \Rightarrow \Pi^{\mathbf{C^\infty Rings}}_{\mathbf{PC^\infty Rings^c}}$ are the identity.

Since $\Pi^{\mathbf{C^\infty Rings}}_{\mathbf{PC^\infty Rings^c_{in}}} = \Pi^{\mathbf{C^\infty Rings}}_{\mathbf{PC^\infty Rings^c}} \circ \mathrm{inc}$, it now follows from Definition 4.11 that $\Pi^{\mathbf{PC^\infty Rings^c_{in}}}_{\mathbf{PC^\infty Rings^c}} \circ F_{\geqslant 0}$ is right adjoint to $\Pi^{\mathbf{C^\infty Rings}}_{\mathbf{PC^\infty Rings^c_{in}}}$. □

The next lemma was used in the previous proof.

Lemma 4.15 *If $f : \mathbb{R}^m \to \mathbb{R}$ is smooth such that $f|_{\mathbb{R}^m_k} = 0$, then there are smooth $f_i : \mathbb{R}^m \to \mathbb{R}$ for $i = 1,\ldots,k$ such that $f_i|_{\mathbb{R}^{i-1}\times[0,\infty)\times\mathbb{R}^{m-i-1}} = 0$ for $i = 1,\ldots,k$, and $f = f_1 + \cdots + f_k$.*

Proof Let f be as in the lemma. Consider the following open subset $U = \mathcal{S}^{k-1}\setminus\{(x_1,\ldots,x_k) : x_i \geqslant 0\}$ of the dimension $k-1$ unit sphere $\mathcal{S}^{k-1} \subset \mathbb{R}^k$.

Define an open cover $U_1,\ldots,U_k$ of U by $U_i=\{(x_1,\ldots,x_k)\in U : x_i<0\}$. Take a partition of unity $\rho_i : U\rightarrow[0,1]$ for $i=1,\ldots,k$, subordinate to $\{U_1,\ldots,U_n\}$, with ρ_i having support on U_i and $\sum_{i=1}^k\rho_i=1$.

Define the f_i as follows:

$$f_i=\begin{cases} f(x_1,\ldots,x_m)\rho_i\left(\frac{(x_1,\ldots,x_k)}{|(x_1,\ldots,x_k)|}\right), & \text{if } x_i<0 \text{ for some } i=1,\ldots,k,\\ 0, & \text{otherwise,}\end{cases}$$

where $|(x_1,\ldots,x_k)|$ is the length of the vector $(x_1,\ldots,x_k)\in\mathbb{R}^k$. It is clear these are smooth where the ρ_i are defined. The ρ_i are not defined in the first quadrant of $\mathbb{R}^k$, where all $x_i\geqslant 0$; however, approaching the boundary of this quadrant, the ρ_i are all constant. As $f|_{\mathbb{R}^m_k}=0$, then all derivatives of f are zero in this quadrant, so the f_i are smooth and identically zero on $\mathbb{R}^m_k$. In addition, $f_i|_{\mathbb{R}^{i-1}\times[0,\infty)\times\mathbb{R}^{m-i-1}}=0$, as the ρ_i are zero outside of U_i. Finally, as $\sum\rho_i=1$ and $f|_{\mathbb{R}^m_k}=0$, then $f=f_1+\cdots+f_k$ as required. □

4.4 C^∞-rings with corners

We can now define C^∞-rings with corners.

Definition 4.16 Let $\boldsymbol{\mathfrak{C}}=(\mathfrak{C},\mathfrak{C}_{\rm ex})$ be a pre C^∞-ring with corners. We call $\boldsymbol{\mathfrak{C}}$ a *C^∞-ring with corners* if any (hence all) of the following hold.

(i) $\Phi_i|_{\mathfrak{C}_{\rm ex}^\times}:\mathfrak{C}_{\rm ex}^\times\rightarrow\mathfrak{C}$ is injective.
(ii) $\Psi_{\exp}:\mathfrak{C}\rightarrow\mathfrak{C}_{\rm ex}^\times$ is surjective.
(iii) $\Psi_{\exp}:\mathfrak{C}\rightarrow\mathfrak{C}_{\rm ex}^\times$ is a bijection.

Here Proposition 4.7 implies that (i) and (ii) are equivalent, and Definition 4.6(e) and Proposition 2.26(a) imply that $\Psi_{\exp}:\mathfrak{C}\rightarrow\mathfrak{C}_{\rm ex}^\times$ is injective, so (ii) and (iii) are equivalent, and therefore (i)–(iii) are equivalent. Write $\mathbf{C^\infty Rings^c}$ for the full subcategory of C^∞-rings with corners in $\mathbf{PC^\infty Rings^c}$.

We call a C^∞-ring with corners $\boldsymbol{\mathfrak{C}}$ *interior* if it is an interior pre C^∞-ring with corners. Write $\mathbf{C^\infty Rings^c_{in}}\subset\mathbf{C^\infty Rings^c}$ for the subcategory of interior C^∞-rings with corners, and interior morphisms between them.

Define a commutative triangle of functors

$$\begin{array}{ccc}\mathbf{C^\infty Rings^c} & \xrightarrow{\Pi^{\mathbf{C^\infty Rings}}_{\mathbf{C^\infty Rings^c}}} & \mathbf{C^\infty Rings}\\ & {\scriptstyle\Pi^{\mathbf{C^\infty Rings^c_{in}}}_{\mathbf{C^\infty Rings^c}}}\searrow \quad \mathbf{C^\infty Rings^c_{in}} \quad \nearrow{\scriptstyle\Pi^{\mathbf{C^\infty Rings}}_{\mathbf{C^\infty Rings^c_{in}}}} & \end{array} \tag{4.9}$$

by restricting the functors in (4.7) to $\mathbf{C^\infty Rings^c},\mathbf{C^\infty Rings^c_{in}}$. It is clear from Definition 4.11 that the restriction of $\Pi_{\mathbf{PC^\infty Rings^c}}^{\mathbf{PC^\infty Rings^c_{in}}}$ to $\mathbf{C^\infty Rings^c}$ maps to $\mathbf{C^\infty Rings^c_{in}}$. Then $\Pi_{\mathbf{PC^\infty Rings^c}}^{\mathbf{PC^\infty Rings^c_{in}}}$ right adjoint to inc in Definition 4.11 implies $\Pi_{\mathbf{C^\infty Rings^c}}^{\mathbf{C^\infty Rings^c_{in}}}$ is right adjoint to $\mathrm{inc} : \mathbf{C^\infty Rings^c_{in}} \hookrightarrow \mathbf{C^\infty Rings^c}$.

Remark 4.17 We can interpret the condition that $\boldsymbol{\mathfrak{C}} = (\mathfrak{C}, \mathfrak{C}_{\rm ex})$ be a C^∞-ring with corners as follows. Imagine there is some 'space with corners' X, such that $\mathfrak{C} = \{\text{smooth maps } c : X \to \mathbb{R}\}$, and $\mathfrak{C}_{\rm ex} = \{\text{exterior maps } c' : X \to [0,\infty)\}$. If $c' \in \mathfrak{C}_{\rm ex}$ is invertible then c' should map $X \to (0,\infty)$, and we require there to exist smooth $c = \log c' : X \to \mathbb{R}$ in $\mathfrak{C}$ with $c' = \exp c$. This makes C^∞-rings and C^∞-schemes with corners behave more like manifolds with corners.

Here is the motivating example, following Example 4.2.

Example 4.18 **(a)** Let X be a manifold with corners. Define a C^∞-ring with corners $\boldsymbol{\mathfrak{C}} = (\mathfrak{C}, \mathfrak{C}_{\rm ex})$ by $\mathfrak{C} = C^\infty(X)$ and $\mathfrak{C}_{\rm ex} = \mathrm{Ex}(X)$, as sets. If $f : \mathbb{R}^m \times [0,\infty)^n \to \mathbb{R}$ is smooth, define the operation $\Phi_f : \mathfrak{C}^m \times \mathfrak{C}_{\rm ex}^n \to \mathfrak{C}$ by

$$\begin{aligned} &\Phi_f(c_1,\ldots,c_m,c'_1,\ldots,c'_n) : \\ &\qquad x \longmapsto f\bigl(c_1(x),\ldots,c_m(x),c'_1(x),\ldots,c'_n(x)\bigr). \end{aligned} \tag{4.10}$$

If $g : \mathbb{R}^m \times [0,\infty)^n \to [0,\infty)$ is exterior, define $\Psi_g : \mathfrak{C}^m \times \mathfrak{C}_{\rm ex}^n \to \mathfrak{C}_{\rm ex}$ by

$$\begin{aligned} &\Psi_g(c_1,\ldots,c_m,c'_1,\ldots,c'_n) : \\ &\qquad x \longmapsto g\bigl(c_1(x),\ldots,c_m(x),c'_1(x),\ldots,c'_n(x)\bigr). \end{aligned} \tag{4.11}$$

It is easy to check that $\boldsymbol{\mathfrak{C}}$ is a C^∞-ring with corners. We write $\boldsymbol{C^\infty}(X) = \boldsymbol{\mathfrak{C}}$.

Suppose $f : X \to Y$ is a smooth map of manifolds with corners, and let $\boldsymbol{\mathfrak{C}}, \boldsymbol{\mathfrak{D}}$ be the C^∞-rings with corners corresponding to X, Y. Write $\boldsymbol{\phi} = (\phi, \phi_{\rm ex})$, where $\phi : \mathfrak{D} \to \mathfrak{C}$ maps $\phi(d) = d \circ f$ and $\phi_{\rm ex} : \mathfrak{D}_{\rm ex} \to \mathfrak{C}_{\rm ex}$ maps $\phi(d') = d' \circ f$. Then $\boldsymbol{\phi} : \boldsymbol{\mathfrak{D}} \to \boldsymbol{\mathfrak{C}}$ is a morphism in $\mathbf{C^\infty Rings^c}$.

Define a functor $F_{\mathbf{Man^c}}^{\mathbf{C^\infty Rings^c}} : \mathbf{Man^c} \to (\mathbf{C^\infty Rings^c})^{\rm op}$ to map $X \mapsto \boldsymbol{\mathfrak{C}}$ on objects, and $f \mapsto \boldsymbol{\phi}$ on morphisms, for $X, Y, \boldsymbol{\mathfrak{C}}, \boldsymbol{\mathfrak{D}}, f, \boldsymbol{\phi}$ as above.

All the above also works for manifolds with g-corners, as in §3.3, giving a functor $F_{\mathbf{Man^{gc}}}^{\mathbf{C^\infty Rings^c}} : \mathbf{Man^{gc}} \to (\mathbf{C^\infty Rings^c})^{\rm op}$.

(b) Similarly, if X is a manifold with (g-)corners, define an interior C^∞-ring with corners $\boldsymbol{\mathfrak{C}} = (\mathfrak{C}, \mathfrak{C}_{\rm ex})$ by $\mathfrak{C} = C^\infty(X)$ and $\mathfrak{C}_{\rm ex} = \mathrm{In}(X) \amalg \{0\}$ as sets, where $0 : X \to [0,\infty)$ is the zero function, with C^∞-operations as in (4.10)–(4.11). We write $\boldsymbol{C^\infty_{\rm in}}(X) = \boldsymbol{\mathfrak{C}}$.

Suppose $f : X \to Y$ is an interior map of manifolds with (g-)corners, and let $\boldsymbol{\mathfrak{C}}, \boldsymbol{\mathfrak{D}}$ be the interior C^∞-rings with corners corresponding to X, Y. Write

$\boldsymbol{\phi} = (\phi, \phi_{\rm ex})$, where $\phi : \mathfrak{D} \to \mathfrak{C}$ maps $\phi(d) = d \circ f$ and $\phi_{\rm ex} : \mathfrak{D}_{\rm ex} \to \mathfrak{C}_{\rm ex}$ maps $\phi(d') = d' \circ f$. Then $\boldsymbol{\phi} : \boldsymbol{\mathfrak{D}} \to \boldsymbol{\mathfrak{C}}$ is a morphism in $\mathbf{C^\infty Rings^c_{in}}$.

Define $F_{\mathbf{Man^c_{in}}}^{\mathbf{C^\infty Rings^c_{in}}} : \mathbf{Man^c_{in}} \to (\mathbf{C^\infty Rings^c_{in}})^{\rm op}$ and $F_{\mathbf{Man^{gc}_{in}}}^{\mathbf{C^\infty Rings^c_{in}}} : \mathbf{Man^{gc}_{in}} \to (\mathbf{C^\infty Rings^c_{in}})^{\rm op}$ to map $X \mapsto \boldsymbol{\mathfrak{C}}$ on objects, and $f \mapsto \boldsymbol{\phi}$ on morphisms, for $X, Y, \boldsymbol{\mathfrak{C}}, \boldsymbol{\mathfrak{D}}, f, \boldsymbol{\phi}$ as above.

Recall from §2.1.3 that if $\mathcal{D}$ is a category and $\mathcal{C} \subset \mathcal{D}$ is a full subcategory, then $\mathcal{C}$ is a *reflective subcategory* if the inclusion inc : $\mathcal{C} \hookrightarrow \mathcal{D}$ has a left adjoint $\Pi : \mathcal{D} \to \mathcal{C}$, which is called a *reflection functor*.

Proposition 4.19 *The inclusion* inc : $\mathbf{C^\infty Rings^c} \hookrightarrow \mathbf{PC^\infty Rings^c}$ *has a left adjoint reflection functor* $\Pi_{\mathbf{PC^\infty Rings^c}}^{\mathbf{C^\infty Rings^c}}$. *Its restriction to* $\mathbf{PC^\infty Rings^c_{in}}$ *is a left adjoint* $\Pi_{\mathbf{PC^\infty Rings^c_{in}}}^{\mathbf{C^\infty Rings^c_{in}}}$ *for* inc : $\mathbf{C^\infty Rings^c_{in}} \hookrightarrow \mathbf{PC^\infty Rings^c_{in}}$.

Proof Let $\boldsymbol{\mathfrak{C}} = (\mathfrak{C}, \mathfrak{C}_{\rm ex})$ be a pre C^∞-ring with corners. We will define a C^∞-ring with corners $\hat{\boldsymbol{\mathfrak{C}}} = (\mathfrak{C}, \hat{\mathfrak{C}}_{\rm ex})$. As a set, define $\hat{\mathfrak{C}}_{\rm ex} = \mathfrak{C}_{\rm ex}/\sim$, where $\sim$ is the equivalence relation on $\mathfrak{C}_{\rm ex}$ given by $c' \sim c''$ if there exists $c''' \in \mathfrak{C}_{\rm ex}^\times$ with $\Phi_i(c''') = 1$ and $c' = c'' \cdot c'''$. That is, $\hat{\mathfrak{C}}_{\rm ex}$ is the quotient of $\mathfrak{C}_{\rm ex}$ by the group $\mathrm{Ker}(\Phi_i|_{\mathfrak{C}_{\rm ex}^\times}) \subseteq \mathfrak{C}_{\rm ex}^\times$. There is a natural surjective projection $\hat{\pi} : \mathfrak{C}_{\rm ex} \to \hat{\mathfrak{C}}_{\rm ex}$.

If $f : \mathbb{R}^m \times [0,\infty)^n \to \mathbb{R}$ is smooth and $g : \mathbb{R}^m \times [0,\infty)^n \to [0,\infty)$ is exterior, it is not difficult to show there exist unique maps $\hat{\Phi}_f : \mathfrak{C}^m \times \hat{\mathfrak{C}}_{\rm ex}^n \to \mathfrak{C}$ and $\hat{\Psi}_g : \mathfrak{C}^m \times \hat{\mathfrak{C}}_{\rm ex}^n \to \hat{\mathfrak{C}}_{\rm ex}$ making the following diagrams commute:

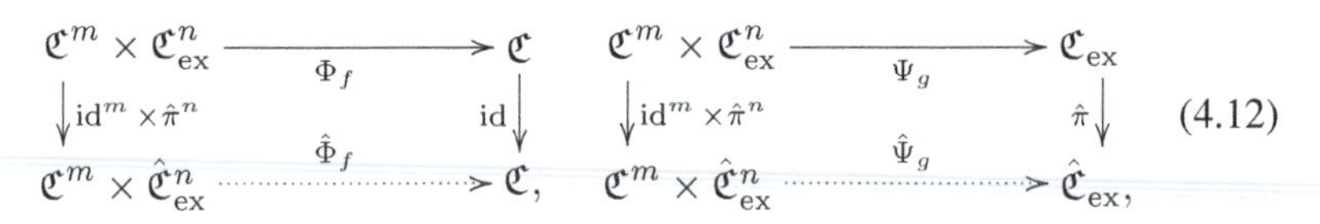

and these make $\hat{\boldsymbol{\mathfrak{C}}}$ into a pre C^∞-ring with corners. Clearly $\hat{\boldsymbol{\mathfrak{C}}}$ satisfies Definition 4.16(i). Therefore $\hat{\boldsymbol{\mathfrak{C}}}$ is a C^∞-ring with corners.

Suppose $\boldsymbol{\phi} = (\phi, \phi_{\rm ex}) : \boldsymbol{\mathfrak{C}} \to \boldsymbol{\mathfrak{D}}$ is a morphism of pre C^∞-rings with corners, and define $\hat{\boldsymbol{\mathfrak{C}}}, \hat{\boldsymbol{\mathfrak{D}}}$ as above. Then by a similar argument to (4.12), we find that there is a unique map $\hat{\phi}_{\rm ex}$ such that the following commutes:

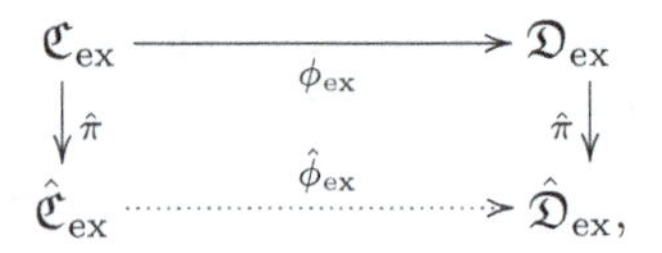

and then $\hat{\boldsymbol{\phi}} = (\phi, \hat{\phi}_{\rm ex}) : \hat{\boldsymbol{\mathfrak{C}}} \to \hat{\boldsymbol{\mathfrak{D}}}$ is a morphism of C^∞-rings with corners. Define a functor $\Pi_{\mathbf{PC^\infty Rings^c}}^{\mathbf{C^\infty Rings^c}} : \mathbf{PC^\infty Rings^c} \to \mathbf{C^\infty Rings^c}$ to map $\boldsymbol{\mathfrak{C}} \mapsto \hat{\boldsymbol{\mathfrak{C}}}$ on objects and $\boldsymbol{\phi} \mapsto \hat{\boldsymbol{\phi}}$ on morphisms.

Now let $\boldsymbol{\mathfrak{C}}$ be a pre C^∞-ring with corners, $\boldsymbol{\mathfrak{D}}$ a C^∞-ring with corners, and

$\phi : \mathfrak{C} \to \mathfrak{D}$ a morphism. Then we have a morphism $\hat\phi : \hat{\mathfrak{C}} \to \mathfrak{D}$, as $\hat{\mathfrak{D}} = \mathfrak{D}$, and $\phi \leftrightarrow \hat\phi$ gives a 1-1 correspondence

$$\begin{aligned}\mathrm{Hom}_{\mathbf{PC^\infty Rings^c}}(\mathfrak{C}, \mathrm{inc}(\mathfrak{D})) &= \mathrm{Hom}_{\mathbf{PC^\infty Rings^c}}(\mathfrak{C}, \mathfrak{D}) \\ &\cong \mathrm{Hom}_{\mathbf{C^\infty Rings^c}}\big(\Pi^{\mathbf{C^\infty Rings^c}}_{\mathbf{PC^\infty Rings^c}}(\mathfrak{C}), \mathfrak{D}\big),\end{aligned}$$

which is functorial in $\mathfrak{C}, \mathfrak{D}$. Hence $\Pi^{\mathbf{C^\infty Rings^c}}_{\mathbf{PC^\infty Rings^c}}$ is left adjoint to inc.

If $\mathfrak{C}, \phi$ above are interior then $\hat{\mathfrak{C}}, \hat\phi$ are interior, so $\Pi^{\mathbf{C^\infty Rings^c}}_{\mathbf{PC^\infty Rings^c}}$ restricts to a functor $\Pi^{\mathbf{C^\infty Rings^c_{in}}}_{\mathbf{PC^\infty Rings^c_{in}}} : \mathbf{PC^\infty Rings^c_{in}} \to \mathbf{C^\infty Rings^c_{in}}$, which is left adjoint to $\mathrm{inc} : \mathbf{C^\infty Rings^c_{in}} \hookrightarrow \mathbf{PC^\infty Rings^c_{in}}$. □

We extend Theorem 4.12, Example 4.13, Proposition 4.14 to $\mathbf{C^\infty Rings^c}$.

Theorem 4.20 **(a)** *All small limits exist in* $\mathbf{C^\infty Rings^c}, \mathbf{C^\infty Rings^c_{in}}$. *The functors* $\Pi_{\rm sm}, \Pi_{\rm ex} : \mathbf{C^\infty Rings^c} \to \mathbf{Sets}$ *and* $\Pi_{\rm sm}, \Pi_{\rm in} : \mathbf{C^\infty Rings^c_{in}} \to \mathbf{Sets}$ *preserve limits.*

(b) *All small colimits exist in* $\mathbf{C^\infty Rings^c}, \mathbf{C^\infty Rings^c_{in}}$, *though in general they are not preserved by* $\Pi_{\rm sm}, \Pi_{\rm ex}$ *and* $\Pi_{\rm sm}, \Pi_{\rm in}$.

(c) *The functors* $\Pi^{\mathbf{C^\infty Rings}}_{\mathbf{C^\infty Rings^c}}, \Pi^{\mathbf{C^\infty Rings}}_{\mathbf{C^\infty Rings^c_{in}}}$ *in* (4.9) *have left and right adjoints, so they preserve limits and colimits.*

(d) *The left adjoint of* $\Pi^{\mathbf{C^\infty Rings^c_{in}}}_{\mathbf{C^\infty Rings^c}}$ *is* $\mathrm{inc} : \mathbf{C^\infty Rings^c_{in}} \hookrightarrow \mathbf{C^\infty Rings^c}$, *so* inc *preserves colimits. However,* inc *does not preserve limits, and has no left adjoint.*

Proof For (a), we claim that $\mathbf{C^\infty Rings^c}, \mathbf{C^\infty Rings^c_{in}}$ are closed under limits in $\mathbf{PC^\infty Rings^c}, \mathbf{PC^\infty Rings^c_{in}}$. To see this, note that $\Pi_{\rm sm}, \Pi_{\rm ex} : \mathbf{PC^\infty Rings^c} \to \mathbf{Sets}$ preserve limits by Theorem 4.12(a), and the conditions Definition 4.16(i)–(iii) are closed under limits of sets $\mathfrak{C} = \Pi_{\rm sm}(\mathfrak{C})$ and $\mathfrak{C}_{\rm ex} = \Pi_{\rm ex}(\mathfrak{C})$. Thus part (a) follows from Theorem 4.12(a).

For (b), let J be a small category and $\boldsymbol{F} : J \to \mathbf{C^\infty Rings^c}$ a functor. Then a colimit $\mathfrak{C} = \varinjlim \boldsymbol{F}$ exists in $\mathbf{PC^\infty Rings^c}$ by Theorem 4.12(b). As $\Pi^{\mathbf{C^\infty Rings^c}}_{\mathbf{PC^\infty Rings^c}}$ is a left adjoint it preserves colimits, so $\Pi^{\mathbf{C^\infty Rings^c}}_{\mathbf{PC^\infty Rings^c}}(\mathfrak{C}) = \varinjlim \Pi^{\mathbf{C^\infty Rings^c}}_{\mathbf{PC^\infty Rings^c}} \circ \boldsymbol{F}$ in $\mathbf{C^\infty Rings^c}$. But $\Pi^{\mathbf{C^\infty Rings^c}}_{\mathbf{PC^\infty Rings^c}}$ is the identity on $\mathbf{C^\infty Rings^c} \subset \mathbf{PC^\infty Rings^c}$, so $\Pi^{\mathbf{C^\infty Rings^c}}_{\mathbf{PC^\infty Rings^c}} \circ \boldsymbol{F} \cong \boldsymbol{F}$, and $\Pi^{\mathbf{C^\infty Rings^c}}_{\mathbf{PC^\infty Rings^c}}(\mathfrak{C}) = \varinjlim \boldsymbol{F}$ in $\mathbf{C^\infty Rings^c}$. The same argument works for $\mathbf{C^\infty Rings^c_{in}}$.

For (c), $\Pi^{\mathbf{C^\infty Rings}}_{\mathbf{C^\infty Rings^c}}$ has a left adjoint by Theorem 4.12(c) and Proposition 4.19, as $\Pi^{\mathbf{C^\infty Rings}}_{\mathbf{C^\infty Rings^c}} = \Pi^{\mathbf{C^\infty Rings}}_{\mathbf{PC^\infty Rings^c}} \circ \mathrm{inc}$. It has a right adjoint as $F_{\geqslant 0}$ in the proof of Proposition 4.14 maps to $\mathbf{C^\infty Rings^c} \subset \mathbf{PC^\infty Rings^c}$. The same works for $\Pi^{\mathbf{C^\infty Rings^c_{in}}}_{\mathbf{C^\infty Rings^c_{in}}}$. For (d), $\Pi^{\mathbf{C^\infty Rings^c_{in}}}_{\mathbf{C^\infty Rings^c}}$ has left adjoint inc by Definition 4.16, and inc does not preserve limits by the analogue of Example 4.13. □

Remark 4.21 Although inc : $\mathbf{C^\infty Rings^c_{in}} \hookrightarrow \mathbf{C^\infty Rings^c}$ has no left adjoint, as in Theorem 4.20(d), we will show in §6.2 that inc : $\mathbf{C^\infty Sch^c_{in}} \hookrightarrow \mathbf{C^\infty Sch^c}$ *does* have a right adjoint, which is the analogous statement for spaces.

4.5 Generators, relations, and localization

We generalize Definition 2.16.

Definition 4.22 If $A, A_{\rm ex}$ are sets then by [1, Rem. 14.12] we can define the *free C^∞-ring with corners* $\mathfrak{F}^{A,A_{\rm ex}} = (\mathfrak{F}^{A,A_{\rm ex}}, \mathfrak{F}^{A,A_{\rm ex}}_{\rm ex})$ *generated by* $(A, A_{\rm ex})$. We may think of $\mathfrak{F}^{A,A_{\rm ex}}$ as $C^\infty(\mathbb{R}^A \times [0,\infty)^{A_{\rm ex}})$, where $\mathbb{R}^A = \{(x_a)_{a\in A} : x_a \in \mathbb{R}\}$ and $[0,\infty)^{A_{\rm ex}} = \{(y_{a'})_{a'\in A_{\rm ex}} : y_{a'} \in [0,\infty)\}$. Explicitly, we define $\mathfrak{F}^{A,A_{\rm ex}}$ to be the set of maps $c : \mathbb{R}^A \times [0,\infty)^{A_{\rm ex}} \to \mathbb{R}$ which depend smoothly on only finitely many variables $x_a, y_{a'}$, and $\mathfrak{F}^{A,A_{\rm ex}}_{\rm ex}$ to be the set of maps $c' : \mathbb{R}^A \times [0,\infty)^{A_{\rm ex}} \to [0,\infty)$ which depend smoothly on only finitely many variables $x_a, y_{a'}$, and operations Φ_f, Ψ_g are defined as in (4.10)–(4.11). Regarding $x_a : \mathbb{R}^A \to \mathbb{R}$ and $y_{a'} : [0,\infty)^{A_{\rm ex}} \to [0,\infty)$ as functions for $a \in A$, $a' \in A_{\rm ex}$, we have $x_a \in \mathfrak{F}^{A,A_{\rm ex}}$ and $y_{a'} \in \mathfrak{F}^{A,A_{\rm ex}}_{\rm ex}$, and we call $x_a, y_{a'}$ the *generators* of $\mathfrak{F}^{A,A_{\rm ex}}$.

Then $\mathfrak{F}^{A,A_{\rm ex}}$ has the property that if $\mathfrak{C} = (\mathfrak{C}, \mathfrak{C}_{\rm ex})$ is any (pre) C^∞-ring with corners then a choice of maps $\alpha : A \to \mathfrak{C}$ and $\alpha_{\rm ex} : A_{\rm ex} \to \mathfrak{C}_{\rm ex}$ uniquely determine a morphism $\phi : \mathfrak{F}^{A,A_{\rm ex}} \to \mathfrak{C}$ with $\phi(x_a) = \alpha(a)$ for $a \in A$ and $\phi_{\rm ex}(y_{a'}) = \alpha_{\rm ex}(a')$ for $a' \in A_{\rm ex}$. We write $\mathfrak{F}^{A,A_{\rm ex}} = \mathfrak{F}^{m,n}$ when $A = \{1,\ldots,m\}$ and $A_{\rm ex} = \{1,\ldots,n\}$, and then $\mathfrak{F}^{m,n} \cong C^\infty(\mathbb{R}^m \times [0,\infty)^n) = C^\infty(\mathbb{R}^{m+n}_n)$.

The analogue of all this also holds in $\mathbf{C^\infty Rings^c_{in}}$, with the same objects $\mathfrak{F}^{m,n}$ and $\mathfrak{F}^{A,A_{\rm in}}$, which are interior C^∞-rings with corners, and the difference that (interior) morphisms $\mathfrak{F}^{m,n} \to \mathfrak{C}$ in $\mathbf{C^\infty Rings^c_{in}}$ or $\mathbf{PC^\infty Rings^c_{in}}$ are uniquely determined by elements $c_1,\ldots,c_m \in \mathfrak{C}$ and $c'_1,\ldots,c'_n \in \mathfrak{C}_{\rm in}$ (rather than $c'_1,\ldots,c'_n \in \mathfrak{C}_{\rm ex}$), and similarly for $\mathfrak{F}^{A,A_{\rm in}}$ with $\alpha_{\rm in} : A_{\rm in} \to \mathfrak{C}_{\rm in}$.

As for Proposition 2.17, every object in $\mathbf{C^\infty Rings^c}, \mathbf{C^\infty Rings^c_{in}}$ can be built out of free C^∞-rings with corners, in a certain sense.

Proposition 4.23 **(a)** *Every object $\mathfrak{C}$ in $\mathbf{C^\infty Rings^c}$ admits a surjective morphism $\phi : \mathfrak{F}^{A,A_{\rm ex}} \to \mathfrak{C}$ from some free C^∞-ring with corners $\mathfrak{F}^{A,A_{\rm ex}}$.*

(b) *Every object $\mathfrak{C}$ in $\mathbf{C^\infty Rings^c}$ fits into a **coequalizer diagram***

$$\mathfrak{F}^{B,B_{\rm ex}} \overset{\alpha}{\underset{\beta}{\rightrightarrows}} \mathfrak{F}^{A,A_{\rm ex}} \xrightarrow{\phi} \mathfrak{C}, \tag{4.13}$$

that is, $\mathfrak{C}$ is the colimit of the diagram $\mathfrak{F}^{B,B_{\rm ex}} \rightrightarrows \mathfrak{F}^{A,A_{\rm ex}}$ in $\mathbf{C^\infty Rings^c}$, where $\phi : \mathfrak{F}^{A,A_{\rm ex}} \to \mathfrak{C}$ is automatically surjective.

The analogues of **(a)** *and* **(b)** *also hold in $\mathbf{C^\infty Rings^c_{in}}$.*

Proof The analogue in $\mathbf{PC^\infty Rings^c}, \mathbf{PC^\infty Rings^c_{in}}$ holds by Adámek et al. [1, Props. 11.26, 11.28, 11.30, Cor. 11.33, and Rem. 14.14], by general facts about algebraic theories. The result for $\mathbf{C^\infty Rings^c}, \mathbf{C^\infty Rings^c_{in}}$ follows by applying the reflection functors $\Pi^{\mathbf{C^\infty Rings^c}}_{\mathbf{PC^\infty Rings^c}}, \Pi^{\mathbf{C^\infty Rings^c_{in}}}_{\mathbf{PC^\infty Rings^c_{in}}}$ of Proposition 4.19. □

Definition 4.24 Let $\mathfrak{C}$ be a C^∞-ring with corners, and $A, A_{\rm ex}$ be sets. We will write $\mathfrak{C}(x_a : a \in A)[y_{a'} : a' \in A_{\rm ex}]$ for the C^∞-ring with corners obtained by adding extra generators x_a for $a \in A$ of type $\mathbb{R}$ and $y_{a'}$ for $a' \in A_{\rm ex}$ of type $[0,\infty)$ to $\mathfrak{C}$. That is, by definition

$$\mathfrak{C}(x_a : a \in A)[y_{a'} : a' \in A_{\rm ex}] := \mathfrak{C} \otimes_\infty \mathfrak{F}^{A,A_{\rm ex}}, \tag{4.14}$$

where $\mathfrak{F}^{A,A_{\rm ex}}$ is the free C^∞-ring with corners from Definition 4.22, with generators x_a for $a \in A$ of type $\mathbb{R}$ and $y_{a'}$ for $a' \in A_{\rm ex}$ of type $[0,\infty)$, and $\otimes_\infty$ is the coproduct in $\mathbf{C^\infty Rings^c}$. As coproducts are a type of colimit, Theorem 4.20(b) implies that $\mathfrak{C}(x_a : a \in A)[y_{a'} : a' \in A_{\rm ex}]$ is well defined. Since $\mathfrak{F}^{A,A_{\rm ex}}$ is interior, if $\mathfrak{C}$ is interior then (4.14) is a coproduct in both $\mathbf{C^\infty Rings^c}$ and $\mathbf{C^\infty Rings^c_{in}}$, so $\mathfrak{C}(x_a : a \in A)[y_{a'} : a' \in A_{\rm ex}]$ is interior by Theorem 4.20(d).

By properties of coproducts and free C^∞-rings with corners, morphisms $\phi : \mathfrak{C}(x_a : a \in A)[y_{a'} : a' \in A_{\rm ex}] \to \mathfrak{D}$ in $\mathbf{C^\infty Rings^c}$ are uniquely determined by a morphism $\psi : \mathfrak{C} \to \mathfrak{D}$ and maps $\alpha : A \to \mathfrak{D}$, $\alpha_{\rm ex} : A_{\rm ex} \to \mathfrak{D}_{\rm ex}$. If $\mathfrak{C}, \mathfrak{D}, \psi$ are interior and $\alpha_{\rm ex}(A_{\rm ex}) \subseteq \mathfrak{D}_{\rm in}$ then ϕ is interior.

Next suppose $B, B_{\rm ex}$ are sets and $f_b \in \mathfrak{C}$ for $b \in B$, $g_{b'}, h_{b'} \in \mathfrak{C}_{\rm ex}$ for $b' \in B_{\rm ex}$. We will write $\mathfrak{C}/(f_b = 0 : b \in B)[g_{b'} = h_{b'} : b' \in B_{\rm ex}]$ for the C^∞-ring with corners obtained by imposing relations $f_b = 0$, $b \in B$ in $\mathfrak{C}$ of type $\mathbb{R}$, and $g_{b'} = h_{b'}$, $b' \in B_{\rm ex}$ in $\mathfrak{C}_{\rm ex}$ of type $[0,\infty)$. That is, we have a coequalizer diagram

$$\mathfrak{F}^{B,B_{\rm ex}} \underset{\beta}{\overset{\alpha}{\rightrightarrows}} \mathfrak{C} \xrightarrow{\pi} \mathfrak{C}/(f_b{=}0{:}b{\in}B)[g_{b'}{=}h_{b'}{:}b'{\in}B_{\rm ex}], \tag{4.15}$$

where α, β are determined uniquely by $\alpha(x_b) = f_b$, $\alpha_{\rm ex}(y_{b'}) = g_{b'}$, $\beta(x_b) = 0$, $\beta_{\rm ex}(y_{b'}) = h_{b'}$ for all $b \in B$ and $b' \in B_{\rm ex}$. As coequalizers are a type of colimit, Theorem 4.20(b) shows that $\mathfrak{C}/(f_b = 0 : b \in B)[g_{b'} = h_{b'} : b' \in B_{\rm ex}]$ is well defined. If $\mathfrak{C}$ is interior and $g_{b'}, h_{b'} \in \mathfrak{C}_{\rm in}$ for all $b' \in B_{\rm ex}$ (that is, $g_{b'}, h_{b'} \neq 0_{\mathfrak{C}_{\rm ex}}$) then (4.14) is also a coequalizer in $\mathbf{C^\infty Rings^c_{in}}$, so Theorem

4.20(d) implies that $\mathfrak{C}/(f_b = 0 : b \in B)[g_{b'} = h_{b'} : b' \in B_{\rm ex}]$ and $\boldsymbol{\pi}$ are interior.

Note that round brackets $(\cdots)$ denote generators or relations of type $\mathbb{R}$, and square brackets $[\cdots]$ denote generators or relations of type $[0,\infty)$. If we add generators or relations of only one type, we use only these brackets.

We construct two explicit examples of quotients in $\mathbf{C^\infty Rings^c}$.

Example 4.25 **(a)** Say we wish to quotient a C^∞-ring with corners $(\mathfrak{C},\mathfrak{C}_{\rm ex})$ by an ideal I in $\mathfrak{C}$. Quotienting the C^∞-ring by the ideal will result in additional relations on the monoid. While this quotient $(\mathfrak{D},\mathfrak{D}_{\rm ex})$ is the coequalizer of a diagram such as (4.15), it is also equivalent to the following construction.

The quotient is a C^∞-ring with corners $(\mathfrak{D},\mathfrak{D}_{\rm ex})$ with a morphism $\boldsymbol{\pi} = (\pi,\pi_{\rm ex}) : (\mathfrak{C},\mathfrak{C}_{\rm ex}) \to (\mathfrak{D},\mathfrak{D}_{\rm ex})$ such that I is contained in the kernel of π, and is universal with respect to this property. That is, if $(\mathfrak{E},\mathfrak{E}_{\rm ex})$ is another C^∞-ring with corners with morphism $\boldsymbol{\pi}' = (\pi',\pi'_{\rm ex}) : (\mathfrak{C},\mathfrak{C}_{\rm ex}) \to (\mathfrak{E},\mathfrak{E}_{\rm ex})$ with I contained in the kernel of π', then there is a unique morphism $\boldsymbol{p} : (\mathfrak{D},\mathfrak{D}_{\rm ex}) \to (\mathfrak{E},\mathfrak{E}_{\rm ex})$ such that $\boldsymbol{p}\circ\boldsymbol{\pi} = \boldsymbol{\pi}'$.

As a coequalizer is a colimit, by Theorem 4.20(c) we have $\mathfrak{D} \cong \mathfrak{C}/I$, the quotient in C^∞-rings. For the monoid, we require that smooth $f : \mathbb{R} \to [0,\infty)$ give well-defined operations $\Psi_f : \mathfrak{D} \to \mathfrak{D}_{\rm ex}$. This means we require that if $a - b \in I$, then $\Psi_f(a) \sim \Psi_f(b) \in \mathfrak{C}_{\rm ex}$, and this needs to generate a monoid equivalence relation on $\mathfrak{C}_{\rm ex}$, so that a quotient by this relation is well defined. If $f : \mathbb{R} \to [0,\infty)$ is identically zero, this follows. If $f : \mathbb{R} \to [0,\infty)$ is non-zero and smooth this means that f is positive, and hence that $\log f$ is well defined with $f = \exp\circ\log f$. By Hadamard's Lemma, if $a - b \in I$, then $\Phi_g(a) - \Phi_g(b) \in I$, for all $g : \mathbb{R} \to \mathbb{R}$ smooth, and therefore $\Phi_{\log f}(a) - \Phi_{\log f}(b) \in I$. Hence in $\mathfrak{C}_{\rm ex}$ we only require that if $a - b \in I$, then $\Psi_{\exp}(a) \sim \Psi_{\exp}(b)$ in $\mathfrak{C}_{\rm ex}$. The monoid equivalence relation that this generates is equivalent to $c_1' \sim_I c_2' \in \mathfrak{C}_{\rm ex}$ if there exists $d \in I$ such that $c_1' = \Psi_{\exp}(d)\cdot c_2'$.

We claim that $(\mathfrak{C}/I, \mathfrak{C}_{\rm ex}/{\sim_I})$ is the required C^∞-ring with corners. If $f : [0,\infty) \to \mathbb{R}$ is smooth, and $c_1' \sim_I c_2' \in \mathfrak{C}_{\rm ex}$, then $\Phi_f(c_1') = \Phi_f(\Psi_{\exp}(d)c_2')$ in $\mathfrak{C}$ for some $d \in I$. Applying Hadamard's Lemma twice, we have

$$\begin{aligned}\Phi_f(c_1') - \Phi_f(c_2') &= \Phi_{(x-y)g(x,y)}(\Psi_{\exp}(d)c_2', c_2')\\ &= \Phi_i(c_2')(\Phi_{\exp}(d) - 1)\Phi_g(\Psi_{\exp}(d)c_2', c_2')\\ &= \Phi_i(c_2')(d-0)\Phi_{h(x,y)}(d,0)\Phi_g(\Psi_{\exp}(d)c_2', c_2'),\end{aligned}$$

for smooth maps $g, h : \mathbb{R}^2 \to \mathbb{R}$. As $d \in I$, then $\Phi_f(c_1') - \Phi_f(c_2') \in I$, and Φ_f is well defined. A similar proof shows all the C^∞-operations are well defined, and so $(\mathfrak{C}/I, \mathfrak{C}_{\rm ex}/{\sim_I})$ is a pre C^∞-ring with corners.

We must show that $\Psi_{\exp} : \mathfrak{C}/I \to \mathfrak{C}_{\rm ex}/{\sim_I}$ has image equal to (not just contained in) the invertible elements $(\mathfrak{C}_{\rm ex}/{\sim_I})^\times$. Say $[c_1'] \in (\mathfrak{C}_{\rm ex}/{\sim_I})^\times$, then there is $[c_2'] \in (\mathfrak{C}_{\rm ex}/{\sim_I})^\times$ such that $[c_1'][c_2'] = [c'd'] = [1]$. So there is $d \in I$ such that $c_1'c_2' = \Psi_{\exp}(d)$. However, $\Psi_{\exp}(d)$ is invertible, so each of c_1', c_2' must be invertible in $\mathfrak{C}_{\rm ex}$, and using that $\Psi_{\exp}$ is surjective onto invertible elements in $\mathfrak{C}_{\rm ex}^\times$ gives the result. Thus $(\mathfrak{C}/I, \mathfrak{C}_{\rm ex}/{\sim_I})$ is a C^∞-ring with corners. The quotient morphisms $\mathfrak{C} \to \mathfrak{C}/I$ and $\mathfrak{C}_{\rm ex} \to \mathfrak{C}_{\rm ex}/{\sim_I}$ give the required map $\boldsymbol{\pi}$.

To show that this satisfies the required universal property of either (4.15) or the beginning of the example, let $(\mathfrak{E}, \mathfrak{E}_{\rm ex})$ be another C^∞-ring with corners with a morphism $\boldsymbol{\pi}' = (\pi', \pi_{\rm ex}') : (\mathfrak{C}, \mathfrak{C}_{\rm ex}) \to (\mathfrak{E}, \mathfrak{E}_{\rm ex})$ such that $I \subset \operatorname{Ker} \pi'$. Then the unique morphism $\boldsymbol{p} : (\mathfrak{D}, \mathfrak{D}_{\rm ex}) \to (\mathfrak{E}, \mathfrak{E}_{\rm ex})$ is defined by $\boldsymbol{p}([c],[c']) = [\pi'(c), \pi'(c')]$. The requirement that $(\mathfrak{E}, \mathfrak{E}_{\rm ex})$ factors through each diagram shows that this morphism is well defined and unique, giving the result.

(b) Say we wish to quotient a C^∞-ring with corners $(\mathfrak{C}, \mathfrak{C}_{\rm ex})$ by an ideal P in the monoid $\mathfrak{C}_{\rm ex}$. By this we mean quotient $\mathfrak{C}_{\rm ex}$ by the equivalence relation $c_1' \sim c_2'$ if $c_1' = c_2'$ or $c_1', c_2' \in P$. This is known as a Rees quotient of semigroups; see Rees [81, p. 389]. Quotienting the monoid by this ideal will result in additional relations on both the monoid and the C^∞-ring, which we will now make explicit. While this quotient $(\mathfrak{D}, \mathfrak{D}_{\rm ex})$ is the coequalizer of a diagram such as (4.15), it is also equivalent to the following construction.

The quotient is a C^∞-ring with corners $(\mathfrak{D}, \mathfrak{D}_{\rm ex})$ with a morphism $\boldsymbol{\pi} = (\pi, \pi_{\rm ex}) : (\mathfrak{C}, \mathfrak{C}_{\rm ex}) \to (\mathfrak{D}, \mathfrak{D}_{\rm ex})$ such that P is contained in the kernel of $\pi_{\rm ex}$, and is universal with respect to this property. That is, if $(\mathfrak{E}, \mathfrak{E}_{\rm ex})$ is another C^∞-ring with corners with morphism $\boldsymbol{\pi}' = (\pi', \pi_{\rm ex}') : (\mathfrak{C}, \mathfrak{C}_{\rm ex}) \to (\mathfrak{E}, \mathfrak{E}_{\rm ex})$ with P contained in the kernel of $\pi_{\rm ex}'$, then there is a unique morphism $\boldsymbol{p} : (\mathfrak{D}, \mathfrak{D}_{\rm ex}) \to (\mathfrak{E}, \mathfrak{E}_{\rm ex})$ such that $\boldsymbol{p} \circ \boldsymbol{\pi} = \boldsymbol{\pi}'$.

Similarly to part (a), we begin by quotienting $\mathfrak{C}_{\rm ex}$ by P, and then require that the C^∞-operations are well defined. As all smooth $f : [0, \infty) \to \mathbb{R}$ are equal to $\hat{f} \circ i : [0, \infty) \to \mathbb{R}$ for a smooth function $\hat{f} : \mathbb{R} \to \mathbb{R}$, we need only require that if $c_1' \sim c_2'$, then $\Phi_i(c_1') \sim \Phi_i(c_2')$. This generates a C^∞-ring equivalence relation on the C^∞-ring $\mathfrak{C}$; such a C^∞-ring equivalence relation is the same data as giving an ideal $I \subset \mathfrak{C}$ such that $c_1 \sim c_2 \in \mathfrak{C}$ whenever $c_1 - c_2 \in I$. Here, this equivalence relation will be given by the ideal $\langle \Phi_i(P) \rangle$, that is, the ideal generated by the image of P under Φ_i. Quotienting $\mathfrak{C}$ by this ideal generates a further condition on the monoid $\mathfrak{C}_{\rm ex}$, as in part (a), that is $c_1' \sim c_2'$ if there is $d \in \langle \Phi_i(P) \rangle$ such that $c_1' = \Psi_{\exp}(d) c_2'$.

The claim then is that we may take $(\mathfrak{D}, \mathfrak{D}_{\rm ex})$ equal to $\big(\mathfrak{C}/\langle \Phi_i(P) \rangle, \mathfrak{C}_{\rm ex}/{\sim_P}\big)$ where $c_1' \sim_P c_2'$ if either $c_1', c_2' \in P$ or there is $d \in \langle \Phi_i(P) \rangle$ such that $c_1' = \Psi_{\exp}(d)(c_2')$. Similar applications of Hadamard's Lemma as in (a) show that

$(\mathfrak{D},\mathfrak{D}_{\rm ex})$ is a pre C^∞-ring with corners, and similar discussions show it is a C^∞-ring with corners, and is isomorphic to the quotient. We will use the notation

$$(\mathfrak{D},\mathfrak{D}_{\rm ex})=(\mathfrak{C},\mathfrak{C}_{\rm ex})/{\sim_P}=(\mathfrak{C}/\langle\Phi_i(P)\rangle,\mathfrak{C}_{\rm ex}/{\sim_P})=(\mathfrak{C}/{\sim_P},\mathfrak{C}_{\rm ex}/{\sim_P})$$

to refer to this quotient later.

Definition 4.26 Let $\boldsymbol{\mathfrak{C}}=(\mathfrak{C},\mathfrak{C}_{\rm ex})$ be a C^∞-ring with corners, and let $A\subseteq\mathfrak{C}$, $A_{\rm ex}\subseteq\mathfrak{C}_{\rm ex}$ be subsets. A *localization* $\boldsymbol{\mathfrak{C}}(a^{-1}:a\in A)[a'^{-1}:a'\in A_{\rm ex}]$ of $\boldsymbol{\mathfrak{C}}$ at $(A,A_{\rm ex})$ is a C^∞-ring with corners $\boldsymbol{\mathfrak{D}}=\boldsymbol{\mathfrak{C}}(a^{-1}:a\in A)[a'^{-1}:a'\in A_{\rm ex}]$ and a morphism $\boldsymbol{\pi}:\boldsymbol{\mathfrak{C}}\to\boldsymbol{\mathfrak{D}}$ such that $\pi(a)$ is invertible in $\mathfrak{D}$ for all $a\in A$ and $\pi_{\rm ex}(a')$ is invertible in $\mathfrak{D}_{\rm ex}$ for all $a'\in A_{\rm ex}$, with the universal property that if $\boldsymbol{\mathfrak{E}}=(\mathfrak{E},\mathfrak{E}_{\rm ex})$ is a C^∞-ring with corners and $\boldsymbol{\phi}:\boldsymbol{\mathfrak{C}}\to\boldsymbol{\mathfrak{E}}$ a morphism with $\phi(a)$ invertible in $\mathfrak{E}$ for all $a\in A$ and $\phi_{\rm ex}(a')$ invertible in $\mathfrak{E}_{\rm ex}$ for all $a'\in A_{\rm ex}$, then there is a unique morphism $\boldsymbol{\psi}:\boldsymbol{\mathfrak{D}}\to\boldsymbol{\mathfrak{E}}$ with $\boldsymbol{\phi}=\boldsymbol{\psi}\circ\boldsymbol{\pi}$.

Localizations $\boldsymbol{\mathfrak{C}}(a^{-1}:a\in A)[a'^{-1}:a'\in A_{\rm ex}]$ always exist, and are unique up to canonical isomorphism. In the notation of Definition 4.24 we may write

$$\begin{aligned}&\boldsymbol{\mathfrak{C}}(a^{-1}:a\in A)[a'^{-1}:a'\in A_{\rm ex}]=\\&\bigl(\boldsymbol{\mathfrak{C}}(x_a:a\in A)[y_{a'}:a'\in A_{\rm ex}]\bigr)/\bigl(a\cdot x_a=1:a\in A\bigr)\bigl[a'\cdot y_{a'}=1:a'\in A_{\rm ex}\bigr].\end{aligned}\tag{4.16}$$

That is, we add an extra generator x_a of type $\mathbb{R}$ and an extra relation $a\cdot x_a=1$ of type $\mathbb{R}$ for each $a\in A$, so that $x_a=a^{-1}$, and similarly for each $a'\in A_{\rm ex}$.

If $\boldsymbol{\mathfrak{C}}$ is interior and $A_{\rm ex}\subseteq\mathfrak{C}_{\rm in}$ then $\boldsymbol{\mathfrak{C}}(a^{-1}:a\in A)[a'^{-1}:a'\in A_{\rm ex}]$ makes sense and exists in $\mathbf{C^\infty Rings^c_{in}}$ as well as in $\mathbf{C^\infty Rings^c}$, and Theorem 4.20(d) implies that the two localizations are the same.

The next lemma follows from Theorem 4.20(c).

Lemma 4.27 *Let $\boldsymbol{\mathfrak{C}}=(\mathfrak{C},\mathfrak{C}_{\rm ex})$ be a C^∞-ring with corners and $c\in\mathfrak{C}$, and write $\boldsymbol{\mathfrak{C}}(c^{-1})=(\mathfrak{D},\mathfrak{D}_{\rm ex})$. Then $\mathfrak{D}\cong\mathfrak{C}(c^{-1})$, the localization of the C^∞-ring.*

4.6 Local C^∞-rings with corners

We now define local C^∞-rings with corners.

Definition 4.28 Let $\boldsymbol{\mathfrak{C}}=(\mathfrak{C},\mathfrak{C}_{\rm ex})$ be a C^∞-ring with corners. We call $\boldsymbol{\mathfrak{C}}$ *local* if there exists a C^∞-ring morphism $\pi:\mathfrak{C}\to\mathbb{R}$ such that each $c\in\mathfrak{C}$ is invertible in $\mathfrak{C}$ if and only if $\pi(c)\neq 0$ in $\mathbb{R}$, and each $c'\in\mathfrak{C}_{\rm ex}$ is invertible in $\mathfrak{C}_{\rm ex}$ if and only if $\pi\circ\Phi_i(c')\neq 0$ in $\mathbb{R}$. Note that if $\boldsymbol{\mathfrak{C}}$ is local then $\pi:\mathfrak{C}\to\mathbb{R}$ is determined uniquely by $\operatorname{Ker}\pi=\{c\in\mathfrak{C}:c\text{ is not invertible}\}$, and $\mathfrak{C}$ is a local C^∞-ring.

Write $\mathbf{C^\infty Rings^c_{lo}} \subset \mathbf{C^\infty Rings^c}$ and $\mathbf{C^\infty Rings^c_{in,lo}} \subset \mathbf{C^\infty Rings^c_{in}}$ for the full subcategories of (interior) local C^∞-rings with corners.

Proposition 4.29 *$\mathbf{C^\infty Rings^c_{lo}}$ is closed under colimits in $\mathbf{C^\infty Rings^c}$. Thus all small colimits exist in $\mathbf{C^\infty Rings^c_{lo}}$ and $\mathbf{C^\infty Rings^c_{in,lo}}$.*

Proof Let J be a category and $\boldsymbol{F} = (F, F_{\rm ex}) : J \to \mathbf{C^\infty Rings^c_{lo}}$ a functor, and suppose a colimit $\boldsymbol{\mathfrak{C}} = (\mathfrak{C}, \mathfrak{C}_{\rm ex}) = \varinjlim \boldsymbol{F}$ exists in $\mathbf{C^\infty Rings^c}$, with projections $\phi_j : \boldsymbol{F}(j) \to \boldsymbol{\mathfrak{C}}$. Then $\mathfrak{C} = \varinjlim F$ in $\mathbf{C^\infty Rings}$ by Theorem 4.20(c). For each $j \in J$ we have a unique C^∞-ring morphism $\pi_j : F(j) \to \mathbb{R}$ as $F(j)$ is local, with $\pi_j = \pi_k \circ F(\alpha)$ for all morphisms $\alpha : j \to k$ in J. Hence by properties of colimits there is a unique morphism $\pi : \mathfrak{C} \to \mathbb{R}$ with $\pi_j = \pi \circ \phi_j$ for all $j \in J$.

Suppose $c \in \mathfrak{C}$ with $\pi(c) \neq 0$. Then $c = \phi_j(f)$ for some $j \in J$ and $f \in F(j)$, with $\pi_j(f) = \pi \circ \phi_j(f) = \pi(c) \neq 0$. Hence as $\boldsymbol{F}(j)$ is local, we have an inverse $f^{-1} \in F(j)$, and $\phi_j(f^{-1}) = c^{-1}$ is the inverse of c in $\mathfrak{C}$. The same argument shows that if $c' \in \mathfrak{C}_{\rm ex}$ with $\pi \circ \Phi_i(c') \neq 0$ then c' is invertible in $\mathfrak{C}_{\rm ex}$. So $\boldsymbol{\mathfrak{C}}$ is local, and $\mathbf{C^\infty Rings^c_{lo}}$ is closed under colimits in $\mathbf{C^\infty Rings^c}$. The last part holds by Theorem 4.20(b), (d). □

Lemma 4.30 *If $\boldsymbol{\mathfrak{C}} = (\mathfrak{C}, \mathfrak{C}_{\rm ex})$ is a (pre) C^∞-ring with corners and $x : \mathfrak{C} \to \mathbb{R}$ a C^∞-ring morphism, then we have a morphism of (pre) C^∞-rings with corners $(x, x_{\rm ex}) : (\mathfrak{C}, \mathfrak{C}_{\rm ex}) \to (\mathbb{R}, [0, \infty))$, where $x_{\rm ex}(c') = x \circ \Phi_i(c')$ for $c' \in \mathfrak{C}_{\rm ex}$.*

Proof Let $\boldsymbol{\mathfrak{C}} = (\mathfrak{C}, \mathfrak{C}_{\rm ex})$ and $x : \mathfrak{C} \to \mathbb{R}$ be as in the statement. Take $c' \in \mathfrak{C}_{\rm ex}$. To show $x_{\rm ex} = x \circ \Phi_i : \mathfrak{C}_{\rm ex} \to [0, \infty)$ is well defined, assume for a contradiction that $x \circ \Phi_i(c') = \epsilon < 0 \in \mathbb{R}$. Let $f : \mathbb{R} \to \mathbb{R}$ be a smooth function such that f is the identity on $[0, \infty)$ and it is zero on $(-\infty, \epsilon/2)$. Then $f \circ i = i$ for $i : [0, \infty) \to \mathbb{R}$ the inclusion. So we have $0 > \epsilon = x \circ \Phi_i(c') = x \circ \Phi_f \circ \Phi_i(c') = f(x \circ \Phi_i(c')) = f(\epsilon) = 0$, and $x_{\rm ex} = x \circ \Phi_i : \mathfrak{C}_{\rm ex} \to [0, \infty)$ is well defined.

For $(x, x_{\rm ex})$ to be a morphism of pre C^∞-rings with corners, it must respect the C^∞-operations. For example, let $f : [0, \infty) \to [0, \infty)$ be smooth, then there is a smooth function $g : \mathbb{R} \to \mathbb{R}$ that extends f, so that $g \circ i = i \circ f$. Then $x_{\rm ex}(\Psi_f(c')) = x \circ \Phi_i(\Psi_f(c')) = \Phi_g(x \circ \Phi_i(c')) = \Phi_g(x_{\rm ex}(c'))$ as required. A similar proof holds for the other C^∞-operations. □

Definition 4.31 Let $\boldsymbol{\mathfrak{C}} = (\mathfrak{C}, \mathfrak{C}_{\rm ex})$ be a C^∞-ring with corners. An $\mathbb{R}$-*point* x of $\boldsymbol{\mathfrak{C}}$ is a C^∞-ring morphism (or equivalently, an $\mathbb{R}$-algebra morphism) $x : \mathfrak{C} \to \mathbb{R}$. Define $\boldsymbol{\mathfrak{C}}_x$ to be the localization

$$\boldsymbol{\mathfrak{C}}_x = \boldsymbol{\mathfrak{C}}\big(c^{-1} : c \in \mathfrak{C},\ x(c) \neq 0\big)\big[c'^{-1} : c' \in \mathfrak{C}_{\rm ex},\ x \circ \Phi_i(c') \neq 0\big], \quad (4.17)$$

with projection $\boldsymbol{\pi}_x : \boldsymbol{\mathfrak{C}} \to \boldsymbol{\mathfrak{C}}_x$. If $\boldsymbol{\mathfrak{C}}$ is interior then $\boldsymbol{\mathfrak{C}}_x$ is interior by Definition 4.26. Theorem 4.32(a) below shows $\boldsymbol{\mathfrak{C}}_x$ is local. Theorem 4.32(c) is the analogue of Proposition 2.24. The point of the proof is to give an alternative construction of $\boldsymbol{\mathfrak{C}}_x$ from $\boldsymbol{\mathfrak{C}}$ by imposing relations, but adding no new generators.

Theorem 4.32 *Let $\boldsymbol{\pi}_x : \boldsymbol{\mathfrak{C}} \to \boldsymbol{\mathfrak{C}}_x$ be as in Definition* 4.31. *Then:*

(a) $\boldsymbol{\mathfrak{C}}_x$ *is a local C^∞-ring with corners;*

(b) $\boldsymbol{\mathfrak{C}}_x = (\mathfrak{C}_x, \mathfrak{C}_{x,\mathrm{ex}})$ *and $\boldsymbol{\pi}_x = (\pi_x, \pi_{x,\mathrm{ex}})$, where $\pi_x : \mathfrak{C} \to \mathfrak{C}_x$ is the local C^∞-ring associated to $x : \mathfrak{C} \to \mathbb{R}$ in Definition* 2.22;

(c) $\pi_x : \mathfrak{C} \to \mathfrak{C}_x$ *and $\pi_{x,\mathrm{ex}} : \mathfrak{C}_{\mathrm{ex}} \to \mathfrak{C}_{x,\mathrm{ex}}$ are surjective.*

Proof Proposition 2.24 says that the local C^∞-ring $\mathfrak{C}_x$ is $\mathfrak{C}/I$, for $I \subset \mathfrak{C}$ the ideal defined in (2.4), with $\pi_x : \mathfrak{C} \to \mathfrak{C}_x$ the projection $\mathfrak{C} \to \mathfrak{C}/I$. Define

$$\mathfrak{D} = \mathfrak{C}_x = \mathfrak{C}/I \quad \text{and} \quad \mathfrak{D}_{\mathrm{ex}} = \mathfrak{C}_{\mathrm{ex}}/\sim,$$

where $\sim$ is the equivalence relation on $\mathfrak{C}_{\mathrm{ex}}$ given by $c' \sim c''$ if there exists $i \in I$ with $c'' = \Psi_{\exp}(i) \cdot c'$, and $\phi : \mathfrak{C} \to \mathfrak{D} = \mathfrak{C}/I$, $\phi_{\mathrm{ex}} : \mathfrak{C}_{\mathrm{ex}} \to \mathfrak{D}_{\mathrm{ex}} = \mathfrak{C}_{\mathrm{ex}}/\sim$ to be the natural surjective projections. Let $f : \mathbb{R}^m \times [0,\infty)^n \to \mathbb{R}$ be smooth and $g : \mathbb{R}^m \times [0,\infty)^n \to [0,\infty)$ be exterior, and write Φ_f, Ψ_g for the operations in $\boldsymbol{\mathfrak{C}}$. Then it is not difficult to show there exist unique maps Φ'_f, Ψ'_g making the following diagrams commute:

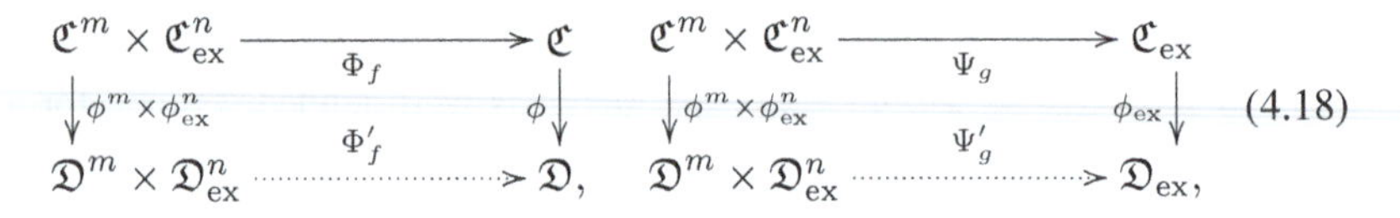

$$\begin{array}{ccc} \mathfrak{C}^m \times \mathfrak{C}^n_{\mathrm{ex}} & \xrightarrow{\Phi_f} & \mathfrak{C} \\ \downarrow \phi^m \times \phi^n_{\mathrm{ex}} & & \phi \downarrow \\ \mathfrak{D}^m \times \mathfrak{D}^n_{\mathrm{ex}} & \xrightarrow{\Phi'_f} & \mathfrak{D}, \end{array} \qquad \begin{array}{ccc} \mathfrak{C}^m \times \mathfrak{C}^n_{\mathrm{ex}} & \xrightarrow{\Psi_g} & \mathfrak{C}_{\mathrm{ex}} \\ \downarrow \phi^m \times \phi^n_{\mathrm{ex}} & & \phi_{\mathrm{ex}} \downarrow \\ \mathfrak{D}^m \times \mathfrak{D}^n_{\mathrm{ex}} & \xrightarrow{\Psi'_g} & \mathfrak{D}_{\mathrm{ex}}, \end{array} \tag{4.18}$$

and these Φ'_f, Ψ'_g make $\boldsymbol{\mathfrak{D}} = (\mathfrak{D}, \mathfrak{D}_{\mathrm{ex}})$ into a C^∞-ring with corners, and $\boldsymbol{\phi} = (\phi, \phi_{\mathrm{ex}}) : \boldsymbol{\mathfrak{C}} \to \boldsymbol{\mathfrak{D}}$ into a surjective morphism.

Suppose that $\boldsymbol{\mathfrak{F}}$ is a C^∞-ring with corners and $\boldsymbol{\chi} = (\chi, \chi_{\mathrm{ex}}) : \boldsymbol{\mathfrak{C}} \to \boldsymbol{\mathfrak{F}}$ a morphism such that $\chi(c)$ is invertible in $\mathfrak{F}$ for all $c \in \mathfrak{C}$ with $x(c) \neq 0$. The definition $\mathfrak{D} = \mathfrak{C}_x = \mathfrak{C}(c^{-1} : c \in \mathfrak{C},\ x(c) \neq 0)$ in $\mathbf{C^\infty Rings}$ in Definition 2.22 implies that $\chi : \mathfrak{C} \to \mathfrak{F}$ factorizes uniquely as $\chi = \xi \circ \phi$ for $\xi : \mathfrak{D} \to \mathfrak{F}$ a morphism in $\mathbf{C^\infty Rings}$. Hence $\chi(i) = 0$ in $\mathfrak{F}$ for all $i \in I$, so $\chi_{\mathrm{ex}}(\Psi_{\exp}(i)) = 1_{\mathfrak{F}_{\mathrm{ex}}}$ for all $i \in I$. Thus if $c', c'' \in \mathfrak{C}_{\mathrm{ex}}$ with $c'' = \Psi_{\exp}(i) \cdot c'$ for $i \in I$ then $\chi_{\mathrm{ex}}(c') = \chi_{\mathrm{ex}}(c'')$. Hence χ_{ex} factorizes uniquely as $\chi_{\mathrm{ex}} = \xi_{\mathrm{ex}} \circ \phi_{\mathrm{ex}}$ for $\xi_{\mathrm{ex}} : \mathfrak{D}_{\mathrm{ex}} \to \mathfrak{F}_{\mathrm{ex}}$.

As $\boldsymbol{\chi}, \boldsymbol{\phi}$ are morphisms in $\mathbf{C^\infty Rings^c}$ with $\boldsymbol{\phi}$ surjective we see that $\boldsymbol{\xi} = (\xi, \xi_{\mathrm{ex}}) : \boldsymbol{\mathfrak{D}} \to \boldsymbol{\mathfrak{F}}$ is a morphism. Therefore $\boldsymbol{\chi} : \boldsymbol{\mathfrak{C}} \to \boldsymbol{\mathfrak{F}}$ factorizes uniquely as $\boldsymbol{\chi} = \boldsymbol{\xi} \circ \boldsymbol{\phi}$. Also $\phi(c)$ is invertible in $\mathfrak{D} = \mathfrak{C}_x$ for all $c \in \mathfrak{C}$ with $x(c) \neq 0$, by definition of $\mathfrak{C}_x$ in Definition 2.22. Therefore we have a canonical isomorphism

$$\mathfrak{D} \cong \mathfrak{C}\big(c^{-1} : c \in \mathfrak{C},\ x(c) \neq 0\big)$$

identifying $\phi : \mathfrak{C} \to \mathfrak{D}$ with the projection $\mathfrak{C} \to \mathfrak{C}\big(c^{-1} : c \in \mathfrak{C},\ x(c) \neq 0\big)$. Note that $x : \mathfrak{C} \to \mathbb{R}$ factorizes as $x = \tilde{x} \circ \phi$ for a unique morphism $\tilde{x} : \mathfrak{D} \to \mathbb{R}$.

Next define

$$\mathfrak{E} = \mathfrak{D} = \mathfrak{C}_x \quad \text{and} \quad \mathfrak{E}_{\mathrm{ex}} = \mathfrak{D}_{\mathrm{ex}} / \approx,$$

where $\approx$ is the monoidal equivalence relation on $\mathfrak{D}_{\mathrm{ex}}$ generated by the conditions that $d' \approx d''$ whenever $d', d'' \in \mathfrak{D}_{\mathrm{ex}}$ with $\Phi'_i(d') = \Phi'_i(d'')$ in $\mathfrak{D}$ and $\tilde{x} \circ \Phi'_i(d') \neq 0$. Write $\psi = \mathrm{id} : \mathfrak{D} \to \mathfrak{E}$, and let $\psi_{\mathrm{ex}} : \mathfrak{D}_{\mathrm{ex}} \to \mathfrak{E}_{\mathrm{ex}}$ be the natural surjective projection. Suppose $f : \mathbb{R}^m \times [0,\infty)^n \to \mathbb{R}$ is smooth and $g : \mathbb{R}^m \times [0,\infty)^n \to [0,\infty)$ is exterior. Then as for (4.18), we claim there are unique maps Φ''_f, Ψ''_g making the following diagrams commute:

$$\begin{array}{ccc} \mathfrak{D}^m \times \mathfrak{D}^n_{\mathrm{ex}} & \xrightarrow{\Phi'_f} & \mathfrak{D} \\ \downarrow \psi^m \times \psi^n_{\mathrm{ex}} & & \psi \downarrow \\ \mathfrak{E}^m \times \mathfrak{E}^n_{\mathrm{ex}} & \xrightarrow{\Phi''_f} & \mathfrak{E}, \end{array} \qquad \begin{array}{ccc} \mathfrak{D}^m \times \mathfrak{D}^n_{\mathrm{ex}} & \xrightarrow{\Psi'_g} & \mathfrak{D}_{\mathrm{ex}} \\ \downarrow \psi^m \times \psi^n_{\mathrm{ex}} & & \psi_{\mathrm{ex}} \downarrow \\ \mathfrak{E}^m \times \mathfrak{E}^n_{\mathrm{ex}} & \xrightarrow{\Psi''_g} & \mathfrak{E}_{\mathrm{ex}}. \end{array} \tag{4.19}$$

To see that Φ''_f in (4.19) is well defined, note that as $\Phi'_i : \mathfrak{D}_{\mathrm{ex}} \to \mathfrak{D}$ is a monoid morphism and $\approx$ is a monoidal equivalence relation generated by $d' \approx d''$ when $\Phi'_i(d') = \Phi'_i(d'')$, we have a factorization $\Phi'_i = \tilde{\Phi}'_i \circ \psi_{\mathrm{ex}}$. We may extend f to smooth $\tilde{f} : \mathbb{R}^{m+n} \to \mathbb{R}$, and then Φ'_f factorizes as

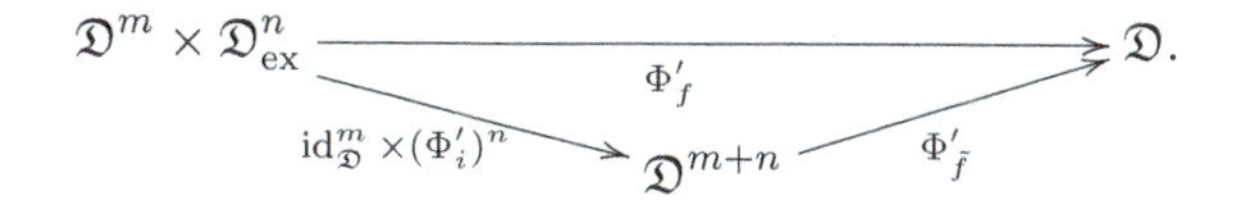

Using $\Phi'_i = \tilde{\Phi}'_i \circ \psi_{\mathrm{ex}}$, we see that Φ''_f in (4.19) exists and is unique.

For Ψ''_g, if $g = 0$ then $\Psi''_g = 0_{\mathfrak{E}_{\mathrm{ex}}} = [0_{\mathfrak{D}_{\mathrm{ex}}}]$ in (4.19). Otherwise we may write g using $a_1, \dots, a_n$ and $h : \mathbb{R}^m \times [0,\infty)^n \to \mathbb{R}$ as in (4.5), and then

$$\begin{aligned} &\Psi_{g'}(d_1, \dots, d_m, d'_1, \dots, d'_n) \\ &= (d'_1)^{a_1} \cdots (d'_n)^{a_n} \cdot \Psi'_{\exp}\big[\Phi'_h(d_1, \dots, d_m, d'_1, \dots, d'_n)\big]. \end{aligned}$$

Since $\psi_{\mathrm{ex}} : \mathfrak{D}_{\mathrm{ex}} \to \mathfrak{E}_{\mathrm{ex}}$ is a monoid morphism, as it is a quotient by a monoidal equivalence relation, we see from this and the previous argument applied to $\Phi_h(d_1, \dots, d_m, d'_1, \dots, d'_n)$ that Ψ''_g in (4.19) exists and is unique. These Φ''_f, Ψ''_g make $\boldsymbol{\mathfrak{E}} = (\mathfrak{E}, \mathfrak{E}_{\mathrm{ex}})$ into a C^∞-ring with corners, and $\boldsymbol{\psi} = (\psi, \psi_{\mathrm{ex}}) : \boldsymbol{\mathfrak{D}} \to \boldsymbol{\mathfrak{E}}$ into a surjective morphism.

We will show that there is a canonical isomorphism $\boldsymbol{\mathfrak{E}} \cong \boldsymbol{\mathfrak{C}}_x$ which identifies $\boldsymbol{\psi} \circ \boldsymbol{\phi} : \boldsymbol{\mathfrak{C}} \to \boldsymbol{\mathfrak{E}}$ with $\boldsymbol{\pi}_x : \boldsymbol{\mathfrak{C}} \to \boldsymbol{\mathfrak{C}}_x$. Firstly, suppose $c \in \mathfrak{C}$ with $x(c) \neq 0$.

Then $\psi \circ \phi(c)$ is invertible in $\mathfrak{E} = \mathfrak{C}_x$ by definition of $\mathfrak{C}_x$ in Definition 2.22. Secondly, suppose $c' \in \mathfrak{C}_{\rm ex}$ with $x \circ \Phi_i(c') \neq 0$. Set $d' = \phi_{\rm ex}(c')$. Then $\Phi'_i(d') = \phi \circ \Phi_i(c')$ is invertible in $\mathfrak{D}$. Now in the proof of Proposition 4.7, we do not actually need c' to be invertible in $\mathfrak{C}_{\rm ex}$, it is enough that $\Phi_i(c')$ is invertible in $\mathfrak{C}$. Thus this proof shows that there exists a unique $d \in \mathfrak{D}$ with $\Phi'_i(d') = \Phi'_{\rm exp}(d) = \Phi'_i \circ \Psi'_{\rm exp}(d)$. But then $d' \approx \Psi'_{\rm exp}(d)$, so $\psi_{\rm ex}(d') = \psi_{\rm ex} \circ \Psi'_{\rm exp}(d) = \Psi''_{\rm exp}(\psi(d))$. Hence $\psi_{\rm ex} \circ \phi_{\rm ex}(c') = \psi_{\rm ex}(d')$ is invertible in $\mathfrak{E}_{\rm ex}$, with inverse $\Psi''_{\rm exp}(-\psi(d))$.

Thirdly, suppose that $\boldsymbol{\mathfrak{G}}$ is a C^∞-ring with corners and $\boldsymbol{\zeta} = (\zeta, \zeta_{\rm ex}) : \boldsymbol{\mathfrak{C}} \to \boldsymbol{\mathfrak{G}}$ a morphism such that $\zeta(c)$ is invertible in $\mathfrak{G}$ for all $c \in \mathfrak{C}$ with $x(c) \neq 0$ and $\zeta_{\rm ex}(c')$ is invertible in $\mathfrak{G}_{\rm ex}$ for all $c' \in \mathfrak{C}_{\rm ex}$ with $x \circ \Phi_i(c') \neq 0$. Then $\boldsymbol{\zeta} = \boldsymbol{\eta} \circ \boldsymbol{\phi}$ for a unique $\boldsymbol{\eta} : \boldsymbol{\mathfrak{D}} \to \boldsymbol{\mathfrak{G}}$, by the universal property of $\boldsymbol{\mathfrak{D}}$. Since $\phi_{\rm ex} : \mathfrak{C}_{\rm ex} \to \mathfrak{D}_{\rm ex}$ is surjective and $x = \tilde{x} \circ \phi$ we see that $\eta_{\rm ex}(d')$ is invertible in $\mathfrak{G}_{\rm ex}$ for all $d' \in \mathfrak{D}_{\rm ex}$ with $\tilde{x} \circ \Phi'_i(d') \neq 0$.

Let $d', d'' \in \mathfrak{D}_{\rm ex}$ with $\Phi'_i(d') = \Phi'_i(d'')$ in $\mathfrak{D}$ and $\tilde{x} \circ \Phi'_i(d') \neq 0$, so that $d' \approx d''$. Then $\eta_{\rm ex}(d'), \eta_{\rm ex}(d'')$ are invertible in $\mathfrak{G}_{\rm ex}$ with $\Phi_i \circ \eta_{\rm ex}(d') = \Phi_i \circ \eta_{\rm ex}(d'')$ in $\mathfrak{G}$, so Definition 4.16(i) for $\boldsymbol{\mathfrak{G}}$ implies that $\eta_{\rm ex}(d') = \eta_{\rm ex}(d'')$. Since $\eta_{\rm ex} : \mathfrak{D}_{\rm ex} \to \mathfrak{G}_{\rm ex}$ is a monoid morphism, and $\approx$ is a monoidal equivalence relation, and $\eta_{\rm ex}(d') = \eta_{\rm ex}(d'')$ for the generating relations $d' \approx d''$, we see that $\eta_{\rm ex}$ factorizes via $\mathfrak{D}_{\rm ex}/\approx$. Thus there exists unique $\theta_{\rm ex} : \mathfrak{E}_{\rm ex} \to \mathfrak{F}_{\rm ex}$ with $\eta_{\rm ex} = \theta_{\rm ex} \circ \psi_{\rm ex}$. Set $\theta = \eta : \mathfrak{E} = \mathfrak{D} \to \mathfrak{G}$. Then $\eta = \theta \circ \psi$ as $\psi = {\rm id}_{\mathfrak{D}}$. As $\boldsymbol{\psi}$ is surjective we see that $\boldsymbol{\theta} = (\theta, \theta_{\rm ex}) : \boldsymbol{\mathfrak{E}} \to \boldsymbol{\mathfrak{F}}$ is a morphism in $\mathbf{C^\infty Rings^c}$, with $\boldsymbol{\eta} = \boldsymbol{\theta} \circ \boldsymbol{\psi}$, so that $\boldsymbol{\zeta} = \boldsymbol{\theta} \circ \boldsymbol{\psi} \circ \boldsymbol{\phi}$.

This proves that $\boldsymbol{\psi} \circ \boldsymbol{\phi} : \boldsymbol{\mathfrak{C}} \to \boldsymbol{\mathfrak{E}}$ satisfies the universal property of $\boldsymbol{\pi}_x : \boldsymbol{\mathfrak{C}} \to \boldsymbol{\mathfrak{C}}_x$ from the localization (4.17), so $\boldsymbol{\mathfrak{E}} \cong \boldsymbol{\mathfrak{C}}_x$ as we claimed. Parts (b) and (c) of the theorem are now immediate, as $\mathfrak{E} = \mathfrak{C}_x$ and $\boldsymbol{\phi}, \boldsymbol{\psi}$ are surjective. For (a), observe that $x : \mathfrak{C} \to \mathbb{R}$ factorizes as $\pi \circ \pi_x$ for $\pi : \mathfrak{C}_x \to \mathbb{R}$ a morphism. If $\bar{c} \in \mathfrak{C}_x$ with $\pi(\bar{c}) \neq 0$ then as $\pi_x : \mathfrak{C} \to \mathfrak{C}_x$ is surjective by (c) we have $\bar{c} = \pi_x(c)$ with $x(c) \neq 0$, so $\bar{c} = \pi_x(c)$ is invertible in $\mathfrak{C}_x$ by (4.17). Similarly, if $\bar{c}' \in \mathfrak{C}_{x,\rm ex}$ with $\pi \circ \Phi_i(\bar{c}') \neq 0$ then as $\pi_{x,\rm ex} : \mathfrak{C}_{\rm ex} \to \mathfrak{C}_{x,\rm ex}$ is surjective we find that $\bar{c}'$ is invertible in $\mathfrak{C}_{x,\rm ex}$. Hence $\boldsymbol{\mathfrak{C}}_x$ is a local C^∞-ring with corners. □

We can characterize the equivalence relations that define $\mathfrak{C}_{x,\rm ex} = \mathfrak{E}_{\rm ex}$ in the previous proof using the following proposition.

Proposition 4.33 *Let $\boldsymbol{\mathfrak{C}} = (\mathfrak{C}, \mathfrak{C}_{\rm ex})$ be a C^∞-ring with corners and $x : \mathfrak{C} \to \mathbb{R}$ an $\mathbb{R}$-point of $\mathfrak{C}$. Let $\boldsymbol{\pi}_x : \boldsymbol{\mathfrak{C}} \to \boldsymbol{\mathfrak{C}}_x$ be as in Definition 4.31, and let I be the ideal defined in (2.4). For any $c'_1, c'_2 \in \mathfrak{C}_{\rm ex}$, then $\pi_{x,\rm ex}(c'_1) = \pi_{x,\rm ex}(c'_2)$ if and only if there are elements $a', b' \in \mathfrak{C}_{\rm ex}$ such that $\Phi_i(a') - \Phi_i(b') \in I$, $x \circ \Phi_i(a') \neq 0$, and $a'c'_1 = b'c'_2$. Hence $\boldsymbol{\mathfrak{C}}_x = (\mathfrak{C}/I, \mathfrak{C}_{\rm ex}/\sim)$ where $c'_1 \sim c'_2 \in$*

$\mathfrak{C}_{\rm ex}$ *if and only if there are elements* $a', b' \in \mathfrak{C}_{\rm ex}$ *such that* $\Phi_i(a') - \Phi_i(b') \in I$, $x \circ \Phi_i(a') \neq 0$, *and* $a'c'_1 = b'c'_2$.

Proof We first show that if $\pi_{x,\rm ex}(c'_1) = \pi_{x,\rm ex}(c'_2)$, then there are a', b' satisfying the conditions. In Theorem 4.32, we constructed C^∞-rings with corners $\boldsymbol{\mathfrak{D}} = (\mathfrak{D}, \mathfrak{D}_{\rm ex})$ and $\boldsymbol{\mathfrak{E}} = (\mathfrak{E}, \mathfrak{E}_{\rm ex})$ and surjective morphisms $\boldsymbol{\phi} = (\phi, \phi_{\rm ex}) : \boldsymbol{\mathfrak{C}} \to \boldsymbol{\mathfrak{D}}$, $\boldsymbol{\psi} = (\psi, \psi_{\rm ex}) : \boldsymbol{\mathfrak{D}} \to \boldsymbol{\mathfrak{E}}$, where

$$\mathfrak{E} = \mathfrak{D} = \mathfrak{C}_x = \mathfrak{C}/I, \quad \mathfrak{D}_{\rm ex} = \mathfrak{C}_{\rm ex}/\sim \quad \text{and} \quad \mathfrak{E}_{\rm ex} = \mathfrak{D}_{\rm ex}/\approx,$$

and $I \subset \mathfrak{C}$ is the ideal defined in (2.4), and $\sim, \approx$ are explicit equivalence relations. Then we showed that there is a unique isomorphism $\boldsymbol{\mathfrak{E}} \cong \boldsymbol{\mathfrak{C}}_x$ identifying $\boldsymbol{\psi} \circ \boldsymbol{\phi} : \boldsymbol{\mathfrak{C}} \to \boldsymbol{\mathfrak{E}}$ with $\boldsymbol{\pi}_x : \boldsymbol{\mathfrak{C}} \to \boldsymbol{\mathfrak{C}}_x$. As $\pi_{x,\rm ex}(c'_1) = \pi_{x,\rm ex}(c'_2)$ we have $\psi_{\rm ex} \circ \phi_{\rm ex}(c'_1) = \psi_{\rm ex} \circ \phi_{\rm ex}(c'_2)$ in $\mathfrak{E}_{\rm ex}$. Thus $\phi_{\rm ex}(c'_1) \approx \phi_{\rm ex}(c'_2)$.

By definition $\approx$ is the monoidal equivalence relation on $\mathfrak{D}_{\rm ex}$ generated by the condition that $d' \approx d''$ whenever $d', d'' \in \mathfrak{D}_{\rm ex}$ with $\Phi'_i(d') = \Phi'_i(d'')$ in $\mathfrak{D}$ and $\tilde{x} \circ \Phi'_i(d') \neq 0$, where $\Phi'_i : \mathfrak{D}_{\rm ex} \to \mathfrak{D}$ is the C^∞-operation from the inclusion $i : [0, \infty) \hookrightarrow \mathbb{R}$, and $x : \mathfrak{C} \to \mathbb{R}$ factorizes as $x = \tilde{x} \circ \phi$ for a unique morphism $\tilde{x} : \mathfrak{D} \to \mathbb{R}$. Hence $\phi_{\rm ex}(c'_1) \approx \phi_{\rm ex}(c'_2)$ means that there is a finite sequence $\phi_{\rm ex}(c'_1) = d'_0, d'_1, d'_2, \ldots, d'_{n-1}, d'_n = \phi_{\rm ex}(c'_2)$ in $\mathfrak{D}_{\rm ex}$, and elements $e'_i, f'_i, g'_i \in \mathfrak{D}_{\rm ex}$ such that $\Phi'_i(e'_i) = \Phi'_i(f'_i)$, $\tilde{x} \circ \Phi'_i(e'_i) \neq 0$, and $d'_{i-1} = e'_i g'_i$, $d'_i = f'_i g'_i$ in $\mathfrak{D}_{\rm ex}$ for $i = 1, \ldots, n$.

As $\phi_{\rm ex} : \mathfrak{C}_{\rm ex} \to \mathfrak{D}_{\rm ex}$ is surjective we can choose $e_i, f_i, g_i \in \mathfrak{C}_{\rm ex}$ with $e'_i = \phi_{\rm ex}(e_i)$, $f'_i = \phi_{\rm ex}(f_i)$, $g'_i = \phi_{\rm ex}(g_i)$ for $i = 1, \ldots, n$. Then the conditions become

$$\begin{gathered}\Phi_i(e_i) - \Phi_i(f_i) \in I, \quad x \circ \Phi_i(e_i) \neq 0 \in \mathbb{R}, \quad i = 1, \ldots, n,\\ c'_1 \sim e_1 g_1, \quad e_{i+1} g_{i+1} \sim f_i g_i, \quad i = 1, \ldots, n-1, \quad c'_2 \sim f_n g_n,\end{gathered}$$

since equality in $\mathfrak{D}_{\rm ex}$ lifts to $\sim$-equivalence in $\mathfrak{C}_{\rm ex}$. By definition of $\sim$, this means that there exist elements $h_0, h_1, \ldots, h_n$ in the ideal $I \subset \mathfrak{C}$ in (2.4) such that

$$\begin{gathered}c'_1 = \Psi_{\rm exp}(h_0) e_1 g_1, \quad c'_2 = \Psi_{\rm exp}(h_n) f_n g_n,\\ \text{and} \quad e_{i+1} g_{i+1} = \Psi_{\rm exp}(h_i) f_i g_i, \quad i = 1, \ldots, n-1.\end{gathered} \tag{4.20}$$

It is not hard to show that for any element $h \in I$, then the conditions $\Phi_i(e_i) - \Phi_i(f_i) \in I$ and $x \circ \Phi_i(e_i) \neq 0 \in \mathbb{R}$ hold if and only if $\Phi_i(\Psi_{\rm exp}(h) e_i) - \Phi_i(f_i) \in I$ and $x \circ \Phi_i(\Psi_{\rm exp}(h) e_i) \neq 0 \in \mathbb{R}$ hold. So we can remove the h_i in (4.20). We have that $\pi_{x,\rm ex}(c'_1) = \pi_{x,\rm ex}(c'_2)$ if and only if there are $e_i, f_i, g_i \in \mathfrak{C}_{\rm ex}$ such that

$$\begin{gathered}c'_1 = e_1 g_1, \quad c'_2 = f_n g_n, \quad e_{i+1} g_{i+1} = f_i g_i, \quad i = 1, \ldots, n-1,\\ \text{and} \quad x \circ \Phi_i(e_i) \neq 0 \in \mathbb{R}, \quad i = 1, \ldots, n.\end{gathered} \tag{4.21}$$

We define $a' = f_1 f_2 \cdots f_n$ and $b' = e_1 e_2 \cdots e_n$. Then using (4.21), we see that $a'c'_1 = b'c'_2$, $\Phi_i(a') - \Phi_i(b') \in I$ and $x \circ \Phi_i(a') \neq 0$ as required.

For the reverse argument, say we have $a', b' \in \mathfrak{C}_{\rm ex}$ with $\Phi_i(a') - \Phi_i(b') \in I$ and $x \circ \Phi_i(a') \neq 0$. Let $n = 1$, $e_1 = a'$, $g_1 = 0$ and $f_1 = b'$ in (4.20) then we see that $\pi_{x,\rm ex}(a') = \pi_{x,\rm ex}(b')$. As $x \circ \Phi_i(a') \neq 0$, then $\pi_{x,\rm ex}(a')$ is invertible in $\mathfrak{C}_{\rm ex}$. If we also have that $a'c'_1 = b'c'_2$, then as $\pi_{x,\rm ex}$ is a morphism, $\pi_{x,\rm ex}(c'_1) = \pi_{x,\rm ex}(c'_2)$ and the result follows. □

Remark 4.34 Suppose $0 \neq c' \in \mathfrak{C}_{\rm ex}$ with $\pi_{x,\rm ex}(c') = 0$ for some $\mathbb{R}$-point $x : \mathfrak{C} \to \mathbb{R}$. Then Proposition 4.33 gives $a' \in \mathfrak{C}_{\rm ex}$ with $x \circ \Phi_i(a') \neq 0$ and $a'c' = 0$. Thus a', c' are zero divisors in $\mathfrak{C}_{\rm ex}$. Conversely, if $\boldsymbol{\mathfrak{C}}$ is interior then $\mathfrak{C}_{\rm ex}$ has no zero divisors, so $\pi_{x,\rm ex}(c') \neq 0$ for all $0 \neq c' \in \mathfrak{C}_{\rm ex}$.

Example 4.35 Let X be a manifold with corners, or a manifold with g-corners, and $x \in X$. Define a C^∞-ring with corners $\boldsymbol{\mathfrak{C}}_x = (\mathfrak{C}, \mathfrak{C}_{\rm ex})$ such that $\mathfrak{C}$ is the set of germs at x of smooth functions $c : X \to \mathbb{R}$, and $\mathfrak{C}_{\rm ex}$ is the set of germs at x of exterior functions $c' : X \to [0, \infty)$.

That is, elements of $\mathfrak{C}$ are $\sim$-equivalence classes $[U, c]$ of pairs (U, c), where U is an open neighbourhood of x in X and $c : U \to \mathbb{R}$ is smooth, and $(U, c) \sim (\tilde{U}, \tilde{c})$ if there exists an open neighbourhood $\hat{U}$ of x in $U \cap \tilde{U}$ with $c|_{\hat{U}} = \tilde{c}|_{\hat{U}}$. Similarly, elements of $\mathfrak{C}_{\rm ex}$ are equivalence classes $[U, c']$, where U is an open neighbourhood of x in X and $c' : U \to [0, \infty)$ is exterior. The C^∞-operations Φ_f, Ψ_g are defined as in (4.10)–(4.11), but for germs.

As the set of germs depends only on the local behaviour, the set of germs at x of exterior functions is equal to the set of germs at x of interior functions and the zero function. Hence $\boldsymbol{\mathfrak{C}}_x$ is an interior C^∞-ring with corners.

There is a morphism $\pi : \mathfrak{C} \to \mathbb{R}$ mapping $\pi : [U, c] \mapsto c(x)$. If $\pi([U, c]) \neq 0$ then $[U, c]$ is invertible in $\mathfrak{C}$, and if $\pi \circ \Phi_i([U, c']) = c'(x) \neq 0$ then $[U, c']$ is invertible in $\mathfrak{C}_{\rm ex}$. Thus $\boldsymbol{\mathfrak{C}}_x$ is a local C^∞-ring with corners. Write $\boldsymbol{C}^\infty_x(X) = \boldsymbol{\mathfrak{C}}_x$.

Example 4.18(a) defines a C^∞-ring with corners $\boldsymbol{C}^\infty(X)$. The localization $\boldsymbol{\pi}_{x_*} : (\boldsymbol{C}^\infty(X))_{x_*}$ of $\boldsymbol{C}^\infty(X)$ at $x_* : C^\infty(X) \to \mathbb{R}$ from Definition 4.31 is also a local C^∞-ring with corners. There is a morphism $\boldsymbol{\pi}_x : \boldsymbol{C}^\infty(X) \to \boldsymbol{C}^\infty_x(X)$ mapping $c \mapsto [X, c]$, $c' \mapsto [X, c']$. By the universal property of $(\boldsymbol{C}^\infty(X))_{x_*}$ we have $\boldsymbol{\pi}_x = \boldsymbol{\lambda}_x \circ \boldsymbol{\pi}_{x_*}$ for a unique morphism $\boldsymbol{\lambda}_x : (\boldsymbol{C}^\infty(X))_{x_*} \to \boldsymbol{C}^\infty_x(X)$.

The proof of [46, Th. 4.41] implies that the C^∞-ring morphism λ_x in $\boldsymbol{\lambda}_x = (\lambda_x, \lambda_{x,\rm ex})$ is always an isomorphism. But $\lambda_{x,\rm ex}$ need not be an isomorphism, as the next proposition shows.

Proposition 4.36 *In Example* 4.35, *suppose X is a manifold with faces, as in Definition* 3.12. *Then $\boldsymbol{\lambda}_x$ is an isomorphism for all $x \in X$.*

If X is a manifold with corners, but not a manifold with faces, then $\boldsymbol{\lambda}_x$ is not an isomorphism for some $x \in X$.

Proof Let X be a manifold with faces, and $x \in X$, giving $\boldsymbol{\lambda}_x = (\lambda_x, \lambda_{x,\mathrm{ex}})$. Then λ_x is an isomorphism as in Example 4.35. We will show $\lambda_{x,\mathrm{ex}}$ is both surjective and injective, so that $\boldsymbol{\lambda}_x$ is an isomorphism.

For surjectivity, we need to show that for any $[U, c'] \in C^\infty_{x,\mathrm{ex}}(X)$ there exists $c'' \in C^\infty_{\mathrm{ex}}(X)$ with $c' = c''$ near x in X, so that $c' = \pi_{x,\mathrm{ex}}(c'') = \lambda_{x,\mathrm{ex}}(\pi_{x_*,\mathrm{ex}}(c''))$. If $[U, c'] = 0$ we take $c' = 0$, so suppose $[U, c'] \neq 0$, and take c'' to be interior. As in Remark 3.15, such c'' has a locally constant map $\mu_{c''} : \partial X \to \mathbb{N}$ such that c'' vanishes to order $\mu_{c''}$ along ∂X locally, and $c'' > 0$ on X°. We choose $\mu_{c''}$ such that $\mu_{c''} = \mu_{c'}$ on any connected component of ∂X which meets x, and $\mu_{c''} = 0$ on other components. Then local choices of interior c'' on X with prescribed vanishing order $\mu_{c''}$ on ∂X can be combined using a partition of unity on X, with the local choice $c'' = c'$ near x, to get c'' as required.

For injectivity, let $c'_1, c'_2 \in C^\infty_{\mathrm{ex}}(X)$ with $\pi_{x,\mathrm{ex}}(c'_1) = \pi_{x,\mathrm{ex}}(c'_2)$ in $C^\infty_{x,\mathrm{ex}}(X)$, which means that $c'_1 = c'_2$ near x in X. We must show that $\pi_{x_*,\mathrm{ex}}(c'_1) = \pi_{x_*,\mathrm{ex}}(c'_2)$ in $C^\infty_{x_*,\mathrm{ex}}(X)$. By Proposition 4.33, this holds if there are $a', b' \in C^\infty_{\mathrm{ex}}(X)$ such that $a' = b'$ near x in X, and $a'(x) \neq 0$, and $a'c'_1 = b'c'_2$.

Let $\tilde{X}$ be the connected component of X containing x. If $c'_1 = 0$ near x then $c'_1|_{\tilde{X}} = c'_2|_{\tilde{X}}$, and we can take $a' = b' = 1$ on $\tilde{X}$ and $a' = b' = 0$ on $X \setminus \tilde{X}$. Otherwise c'_1, c'_2 are interior on $\tilde{X}$, so $\mu_{c'_1}, \mu_{c'_2}$ map $\partial\tilde{X} \to \mathbb{N}$. Define a', b' to be zero on $X \setminus \tilde{X}$. On $\tilde{X}$, define locally constant $\mu_{a'}, \mu_{b'} : \partial\tilde{X} \to \mathbb{N}$ by $\mu_{a'} = \max(\mu_{c'_2} - \mu_{c'_1}, 0)$ and $\mu_{a'} = \max(\mu_{c'_1} - \mu_{c'_2}, 0)$, so that $\mu_{a'} + \mu_{c'_1} = \mu_{b'} + \mu_{c'_2}$ on $\partial\tilde{X}$, and $\mu_{a'}, \mu_{b'} = 0$ near $\Pi_{\tilde{X}}^{-1}(x)$. Let $a'|_{\tilde{X}} : \tilde{X} \to [0, \infty)$ be arbitrary interior with order of vanishing $\mu_{a'}$ on $\partial\tilde{X}$. Then there is a unique interior $b'|_{\tilde{X}} : \tilde{X} \to [0, \infty)$ with $a'c'_1 = b'c'_2$, defined by $b' = a'c'_1(c'_2)^{-1}$ on $\tilde{X}^\circ$, and extended continuously over $\tilde{X}$. These a', b' show $\lambda_{x,\mathrm{ex}}$ is injective.

If X is a manifold with corners, but not a manifold with faces, then there exist $x \in X$ and distinct $x'_1, x'_2 \in \Pi_X^{-1}(x)$ which lie in the same connected component $\partial_i X$ of ∂X. Then any $c'' \in C^\infty_{\mathrm{ex}}(X)$ has $\mu_{c''}(x'_1) = \mu_{c''}(x'_2)$, but there exists $[U, c'] \in C^\infty_{x,\mathrm{ex}}(X)$ with $\mu_{c'}(x'_1) \neq \mu_{c'}(x'_2)$, so $[U, c']$ does not lie in the image of $\pi_{x,\mathrm{ex}} = \lambda_{x,\mathrm{ex}} \circ \pi_{x_*,\mathrm{ex}}$, and hence not in the image of $\lambda_{x,\mathrm{ex}}$, as $\pi_{x_*,\mathrm{ex}}$ is surjective. Thus $\lambda_{x,\mathrm{ex}}$ is not surjective, and $\boldsymbol{\lambda}_x$ not an isomorphism. □

The proposition justifies the following definition in the g-corners case.

Definition 4.37 A manifold with g-corners X is called a *manifold with g-faces* if $\boldsymbol{\lambda}_x : (\boldsymbol{C}^\infty(X))_{x_*} \to \boldsymbol{C}_x^\infty(X)$ in Example 4.35 is an isomorphism for all $x \in X$.

In §5.4 we will show that a manifold with g-corners defines an *affine* C^∞-scheme with corners if and only if it is a manifold with g-faces.

Example 4.38 For any weakly toric monoid P, one can show using an argument similar to the proof of Proposition 4.36 that X_P in Definition 3.5 is a manifold with g-faces. Since every manifold with g-corners X can be covered by open neighbourhoods diffeomorphic to X_P, it follows that every manifold with g-corners can be covered by open subsets which are manifolds with g-faces.

Remark 4.39 Let us compare Definitions 3.13 and 4.37. For a manifold with g-corners X to be a manifold with g-faces it is necessary, but not sufficient, that $\Pi_X|_{\partial_i X} : \partial_i X \to X$ is injective for each connected component $\partial_i X$ of ∂X. This is because not every locally constant map $\mu_{c'} : \partial X \to \mathbb{N}$ can be the vanishing multiplicity map of an interior morphism $c' : X \to [0, \infty)$, as g-corner strata of codimension $\geqslant 3$ can impose extra consistency conditions on $\mu_{c'}$. The first author [25, Ex. 5.5.4] gives an example of a manifold with g-corners with $\Pi_X|_{\partial_i X}$ injective for all components $\partial_i X \subset \partial X$, which is not a manifold with g-faces.

4.7 Special classes of C^∞-rings with corners

We define and study several classes of C^∞-rings with corners. Throughout, if $\boldsymbol{\mathfrak{C}} = (\mathfrak{C}, \mathfrak{C}_{\rm ex})$ is an (interior) C^∞-ring with corners, then we regard $\mathfrak{C}_{\rm ex}$ (and $\mathfrak{C}_{\rm in}$) as monoids under multiplication, as in §3.2.

Definition 4.40 Let $\boldsymbol{\mathfrak{C}}$ be a C^∞-ring with corners, and see Proposition 4.23. We call $\boldsymbol{\mathfrak{C}}$ *finitely generated* if there exists a surjective morphism $\boldsymbol{\phi} : \boldsymbol{\mathfrak{F}}^{A,A_{\rm ex}} \to \boldsymbol{\mathfrak{C}}$ from some free C^∞-ring with corners $\boldsymbol{\mathfrak{F}}^{A,A_{\rm ex}}$ with $A, A_{\rm ex}$ finite sets. Equivalently, $\boldsymbol{\mathfrak{C}}$ is finitely generated if it fits into a coequalizer diagram (4.13) with $A, A_{\rm ex}$ finite. We call $\boldsymbol{\mathfrak{C}}$ *finitely presented* if it fits into a coequalizer diagram (4.13) with $A, A_{\rm ex}, B, B_{\rm ex}$ finite. Finitely presented implies finitely generated.

Write $\mathbf{C^\infty Rings^c_{fp}} \subset \mathbf{C^\infty Rings^c_{fg}} \subset \mathbf{C^\infty Rings^c}$ for the full subcategories of finitely presented, and finitely generated, objects in $\mathbf{C^\infty Rings^c}$.

Proposition 4.41 *Suppose we are given a pushout diagram in* $\mathbf{C^\infty Rings^c}$*:*

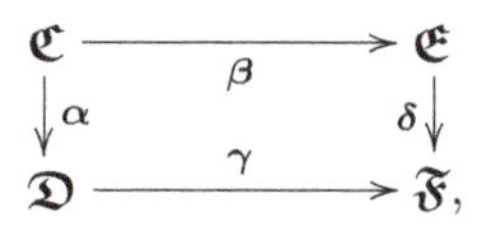

so $\mathfrak{F} = \mathfrak{D} \amalg_{\mathfrak{C}} \mathfrak{E}$*, and* $\mathfrak{C},\mathfrak{D},\mathfrak{E}$ *fit into coequalizer diagrams in* $\mathbf{C^\infty Rings^c}$*:*

$$\mathfrak{F}^{B,B_{\rm ex}} \underset{\zeta}{\overset{\epsilon}{\rightrightarrows}} \mathfrak{F}^{A,A_{\rm ex}} \xrightarrow{\phi} \mathfrak{C}, \tag{4.22}$$

$$\mathfrak{F}^{D,D_{\rm ex}} \underset{\theta}{\overset{\eta}{\rightrightarrows}} \mathfrak{F}^{C,C_{\rm ex}} \xrightarrow{\chi} \mathfrak{D}, \tag{4.23}$$

$$\mathfrak{F}^{F,F_{\rm ex}} \underset{\kappa}{\overset{\iota}{\rightrightarrows}} \mathfrak{F}^{E,E_{\rm ex}} \xrightarrow{\psi} \mathfrak{E}. \tag{4.24}$$

Then $\mathfrak{F}$ *fits into a coequalizer diagram in* $\mathbf{C^\infty Rings^c}$*:*

$$\mathfrak{F}^{H,H_{\rm ex}} \underset{\mu}{\overset{\lambda}{\rightrightarrows}} \mathfrak{F}^{G,G_{\rm ex}} \xrightarrow{\omega} \mathfrak{F}, \qquad \textit{with} \tag{4.25}$$

$$G=C\amalg E,\; G_{\rm ex}=C_{\rm ex}\amalg E_{\rm ex},\; H=A\amalg D\amalg F,\; H_{\rm ex}=A_{\rm ex}\amalg D_{\rm ex}\amalg F_{\rm ex}.$$

The analogue holds in $\mathbf{C^\infty Rings^c_{in}}$ *rather than* $\mathbf{C^\infty Rings^c}$.

Proof As coproducts exist in $\mathbf{C^\infty Rings^c}$, by general properties of colimits we may rewrite the pushout $\mathfrak{F} = \mathfrak{D} \amalg_{\mathfrak{C}} \mathfrak{E}$ as a coequalizer diagram

$$\mathfrak{C} \underset{\iota_{\mathfrak{E}}\circ\beta}{\overset{\iota_{\mathfrak{D}}\circ\alpha}{\rightrightarrows}} \mathfrak{D} \otimes_\infty \mathfrak{E} \xrightarrow{(\gamma,\delta)} \mathfrak{F}. \tag{4.26}$$

Now in a diagram $\mathfrak{H} \rightrightarrows \mathfrak{G}$, the coequalizer is the quotient of $\mathfrak{G}$ by relations from each element of $\mathfrak{H},\mathfrak{H}_{\rm ex}$. Given a surjective morphism $\mathfrak{I} \twoheadrightarrow \mathfrak{H}$, the coequalizers of $\mathfrak{H} \rightrightarrows \mathfrak{G}$ and $\mathfrak{I} \rightrightarrows \mathfrak{G}$ are the same, as we quotient $\mathfrak{G}$ by the same set of relations. Hence using (4.22), in (4.26) we can replace $\mathfrak{C}$ by $\mathfrak{F}^{A,A_{\rm ex}}$. Also using (4.23)–(4.24) gives a coequalizer diagram

$$\mathfrak{F}^{A,A_{\rm ex}} \rightrightarrows \mathrm{Coeq}\bigl(\mathfrak{F}^{D,D_{\rm ex}} \rightrightarrows \mathfrak{F}^{C,C_{\rm ex}}\bigr) \otimes_\infty \mathrm{Coeq}\bigl(\mathfrak{F}^{F,F_{\rm ex}} \rightrightarrows \mathfrak{F}^{E,E_{\rm ex}}\bigr) \longrightarrow \mathfrak{F}.$$

As coproducts commute with coequalizers, this is equivalent to

$$\mathfrak{F}^{A,A_{\rm ex}} \rightrightarrows \mathrm{Coeq}\bigl((\mathfrak{F}^{D,D_{\rm ex}} \otimes_\infty \mathfrak{F}^{F,F_{\rm ex}}) \rightrightarrows (\mathfrak{F}^{C,C_{\rm ex}} \otimes_\infty \mathfrak{F}^{E,E_{\rm ex}})\bigr) \longrightarrow \mathfrak{F}.$$

Since $\mathfrak{F}^{A,A_{\rm ex}}$ is free, the two morphisms $\mathfrak{F}^{A,A_{\rm ex}} \to \mathrm{Coeq}(\cdots)$ factor through the surjective morphism $\mathfrak{F}^{C,C_{\rm ex}} \otimes_\infty \mathfrak{F}^{E,E_{\rm ex}} \to \mathrm{Coeq}(\cdots)$. Thus we may combine the two coequalizers into one, giving a coequalizer diagram

$$\mathfrak{F}^{A,A_{\rm ex}} \otimes_\infty \mathfrak{F}^{D,D_{\rm ex}} \otimes_\infty \mathfrak{F}^{F,F_{\rm ex}} \rightrightarrows \mathfrak{F}^{C,C_{\rm ex}} \otimes_\infty \mathfrak{F}^{E,E_{\rm ex}} \longrightarrow \mathfrak{F}. \quad (4.27)$$

But coproducts of free C^∞-rings with corners are free over the disjoint unions of the generating sets. Thus (4.27) is equivalent to (4.25). The same argument works in $\mathbf{C^\infty Rings^c_{in}}$. □

Proposition 4.42 $\mathbf{C^\infty Rings^c_{fg}}$ *and* $\mathbf{C^\infty Rings^c_{fp}}$ *are closed under finite colimits in* $\mathbf{C^\infty Rings^c}$.

Proof $C^\infty(*)=(\mathbb{R},[0,\infty))$ is an initial object in $\mathbf{C^\infty Rings^c_{fg}}, \mathbf{C^\infty Rings^c_{fp}}$. As $\mathbf{C^\infty Rings^c_{fg}}, \mathbf{C^\infty Rings^c_{fp}}$ have an initial object, finite colimits may be written as iterated pushouts, so it is enough that $\mathbf{C^\infty Rings^c_{fg}}, \mathbf{C^\infty Rings^c_{fp}}$ are closed under pushouts. This follows from Proposition 4.41, noting that if $A, A_{\rm ex}, C, C_{\rm ex}, E, E_{\rm ex}$ are finite then $G, G_{\rm ex}$ are finite, and if $A,\ldots,F_{\rm ex}$ are finite then $G,\ldots,H_{\rm ex}$ are finite. □

Definition 4.43 We call a C^∞-ring with corners $\mathfrak{C}=(\mathfrak{C},\mathfrak{C}_{\rm ex})$ *firm* if the sharpening $\mathfrak{C}^\sharp_{\rm ex}$ is a finitely generated monoid. We denote by $\mathbf{C^\infty Rings^c_{fi}}$ the full subcategory of $\mathbf{C^\infty Rings^c}$ consisting of firm C^∞-rings with corners.

If $\mathfrak{C}$ is firm, then there are $c'_1,\ldots,c'_n$ in $\mathfrak{C}_{\rm ex}$ whose images under the quotient $\mathfrak{C}_{\rm ex} \to \mathfrak{C}^\sharp_{\rm ex}$ generate $\mathfrak{C}^\sharp_{\rm ex}$. This implies that each element in $\mathfrak{C}_{\rm ex}$ can be written as $\Psi_{\exp}(c)c_1'^{a_1}\cdots c_n'^{a_n}$ for some $c\in\mathfrak{C}$ and $a_i \geqslant 0$. If there exists a surjective morphism $\phi:\mathfrak{F}^{A,A_{\rm ex}} \to \mathfrak{C}$ with $A_{\rm ex}$ finite then $\mathfrak{C}$ is firm, as $[\phi_{\rm ex}(y_{a'})]$ for $a' \in A_{\rm ex}$ generate $\mathfrak{C}^\sharp_{\rm ex}$. Hence $\mathbf{C^\infty Rings^c_{fp}} \subset \mathbf{C^\infty Rings^c_{fg}} \subset \mathbf{C^\infty Rings^c_{fi}}$.

Proposition 4.44 $\mathbf{C^\infty Rings^c_{fi}}$ *is closed under finite colimits in* $\mathbf{C^\infty Rings^c}$.

Proof As in the proof of Proposition 4.42, it is enough to show $\mathbf{C^\infty Rings^c_{fi}}$ is closed under pushouts in $\mathbf{C^\infty Rings^c}$. Take $\mathfrak{C},\mathfrak{D},\mathfrak{E} \in \mathbf{C^\infty Rings^c_{fi}}$ with morphisms $\mathfrak{C}\to\mathfrak{D}$ and $\mathfrak{C}\to\mathfrak{E}$, and consider the pushout $\mathfrak{D}\amalg_{\mathfrak{C}}\mathfrak{E}$, with its morphisms $\phi:\mathfrak{D}\to\mathfrak{D}\amalg_{\mathfrak{C}}\mathfrak{E}$ and $\psi:\mathfrak{E}\to\mathfrak{D}\amalg_{\mathfrak{C}}\mathfrak{E}$. Then every element of $(\mathfrak{D}\amalg_{\mathfrak{C}}\mathfrak{E})_{\rm ex}$ is of the form

$$\Psi_f(\phi(d_1),\ldots,\phi(d_m),\psi(e_1),\ldots,\psi(e_n),$$
$$\phi_{\rm ex}(d'_1),\ldots,\phi_{\rm ex}(d'_k),\psi_{\rm ex}(e'_1),\ldots,\psi_{\rm ex}(e'_l)),$$

for smooth $f:\mathbb{R}^{m+n}\times[0,\infty)^{k+l}\to[0,\infty)$, where $d_i\in\mathfrak{D}, d'_i\in\mathfrak{D}_{\rm ex}, e_i\in\mathfrak{E}, e'_i\in\mathfrak{E}_{\rm ex}$, and d'_i are generators of the sharpening of $\mathfrak{D}_{\rm ex}$, and e'_i generate the sharpening of $\mathfrak{E}_{\rm ex}$. If $f=0$ this gives 0, otherwise we may write

$$f(x_1,\ldots,x_{m+n},y_1,\ldots,y_{k+l}) = y_1^{a_1}\cdots y_{k+l}^{a_{k+l}}\,e^{F(x_1,\ldots,x_{m+n},y_1,\ldots,y_{k+l})},$$

for $F : \mathbb{R}^{m+n} \times [0,\infty)^{k+l} \to \mathbb{R}$ smooth. In $(\mathfrak{D} \amalg_{\mathfrak{C}} \mathfrak{E})^\sharp_{\rm ex}$, the above element maps to

$$[\phi_{\rm ex}(d'_1)]^{a_1} \cdots [\phi_{\rm ex}(d'_k)]^{a_k} [\psi_{\rm ex}(e'_1)]^{a_{k+1}} \cdots [\psi_{\rm ex}(e'_l)]^{a_{k+l}}.$$

Hence $(\mathfrak{D} \amalg_{\mathfrak{C}} \mathfrak{E})^\sharp_{\rm ex}$ is generated by the images of the generators of $\mathfrak{D}^\sharp_{\rm ex}$ and $\mathfrak{E}^\sharp_{\rm ex}$, and zero, and so $(\mathfrak{D} \amalg_{\mathfrak{C}} \mathfrak{E})^\sharp_{\rm ex}$ is finitely generated. Thus $\mathbf{C^\infty Rings^c_{fi}}$ is closed under pushouts. □

Definition 4.45 Suppose $\mathfrak{C} = (\mathfrak{C}, \mathfrak{C}_{\rm ex})$ is an interior C^∞-ring with corners, and let $\mathfrak{C}_{\rm in} \subseteq \mathfrak{C}_{\rm ex}$ be the submonoid of §4.2, and $\mathfrak{C}^\sharp_{\rm in}$ its sharpening. Then:

(i) We call $\mathfrak{C}$ *integral* if $\mathfrak{C}_{\rm in}$ is an integral monoid.
(ii) We call $\mathfrak{C}$ *torsion-free* if $\mathfrak{C}_{\rm in}$ is an integral, torsion-free monoid.
(iii) We call $\mathfrak{C}$ *saturated* if it is integral, and $\mathfrak{C}_{\rm in}$ is a saturated monoid. Note that $\mathfrak{C}^\times_{\rm in} \cong \mathfrak{C}$ as abelian groups since $\mathfrak{C}$ is a C^∞-ring with corners, so $\mathfrak{C}^\times_{\rm in}$ is torsion-free. Therefore $\mathfrak{C}$ saturated implies that $\mathfrak{C}$ is torsion-free.
(iv) We call $\mathfrak{C}$ *toric* if it is saturated and firm. This implies that $\mathfrak{C}^\sharp_{\rm in}$ is a toric monoid.
(v) We call $\mathfrak{C}$ *simple* if it is toric and $\mathfrak{C}^\sharp_{\rm in} \cong \mathbb{N}^k$ for some $k \in \mathbb{N}$.

We write $\mathbf{C^\infty Rings^c_{si}} \subset \mathbf{C^\infty Rings^c_{to}} \subset \mathbf{C^\infty Rings^c_{sa}} \subset \mathbf{C^\infty Rings^c_{tf}} \subset \mathbf{C^\infty Rings^c_{\mathbb{Z}}} \subset \mathbf{C^\infty Rings^c_{in}}$ for the full subcategories of simple, toric, saturated, torsion-free, and integral objects in $\mathbf{C^\infty Rings^c_{in}}$. In the first author [25], simple C^∞-rings with corners were called 'simplicial', but we change to simple to avoid confusion with 'simplicial C^∞-rings' as used in Derived Differential Geometry, as in Spivak [87]; see §2.9.2.

Example 4.46 Let X be a manifold with corners, and $\boldsymbol{C^\infty_{\rm in}}(X)$ be the interior C^∞-ring with corners from Example 4.18(b). Let S be the set of connected components of ∂X. For each $F \in S$, we choose an interior map $c_F : X \to [0,\infty)$ which vanishes to order 1 on F, and to order zero on $\partial X \setminus F$, such that $c_F = 1$ outside a small neighbourhood U_F of $\Pi_X(F)$ in X, where we choose $\{U_F : F \in S\}$ to be locally finite in X. Then every interior map $g : X \to [0,\infty)$ may be written uniquely as $g = \exp(f) \cdot \prod_{F\in S} c_F^{a_F}$, for $f \in C^\infty(X)$ and $a_F \in \mathbb{N}$, $F \in S$.

Hence as monoids we have $\mathrm{In}(X) \cong C^\infty(X) \times \mathbb{N}^S$. Therefore $\boldsymbol{C^\infty_{\rm in}}(X)$ is integral, torsion-free, and saturated, and it is simple and toric if and only if ∂X has finitely many connected components. A more complicated proof shows that if X is a manifold with g-corners then $\boldsymbol{C^\infty_{\rm in}}(X)$ is integral, torsion-free, and saturated, and it is toric if ∂X has finitely many connected components.

The functor $\Pi_{\rm in} : \mathbf{C^\infty Rings^c_{in}} \to \mathbf{Sets}$ mapping $\mathfrak{C} \mapsto \mathfrak{C}_{\rm in}$ may be enhanced to a functor $\bar\Pi_{\rm in} : \mathbf{C^\infty Rings^c_{in}} \to \mathbf{Mon}$ by regarding $\mathfrak{C}_{\rm in}$ as a monoid, and then Theorem 4.20(a) implies that $\bar\Pi_{\rm in}$ preserves limits and directed colimits. Write $\mathbf{Mon_{sa,tf,\mathbb{Z}}} \subset \mathbf{Mon_{tf,\mathbb{Z}}} \subset \mathbf{Mon_{\mathbb{Z}}} \subset \mathbf{Mon}$ for the full subcategories of saturated, torsion-free integral, and torsion-free integral, and integral monoids. They are closed under limits and directed colimits in $\mathbf{Mon}$. Thus we deduce the following.

Proposition 4.47 $\mathbf{C^\infty Rings^c_{sa}}, \mathbf{C^\infty Rings^c_{tf}}$, *and* $\mathbf{C^\infty Rings^c_{\mathbb{Z}}}$ *are closed under limits and under directed colimits in* $\mathbf{C^\infty Rings^c_{in}}$. *Thus, all small limits and directed colimits exist in* $\mathbf{C^\infty Rings^c_{sa}}, \mathbf{C^\infty Rings^c_{tf}}, \mathbf{C^\infty Rings^c_{\mathbb{Z}}}$.

In a similar way to Proposition 4.19 we prove the following.

Theorem 4.48 *There are reflection functors* $\Pi_{\rm in}^{\mathbb{Z}}, \Pi_{\mathbb{Z}}^{\rm tf}, \Pi_{\rm tf}^{\rm sa}, \Pi_{\rm in}^{\rm sa}$ *in a diagram*

$$\mathbf{C^\infty Rings^c_{sa}} \underset{\rm inc}{\overset{\Pi_{\rm tf}^{\rm sa}}{\rightleftarrows}} \mathbf{C^\infty Rings^c_{tf}} \underset{\rm inc}{\overset{\Pi_{\mathbb{Z}}^{\rm tf}}{\rightleftarrows}} \mathbf{C^\infty Rings^c_{\mathbb{Z}}} \underset{\rm inc}{\overset{\Pi_{\rm in}^{\mathbb{Z}}}{\rightleftarrows}} \mathbf{C^\infty Rings^c_{in}}, \qquad \Pi_{\rm in}^{\rm sa} : \mathbf{C^\infty Rings^c_{in}} \to \mathbf{C^\infty Rings^c_{sa}}$$

such that each of $\Pi_{\rm in}^{\mathbb{Z}}, \Pi_{\mathbb{Z}}^{\rm tf}, \Pi_{\rm tf}^{\rm sa}, \Pi_{\rm in}^{\rm sa}$ *is left adjoint to the corresponding inclusion functor* inc.

Proof Let $\mathfrak{C}$ be an object in $\mathbf{C^\infty Rings^c_{in}}$. We will construct an object $\mathfrak{D} = \Pi_{\rm in}^{\mathbb{Z}}(\mathfrak{C})$ in $\mathbf{C^\infty Rings^c_{\mathbb{Z}}}$ and a projection $\boldsymbol{\pi} : \mathfrak{C} \to \mathfrak{D}$, with the property that if $\boldsymbol{\phi} : \mathfrak{C} \to \mathfrak{E}$ is a morphism in $\mathbf{C^\infty Rings^c_{in}}$ with $\mathfrak{E} \in \mathbf{C^\infty Rings^c_{\mathbb{Z}}}$ then $\boldsymbol{\phi} = \boldsymbol{\psi} \circ \boldsymbol{\pi}$ for a unique morphism $\boldsymbol{\psi} : \mathfrak{D} \to \mathfrak{E}$. Consider the diagram:

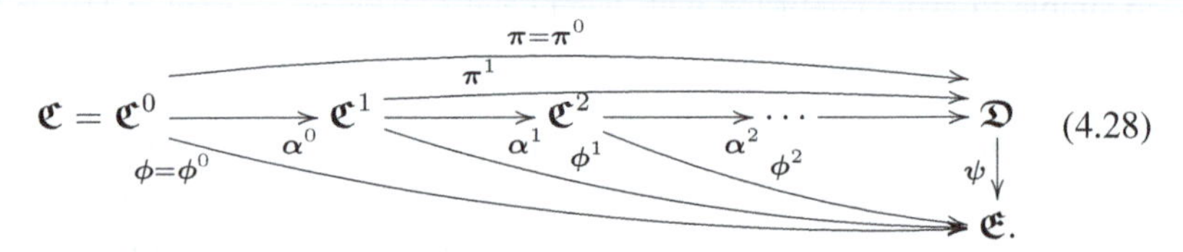

Define $\mathfrak{C}^0 = \mathfrak{C}$ and $\boldsymbol{\phi}^0 = \boldsymbol{\phi}$. By induction on $n = 0, 1, \ldots$, if $\mathfrak{C}^n, \boldsymbol{\phi}^n$ are defined, define an object $\mathfrak{C}^{n+1} \in \mathbf{C^\infty Rings^c_{in}}$ and morphisms $\boldsymbol{\alpha}^n : \mathfrak{C}^n \to \mathfrak{C}^{n+1}$, $\boldsymbol{\phi}^{n+1} : \mathfrak{C}^{n+1} \to \mathfrak{E}$ as follows. We have a monoid $\mathfrak{C}^n_{\rm in}$, which as in §3.2 has an abelian group $(\mathfrak{C}^n_{\rm in})^{\rm gp}$ with projection $\pi^{\rm gp} : \mathfrak{C}^n_{\rm in} \to (\mathfrak{C}^n_{\rm in})^{\rm gp}$, where $\mathfrak{C}^n_{\rm in}, \mathfrak{C}^n$ are integral if $\pi^{\rm gp}$ is injective. Using the notation of Definition 4.24, we define

$$\mathfrak{C}^{n+1} = \mathfrak{C}^n / \big[c' = c'' \text{ if } c', c'' \in \mathfrak{C}^n_{\rm in} \text{ with } \pi^{\rm gp}(c') = \pi^{\rm gp}(c'')\big]. \tag{4.29}$$

Write $\boldsymbol{\alpha}^n : \mathfrak{C}^n \to \mathfrak{C}^{n+1}$ for the natural surjective projection. Then $\mathfrak{C}^{n+1}, \boldsymbol{\alpha}^n$ are both interior, since the relations $c' = c''$ in (4.29) are all interior.

We have a morphism $\boldsymbol\phi^n : \mathfrak{C}^n \to \mathfrak{E}$ with $\mathfrak{E}$ integral, so by considering the diagram with bottom morphism injective

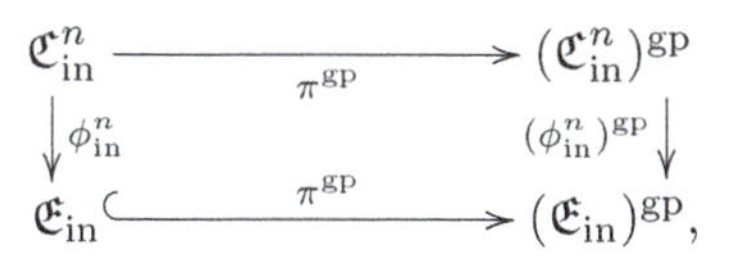

we see that if $c', c'' \in \mathfrak{C}^n_{\rm in}$ with $\pi^{\rm gp}(c') = \pi^{\rm gp}(c'')$ then $\phi^n_{\rm in}(c') = \phi^n_{\rm in}(c'')$. Thus by the universal property of (4.29), there is a unique morphism $\boldsymbol\phi^{n+1} : \mathfrak{C}^{n+1} \to \mathfrak{E}$ with $\boldsymbol\phi^n = \boldsymbol\phi^{n+1} \circ \boldsymbol\alpha^n$. This completes the inductive step, so we have defined $\mathfrak{C}^n, \boldsymbol\alpha^n, \boldsymbol\phi^n$ for all $n = 0, 1, \ldots$, where $\mathfrak{C}^n, \boldsymbol\alpha^n$ are independent of $\mathfrak{E}, \boldsymbol\phi$.

Now define $\mathfrak{D}$ to be the directed colimit $\mathfrak{D} = \varinjlim_{n=0}^\infty \mathfrak{C}^n$ in $\mathbf{C^\infty Rings^c_{in}}$, using the morphisms $\boldsymbol\alpha^n : \mathfrak{C}^n \to \mathfrak{C}^{n+1}$. This exists by Theorem 4.20(a), and commutes with $\Pi_{\rm in} : \mathbf{C^\infty Rings^c_{in}} \to \mathbf{Sets}$, where we can think of $\Pi_{\rm in}$ as mapping to monoids. It has a natural projection $\boldsymbol\pi : \mathfrak{C} \to \mathfrak{D}$, and also projections $\boldsymbol\pi^n : \mathfrak{C}^n \to \mathfrak{D}$ for all n. By the universal property of colimits, there is a unique morphism $\boldsymbol\psi$ in $\mathbf{C^\infty Rings^c_{in}}$ making (4.28) commute.

The purpose of the quotient (4.29) is to modify $\mathfrak{C}^n$ to make it integral, since if $\mathfrak{C}^n$ were integral then $\pi^{\rm gp}(c') = \pi^{\rm gp}(c'')$ implies $c' = c''$. It is not obvious that $\mathfrak{C}^{n+1}$ in (4.29) is integral, as the quotient modifies $(\mathfrak{C}^n_{\rm in})^{\rm gp}$. However, the direct limit $\mathfrak{D}$ is integral. To see this, suppose $d', d'' \in \mathfrak{D}_{\rm in}$ with $\pi^{\rm gp}(d') = \pi^{\rm gp}(d'')$ in $(\mathfrak{D}_{\rm in})^{\rm gp}$. Since $\mathfrak{D}_{\rm in} = \varinjlim_{m=0}^\infty \mathfrak{C}^m_{\rm in}$ in $\mathbf{Mon}$, for $m \gg 0$ we may write $d' = \pi^m_{\rm in}(c')$, $d'' = \pi^m_{\rm in}(c'')$ for $c', c'' \in \mathfrak{C}^m_{\rm in}$. As $(\mathfrak{D}_{\rm in})^{\rm gp} = \varinjlim_{n=0}^\infty (\mathfrak{C}^n_{\rm in})^{\rm gp}$ and $\pi^{\rm gp}(d') = \pi^{\rm gp}(d'')$, for some $n \gg m$ we have

$$\pi^{\rm gp} \circ \alpha^{n-1}_{\rm in} \circ \cdots \circ \alpha^m_{\rm in}(c') = \pi^{\rm gp} \circ \alpha^{n-1}_{\rm in} \circ \cdots \circ \alpha^m_{\rm in}(c'') \quad \text{in } (\mathfrak{C}^n_{\rm in})^{\rm gp}.$$

But then (4.29) implies that $\alpha^n_{\rm in} \circ \cdots \circ \alpha^m_{\rm in}(c') = \alpha^n_{\rm in} \circ \cdots \circ \alpha^m_{\rm in}(c'')$ in $\mathfrak{C}^{n+1}_{\rm in}$, so $d' = d''$. Therefore $\pi^{\rm gp} : \mathfrak{D}_{\rm in} \to (\mathfrak{D}_{\rm in})^{\rm gp}$ is injective, and $\mathfrak{D}$ is integral.

Set $\Pi^{\mathbb{Z}}_{\rm in}(\mathfrak{C}) = \mathfrak{D}$. If $\boldsymbol\xi : \mathfrak{C} \to \mathfrak{C}'$ is a morphism in $\mathbf{C^\infty Rings^c_{in}}$, by taking $\mathfrak{E} = \Pi^{\mathbb{Z}}_{\rm in}(\mathfrak{C}')$ and $\boldsymbol\phi = \boldsymbol\pi' \circ \boldsymbol\xi$ in (4.28) we see that there is a unique morphism $\Pi^{\mathbb{Z}}_{\rm in}(\boldsymbol\xi)$ in $\mathbf{C^\infty Rings^c_{\mathbb{Z}}}$ making the following commute:

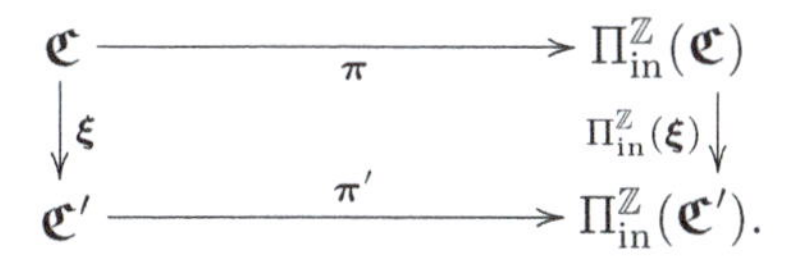

This defines the functor $\Pi^{\mathbb{Z}}_{\rm in}$. For any $\mathfrak{E} \in \mathbf{C^\infty Rings^c_{\mathbb{Z}}}$, the correspondence between $\boldsymbol\phi$ and $\boldsymbol\psi$ in (4.28) implies that we have a natural bijection

$$\mathrm{Hom}_{\mathbf{C^\infty Rings^c_{in}}}\bigl(\mathfrak{C}, \mathrm{inc}(\mathfrak{E})\bigr) \cong \mathrm{Hom}_{\mathbf{C^\infty Rings^c_{\mathbb{Z}}}}\bigl(\Pi^{\mathbb{Z}}_{\rm in}(\mathfrak{C}), \mathfrak{E}\bigr).$$

This is functorial in $\mathfrak{C},\mathfrak{E}$, and so $\Pi_{\rm in}^{\mathbb{Z}}$ is left adjoint to $\mathrm{inc}:\mathbf{C^\infty Rings^c_{\mathbb{Z}}}\hookrightarrow\mathbf{C^\infty Rings^c_{in}}$, as we have to prove.

The constructions of $\Pi_{\mathbb{Z}}^{\rm tf},\Pi_{\rm tf}^{\rm sa}$ are very similar. For $\Pi_{\mathbb{Z}}^{\rm tf}$, if $\mathfrak{C}$ is an object in $\mathbf{C^\infty Rings^c_{\mathbb{Z}}}$, the analogue of (4.29) is

$$\mathfrak{C}^{n+1}=\mathfrak{C}^n/\bigl[c'=c''\text{ if }c',c''\in\mathfrak{C}^n_{\rm in}\text{ with }\pi^{\rm tf}(c')=\pi^{\rm tf}(c'')\bigr],$$

where $\pi^{\rm tf}:\mathfrak{C}^n_{\rm in}\to(\mathfrak{C}^n_{\rm in})^{\rm gp}/\text{torsion}$ is the natural projection. For $\Pi_{\rm tf}^{\rm sa}$, if $\mathfrak{C}$ is an object in $\mathbf{C^\infty Rings^c_{tf}}$, the analogue of (4.29) is

$$\mathfrak{C}^{n+1}=\mathfrak{C}^n\bigl[s_{c'}:c'\in\mathfrak{C}^n_{\rm in}\subseteq(\mathfrak{C}^n_{\rm in})^{\rm gp}\text{ and there exists }c''\in(\mathfrak{C}^n_{\rm in})^{\rm gp}\setminus\mathfrak{C}^n_{\rm in}\\ \text{with }c'=n_{c'}\cdot c'',n'_c=2,3,\ldots\bigr]/\bigl[n_{c'}\cdot s_{c'}=c',\text{ all }c',n_{c'},s_{c'}\bigr].$$

Finally we set $\Pi_{\rm in}^{\rm sa}=\Pi_{\rm tf}^{\rm sa}\circ\Pi_{\mathbb{Z}}^{\rm tf}\circ\Pi_{\rm in}^{\mathbb{Z}}$. This completes the proof. □

As for Theorem 4.20(b), we deduce the following.

Corollary 4.49 *All small colimits exist in* $\mathbf{C^\infty Rings^c_{sa}},\mathbf{C^\infty Rings^c_{tf}}$, *and* $\mathbf{C^\infty Rings^c_{\mathbb{Z}}}$.

In Definition 4.24 we explained how to modify a C^∞-ring with corners $\mathfrak{C}$ by adding generators $\mathfrak{C}(x_a:a\in A)[y_{a'}:a'\in A_{\rm ex}]$ and imposing relations $\mathfrak{C}/(f_b=0:b\in B)[g_{b'}=h_{b'}:b'\in B_{\rm ex}]$. This is just notation for certain small colimits in $\mathbf{C^\infty Rings^c}$ or $\mathbf{C^\infty Rings^c_{in}}$. Corollary 4.49 implies that we can also add generators and relations in $\mathbf{C^\infty Rings^c_{sa}},\mathbf{C^\infty Rings^c_{tf}}$, and $\mathbf{C^\infty Rings^c_{\mathbb{Z}}}$, provided the relations $g_{b'}=h_{b'}$ are interior, that is, $g_{b'},h_{b'}\neq 0_{\mathfrak{C}_{\rm ex}}$.

Proposition 4.50 $\mathbf{C^\infty Rings^c_{to}}$ *is closed under finite colimits in* $\mathbf{C^\infty Rings^c_{sa}}$.

Proof Let J be a finite category and $\boldsymbol{F}:J\to\mathbf{C^\infty Rings^c_{to}}$ a functor. Write $\mathfrak{C}=\varinjlim\boldsymbol{F}$ for the colimit in $\mathbf{C^\infty Rings^c}$. As $\mathbf{C^\infty Rings^c_{to}}\subset\mathbf{C^\infty Rings^c_{fi}}$, Proposition 4.44 says $\mathfrak{C}$ is firm. Now $\Pi_{\rm in}^{\rm sa}(\mathfrak{C})=\varinjlim\boldsymbol{F}$ in $\mathbf{C^\infty Rings^c_{sa}}$, as $\Pi_{\rm in}^{\rm sa}$ is a reflection functor. By construction in the proof of Theorem 4.48 we see that $\Pi_{\rm in}^{\rm sa}$ takes firm objects to firm objects, so $\Pi_{\rm in}^{\rm sa}(\mathfrak{C})$ is firm, and hence toric, as toric means saturated and firm. □

5

C^∞-schemes with corners

We extend the theory of C^∞-schemes in Chapter 2 to C^∞-schemes with corners. We start by defining *(local) C^∞-ringed spaces with corners* in §5.1, using local C^∞-rings with corners from §4.6, and discuss some some special classes of local C^∞-ringed spaces with corners in §5.2. We then construct the *spectrum functor* $\mathrm{Spec^c} : (\mathbf{C^\infty Rings^c})^{\mathrm{op}} \to \mathbf{LC^\infty RS^c}$ in §5.3, and use it to define (affine) C^∞-schemes with corners in §5.4. The spectrum functor extends the spectrum functor for C^∞-rings in a natural way, and is right adjoint to the global sections functor $\Gamma^{\mathrm{c}} : \mathbf{LC^\infty RS^c} \to (\mathbf{C^\infty Rings^c})^{\mathrm{op}}$.

In Chapter 2 we saw that $\mathrm{Spec} : \mathbf{C^\infty Rings}^{\mathrm{op}} \to \mathbf{LC^\infty RS}$ restricts to an equivalence from the subcategory $(\mathbf{C^\infty Rings_{co}})^{\mathrm{op}} \subset \mathbf{C^\infty Rings}^{\mathrm{op}}$ of *complete* C^∞-rings to the subcategory $\mathbf{AC^\infty Sch} \subset \mathbf{LC^\infty RS}$ of affine C^∞-schemes. Unfortunately, no analogous notion of complete C^∞-rings with corners exists. But in §5.5 we study *semi-complete* C^∞-rings with corners, that have some of the good properties of complete C^∞-rings, and will be useful in later proofs. We introduce many special classes of C^∞-schemes with corners in §5.6, and then study fibre products of C^∞-schemes with corners in §5.7, where we show they exist under mild conditions.

5.1 (Local) C^∞-ringed spaces with corners

Definition 5.1 A *C^∞-ringed space with corners* $\boldsymbol{X} = (X, \boldsymbol{\mathcal{O}}_X)$ is a topological space X with a sheaf $\boldsymbol{\mathcal{O}}_X$ of C^∞-rings with corners on X. That is, for each open set $U \subseteq X$, then $\boldsymbol{\mathcal{O}}_X(U) = (\mathcal{O}_X(U), \mathcal{O}_X^{\mathrm{ex}}(U))$ is a C^∞-ring with corners and $\boldsymbol{\mathcal{O}}_X$ satisfies the sheaf axioms in §2.4.

A *morphism* $\boldsymbol{f} = (f, \boldsymbol{f}^\sharp) : (X, \boldsymbol{\mathcal{O}}_X) \to (Y, \boldsymbol{\mathcal{O}}_Y)$ of C^∞-ringed spaces with corners is a continuous map $f : X \to Y$ and a morphism $\boldsymbol{f}^\sharp = (f^\sharp, f^\sharp_{\mathrm{ex}}) : f^{-1}(\boldsymbol{\mathcal{O}}_Y) \to \boldsymbol{\mathcal{O}}_X$ of sheaves of C^∞-rings with corners on X, for $f^{-1}(\boldsymbol{\mathcal{O}}_Y) =$

$(f^{-1}(\mathcal{O}_Y), f^{-1}(\mathcal{O}_Y^{\mathrm{ex}}))$ as in Definition 2.38. Note that $\boldsymbol{f}^\sharp$ is adjoint to a morphism $\boldsymbol{f}_\sharp : \boldsymbol{\mathcal{O}}_Y \to f_*(\boldsymbol{\mathcal{O}}_X)$ on Y as in (2.9).

A *local C^∞-ringed space with corners* $\boldsymbol{X} = (X, \boldsymbol{\mathcal{O}}_X)$ is a C^∞-ringed space for which the stalks $\boldsymbol{\mathcal{O}}_{X,x} = (\mathcal{O}_{X,x}, \mathcal{O}_{X,x}^{\mathrm{ex}})$ of $\boldsymbol{\mathcal{O}}_X$ at x are local C^∞-rings with corners for all $x \in X$. We define morphisms of local C^∞-ringed spaces with corners $(X, \boldsymbol{\mathcal{O}}_X), (Y, \boldsymbol{\mathcal{O}}_Y)$ to be morphisms of C^∞-ringed spaces with corners, without any additional locality condition.

Write $\mathbf{C^\infty RS^c}$ for the category of C^∞-ringed spaces with corners, and write $\mathbf{LC^\infty RS^c}$ for the full subcategory of local C^∞-ringed spaces with corners.

For brevity, we will use the notation that bold upper-case letters $\boldsymbol{X}, \boldsymbol{Y}, \boldsymbol{Z}, \ldots$ represent C^∞-ringed spaces with corners $(X, \boldsymbol{\mathcal{O}}_X)$, $(Y, \boldsymbol{\mathcal{O}}_Y)$, $(Z, \boldsymbol{\mathcal{O}}_Z)$, $\ldots$, and bold lower-case letters $\boldsymbol{f}, \boldsymbol{g}, \ldots$ represent morphisms of C^∞-ringed spaces with corners $(f, \boldsymbol{f}^\sharp), (g, \boldsymbol{g}^\sharp), \ldots$. When we write '$x \in \boldsymbol{X}$' we mean that $\boldsymbol{X} = (X, \boldsymbol{\mathcal{O}}_X)$ and $x \in X$. When we write '$\boldsymbol{U}$ *is open in* $\boldsymbol{X}$' we mean that $\boldsymbol{U} = (U, \boldsymbol{\mathcal{O}}_U)$ and $\boldsymbol{X} = (X, \boldsymbol{\mathcal{O}}_X)$ with $U \subseteq X$ an open set and $\boldsymbol{\mathcal{O}}_U = \boldsymbol{\mathcal{O}}_X|_U$.

Let $\boldsymbol{X} = (X, \mathcal{O}_X, \mathcal{O}_X^{\mathrm{ex}}) \in \mathbf{LC^\infty RS^c}$, and let U be open in X. Take elements $s \in \mathcal{O}_X(U)$ and $s' \in \mathcal{O}_X^{\mathrm{ex}}(U)$. Then s and s' induce functions $s : U \to \mathbb{R}$, $s' : U \to [0, \infty)$, that at each $x \in U$ are the compositions

$$\mathcal{O}_X(U) \xrightarrow{\rho_{X,x}} \mathcal{O}_{X,x} \xrightarrow{\pi_x} \mathbb{R}, \text{ and } \mathcal{O}_X^{\mathrm{ex}}(U) \xrightarrow{\rho_{X,x}^{\mathrm{ex}}} \mathcal{O}_{X,x}^{\mathrm{ex}} \xrightarrow{\pi_x^{\mathrm{ex}}} [0, \infty).$$

Here, $\rho_{X,x}, \rho_{X,x}^{\mathrm{ex}}$ are the restriction morphism to the stalks, and $\pi_x, \pi_x^{\mathrm{ex}}$ are the unique morphisms that exist as $\mathcal{O}_{X,x}$ is local for each $x \in X$, as in Definition 4.28 and Lemma 4.30. We denote by $s(x)$ and $s'(x)$ the values of $s : U \to \mathbb{R}$ and $s' : U \to [0, \infty)$ respectively at the point $x \in U$. We denote by $s_x \in \mathcal{O}_{X,x}$ and $s'_x \in \mathcal{O}_{X,x}^{\mathrm{ex}}$ the values of s and s' under the restriction morphisms to the stalks $\rho_{X,x}$ and $\rho_{X,x}^{\mathrm{ex}}$ respectively.

Definition 5.2 Let $\boldsymbol{X} = (X, \boldsymbol{\mathcal{O}}_X)$ be a C^∞-ringed space with corners. We call $\boldsymbol{X}$ an *interior* C^∞-ringed space with corners if one (hence all) of the following conditions hold, which are equivalent by Lemma 5.3.

(a) For all open $U \subseteq X$ and each $s' \in \mathcal{O}_X^{\mathrm{ex}}(U)$, then $U_{s'} = \{x \in U : s'_x \neq 0 \in \mathcal{O}_{X,x}^{\mathrm{ex}}\}$, which is always closed in U, is open in U, and the stalks $\boldsymbol{\mathcal{O}}_{X,x}$ are interior C^∞-rings with corners.
(b) For all open $U \subseteq X$ and each $s' \in \mathcal{O}_X^{\mathrm{ex}}(U)$, then $U \setminus U_{s'} = \hat{U}_{s'} = \{x \in U : s'_x = 0 \in \mathcal{O}_{X,x}^{\mathrm{ex}}\}$, which is always open in U, is closed in U, and the stalks $\boldsymbol{\mathcal{O}}_{X,x}$ are interior C^∞-rings with corners.
(c) $\mathcal{O}_X^{\mathrm{ex}}$ is the sheafification of a presheaf of the form $\mathcal{O}_X^{\mathrm{in}} \amalg \{0\}$, where $\mathcal{O}_X^{\mathrm{in}}$ is a sheaf of monoids, such that $\boldsymbol{\mathcal{O}}_X^{\mathrm{in}}(U) = (\mathcal{O}_X(U), \mathcal{O}_X^{\mathrm{in}}(U) \amalg \{0\})$ is an interior C^∞-ring with corners for each open $U \subseteq X$.

In each case, we can define a sheaf of monoids $\mathcal{O}_X^{\rm in}$, such that $\mathcal{O}_X^{\rm in}(U) = \{s' \in \mathcal{O}_X^{\rm ex}(U) : s'_x \neq 0 \in \mathcal{O}_{X,x}^{\rm ex}$ for all $x \in U\}$.

We call $\boldsymbol{X}$ an *interior* local C^∞-ringed space with corners if $\boldsymbol{X}$ is both a local C^∞-ringed space with corners and an interior C^∞-ringed space with corners.

If $\boldsymbol{X}, \boldsymbol{Y}$ are interior (local) C^∞-ringed spaces with corners, a morphism $\boldsymbol{f} : \boldsymbol{X} \to \boldsymbol{Y}$ is called *interior* if the induced maps on stalks $\boldsymbol{f}_x^\sharp : \boldsymbol{\mathcal{O}}_{Y,f(x)} \to \boldsymbol{\mathcal{O}}_{X,x}$ are interior morphisms of interior C^∞-rings with corners for all $x \in X$. This gives a morphism of sheaves $f^{-1}(\mathcal{O}_Y^{\rm in}) \to \mathcal{O}_X^{\rm in}$. Write $\mathbf{C^\infty RS^c_{in}} \subset \mathbf{C^\infty RS^c}$ (and $\mathbf{LC^\infty RS^c_{in}} \subset \mathbf{LC^\infty RS^c}$) for the non-full subcategories of interior (local) C^∞-ringed spaces with corners and interior morphisms.

Lemma 5.3 *Parts* (a)–(c) *in Definition* 5.2 *are equivalent.*

Proof Parts (a) and (b) are equivalent by definition. The set $\hat{U}_{s'}$ is open, as the requirement that an element is zero in the stalk is a local requirement. That is, $s'_x = 0$ if and only if $s'|_V = 0 \in \mathcal{O}_X^{\rm ex}(V)$ for some $x \in V \subseteq U$.

Suppose (a) and (b) hold. Write $\mathcal{O}_X^{\rm in}(U) = \{s' \in \mathcal{O}_X^{\rm ex}(U) : s'_x \neq 0 \in \mathcal{O}_{X,x}^{\rm ex}$ for all $x \in U\}$. If $s'_1, s'_2 \in \mathcal{O}_X^{\rm in}(U)$ then $s'_{1,x}, s'_{2,x} \neq 0$ for $x \in U$, and as the stalks are interior $s'_{1,x} \cdot s'_{2,x} \neq 0 \in \mathcal{O}_{X,x}^{\rm ex}$. So $s'_1 \cdot s'_2 \in \mathcal{O}_X^{\rm in}(U)$, and $\mathcal{O}_X^{\rm in}$ is a monoid. Thus $(\mathcal{O}_X(U), \mathcal{O}_X^{\rm in}(U) \amalg \{0\})$ is a pre C^∞-ring with corners, where the C^∞-operations come from restriction from $\boldsymbol{\mathcal{O}}_X(U)$. As the invertible elements of the monoid and the C^∞-rings of $(\mathcal{O}_X(U), \mathcal{O}_X^{\rm in}(U) \amalg \{0\})$ are the same as those from $\boldsymbol{\mathcal{O}}_X(U)$, it is a C^∞-ring with corners. Let $\hat{\mathcal{O}}_X^{\rm ex}$ be the sheafification of $\mathcal{O}_X^{\rm in} \amalg \{0\}$, which is a subsheaf of $\mathcal{O}_X^{\rm ex}$. Note that $\mathcal{O}_X^{\rm in}(U) \amalg \{0\}$ already satisfies uniqueness, so the sheafification process means $\hat{\mathcal{O}}_X^{\rm ex}$ now satisfies gluing. Then $(\mathcal{O}_X, \hat{\mathcal{O}}_X^{\rm ex})$ is a sheaf of C^∞-rings with corners.

There is a morphism $(\mathrm{id}, \mathrm{id}^\sharp, \iota_{\rm ex}^\sharp) : (X, \mathcal{O}_X, \mathcal{O}_X^{\rm ex}) \to (X, \mathcal{O}_X, \hat{\mathcal{O}}_X^{\rm ex})$. This is the identity on X, $\mathcal{O}_X$. On the sheaves of monoids, we have an inclusion $\iota_{\rm ex}^\sharp(U) : \hat{\mathcal{O}}_X^{\rm ex}(U) \to \mathcal{O}_X^{\rm ex}(U)$. On stalks, any non-zero element of $\mathcal{O}_{X,x}^{\rm ex}$ is an equivalence class represented by a section $s' \in \mathcal{O}_X^{\rm ex}(U)$. As (a) holds, we can choose $U \ni x$ so that $s'_y \neq 0$ for all $y \in U$. Then $s' \in \mathcal{O}_X^{\rm in}(U)$, so there is $s'' \in \hat{\mathcal{O}}_X^{\rm ex}(U)$ mapping to s' under $\iota_{\rm ex}^\sharp(U)$. So $s''_x \mapsto s'_x$, and $\iota_{\rm ex}^\sharp$ is surjective on stalks as $0 \mapsto 0$. As $\hat{\mathcal{O}}_X^{\rm ex}$ is a subsheaf of $\mathcal{O}_X^{\rm ex}$, $\iota_{\rm ex}^\sharp$ is injective on stalks. Hence $(\mathrm{id}, \mathrm{id}^\sharp, \iota_{\rm ex}^\sharp)$ is an isomorphism, and (c) holds. Thus (a) and (b) imply (c).

Now suppose (c) holds. If $s' \in \mathcal{O}_X^{\rm ex}(U)$, where $\mathcal{O}_X^{\rm ex}$ is the sheafification of $\mathcal{O}_X^{\rm in} \amalg \{0\}$, then if $s'_x \neq 0 \in \mathcal{O}_{X,x}^{\rm ex}$ there is an open $x \in V \subseteq U$ and $s'' \in \mathcal{O}_X^{\rm in}(V)$ representing $s'|_V$, and therefore $s'_{x'} \neq 0 \in \mathcal{O}_{X,x'}^{\rm ex}$ for all $x' \in V$, and the $U_{s'}$ defined in (a) is open. If $s'_1, s'_2 \in \mathcal{O}_X^{\rm ex}(U)$, and $s'_{1,x}, s'_{2,x} \neq 0 \in \mathcal{O}_{X,x}^{\rm ex}$, then there is an open set open $x \in V \subseteq U$ and $s''_1, s''_2 \in \mathcal{O}_X^{\rm in}(V)$

representing $s_1'|_V, s_2'|_V$. Then $s''_{1,x} \cdot s''_{2,x} \in \mathcal{O}^{\rm in}_{X,x} \not\ni 0$, so the stalk $\boldsymbol{\mathcal{O}}_{X,x} = (\mathcal{O}_{X,x}, \mathcal{O}^{\rm in}_{X,x} \amalg \{0\})$ is interior, and (a) holds. Hence (c) implies (a). □

Remark 5.4 In §4.1 we defined interior C^∞-rings with corners $\boldsymbol{\mathfrak{C}} = (\mathfrak{C}, \mathfrak{C}_{\rm ex})$ as special examples of C^∞-rings of corners, with $\mathfrak{C}_{\rm ex} = \mathfrak{C}_{\rm in} \amalg \{0\}$, so that $0 \in \mathfrak{C}_{\rm ex}$ plays a special, somewhat artificial rôle in the interior theory; really one would prefer to write $\boldsymbol{\mathfrak{C}} = (\mathfrak{C}, \mathfrak{C}_{\rm in})$ and exclude 0 altogether.

One place this artificiality appears is in Definition 5.2(a)–(c). We want interior (local) C^∞-ringed spaces with corners $\boldsymbol{X} = (X, \boldsymbol{\mathcal{O}}_X)$ to be special examples of (local) C^∞-ringed spaces with corners. At the same time, we naïvely expect that if $\boldsymbol{X}$ is interior then $\boldsymbol{\mathcal{O}}_X$ should be a sheaf of interior C^∞-rings with corners. But this is false: $\boldsymbol{\mathcal{O}}_X$ is *never* a sheaf valued in $\mathbf{C^\infty Rings^c_{in}}$ in the sense of Definition 2.34. For example, $\boldsymbol{\mathcal{O}}_X(\emptyset) = (\{0\}, \{0\})$, where $(\{0\}, \{0\})$ is the final object in $\mathbf{C^\infty Rings^c}$, and it is not equal to $(\{0\}, \{0, 1\})$, the final object in $\mathbf{C^\infty Rings^c_{in}}$, contradicting Definition 2.34(iii).

A sheaf $\boldsymbol{\mathcal{O}}^{\rm in}_X$ on X valued in $\mathbf{C^\infty Rings^c_{in}}$ is also a presheaf (but not a sheaf) valued in $\mathbf{C^\infty Rings^c}$, so it has a sheafification $(\boldsymbol{\mathcal{O}}^{\rm in}_X)^{\rm sh}$ valued in $\mathbf{C^\infty Rings^c}$. Conversely, given a sheaf $\boldsymbol{\mathcal{O}}_X$ on X valued in $\mathbf{C^\infty Rings^c}$, we can ask whether it is the sheafification of a (necessarily unique) sheaf $\boldsymbol{\mathcal{O}}^{\rm in}_X$ valued in $\mathbf{C^\infty Rings^c_{in}}$. Definition 5.2(a)–(c) characterize when this holds, and this is the correct notion of 'sheaf of interior C^∞-rings with corners' in our context. The extra conditions $U_{s'}$ open and $\hat U_{s'}$ closed in U in Definition 5.2(a) and (b) are surprising, but they will be essential in the construction of the corner functor in §6.1.

The proof of the next theorem is essentially the same as showing ordinary ringed spaces have all small limits. An explicit proof can be found in the first author [25, Th. 5.1.10]. Existence of small limits in $\mathbf{C^\infty RS^c}, \mathbf{C^\infty RS^c_{in}}$ uses Theorem 4.20(b); and $\mathbf{LC^\infty RS^c}, \mathbf{LC^\infty RS^c_{in}}$ being closed under limits in $\mathbf{C^\infty RS^c}, \mathbf{C^\infty RS^c_{in}}$ uses Proposition 4.29; and $\mathbf{LC^\infty RS^c_{in}}, \mathbf{C^\infty RS^c_{in}}$ being closed under limits in $\mathbf{LC^\infty RS^c}, \mathbf{C^\infty RS^c}$ uses Theorem 4.20(d).

Theorem 5.5 *The categories* $\mathbf{C^\infty RS^c}$, $\mathbf{LC^\infty RS^c}$, $\mathbf{C^\infty RS^c_{in}}$, $\mathbf{LC^\infty RS^c_{in}}$ *have all small limits. Small limits commute with the inclusion and forgetful functors in the following diagram:*

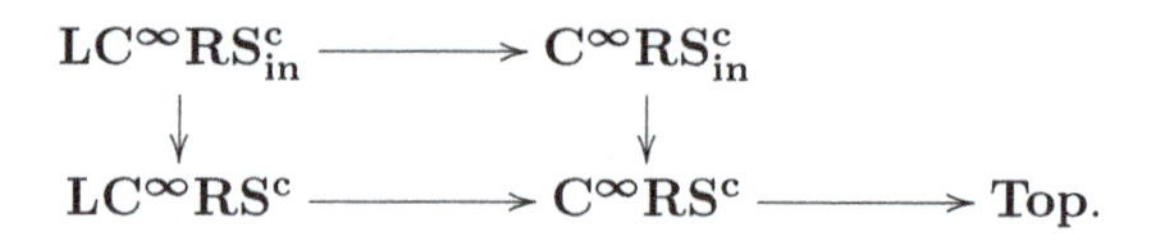
$$\begin{array}{ccccc} \mathbf{LC^\infty RS^c_{in}} & \longrightarrow & \mathbf{C^\infty RS^c_{in}} & & \\ \downarrow & & \downarrow & & \\ \mathbf{LC^\infty RS^c} & \longrightarrow & \mathbf{C^\infty RS^c} & \longrightarrow & \mathbf{Top}. \end{array}$$

In contrast to Theorem 4.20(d), in Theorem 6.3 below we show that $\mathrm{inc} : \mathbf{LC^\infty RS^c_{in}} \hookrightarrow \mathbf{LC^\infty RS^c}$ has a right adjoint C. This implies the following.

Corollary 5.6 *The inclusion* inc : $\mathbf{LC^\infty RS^c_{in}} \hookrightarrow \mathbf{LC^\infty RS^c}$ *preserves colimits.*

5.2 Special classes of local C^∞-ringed spaces with corners

We extend parts of §4.7 to local C^∞-ringed spaces with corners.

Definition 5.7 Let $\boldsymbol{X} = (X, \mathcal{O}_X)$ be an interior local C^∞-ringed space with corners. We call $\boldsymbol{X}$ *integral*, *torsion-free*, or *saturated*, if for each $x \in \boldsymbol{X}$ the stalk $\mathcal{O}_{X,x}$ is an integral, torsion-free, or saturated, interior local C^∞-ring with corners, respectively, in the sense of Definition 4.45.

For each open $U \subseteq X$, the monoid $\mathcal{O}^{\rm in}_X(U)$ is a submonoid of $\prod_{x\in U} \mathcal{O}^{\rm in}_{X,x}$. If $\boldsymbol{X}$ is integral then each $\mathcal{O}^{\rm in}_{X,x}$ is an integral monoid, so it is a submonoid of a group $(\mathcal{O}^{\rm in}_{X,x})^{\rm gp}$, and $\mathcal{O}^{\rm in}_X(U)$ is a submonoid of a group $\prod_{x\in U}(\mathcal{O}^{\rm in}_{X,x})^{\rm gp}$, so it is integral, and $\boldsymbol{\mathcal{O}}^{\rm in}_X(U)$ is an integral C^∞-ring with corners. Conversely, if $\boldsymbol{\mathcal{O}}^{\rm in}_X(U)$ is integral for all open $U \subseteq X$ one can show the stalks $\boldsymbol{\mathcal{O}}_{X,x}$ are integral, so $\boldsymbol{X}$ is integral. Thus, an alternative definition is that $\boldsymbol{X}$ is integral if $\boldsymbol{\mathcal{O}}^{\rm in}_X(U)$ is integral for all open $U \subseteq X$. The analogues hold for torsion-free and saturated.

Write $\mathbf{LC^\infty RS^c_{sa}} \subset \mathbf{LC^\infty RS^c_{tf}} \subset \mathbf{LC^\infty RS^c_{\mathbb{Z}}} \subset \mathbf{LC^\infty RS^c_{in}}$ for the full subcategories of saturated, torsion-free, and integral objects in $\mathbf{LC^\infty RS^c_{in}}$. Also write $\mathbf{LC^\infty RS^c_{sa,ex}} \subset \mathbf{LC^\infty RS^c_{tf,ex}} \subset \mathbf{LC^\infty RS^c_{\mathbb{Z},ex}} \subset \mathbf{LC^\infty RS^c_{in,ex}}$ $\subset \mathbf{LC^\infty RS^c}$ for the full subcategories of saturated, …, interior objects in $\mathbf{LC^\infty RS^c}$, but with exterior (i.e. all) rather than interior morphisms.

Here is the analogue of Theorem 4.48.

Theorem 5.8 *There are coreflection functors* $\Pi^{\mathbb{Z}}_{\rm in}, \Pi^{\rm tf}_{\mathbb{Z}}, \Pi^{\rm sa}_{\rm tf}, \Pi^{\rm sa}_{\rm in}$ *in a diagram*

$$\mathbf{LC^\infty RS^c_{sa}} \underset{\rm inc}{\overset{\Pi^{\rm sa}_{\rm tf}}{\rightleftarrows}} \mathbf{LC^\infty RS^c_{tf}} \underset{\rm inc}{\overset{\Pi^{\rm tf}_{\mathbb{Z}}}{\rightleftarrows}} \mathbf{LC^\infty RS^c_{\mathbb{Z}}} \underset{\rm inc}{\overset{\Pi^{\mathbb{Z}}_{\rm in}}{\rightleftarrows}} \mathbf{LC^\infty RS^c_{in}}, \qquad \Pi^{\rm sa}_{\rm in}: \mathbf{LC^\infty RS^c_{in}} \to \mathbf{LC^\infty RS^c_{sa}}$$

such that each of $\Pi^{\mathbb{Z}}_{\rm in}, \Pi^{\rm tf}_{\mathbb{Z}}, \Pi^{\rm sa}_{\rm tf}, \Pi^{\rm sa}_{\rm in}$ *is right adjoint to the corresponding inclusion functor* inc.

Proof We explain the functor $\Pi^{\mathbb{Z}}_{\rm in}$; the arguments for $\Pi^{\rm tf}_{\mathbb{Z}}, \Pi^{\rm sa}_{\rm tf}, \Pi^{\rm sa}_{\rm in}$ are the same. Let $\boldsymbol{X} = (X, \mathcal{O}_X)$ be an object in $\mathbf{LC^\infty RS^c_{in}}$. We will construct an object $\tilde{\boldsymbol{X}} = (X, \tilde{\mathcal{O}}_X)$ in $\mathbf{LC^\infty RS^c_{\mathbb{Z}}}$, and set $\Pi^{\mathbb{Z}}_{\rm in}(\boldsymbol{X}) = \tilde{\boldsymbol{X}}$. The topological spaces X in $\tilde{\boldsymbol{X}}, \boldsymbol{X}$ are the same. Define a presheaf $\mathcal{P}\tilde{\mathcal{O}}_X$ of C^∞-rings with corners on X by $\mathcal{P}\tilde{\mathcal{O}}_X(U) = \Pi^{\mathbb{Z}}_{\rm in}(\boldsymbol{\mathcal{O}}^{\rm in}_X(U))$ for each open $U \subseteq X$, where

$\mathcal{O}_X^{\rm in}(U) \in \mathbf{C^\infty Rings^c_{in}}$ is as in Definition 5.2(c), and $\Pi_{\rm in}^{\mathbb{Z}} : \mathbf{C^\infty Rings^c_{in}} \to \mathbf{C^\infty Rings^c_{\mathbb{Z}}}$ is as in Theorem 4.48. For open $V \subseteq U \subseteq X$, the restriction morphism ρ_{UV} in $\mathcal{P}\tilde{\mathcal{O}}_X$ is $\Pi_{\rm in}^{\mathbb{Z}}$ applied to $\rho_{UV} : \mathcal{O}_X^{\rm in}(U) \to \mathcal{O}_X^{\rm in}(V)$. Let $\tilde{\mathcal{O}}_X$ be the sheafification of $\mathcal{P}\tilde{\mathcal{O}}_X$.

Definition 5.2(c) implies that $\tilde{\boldsymbol{X}} = (X, \tilde{\mathcal{O}}_X)$ is an object in $\mathbf{LC^\infty RS^c_{in}}$. As stalks are given by direct limits as in Definition 2.35, and $\Pi_{\rm in}^{\mathbb{Z}} : \mathbf{C^\infty Rings^c_{in}} \to \mathbf{C^\infty Rings^c_{\mathbb{Z}}}$ preserves direct limits since it is a left adjoint, we have

$$\tilde{\mathcal{O}}_{X,x} \cong \mathcal{P}\tilde{\mathcal{O}}_{X,x} \cong \Pi_{\rm in}^{\mathbb{Z}}(\mathcal{O}_{X,x}) \qquad \text{for all } x \in X.$$

Hence $\tilde{\boldsymbol{X}}$ is an object of $\mathbf{LC^\infty RS^c_{\mathbb{Z}}}$ by Definition 5.7.

If $\boldsymbol{f} = (f, f^\sharp) : \boldsymbol{X} \to \boldsymbol{Y}$ is a morphism in $\mathbf{LC^\infty RS^c_{in}}$ and $\tilde{\boldsymbol{X}}, \tilde{\boldsymbol{Y}}$ are as above, in an obvious way we define a morphism $\tilde{\boldsymbol{f}} = (f, \tilde{f}^\sharp) : \tilde{\boldsymbol{X}} \to \tilde{\boldsymbol{Y}}$ in $\mathbf{LC^\infty RS^c_{\mathbb{Z}}}$ such that if $x \in X$ with $f(x) = y \in Y$ then $\tilde{f}^\sharp$ acts on stalks by

$$\tilde{f}_x^\sharp \cong \Pi_{\rm in}^{\mathbb{Z}}(f_x^\sharp) : \tilde{\mathcal{O}}_{Y,y} \cong \Pi_{\rm in}^{\mathbb{Z}}(\mathcal{O}_{Y,y}) \longrightarrow \tilde{\mathcal{O}}_{X,x} \cong \Pi_{\rm in}^{\mathbb{Z}}(\mathcal{O}_{X,x}).$$

We define $\Pi_{\rm in}^{\mathbb{Z}}(\boldsymbol{f}) = \tilde{\boldsymbol{f}}$. This gives a functor $\Pi_{\rm in}^{\mathbb{Z}} : \mathbf{LC^\infty RS^c_{in}} \to \mathbf{LC^\infty RS^c_{\mathbb{Z}}}$. We can deduce that $\Pi_{\rm in}^{\mathbb{Z}}$ is right adjoint to inc : $\mathbf{LC^\infty RS^c_{\mathbb{Z}}} \hookrightarrow \mathbf{LC^\infty RS^c_{in}}$ from the fact that $\Pi_{\rm in}^{\mathbb{Z}}$ is left adjoint to inc in Theorem 4.48, applied to stalks $\mathcal{O}_{X,x}, \mathcal{O}_{Y,y}$ as above for all $x \in X$ with $f(x) = y$ in Y. □

Remark 5.9 The analogue of Theorem 5.8 does not work for the categories $\mathbf{LC^\infty RS^c_{sa,ex}} \subset \cdots \subset \mathbf{LC^\infty RS^c_{in,ex}}$. This is because the functors $\Pi_{\rm in}^{\mathbb{Z}}, \ldots, \Pi_{\rm in}^{\rm sa}$ in Theorem 4.48 do not extend to exterior morphisms.

Since right adjoints preserve limits, Theorems 5.5 and 5.8 imply the following.

Corollary 5.10 *All small limits exist in* $\mathbf{LC^\infty RS^c_{sa}}, \mathbf{LC^\infty RS^c_{tf}}, \mathbf{LC^\infty RS^c_{\mathbb{Z}}}$.

5.3 The spectrum functor

We now define a spectrum functor for C^∞-rings with corners, in a similar way to Definition 2.43.

Definition 5.11 Let $\boldsymbol{\mathfrak{C}} = (\mathfrak{C}, \mathfrak{C}_{\rm ex})$ be a C^∞-ring with corners, and use the notation of Definition 4.31. As in Definition 2.42, write $X_{\mathfrak{C}}$ for the set of $\mathbb{R}$-points of $\mathfrak{C}$ with topology $\mathcal{T}_{\mathfrak{C}}$. For each open $U \subseteq X_{\mathfrak{C}}$, define $\boldsymbol{\mathcal{O}}_{X_{\mathfrak{C}}}(U) = (\mathcal{O}_{X_{\mathfrak{C}}}(U), \mathcal{O}_{X_{\mathfrak{C}}}^{\rm ex}(U))$. Here $\mathcal{O}_{X_{\mathfrak{C}}}(U)$ is the set of functions $s : U \to \coprod_{x \in U} \mathfrak{C}_x$ (where we write s_x for its value at the point $x \in U$) such that $s_x \in \mathfrak{C}_x$ for all $x \in U$, and such that U may be covered by open $W \subseteq U$ for which there exist

$c \in \mathfrak{C}$ with $s_x = \pi_x(c)$ in $\mathfrak{C}_x$ for all $x \in W$. Similarly, $\mathcal{O}^{\rm ex}_{X_{\mathfrak{C}}}(U)$ is the set of $s' : U \to \coprod_{x\in U} \mathfrak{C}_{x,\rm ex}$ with $s'_x \in \mathfrak{C}_{x,\rm ex}$ for all $x \in U$, and such that U may be covered by open $W \subseteq U$ for which there exist $c' \in \mathfrak{C}_{\rm ex}$ with $s'_x = \pi_{x,\rm ex}(c')$ in $\mathfrak{C}_{x,\rm ex}$ for all $x \in W$.

Define operations Φ_f and Ψ_g on $\boldsymbol{\mathcal{O}}_{X_{\mathfrak{C}}}(U)$ pointwise in $x \in U$ using the operations Φ_f and Ψ_g on $\mathfrak{C}_x$. This makes $\boldsymbol{\mathcal{O}}_{X_{\mathfrak{C}}}(U)$ into a C^∞-ring with corners. If $V \subseteq U \subseteq X_{\mathfrak{C}}$ are open, the restriction maps $\boldsymbol{\rho}_{UV} = (\rho_{UV}, \rho_{UV,\rm ex}) : \boldsymbol{\mathcal{O}}_{X_{\mathfrak{C}}}(U) \to \boldsymbol{\mathcal{O}}_{X_{\mathfrak{C}}}(V)$ mapping $\rho_{UV} : s \mapsto s|_V$ and $\rho_{UV,\rm ex} : s' \mapsto s'|_V$ are morphisms of C^∞-rings with corners.

The local nature of the definition implies that $\boldsymbol{\mathcal{O}}_{X_{\mathfrak{C}}} = (\mathcal{O}_{X_{\mathfrak{C}}}, \mathcal{O}^{\rm ex}_{X_{\mathfrak{C}}})$ is a sheaf of C^∞-rings with corners on $X_{\mathfrak{C}}$. In fact, $\mathcal{O}_{X_{\mathfrak{C}}}$ is the sheaf of C^∞-rings in Definition 2.43. By Proposition 5.12 below, the stalk $\boldsymbol{\mathcal{O}}_{X_{\mathfrak{C}},x}$ at $x \in X_{\mathfrak{C}}$ is naturally isomorphic to $\mathfrak{C}_x$, which is a local C^∞-ring with corners by Theorem 4.32(a). Hence $(X_{\mathfrak{C}}, \boldsymbol{\mathcal{O}}_{X_{\mathfrak{C}}})$ is a local C^∞-ringed space with corners, which we call the *spectrum* of $\mathfrak{C}$, and write as $\mathrm{Spec^c}\,\mathfrak{C}$.

Now let $\boldsymbol{\phi} = (\phi, \phi_{\rm ex}) : \mathfrak{C} \to \mathfrak{D}$ be a morphism of C^∞-rings with corners. As in Definition 2.43, define the continuous function $f_\phi : X_{\mathfrak{D}} \to X_{\mathfrak{C}}$ by $f_\phi(x) = x \circ \phi$. For open $U \subseteq X_{\mathfrak{C}}$ define $(\boldsymbol{f}_\phi)_\sharp(U) : \boldsymbol{\mathcal{O}}_{X_{\mathfrak{C}}}(U) \to \boldsymbol{\mathcal{O}}_{X_{\mathfrak{D}}}(f_\phi^{-1}(U))$ to act by $\boldsymbol{\phi}_x : \mathfrak{C}_{f_\phi(x)} \to \mathfrak{D}_x$ on stalks at each $x \in f_\phi^{-1}(U)$, where $\boldsymbol{\phi}_x$ is the induced morphism of local C^∞-rings with corners. Then $(\boldsymbol{f}_\phi)_\sharp : \boldsymbol{\mathcal{O}}_{X_{\mathfrak{C}}} \to (f_\phi)_*(\boldsymbol{\mathcal{O}}_{X_{\mathfrak{D}}})$ is a morphism of sheaves of C^∞-rings with corners on $X_{\mathfrak{C}}$.

Let $\boldsymbol{f}^\sharp_\phi : f_\phi^{-1}(\boldsymbol{\mathcal{O}}_{X_{\mathfrak{C}}}) \to \boldsymbol{\mathcal{O}}_{X_{\mathfrak{D}}}$ be the corresponding morphism of sheaves of C^∞-rings with corners on $X_{\mathfrak{D}}$ under (2.9). The stalk map $\boldsymbol{f}^\sharp_{\phi,x} : \boldsymbol{\mathcal{O}}_{X_{\mathfrak{C}},f_\phi(x)} \to \boldsymbol{\mathcal{O}}_{X_{\mathfrak{D}},x}$ of $\boldsymbol{f}^\sharp_\phi$ at $x \in X_{\mathfrak{D}}$ is identified with $\boldsymbol{\phi}_x : \mathfrak{C}_{f_\phi(x)} \to \mathfrak{D}_x$ under the isomorphisms $\boldsymbol{\mathcal{O}}_{X_{\mathfrak{C}},f_\phi(x)} \cong \mathfrak{C}_{f_\phi(x)}$, $\boldsymbol{\mathcal{O}}_{X_{\mathfrak{D}},x} \cong \mathfrak{D}_x$ in Proposition 5.12. Then $\boldsymbol{f}_\phi = (f_\phi, \boldsymbol{f}^\sharp_\phi) : (X_{\mathfrak{D}}, \boldsymbol{\mathcal{O}}_{X_{\mathfrak{D}}}) \to (X_{\mathfrak{C}}, \boldsymbol{\mathcal{O}}_{X_{\mathfrak{C}}})$ is a morphism of local C^∞-ringed spaces with corners. Define $\mathrm{Spec^c}\,\boldsymbol{\phi} : \mathrm{Spec^c}\,\mathfrak{D} \to \mathrm{Spec^c}\,\mathfrak{C}$ by $\mathrm{Spec^c}\,\boldsymbol{\phi} = \boldsymbol{f}_\phi$. Then $\mathrm{Spec^c}$ is a functor $(\mathbf{C^\infty Rings^c})^{\rm op} \to \mathbf{LC^\infty RS^c}$, the *spectrum functor*.

Proposition 5.12 *In Definition* 5.11, *the stalk $\boldsymbol{\mathcal{O}}_{X_{\mathfrak{C}},x}$ of $\boldsymbol{\mathcal{O}}_{X_{\mathfrak{C}}}$ at $x \in X_{\mathfrak{C}}$ is naturally isomorphic to $\mathfrak{C}_x$.*

Proof We have $\boldsymbol{\mathcal{O}}_{X_{\mathfrak{C}},x} = (\mathcal{O}_{X_{\mathfrak{C}},x}, \mathcal{O}^{\rm ex}_{X_{\mathfrak{C}},x})$ where elements $[U,s] \in \mathcal{O}_{X_{\mathfrak{C}},x}$ and $[U,s'] \in \mathcal{O}^{\rm ex}_{X_{\mathfrak{C}},x}$ are $\sim$-equivalence classes of pairs (U,s) and (U,s'), where U is an open neighbourhood of x in $X_{\mathfrak{C}}$ and $s \in \mathcal{O}_{X_{\mathfrak{C}}}(U)$, $s' \in \mathcal{O}^{\rm ex}_{X_{\mathfrak{C}}}(U)$, and $(U,s) \sim (V,t)$, $(U,s') \sim (V,t')$ if there exists open $x \in W \subseteq U \cap V$ with $s|_W = t|_W$ in $\mathcal{O}_{X_{\mathfrak{C}}}(W)$ and $s'|_W = t'|_W$ in $\mathcal{O}^{\rm ex}_{X_{\mathfrak{C}}}(W)$. Define a morphism of C^∞-rings with corners $\boldsymbol{\Pi} = (\Pi, \Pi_{\rm ex}) : \boldsymbol{\mathcal{O}}_{X_{\mathfrak{C}},x} \to \boldsymbol{\mathfrak{C}}_x$ by $\Pi : [U, s] \mapsto s_x \in \mathfrak{C}_x$ and $\Pi_{\rm ex} : [U, s'] \mapsto s'_x \in \mathfrak{C}_{x,\rm ex}$.

Suppose $c_x \in \mathfrak{C}_x$ and $c'_x \in \mathfrak{C}_{x,\rm ex}$. Then $c_x = \pi_x(c)$ for $c \in \mathfrak{C}_x$ and $c'_x = \pi_{x,\rm ex}(c')$ for $c' \in \mathfrak{C}_{x,\rm ex}$ by Theorem 4.32(c). Define $s : X_{\mathfrak{C}} \to \coprod_{y\in X_{\mathfrak{C}}} \mathfrak{C}_y$ and $s' : X_{\mathfrak{C}} \to \coprod_{y\in X_{\mathfrak{C}}} \mathfrak{C}_{y,\rm ex}$ by $s_y = \pi_y(c)$ and $s'_y = \pi_{y,\rm ex}(c')$. Then $s \in \mathcal{O}_{X_{\mathfrak{C}}}(X_{\mathfrak{C}})$, so that $[X_{\mathfrak{C}}, s] \in \mathcal{O}_{X_{\mathfrak{C}},x}$ with $\Pi([X_{\mathfrak{C}}, s]) = s_x = \pi_x(c) = c_x$, and similarly $s' \in \mathcal{O}^{\rm ex}_{X_{\mathfrak{C}}}(X_{\mathfrak{C}})$ with $\Pi_{\rm ex}([X_{\mathfrak{C}}, s']) = c'_x$. Hence $\Pi : \mathcal{O}_{X_{\mathfrak{C}},x} \to \mathfrak{C}_x$ and $\Pi_{\rm ex} : \mathcal{O}^{\rm ex}_{X_{\mathfrak{C}},x} \to \mathfrak{C}_{x,\rm ex}$ are surjective.

Let $[U_1, s_1], [U_2, s_2] \in \mathcal{O}_{X_{\mathfrak{C}},x}$ with $\Pi([U_1, s_1]) = s_{1,x} = s_{2,x} = \Pi([U_2, s_2])$. Then by definition of $\boldsymbol{\mathcal{O}}_{X_{\mathfrak{C}}}(U_1), \boldsymbol{\mathcal{O}}_{X_{\mathfrak{C}}}(U_2)$ there exists an open neighbourhood V of x in $U_1 \cap U_2$ and $c_1, c_2 \in \mathfrak{C}$ with $s_{1,v} = \pi_v(c_1)$ and $s_{2,v} = \pi_v(c_2)$ for all $v \in V$. Thus $\pi_x(c_1) = \pi_x(c_2)$ as $s_{1,x} = s_{2,x}$. Hence $c_1 - c_2$ lies in the ideal I in (2.4) by Proposition 2.24. Thus there exists $d \in \mathfrak{C}$ with $x(d) \neq 0 \in \mathbb{R}$ and $d \cdot (c_1 - c_2) = 0 \in \mathfrak{C}$.

Making V smaller we can suppose that $v(d) \neq 0$ for all $v \in V$, as this is an open condition. Then $\pi_v(c_1) = \pi_v(c_2) \in \mathfrak{C}_v$ for $v \in V$, since $\pi_v(d) \cdot \pi_v(c_1) = \pi_v(d) \cdot \pi_v(c_2)$ as $d \cdot c_1 = d \cdot c_2$ and $\pi_v(d)$ is invertible in $\mathfrak{C}_v$. Thus $s_{1,v} = \pi_v(c_1) = \pi_v(c_2) = s_{2,v}$ for $v \in V$, so $s_1|_V = s_2|_V$, and $[U_1, s_1] = [V, s_1|_V] = [V, s_2|_V] = [U_2, s_2]$. Therefore $\Pi : \mathcal{O}_{X_{\mathfrak{C}},x} \to \mathfrak{C}_x$ is injective, and an isomorphism.

Let $[U_1, s'_1], [U_2, s'_2]$ lie in $\mathcal{O}^{\rm ex}_{X_{\mathfrak{C}},x}$ with $\Pi_{\rm ex}([U_1, s'_1]) = s'_{1,x} = s'_{2,x} = \Pi([U_2, s'_2])$. As above there exist an open neighbourhood V of x in $U_1 \cap U_2$ and $c'_1, c'_2 \in \mathfrak{C}_{\rm ex}$ with $s'_{1,v} = \pi_{v,\rm ex}(c'_1)$ and $s'_{2,v} = \pi_{v,\rm ex}(c'_2)$ for all $v \in V$. At this point we can use Proposition 4.33, which says that $\pi_{x,\rm ex}(x'_2) = \pi_{x,\rm ex}(c'_1)$ if and only if there are $a, b \in \mathfrak{C}_{\rm ex}$ such that $\Phi_i(a) - \Phi_i(b) \in I_x$, $x \circ \Phi_i(a) \neq 0$ and $ac_1 = bc_2$, where I_x is the ideal in (2.4). The third condition does not depend on x, whereas the first two conditions are open conditions in x; that is, if $\Phi_i(a) - \Phi_i(b) \in I_x$, $x \circ \Phi_i(a) \neq 0$, then there is an open neighbourhood of X such that $\Phi_i(a) - \Phi_i(b) \in I_v$, $v \circ \Phi_i(a) \neq 0$ for all v in that neighbourhood.

Making V above smaller if necessary, we can suppose that these conditions hold in V and thus that $\pi_{v,\rm ex}(c'_1) = \pi_{v,\rm ex}(c'_2)$ for all $v \in V$. Hence $s'_{1,v} = s'_{2,v}$ for all $v \in V$, and $s'_1|_V = s'_2|_V$, so that $[U_1, s'_1] = [V, s'_1|_V] = [V, s'_2|_V] = [U_2, s'_2]$. Therefore $\Pi_{\rm ex} : \mathcal{O}^{\rm ex}_{X_{\mathfrak{C}},x} \to \mathfrak{C}_{x,\rm ex}$ is injective, and an isomorphism. So $\boldsymbol{\Pi} = (\Pi, \Pi_{\rm ex}) : \boldsymbol{\mathcal{O}}_{X_{\mathfrak{C}},x} \to \boldsymbol{\mathfrak{C}}_x$ is an isomorphism, as we have to prove. □

Definition 5.13 As $\mathbf{C^\infty Rings^c_{in}}$ is a subcategory of $\mathbf{C^\infty Rings^c}$ we can define the functor $\mathrm{Spec}^{\rm c}_{\rm in}$ by restricting $\mathrm{Spec}^{\rm c}$ to $(\mathbf{C^\infty Rings^c_{in}})^{\rm op}$. Let $\boldsymbol{\mathfrak{C}} = (\mathfrak{C}, \mathfrak{C}_{\rm ex})$ be an interior C^∞-ring with corners, and $\boldsymbol{X} = \mathrm{Spec}^{\rm c}_{\rm in}\, \boldsymbol{\mathfrak{C}} = (X, \boldsymbol{\mathcal{O}}_X)$. Then Definition 4.26 implies the localizations $\boldsymbol{\mathfrak{C}}_x$ are interior C^∞-rings with corners, and $\boldsymbol{\mathfrak{C}}_x \cong \boldsymbol{\mathcal{O}}_{X,x}$ by Proposition 5.12.

If $s' \in \mathcal{O}_X^{\rm ex}(U)$ with $s'_x \neq 0$ in $\mathcal{O}_{X,x}^{\rm ex}$ at $x \in X$, then $s'_{x'} = \pi_{x',{\rm ex}}(c')$ in $\mathfrak{C}_{x',{\rm ex}} \cong \mathcal{O}_{X,x'}^{\rm ex}$ for some $c' \in \mathfrak{C}_{\rm ex}$ and all x' in an open neighbourhood V of x in U. Then $c' \neq 0$ in $\mathfrak{C}_{\rm ex}$, so c' is non-zero in every stalk by Remark 4.34 as $\mathfrak{C}$ is interior, and s' is non-zero at every point in V. Therefore $\boldsymbol{X}$ is an interior local C^∞-ringed space with corners by Definition 5.2(a).

If $\boldsymbol{\phi} : \mathfrak{C} \to \mathfrak{D}$ is a morphism of interior C^∞-rings with corners, then $\operatorname{Spec}^{\rm c}_{\rm in} \boldsymbol{\phi} = (f, \boldsymbol{f}^\sharp)$ has stalk map $\boldsymbol{f}^\sharp_x = \boldsymbol{\phi}_x : \mathfrak{C}_{f_\phi(x)} \to \mathfrak{D}_x$. This map fits into the commutative diagram

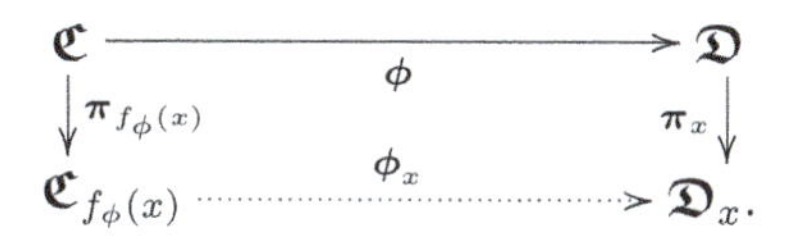

As $\boldsymbol{\phi}$ is interior, and the maps $\boldsymbol{\pi}_{f_\phi(x)}, \boldsymbol{\pi}_x$ are interior and surjective, then $\boldsymbol{f}^\sharp_x$ is interior. This implies that $\operatorname{Spec}^{\rm c}_{\rm in} \boldsymbol{\phi}$ is an interior morphism of interior local C^∞-ringed spaces with corners. Hence $\operatorname{Spec}^{\rm c}_{\rm in} : (\mathbf{C^\infty Rings^c_{in}})^{\rm op} \to \mathbf{LC^\infty RS^c_{in}}$ is a well-defined functor, which we call the *interior spectrum functor*.

Definition 5.14 The *global sections functor*

$$\Gamma^{\rm c} : \mathbf{LC^\infty RS^c} \longrightarrow (\mathbf{C^\infty Rings^c})^{\rm op}$$

takes objects $(X, \boldsymbol{\mathcal{O}}_X) \in \mathbf{LC^\infty RS^c}$ to $\boldsymbol{\mathcal{O}}_X(X)$ and takes morphisms $(f, \boldsymbol{f}^\sharp) : (X, \boldsymbol{\mathcal{O}}_X) \to (Y, \boldsymbol{\mathcal{O}}_Y)$ to $\Gamma^{\rm c} : (f, \boldsymbol{f}^\sharp) \mapsto \boldsymbol{f}_\sharp(Y)$. Here $\boldsymbol{f}_\sharp : \boldsymbol{\mathcal{O}}_Y \to f_*(\boldsymbol{\mathcal{O}}_X)$ corresponds to $\boldsymbol{f}^\sharp$ under (2.9).

The composition $\Gamma^{\rm c} \circ \operatorname{Spec}^{\rm c}$ maps $(\mathbf{C^\infty Rings^c})^{\rm op} \to (\mathbf{C^\infty Rings^c})^{\rm op}$, or equivalently maps $\mathbf{C^\infty Rings^c} \to \mathbf{C^\infty Rings^c}$. For each C^∞-ring with corners $\mathfrak{C}$ we define a morphism $\boldsymbol{\Xi}_{\mathfrak{C}} = (\Xi, \Xi_{\rm ex}) : \mathfrak{C} \to \Gamma^{\rm c} \circ \operatorname{Spec}^{\rm c} \mathfrak{C}$ in $\mathbf{C^\infty Rings^c}$ by $\Xi(c) : X_{\mathfrak{C}} \to \coprod_{x \in X_{\mathfrak{C}}} \mathcal{O}_{X_{\mathfrak{C}},x}$, $\Xi(c) : x \mapsto \pi_x(c)$ in $\mathfrak{C}_x \cong \mathcal{O}_{X_{\mathfrak{C}},x}$ for $c \in \mathfrak{C}$, and $\Xi_{\rm ex}(c') : X_{\mathfrak{C}} \to \coprod_{x \in X_{\mathfrak{C}}} \mathcal{O}^{\rm ex}_{X_{\mathfrak{C}},x}$, $\Xi_{\rm ex}(c') : x \mapsto \pi_{x,{\rm ex}}(c')$ in $\mathfrak{C}_{x,{\rm ex}} \cong \mathcal{O}^{\rm ex}_{X_{\mathfrak{C}},x}$ for $c' \in \mathfrak{C}_{\rm ex}$. This is functorial in $\mathfrak{C}$, so that the $\boldsymbol{\Xi}_{\mathfrak{C}}$ for all $\mathfrak{C}$ define a natural transformation $\boldsymbol{\Xi} : \mathrm{Id}_{\mathbf{C^\infty Rings^c}} \Rightarrow \Gamma^{\rm c} \circ \operatorname{Spec}^{\rm c}$ of functors $\mathrm{Id}_{\mathbf{C^\infty Rings^c}}, \Gamma^{\rm c} \circ \operatorname{Spec}^{\rm c} : \mathbf{C^\infty Rings^c} \to \mathbf{C^\infty Rings^c}$.

Here is the analogue of Lemma 2.45.

Lemma 5.15 *Let $\mathfrak{C}$ be a C^∞-ring with corners, and $\boldsymbol{X} = \operatorname{Spec}^{\rm c} \mathfrak{C}$. For any $c \in \mathfrak{C}$, let $U_c \subseteq X$ be as in Definition 2.42. Then $\boldsymbol{U}_c \cong \operatorname{Spec}^{\rm c}(\mathfrak{C}(c^{-1}))$. If $\mathfrak{C}$ is firm, or interior, then so is $\mathfrak{C}(c^{-1})$.*

Proof Write $\mathfrak{C}(c^{-1}) = (\mathfrak{D}, \mathfrak{D}_{\rm ex})$. By Lemma 4.27 we have $\mathfrak{D} \cong \mathfrak{C}(c^{-1})$. By Lemma 2.45, we need only show there is an isomorphism of stalks $\mathfrak{C}_{x,{\rm ex}} \to$

$\mathfrak{D}_{\hat{x},\mathrm{ex}}$. However, using the universal properties of $\mathfrak{C}_x$, $\mathfrak{C}(c^{-1})$, and $\mathfrak{C}(c^{-1})_{\hat{x}}$ this follows by the same reasoning as Lemma 2.45. If $\mathfrak{C}$ is firm, then so is $\mathfrak{C}(c^{-1})$, as $\mathfrak{C}(c^{-1})^\sharp_{\mathrm{ex}}$ is generated by the image of $\mathfrak{C}^\sharp_{\mathrm{ex}}$ under the morphism $\mathfrak{C} \to \mathfrak{C}(c^{-1})$. If $\mathfrak{C}$ is interior, then $\mathfrak{C}(c^{-1})$ is also interior, as otherwise zero divisors in $\mathfrak{C}(c^{-1})_{\mathrm{ex}}$ would have to come from zero divisors in $\mathfrak{C}_{\mathrm{ex}}$. $\square$

Theorem 5.16 *The functor* $\mathrm{Spec^c} : (\mathbf{C^\infty Rings^c})^{\mathrm{op}} \to \mathbf{LC^\infty RS^c}$ *is* ***right adjoint*** *to* $\Gamma^{\mathrm{c}} : \mathbf{LC^\infty RS^c} \to (\mathbf{C^\infty Rings^c})^{\mathrm{op}}$. *Thus for all* $\mathfrak{C}$ *in* $\mathbf{C^\infty Rings^c}$ *and all* $\boldsymbol{X}$ *in* $\mathbf{LC^\infty RS^c}$ *there are inverse bijections*

$$\mathrm{Hom}_{\mathbf{C^\infty Rings^c}}(\mathfrak{C}, \Gamma^{\mathrm{c}}(\boldsymbol{X})) \underset{R_{\mathfrak{C},\boldsymbol{X}}}{\overset{L_{\mathfrak{C},\boldsymbol{X}}}{\rightleftarrows}} \mathrm{Hom}_{\mathbf{LC^\infty RS^c}}(\boldsymbol{X}, \mathrm{Spec^c}\,\mathfrak{C}). \tag{5.1}$$

If we let $\boldsymbol{X} = \mathrm{Spec^c}\,\mathfrak{C}$ *then* $\Xi_{\mathfrak{C}} = R_{\mathfrak{C},\boldsymbol{X}}(\mathbf{id}_{\boldsymbol{X}})$, *and* Ξ *is the unit of the adjunction between* Γ^{c} *and* $\mathrm{Spec^c}$.

Proof We follow the proof of [49, Th. 4.20]. Take $\boldsymbol{X} \in \mathbf{LC^\infty RS^c}$ and $\mathfrak{C} \in \mathbf{C^\infty Rings^c}$, and let $\boldsymbol{Y} = (Y, \boldsymbol{\mathcal{O}}_Y) = \mathrm{Spec^c}\,\mathfrak{C}$. Define a functor $R_{\mathfrak{C},\boldsymbol{X}}$ in (5.1) by taking $R_{\mathfrak{C},\boldsymbol{X}}(\boldsymbol{f}) : \mathfrak{C} \to \Gamma^{\mathrm{c}}(\boldsymbol{X})$ to be the composition

$$\mathfrak{C} \xrightarrow{\Xi_{\mathfrak{C}}} \Gamma^{\mathrm{c}} \circ \mathrm{Spec^c}\,\mathfrak{C} = \Gamma^{\mathrm{c}}(\boldsymbol{Y}) \xrightarrow{\Gamma^{\mathrm{c}}(\boldsymbol{f})} \Gamma^{\mathrm{c}}(\boldsymbol{X})$$

for each morphism $\boldsymbol{f} : \boldsymbol{X} \to \boldsymbol{Y}$ in $\mathbf{LC^\infty RS^c}$. If $\boldsymbol{X} = \mathrm{Spec^c}\,\mathfrak{C}$ then we have $\Xi_{\mathfrak{C}} = R_{\mathfrak{C},\boldsymbol{X}}(\mathbf{id}_{\boldsymbol{X}})$. We see that $R_{\mathfrak{C},\boldsymbol{X}}$ is an extension of the functor $R_{\mathfrak{C},X}$ constructed in [49, Th. 4.20] for the adjunction between Spec and Γ. This will also occur for $L_{\mathfrak{C},\boldsymbol{X}}$.

Given a morphism $\boldsymbol{\phi} = (\phi, \phi_{\mathrm{ex}}) : \mathfrak{C} \to \Gamma^{\mathrm{c}}(\boldsymbol{X})$ in $\mathbf{C^\infty Rings^c}$ we define $L_{\mathfrak{C},\boldsymbol{X}}(\boldsymbol{\phi}) = \boldsymbol{g} = (g, g^\sharp, g^\sharp_{\mathrm{ex}})$ where $(g, g^\sharp) = L_{\mathfrak{C},X}(\phi)$ with $L_{\mathfrak{C},X}$ constructed in [49, Th. 4.20]. Here, g acts by $x \mapsto x_* \circ \phi$, where $x_* : \mathcal{O}_X(X) \to \mathbb{R}$ is the composition of the $\sigma_x : \mathcal{O}_X(X) \to \mathcal{O}_{X,x}$ with the unique morphism $\pi : \mathcal{O}_{X,x} \to \mathbb{R}$, as $\boldsymbol{\mathcal{O}}_{X,x}$ is a local C^∞-ring with corners. The morphisms $g^\sharp, g^\sharp_{\mathrm{ex}}$ are constructed as $g^\sharp$ is in [49, Th. 4.20], and we explain this explicitly now.

For $x \in X$ and $g(x) = y \in Y$, take the stalk map $\boldsymbol{\sigma}_x = (\sigma_x, \sigma_x^{\mathrm{ex}}) : \boldsymbol{\mathcal{O}}_X(X) \to \boldsymbol{\mathcal{O}}_{X,x}$. This gives the following diagram of C^∞-rings with corners

$$\begin{array}{ccccc} \mathfrak{C} & \xrightarrow{\quad\boldsymbol{\phi}\quad} & \Gamma^{\mathrm{c}}(\boldsymbol{X}) & & \\ \downarrow{\scriptstyle \boldsymbol{\pi}_y} & & {\scriptstyle \boldsymbol{\sigma}_x}\downarrow & & \\ \mathfrak{C}_y \cong \boldsymbol{\mathcal{O}}_{Y,y} & \xrightarrow{\boldsymbol{\phi}_x} & \boldsymbol{\mathcal{O}}_{X,x} & \xrightarrow{\quad\boldsymbol{\pi}\quad} & \mathbb{R}. \end{array} \tag{5.2}$$

We know $\mathfrak{C}_y \cong \boldsymbol{\mathcal{O}}_{Y,y}$ by Proposition 5.12 and $\pi : \mathcal{O}_{X,x} \to \mathbb{R}$ is the unique local morphism. If we have $(c, c') \in \mathfrak{C}$ with $y(c) \neq 0$, and $y \circ \Phi_i(c') \neq 0$ then

$\boldsymbol{\sigma}_x \circ \boldsymbol{\phi}(c, c') \in \boldsymbol{\mathcal{O}}_{X,x}$ with $\pi(\sigma_x \circ \phi(c)) \neq 0$ and

$$\pi\big(\Phi_i \circ \sigma_x^{\rm ex} \circ \phi_{\rm ex}(c)\big) = \pi(\sigma_x \circ \phi \circ \Phi_i(c)) \neq 0.$$

As $\boldsymbol{\mathcal{O}}_{X,x}$ is a local C^∞-ring with corners then $\boldsymbol{\sigma}_x \circ \boldsymbol{\phi}(c, c')$ is invertible in $\boldsymbol{\mathcal{O}}_{X,x}$. The universal property of $\boldsymbol{\pi}_y : \mathfrak{C} \to \mathfrak{C}_y$ gives a unique morphism $\boldsymbol{\phi}_x : \boldsymbol{\mathcal{O}}_{Y,y} \to \boldsymbol{\mathcal{O}}_{X,x}$ that makes (5.2) commute.

We define

$$\boldsymbol{g}_\sharp(V) = (g_\sharp(V), g_\sharp^{\rm ex}(V)) : \boldsymbol{\mathcal{O}}_Y(V) \to g_*(\boldsymbol{\mathcal{O}}_X)(V) = \boldsymbol{\mathcal{O}}_X(U)$$

for each open $V \subseteq Y$ with $U = g^{-1}(V) \subseteq X$ to act by $\boldsymbol{\phi}_x = (\phi_x, \phi_{x,\rm ex})$ on stalks at each $x \in U$. We can identify elements $s \in \mathcal{O}_X(U)$, $s' \in \mathcal{O}_X^{\rm ex}(U)$ with maps $s : U \to \coprod_{x \in U} \mathcal{O}_{X,x}$ and $s' : U \to \coprod_{x \in U} \mathcal{O}_{X,x}^{\rm ex}$, such that s, s' are locally of the form $s : x \mapsto \pi_x(c)$ in $\mathfrak{C}_x \cong \mathcal{O}_{X,x}$ for $c \in \mathfrak{C}$ and $s' : x \mapsto \pi_{x,\rm ex}(c')$ in $\mathfrak{C}_{x,\rm ex} \cong \mathcal{O}_{X,x}^{\rm ex}$ for $c' \in \mathfrak{C}_{\rm ex}$, and similarly for $t \in \mathcal{O}_Y(V)$, $t' \in \mathcal{O}_Y^{\rm ex}(V)$. If t is locally of the form $t : y \mapsto \pi_y(d)$ in $\mathfrak{D}_y \cong \mathcal{O}_{Y,c}$ near $\tilde{y} = g(\tilde{x})$ for $d \in \mathfrak{D}$, then $s = g_\sharp(V)(t)$ is locally of the form $s : x \mapsto \pi_x(c)$ in $\mathfrak{C}_x \cong \mathcal{O}_{X,x}$ near $\tilde{x}$ for $c = \phi(d) \in \mathfrak{C}$, so $g_\sharp(V)$ does map $\mathcal{O}_Y(V) \to \mathcal{O}_X(U)$, and similarly $g_\sharp^{\rm ex}(V)$ maps $\mathcal{O}_Y^{\rm ex}(V) \to \mathcal{O}_X^{\rm ex}(U)$. Hence $\boldsymbol{g}_\sharp(V)$ is well defined.

This defines a morphism $\boldsymbol{g}_\sharp : \boldsymbol{\mathcal{O}}_Y \to g_*(\boldsymbol{\mathcal{O}}_X)$ of sheaves of C^∞-rings with corners on Y, and we write $\boldsymbol{g}^\sharp : g^{-1}(\boldsymbol{\mathcal{O}}_Y) \to \boldsymbol{\mathcal{O}}_X$ for the corresponding morphism of sheaves of C^∞-rings with corners on X under (2.9). At a point $x \in X$ with $g(x) = y \in Y$, the stalk map $\boldsymbol{g}_x^\sharp : \boldsymbol{\mathcal{O}}_{Y,y} \to \boldsymbol{\mathcal{O}}_{X,x}$ is $\boldsymbol{\phi}_x$. Then $\boldsymbol{g} = (g, \boldsymbol{g}^\sharp)$ is a morphism in $\mathbf{LC^\infty RS^c}$, and $L_{\mathfrak{C},\boldsymbol{X}}(\boldsymbol{\phi}) = \boldsymbol{g}$. It remains to show that these define natural bijections, but this follows in a very similar way to [49, Th. 4.20]. □

Definition 5.17 Define the *interior global sections functor* $\Gamma_{\rm in}^{\rm c} : \mathbf{LC^\infty RS^c_{in}} \to (\mathbf{C^\infty Rings^c_{in}})^{\rm op}$ to act on objects $(X, \boldsymbol{\mathcal{O}}_X)$ by $\Gamma_{\rm in}^{\rm c} : (X, \boldsymbol{\mathcal{O}}_X) \mapsto (\mathfrak{C}, \mathfrak{C}_{\rm ex})$, where $\mathfrak{C} = \mathcal{O}_X(X)$ and $\mathfrak{C}_{\rm ex}$ is the set containing the zero element of $\mathcal{O}_X^{\rm ex}(X)$ and the elements of $\mathcal{O}_X^{\rm ex}(X)$ that are non-zero in every stalk. That is,

$$\begin{aligned}\mathfrak{C}_{\rm ex} = \{c' \in \mathcal{O}_X^{\rm ex}(X) : c' = 0 \in \mathcal{O}_X^{\rm ex}(X)\\ \text{or } \sigma_x^{\rm ex}(c') \neq 0 \in \mathcal{O}_{X,x}^{\rm ex}\ \forall x \in X\},\end{aligned} \tag{5.3}$$

where $\sigma_x^{\rm ex}$ is the stalk map $\sigma_x^{\rm ex} : \mathcal{O}_X^{\rm ex} \to \mathcal{O}_{X,x}^{\rm ex}$. This is an interior C^∞-ring with corners, where the C^∞-ring with corners structure is given by restriction from $(\mathcal{O}_X(X), \mathcal{O}_X^{\rm ex}(X))$. We define $\Gamma_{\rm in}^{\rm c}$ to act on morphisms $(f, \boldsymbol{f}^\sharp) : (X, \boldsymbol{\mathcal{O}}_X) \to (Y, \boldsymbol{\mathcal{O}}_Y)$ by $\Gamma_{\rm in}^{\rm c} : (f, \boldsymbol{f}^\sharp) \mapsto \boldsymbol{f}_\sharp(Y)|_{(\mathfrak{C},\mathfrak{C}_{\rm ex})}$ for $\boldsymbol{f}_\sharp : \boldsymbol{\mathcal{O}}_Y \to f_*(\boldsymbol{\mathcal{O}}_X)$ corresponding to $\boldsymbol{f}^\sharp$ under (2.9).

As in Definition 5.14, for each interior C^∞-ring with corners $\boldsymbol{\mathfrak{C}}$, we define a morphism $\boldsymbol{\Xi}_{\boldsymbol{\mathfrak{C}}}^{\rm in} = (\Xi^{\rm in}, \Xi_{\rm ex}^{\rm in}) : \boldsymbol{\mathfrak{C}} \to \Gamma_{\rm in}^{\rm c} \circ \mathrm{Spec}_{\rm in}^{\rm c} \boldsymbol{\mathfrak{C}}$ in $\mathbf{C^\infty Rings^c_{in}}$ by

$\Xi^{\rm in}(c) : X_{\mathfrak C} \to \coprod_{x\in X_{\mathfrak C}} \mathcal{O}_{X_{\mathfrak C},x}$, $\Xi^{\rm in}(c) : x \mapsto \pi_x(c)$ in $\mathfrak{C}_x \cong \mathcal{O}_{X_{\mathfrak C},x}$ for $c \in \mathfrak{C}$, and $\Xi^{\rm in}_{\rm ex}(c') : X_{\mathfrak C} \to \coprod_{x\in X_{\mathfrak C}} \mathcal{O}^{\rm ex}_{X_{\mathfrak C},x}$, $\Xi^{\rm in}_{\rm ex}(c') : x \mapsto \pi_{x,{\rm ex}}(c')$ in $\mathfrak{C}_{x,{\rm ex}} \cong \mathcal{O}^{\rm ex}_{X_{\mathfrak C},x}$ for $c' \in \mathfrak{C}_{\rm ex}$. We need to check $\Xi^{\rm in}_{\rm ex}(c')$ lies in (5.3), but this is immediate as $\boldsymbol{\pi}_x = (\pi_x, \pi_{x,{\rm ex}}) : \mathfrak{C} \to \mathfrak{C}_x$ is interior. The $\Xi^{\rm in}_{\mathfrak C}$ for all $\mathfrak{C}$ define a natural transformation $\Xi^{\rm in} : {\rm Id}_{\mathbf{C^\infty Rings^c_{in}}} \Rightarrow \Gamma^{\rm c}_{\rm in} \circ {\rm Spec}^{\rm c}_{\rm in}$ of functors ${\rm Id}_{\mathbf{C^\infty Rings^c_{in}}}, \Gamma^{\rm c}_{\rm in} \circ {\rm Spec}^{\rm c}_{\rm in} : \mathbf{C^\infty Rings^c_{in}} \to \mathbf{C^\infty Rings^c_{in}}$.

Theorem 5.18 *The functor* ${\rm Spec}^{\rm c}_{\rm in} : (\mathbf{C^\infty Rings^c_{in}})^{\rm op} \to \mathbf{LC^\infty RS^c_{in}}$ *is right adjoint to* $\Gamma^{\rm c}_{\rm in} : \mathbf{LC^\infty RS^c_{in}} \to (\mathbf{C^\infty Rings^c_{in}})^{\rm op}$.

Proof This proof is identical to that of Theorem 5.16. We need only check that the definition of $\Gamma^{\rm c}_{\rm in}(\boldsymbol{X})$, which may not be equal to $\mathcal{O}_X(X)$, gives well-defined maps $\boldsymbol{\sigma}^{\rm in}_x : \Gamma^{\rm c}_{\rm in}(\boldsymbol{X}) \to \mathcal{O}_{X,x}$. As $\Gamma^{\rm c}_{\rm in}(\boldsymbol{X})$ is a subobject of $\mathcal{O}_X(X)$, these maps are the restriction of the stalk maps $\boldsymbol{\sigma}_x : \mathcal{O}_X(X) \to \mathcal{O}_{X,x}$ to $\Gamma^{\rm c}_{\rm in}(\boldsymbol{X})$. The definition of $\Gamma^{\rm c}_{\rm in}(\boldsymbol{X})$ implies these maps are interior. □

5.4 C^∞-schemes with corners

Definition 5.19 A local C^∞-ringed space with corners that is isomorphic in $\mathbf{LC^\infty RS^c}$ to ${\rm Spec}^{\rm c}\,\mathfrak{C}$ for some C^∞-ring with corners $\mathfrak{C}$ is called an *affine C^∞-scheme with corners*. We define the category $\mathbf{AC^\infty Sch^c}$ to be the full subcategory of affine C^∞-schemes with corners in $\mathbf{LC^\infty RS^c}$.

Let $\boldsymbol{X} = (X, \mathcal{O}_X)$ be a local C^∞-ringed space with corners. We call $\boldsymbol{X}$ a *C^∞-scheme with corners* if X can be covered by open sets $U \subseteq X$ such that $(U, \mathcal{O}_X|_U)$ is a affine C^∞-scheme with corners. We define the category $\mathbf{C^\infty Sch^c}$ of C^∞-schemes with corners to be the full subcategory of C^∞-schemes with corners in $\mathbf{LC^\infty RS^c}$. Then $\mathbf{AC^\infty Sch^c}$ is a full subcategory of $\mathbf{C^\infty Sch^c}$.

A local C^∞-ringed space with corners that is isomorphic in $\mathbf{LC^\infty RS^c_{in}}$ to ${\rm Spec}^{\rm c}_{\rm in}\,\mathfrak{C}$ for some interior C^∞-ring with corners $\mathfrak{C}$ is called an *interior affine C^∞-scheme with corners*. We define the category $\mathbf{AC^\infty Sch^c_{in}}$ of interior affine C^∞-schemes with corners to be the full subcategory of interior affine C^∞-schemes with corners in $\mathbf{LC^\infty RS^c_{in}}$, so $\mathbf{AC^\infty Sch^c_{in}}$ is a non-full subcategory of $\mathbf{AC^\infty Sch^c}$. We call an object $\boldsymbol{X} \in \mathbf{LC^\infty RS^c_{in}}$ an *interior C^∞-scheme with corners* if it can be covered by open sets $U \subseteq X$ such that $(U, \mathcal{O}_X|_U)$ is an interior affine C^∞-scheme with corners. We define the category $\mathbf{C^\infty Sch^c_{in}}$ of interior C^∞-schemes with corners to be the full subcategory of interior C^∞-schemes with corners in $\mathbf{LC^\infty RS^c_{in}}$. This implies that $\mathbf{AC^\infty Sch^c_{in}}$ is a full subcategory of $\mathbf{C^\infty Sch^c_{in}}$. Clearly $\mathbf{C^\infty Sch^c_{in}} \subset \mathbf{C^\infty Sch^c}$.

In Theorem 5.28(b) we will show that an object $\boldsymbol{X}$ in $\mathbf{C^\infty Sch^c}$ lies in $\mathbf{C^\infty Sch^c_{in}}$ if and only if it lies in $\mathbf{LC^\infty RS^c_{in}}$. That is, the two natural definitions of 'interior C^∞-scheme with corners' turn out to be equivalent.

Lemma 5.20 *Let $\boldsymbol{X}$ be an (interior) C^∞-scheme with corners, and $\boldsymbol{U} \subseteq \boldsymbol{X}$ be open. Then $\boldsymbol{U}$ is also an (interior) C^∞-scheme with corners.*

Proof This follows from Lemma 5.15 and the fact in Definition 2.42 that the topology on $\operatorname{Spec^c} \mathfrak{C}$ is generated by subsets U_c. □

We explain the relation with manifolds with (g-)corners in Chapter 3.

Definition 5.21 Define a functor $F_{\mathbf{Man^c}}^{\mathbf{C^\infty Sch^c}} : \mathbf{Man^c} \to \mathbf{C^\infty Sch^c}$ that acts on objects $X \in \mathbf{Man^c}$ by $F_{\mathbf{Man^c}}^{\mathbf{C^\infty Sch^c}}(X) = (X, \boldsymbol{\mathcal{O}}_X)$, where $\boldsymbol{\mathcal{O}}_X(U) = \boldsymbol{C^\infty}(U) = (C^\infty(U), \mathrm{Ex}(U))$ from Example 4.18(a) for each open subset $U \subseteq X$. If $V \subseteq U \subseteq X$ are open we define $\boldsymbol{\rho}_{UV} = (\rho_{UV}, \rho_{UV}^{\mathrm{ex}}) : \boldsymbol{C^\infty}(U) \to \boldsymbol{C^\infty}(V)$ by $\rho_{UV} : c \mapsto c|_V$ and $\rho_{UV}^{\mathrm{ex}} : c' \mapsto c'|_V$.

It is easy to verify that $\boldsymbol{\mathcal{O}}_X$ is a sheaf of C^∞-rings with corners on X, so $\boldsymbol{X} = (X, \boldsymbol{\mathcal{O}}_X)$ is a C^∞-ringed space with corners. We show in Theorem 5.22(b) that $\boldsymbol{X}$ is a C^∞-scheme with corners, and it is also interior.

Let $f : X \to Y$ be a morphism in $\mathbf{Man^c}$. Writing $F_{\mathbf{Man^c}}^{\mathbf{C^\infty Sch^c}}(X) = (X, \boldsymbol{\mathcal{O}}_X)$ and $F_{\mathbf{Man^c}}^{\mathbf{C^\infty Sch^c}}(Y) = (Y, \boldsymbol{\mathcal{O}}_Y)$, for all open $U \subseteq Y$, we define

$$\boldsymbol{f}_\sharp(U) : \boldsymbol{\mathcal{O}}_Y(U) = \boldsymbol{C^\infty}(U) \longrightarrow f_*(\boldsymbol{\mathcal{O}}_x)(U) = \boldsymbol{\mathcal{O}}_X(f^{-1}(U)) = \boldsymbol{C^\infty}(f^{-1}(U))$$

by $f_\sharp(U) : c \mapsto c \circ f$ for all $c \in C^\infty(U)$ and $f_\sharp^{\mathrm{ex}}(U) : c' \mapsto c' \circ f$ for all $c' \in \mathrm{Ex}(U)$. Then $\boldsymbol{f}_\sharp(U)$ is a morphism of C^∞ rings with corners, and $\boldsymbol{f}_\sharp : \boldsymbol{\mathcal{O}}_Y \to f_*(\boldsymbol{\mathcal{O}}_X)$ is a morphism of sheaves of C^∞-rings with corners on Y. Let $\boldsymbol{f}^\sharp : f^{-1}(\boldsymbol{\mathcal{O}}_Y) \to \boldsymbol{\mathcal{O}}_X$ correspond to $\boldsymbol{f}_\sharp$ under (2.9). Then $F_{\mathbf{Man^c}}^{\mathbf{C^\infty Sch^c}}(f) = \boldsymbol{f} = (f, \boldsymbol{f}^\sharp) : (X, \boldsymbol{\mathcal{O}}_X) \to (Y, \boldsymbol{\mathcal{O}}_Y)$ is a morphism in $\mathbf{C^\infty Sch^c}$. Define a functor $F_{\mathbf{Man^c_{in}}}^{\mathbf{C^\infty Sch^c_{in}}} : \mathbf{Man^c_{in}} \to \mathbf{C^\infty Sch^c_{in}}$ by restriction of $F_{\mathbf{Man^c}}^{\mathbf{C^\infty Sch^c}}$ to $\mathbf{Man^c_{in}}$.

All the above generalizes immediately from manifolds with corners to manifolds with g-corners in §3.3, giving functors $F_{\mathbf{Man^{gc}}}^{\mathbf{C^\infty Sch^c}} : \mathbf{Man^{gc}} \to \mathbf{C^\infty Sch^c}$ and $F_{\mathbf{Man^{gc}_{in}}}^{\mathbf{C^\infty Sch^c_{in}}} : \mathbf{Man^{gc}_{in}} \to \mathbf{C^\infty Sch^c_{in}}$. It also generalizes to manifolds with (g-)corners of mixed dimension $\check{\mathbf{M}}\mathbf{an^c}, \ldots, \check{\mathbf{M}}\mathbf{an^{gc}_{in}}$.

The functors $F_{\mathbf{Man^c}}^{\mathbf{C^\infty Sch^c}}, \ldots, F_{\mathbf{Man^{gc}_{in}}}^{\mathbf{C^\infty Sch^c_{in}}}$ map to various subcategories of $\mathbf{C^\infty Sch^c}, \mathbf{C^\infty Sch^c_{in}}$ defined in §5.6, including $\mathbf{C^\infty Sch^c_{to,ex}}, \mathbf{C^\infty Sch^c_{to}}$ of *toric* C^∞-schemes with corners. We will write $F_{\mathbf{Man^c}}^{\mathbf{C^\infty Sch^c_{to}}}, \ldots$ for $F_{\mathbf{Man^c}}^{\mathbf{C^\infty Sch^c}}, \ldots$ considered as mapping to these subcategories.

We say that a C^∞-scheme with corners $\boldsymbol{X}$ *is a manifold with corners* (or *is a manifold with g-corners)* if $\boldsymbol{X} \cong F_{\mathbf{Man^c}}^{\mathbf{C^\infty Sch^c}}(X)$ for some $X \in \mathbf{Man^c}$ (or $\boldsymbol{X} \cong F_{\mathbf{Man^{gc}}}^{\mathbf{C^\infty Sch^c}}(X)$ for some $X \in \mathbf{Man^{gc}}$).

Theorem 5.22 *Let X be a manifold with corners, or a manifold with g-corners, and $\boldsymbol{X} = F_{\mathbf{Man^c}}^{\mathbf{C^\infty Sch^c}}(X)$ or $\boldsymbol{X} = F_{\mathbf{Man^{gc}}}^{\mathbf{C^\infty Sch^c}}(X)$. Then we have the following.*

(a) *If X is a manifold with faces or g-faces, then $\boldsymbol{X}$ is an affine C^∞-scheme with corners, and is isomorphic to* $\mathrm{Spec^c}\, \boldsymbol{C}^\infty(X)$. *It is also an interior affine C^∞-scheme with corners, with $\boldsymbol{X} \cong \mathrm{Spec^c_{in}}\, \boldsymbol{C}^\infty_{\mathrm{in}}(X)$.*

If X is a manifold with corners, but not with faces, then $\boldsymbol{X}$ is not affine.

(b) *In general, $\boldsymbol{X}$ is an interior C^∞-scheme with corners.*

(c) *The functors $F_{\mathbf{Man^c}}^{\mathbf{C^\infty Sch^c}}, F_{\mathbf{Man^c_{in}}}^{\mathbf{C^\infty Sch^c_{in}}}, F_{\mathbf{Man^{gc}}}^{\mathbf{C^\infty Sch^c}}, F_{\mathbf{Man^{gc}_{in}}}^{\mathbf{C^\infty Sch^c_{in}}}$ are fully faithful.*

Proof For (a), let X be a manifold with corners or a manifold with g-corners, and write $\boldsymbol{X} = (X, \boldsymbol{\mathcal{O}}_X)$. Note that $\Gamma^{\mathrm{c}}(\boldsymbol{X}) = \boldsymbol{\mathcal{O}}_X(X) = \boldsymbol{C}^\infty(X)$ by the definition of $\boldsymbol{\mathcal{O}}_X$. Consider the map $L_{\mathfrak{C},\boldsymbol{X}}$ in (5.1). In the notation of Theorem 5.16, if we let $\mathfrak{C} = \Gamma^{\mathrm{c}}(\boldsymbol{X}) = \boldsymbol{C}^\infty(X)$, then $L_{\boldsymbol{C}^\infty(X),\boldsymbol{X}}$ is a bijection

$$\begin{aligned} L_{\boldsymbol{C}^\infty(X),\boldsymbol{X}} : \mathrm{Hom}_{\mathbf{C^\infty Rings^c}}&(\boldsymbol{C}^\infty(X), \boldsymbol{C}^\infty(X)) \\ &\longrightarrow \mathrm{Hom}_{\mathbf{LC^\infty RS^c}}(\boldsymbol{X}, \mathrm{Spec^c}\, \boldsymbol{C}^\infty(X)). \end{aligned}$$

Write $\boldsymbol{Y} = \mathrm{Spec^c}\, \boldsymbol{C}^\infty(X)$, and define a morphism $\boldsymbol{g} = (g, \boldsymbol{g}^\sharp) : \boldsymbol{X} \to \boldsymbol{Y}$ in $\mathbf{LC^\infty RS^c}$ by $\boldsymbol{g} = L_{\boldsymbol{C}^\infty(X),\boldsymbol{X}}(\mathrm{id}_{\boldsymbol{C}^\infty(X)})$.

The continuous map $g : X \to Y$ is defined in the proof of Theorem 5.16 by $g(x) = x_* \circ \mathrm{id}_{\boldsymbol{C}^\infty(X)}$, where x_* is the evaluation map at the point $x \in X$. This is a homeomorphism of topological spaces as in the proof of [49, Th. 4.41]. On stalks at each $x \in X$ we have $\boldsymbol{g}^\sharp_x = \boldsymbol{\lambda}_x : (\boldsymbol{C}^\infty(X))_{x_*} \to \boldsymbol{C}^\infty_x(X)$, where $\boldsymbol{\lambda}_x$ is as in Example 4.35. If X has faces then $\boldsymbol{\lambda}_x$ is an isomorphism by Proposition 4.36, and if X has g-faces then $\boldsymbol{\lambda}_x$ is an isomorphism by Definition 4.37. Hence if X has (g-)faces then $g, \boldsymbol{g}^\sharp$ and $\boldsymbol{g}$ are isomorphisms, and $\boldsymbol{X} \cong \mathrm{Spec^c}\, \boldsymbol{C}^\infty(X)$.

Essentially the same proof for the interior case, using $\mathrm{Spec^c_{in}}, \Gamma^{\mathrm{c}}_{\mathrm{in}}$ adjoint by Theorem 5.18, shows that $\boldsymbol{X} \cong \mathrm{Spec^c_{in}}\, \boldsymbol{C}^\infty_{\mathrm{in}}(X)$ if X has (g-)faces.

If X has corners, but not faces, then Proposition 4.36 gives $x \in X$ such that $\boldsymbol{g}^\sharp_x = \boldsymbol{\lambda}_x$ is not an isomorphism. The proof shows that $\lambda_{x,\mathrm{ex}}$ is not surjective. If $\boldsymbol{X} \cong \mathrm{Spec^c}\, \mathfrak{C}$ for some C^∞-ring with corners $\mathfrak{C}$ then $\mathfrak{C}_x \cong \boldsymbol{\mathcal{O}}_{X,x} = \boldsymbol{C}^\infty_x(X)$, and the localization morphism $\mathfrak{C} \to \mathfrak{C}_x$ is surjective. But $\mathfrak{C} \to \mathfrak{C}_x \cong \boldsymbol{C}^\infty_x(X)$ factors as $\mathfrak{C} \to \boldsymbol{C}^\infty(X) \xrightarrow{\boldsymbol{\lambda}_x} \boldsymbol{C}^\infty_x(X)$, and $\lambda_{x,\mathrm{ex}}$ is not surjective, a contradiction. Hence $\boldsymbol{X}$ is not affine, completing part (a).

For (b), for any point $x \in X$, we can find an open neighbourhood U of x which is a manifold with (g-)faces, as in Example 4.38 in the g-corners case. Then $\boldsymbol{U} \cong \mathrm{Spec^c}\, \boldsymbol{C}^\infty(U) \cong \mathrm{Spec^c_{in}}\, \boldsymbol{C}^\infty_{\mathrm{in}}(U)$ by (a). So $\boldsymbol{X}$ can be covered by open $U \subseteq X$ which are (interior) affine C^∞-schemes with corners, and $\boldsymbol{X}$ is an (interior) C^∞-scheme with corners.

For (c), if $f,g:X\to Y$ are morphisms in $\mathbf{Man^c}$ and $(f,\boldsymbol{f}^\sharp)=F_{\mathbf{Man^c}}^{\mathbf{C^\infty Sch^c}}(f)$ $=F_{\mathbf{Man^c}}^{\mathbf{C^\infty Sch^c}}(g)=(g,\boldsymbol{g}^\sharp)$, we see that $f=g$, so $F_{\mathbf{Man^c}}^{\mathbf{C^\infty Sch^c}}$ is faithful. Suppose $\boldsymbol{h}=(h,\boldsymbol{h}^\sharp):\boldsymbol{X}\to\boldsymbol{Y}$ is a morphism in $\mathbf{C^\infty Sch^c}$, and $x\in X$ with $h(y)=y$ in Y. We may choose open $y\in V\subseteq Y$ and $x\in U\subseteq h^{-1}(V)\subseteq X$ such that $\boldsymbol{U}=(U,\boldsymbol{\mathcal{O}}_X|_U)$ and $\boldsymbol{V}=(V,\boldsymbol{\mathcal{O}}_Y|_V)$ are affine. Then $\boldsymbol{h}^\sharp$ induces a morphism $\boldsymbol{h}^\sharp_{UV}:\boldsymbol{\mathcal{O}}_Y(V)=\boldsymbol{C^\infty}(V)\to\boldsymbol{\mathcal{O}}_X(U)=\boldsymbol{C^\infty}(U)$ in $\mathbf{C^\infty Rings^c}$.

By considering restriction to points in U,V we see that $\boldsymbol{h}^\sharp_{UV}$ maps $C^\infty(V)$ $\to C^\infty(U)$ and $C^\infty_{\rm ex}(V)\to C^\infty_{\rm ex}(U)$ by $c\mapsto c\circ h|_U$ for c in $C^\infty(V)$ or $C^\infty_{\rm ex}(V)$. Thus, if $c:V\to\mathbb{R}$ is smooth, or $c:V\to[0,\infty)$ is exterior, then $c\circ h|_U:U\to$ $\mathbb{R}$ is smooth, or $c\circ h|_U:U\to[0,\infty)$ is exterior. Since V is affine, this implies that $h|_U:U\to V$ is smooth, and $\boldsymbol{h}|_U=F_{\mathbf{Man^c}}^{\mathbf{C^\infty Sch^c}}(h|_U)$. As we can cover X,Y by such U,V, we see that h is smooth and $\boldsymbol{h}=F_{\mathbf{Man^c}}^{\mathbf{C^\infty Sch^c}}(h)$. Hence $F_{\mathbf{Man^c}}^{\mathbf{C^\infty Sch^c}}$ is full. The argument for $F_{\mathbf{Man^c_{in}}}^{\mathbf{C^\infty Sch^c_{in}}},\ldots,F_{\mathbf{Man^{gc}_{in}}}^{\mathbf{C^\infty Sch^c_{in}}}$ is essentially the same. □

Theorem 5.22 shows we can regard $\mathbf{Man^c},\mathbf{Man^{gc}}$ (or $\mathbf{Man^c_{in}},\mathbf{Man^{gc}_{in}}$) as full subcategories of $\mathbf{C^\infty Sch^c}$ (or $\mathbf{C^\infty Sch^c_{in}}$), and regard C^∞-schemes with corners as generalizations of manifolds with (g-)corners.

5.5 Semi-complete C^∞-rings with corners

Proposition 2.51 says $\operatorname{Spec}\Xi_{\mathfrak{C}}:\operatorname{Spec}\circ\Gamma\circ\operatorname{Spec}\mathfrak{C}\to\operatorname{Spec}\mathfrak{C}$ is an isomorphism for any C^∞-ring $\mathfrak{C}$. This was used in §2.6 to define the category $\mathbf{C^\infty Rings_{co}}$ of *complete* C^∞-rings (those isomorphic to $\Gamma\circ\operatorname{Spec}\mathfrak{C}$), such that $(\mathbf{C^\infty Rings_{co}})^{\rm op}$ is equivalent to the category of affine C^∞-schemes $\mathbf{AC^\infty Sch}$. This can be used to prove that fibre products and finite limits exist in $\mathbf{C^\infty Sch}$.

We now consider to what extent these results generalize to the corners case. Firstly, it turns out that $\operatorname{Spec^c}\boldsymbol{\Xi}_{\boldsymbol{\mathfrak{C}}}$ need not be an isomorphism for C^∞-rings with corners $\boldsymbol{\mathfrak{C}}$. The next remark and example explain what can go wrong.

Remark 5.23 Let $\boldsymbol{\mathfrak{C}}=(\mathfrak{C},\mathfrak{C}_{\rm ex})$ be a C^∞-ring with corners. Write $\boldsymbol{X}=$ $(X,\boldsymbol{\mathcal{O}}_X)=\operatorname{Spec^c}\boldsymbol{\mathfrak{C}}$, an affine C^∞-scheme with corners. The global sections $\Gamma^{\rm c}(\boldsymbol{X})=\boldsymbol{\mathcal{O}}_X(X)$ is a C^∞-ring with corners, with a morphism $\boldsymbol{\Xi}_{\boldsymbol{\mathfrak{C}}}:\boldsymbol{\mathfrak{C}}\to$ $\boldsymbol{\mathcal{O}}_X(X)$, so we have a morphism of C^∞-schemes with corners

$$\operatorname{Spec^c}\boldsymbol{\Xi}_{\boldsymbol{\mathfrak{C}}}=\operatorname{Spec^c}(\Xi,\Xi_{\rm ex}):\operatorname{Spec^c}\circ\Gamma^{\rm c}\circ\operatorname{Spec^c}\boldsymbol{\mathfrak{C}}\longrightarrow\operatorname{Spec^c}\boldsymbol{\mathfrak{C}}. \tag{5.4}$$

We want to know how close (5.4) is to being an isomorphism.

First note that as the underlying C^∞-scheme of $\boldsymbol{X}$ is $\underline{X} = \operatorname{Spec} \mathfrak{C}$, Proposition 2.51 shows (5.4) is an isomorphism on the level of C^∞-schemes, and hence of topological spaces. Thus (5.4) is an isomorphism if Ξ_{ex} induces an isomorphism of sheaves of monoids. It is sufficient to check this on stalks. Therefore, (5.4) is an isomorphism if the following is an isomorphism for all $x \in X$:

$$\Xi_{x,\mathrm{ex}} : \mathfrak{C}_{x,\mathrm{ex}} \cong \mathcal{O}^{\mathrm{ex}}_{X,x} \longrightarrow (\mathcal{O}^{\mathrm{ex}}_X(X))_{x_*}.$$

Consider the diagram of monoids for $x \in X$:

$$\begin{array}{ccc} \mathfrak{C}_{\mathrm{ex}} & \xrightarrow{\pi_{x,\mathrm{ex}}} & \mathfrak{C}_{x,\mathrm{ex}} \cong \mathcal{O}^{\mathrm{ex}}_{X,x} \\ \Big\downarrow \Xi_{\mathrm{ex}} & \overset{\rho^{\mathrm{ex}}_{X,x}}{\nearrow} & \Big\downarrow \Xi_{x,\mathrm{ex}} \\ \mathcal{O}^{\mathrm{ex}}_X(X) & \xrightarrow[\hat\pi_{x,\mathrm{ex}}]{} & (\mathcal{O}^{\mathrm{ex}}_X(X))_{x_*}. \end{array} \tag{5.5}$$

Here $\rho^{\mathrm{ex}}_{X,x}$ takes an element $s' \in \mathcal{O}^{\mathrm{ex}}_X(X)$ to its value in the stalk $\mathcal{O}^{\mathrm{ex}}_{X,x}$. It is not difficult to show that the upper left triangle and the outer rectangle in (5.5) commute, $\pi_{x,\mathrm{ex}}, \hat\pi_{x,\mathrm{ex}}$ are surjective, and $\Xi_{x,\mathrm{ex}}$ is injective. However, as Example 5.24 shows, $\Xi_{x,\mathrm{ex}}$ need not be surjective (in which case (5.4) is not an isomorphism), and the bottom right triangle of (5.5) need not commute.

That is, if two elements of $\mathcal{O}^{\mathrm{ex}}_X(X)$ agree locally, then while they have the same image in the stalk $\mathcal{O}^{\mathrm{ex}}_{X,x}$ they do not necessarily have the same value in the localization of $\mathcal{O}^{\mathrm{ex}}_X(X)$ at x_*. This is because for $c', d' \in \mathfrak{C}_{\mathrm{ex}}$, equality in $\mathfrak{C}_{\mathrm{ex},x}$ requires a global equality. That is, there need to be $a', b' \in \mathfrak{C}_{\mathrm{ex}}$ such that $a'c' = b'd' \in \mathfrak{C}_{\mathrm{ex}}$ with a', b' satisfying additional conditions as in Proposition 4.33. In the C^∞-ring $\mathfrak{C}$, this equality is only a local equality, as the a' and b' can come from bump functions. However, in the monoid, bump functions do not necessarily exist, meaning this condition is stronger and harder to satisfy.

In the following example, $\mathfrak{C}$ is interior, and the previous discussion also holds with $\mathrm{Spec}^{\mathrm{c}}_{\mathrm{in}}, \Gamma^{\mathrm{c}}_{\mathrm{in}}$ in place of $\mathrm{Spec}^{\mathrm{c}}, \Gamma^{\mathrm{c}}$.

Example 5.24 Define open subsets $U, V \subset \mathbb{R}^2$ by $U = (-1,1) \times (-1,\infty)$ and $V = (-1,1) \times (-\infty,1)$. The important properties for our purposes are that $\mathbb{R}^2 \setminus U, \mathbb{R}^2 \setminus V$ are connected, but $\mathbb{R}^2 \setminus (U \cup V)$ is disconnected, being divided into connected components $(-\infty,-1] \times \mathbb{R}$ and $[1,\infty) \times \mathbb{R}$.

Choose smooth $\varphi, \psi : \mathbb{R}^2 \to \mathbb{R}$ such that $\varphi|_U > 0$, $\varphi|_{\mathbb{R}^2 \setminus U} = 0$, $\psi|_V > 0$, $\psi|_{\mathbb{R}^2 \setminus V} = 0$, so φ, ψ map to $[0,\infty)$. Define a C^∞-ring with corners $\mathfrak{C}$ by

$$\begin{aligned} \mathfrak{C} = \mathbb{R}(x_1,x_2)[y_1,\ldots,y_6]/\big(&\Phi_i(y_1) = \Phi_i(y_2) = 0,\ \Phi_i(y_3) = \Phi_i(y_4) \\ &= \varphi(x_1,x_2),\ \Phi_i(y_5) = \Phi_i(y_6) = \psi(x_1,x_2)\big)\big[y_1y_3 = y_2y_4,\ y_1y_5 = y_2y_6\big], \end{aligned}$$

using the notation of Definition 4.24. Let $\boldsymbol{X} = (X, \mathcal{O}_X) = \mathrm{Spec}^{\mathrm{c}}\, \mathfrak{C}$, an affine

C^∞-scheme with corners. As a topological space we write

$$X = \{(x_1, x_2, y_1, \ldots, y_6) \in \mathbb{R}^2 \times [0,\infty)^6 : y_1 = y_2 = 0,$$
$$y_3 = y_4 = \varphi(x_1, x_2),\ y_5 = y_6 = \psi(x_1, x_2)\}.$$

The projection $X \to \mathbb{R}^2$ mapping $(x_1, x_2, y_1, \ldots, y_6) \mapsto (x_1, x_2)$ is a homeomorphism, and we use this to identify $X \cong \mathbb{R}^2$.

We will show that (5.4) is *not* an isomorphism, in contrast to Proposition 2.51 for C^∞-rings and C^∞-schemes.

For each $x \in X$ we have a monoid $\mathcal{O}_{X,x,\mathrm{ex}} \cong \mathfrak{C}_{x,\mathrm{ex}}$ by Proposition 5.12, containing elements $\pi_{x,\mathrm{ex}}(y_j)$ for $j = 1, \ldots, 6$. If $x \in U$ then using Proposition 4.33 and the relations $\Phi_i(y_3) = \Phi_i(y_4) = \varphi(x_1, x_2)$, $y_1 y_3 = y_2 y_4$, and $\varphi(x) > 0$ we see that $\pi_{x,\mathrm{ex}}(y_1) = \pi_{x,\mathrm{ex}}(y_2)$. Similarly if $x \in V$ then $\Phi_i(y_5) = \Phi_i(y_6) = \psi(x_1, x_2)$, $y_1 y_5 = y_2 y_6$, and $\psi(x) > 0$ give $\pi_{x,\mathrm{ex}}(y_1) = \pi_{x,\mathrm{ex}}(y_2)$. Thus as $U \cup V = (-1,1) \times \mathbb{R}$ we may define an element $z \in \mathcal{O}_X^{\mathrm{ex}}(X)$ by

$$z(x_1, x_2) = \begin{cases} \pi_{x,\mathrm{ex}}(y_1), & x \leqslant -1, \\ \pi_{x,\mathrm{ex}}(y_1) = \pi_{x,\mathrm{ex}}(y_2), & x \in (-1,1), \\ \pi_{x,\mathrm{ex}}(y_2), & x \geqslant 1. \end{cases}$$

We will show that for $x = (-2, 0)$ in X there does not exist $c' \in \mathfrak{C}_{\mathrm{ex}}$ with $\Xi_{x,\mathrm{ex}} \circ \pi_{x,\mathrm{ex}}(c') = \hat{\pi}_{x,\mathrm{ex}}(z)$ in $(\mathcal{O}_X^{\mathrm{ex}}(X))_{x_*}$. Thus $\Xi_{x,\mathrm{ex}} : \mathcal{O}_{X,x}^{\mathrm{ex}} \to (\mathcal{O}_X^{\mathrm{ex}}(X))_{x_*}$ is not surjective. This implies that the bottom right triangle of (5.5) does not commute at $z \in \mathcal{O}_X^{\mathrm{ex}}(X)$. Also $\mathrm{Spec^c}\, \boldsymbol{\Xi}_{\mathfrak{C}}$ in (5.4) is not an isomorphism, as $\Xi_{x,\mathrm{ex}}$ is part of the action of $\mathrm{Spec^c}\, \boldsymbol{\Xi}_{\mathfrak{C}}$ on stalks at x, and is not an isomorphism.

Suppose for a contradiction that such c' does exist. Then $\hat{\pi}_{x,\mathrm{ex}} \circ \Xi_{\mathrm{ex}}(c') = \hat{\pi}_{x,\mathrm{ex}}(z)$ as the outer rectangle of (5.5) commutes, so by Proposition 4.33 there exist $a', b' \in \mathcal{O}_X^{\mathrm{ex}}(X)$ with $a'(x) = b'(x) \neq 0$ and $a' = b'$ near x, such that $a' z = b' \Xi_{\mathrm{ex}}(c')$. Locally on X, a', b' are of the form $\exp(f(x_1, x_2, y_1, \ldots, y_6)) y_1^{a_1} \cdots y_6^{a_6}$ or zero, where the transitions between different representations of this form (e.g. changing $a_1, \ldots, a_6$) satisfy strict rules. We will consider a', b' in the regions $x_1 \leqslant -1$, $x_1 \geqslant 1$, $U \setminus V$, $V \setminus U$. In each region, a', b' must admit a global representation $\exp(f) y_1^{a_1} \cdots y_6^{a_6}$.

Since $a'(x) = b'(x) \neq 0$, we see that a', b' are of the form $\exp(f)$ in $x_1 \leqslant -1$. On crossing the wall $x = -1$, $y \leqslant -1$ between $x \leqslant -1$ and $U \setminus V$, y_3, y_4 become invertible, so a', b' may be of the form $\exp(f) y_3^{a_3} y_4^{a_4}$ in $U \setminus V$. On crossing the wall $x = 1$, $y \leqslant -1$ between $U \setminus V$ and $x \geqslant 1$, we see that a', b' must be of the form $\exp(f) y_3^{a_3} y_4^{a_4}$ in $x_1 \geqslant 1$. Similarly, on crossing the wall $x = -1$, $y \geqslant 1$ between $x \leqslant -1$ and $V \setminus U$, y_5, y_6 become invertible, so a', b' may be of the form $\exp(f) y_5^{a_5} y_6^{a_6}$ in $V \setminus U$. On crossing the wall

$x = 1$, $y \geqslant 1$ we see that a', b' must be of the form $\exp(f) y_5^{a_5} y_6^{a_6}$ in $x_1 \geqslant 1$. Comparing this with the previous statement gives $a_5 = a_6 = 0$. Hence a', b' are invertible at $(2, 0)$.

Now c' is either zero, or of the form $\exp(f) y_1^{a_1} \ldots y_6^{a_6}$. Specializing $a'z = b'\Xi_{\rm ex}(c')$ at $(-2, 0)$ and using a', b' invertible there gives $c' \neq 0$ with $a_1 = 0$ and $a_i = 0$ for $i \neq 1$. But specializing at $(2, 0)$ and using a', b' invertible there gives $a_2 = 1$ and $a_i = 0$ for $i \neq 2$, a contradiction. So no such c' exists.

One can prove that $\mathfrak{C}$ above is toric, and hence also firm, saturated, torsion-free, integral, and interior, in the notation of §4.7. There are simpler examples of $\mathfrak{C}$ lacking these properties for which (5.4) is not an isomorphism and (5.5) does not commute. We chose this example to show that we cannot easily avoid the problem by restricting to nicer classes of C^∞-rings with corners.

Example 5.24 implies that we cannot generalize complete C^∞-rings in §2.6 to the corners case, such that the analogue of Theorem 2.53(a) holds. But we will use the following proposition to define *semi-complete* C^∞-rings with corners, which have some of the good properties of complete C^∞-rings. Note that Example 5.24 gives an example where no choice of $\mathfrak{D}_{\rm ex}$ can make the canonical map $(\mathfrak{D}, \mathfrak{D}_{\rm ex}) \to \Gamma^{\rm c} \circ \operatorname{Spec}^{\rm c}(\mathfrak{D}, \mathfrak{D}_{\rm ex})$ surjective on the monoids.

Proposition 5.25 *Let $(\mathfrak{C}, \mathfrak{C}_{\rm ex})$ be a C^∞-ring with corners and let $\boldsymbol{X} = \operatorname{Spec}^{\rm c}(\mathfrak{C}, \mathfrak{C}_{\rm ex})$ be the corresponding C^∞-scheme with corners. Then there is a C^∞-ring with corners $(\mathfrak{D}, \mathfrak{D}_{\rm ex})$ with $\mathfrak{D} \cong \Gamma \circ \operatorname{Spec} \mathfrak{C}$ a complete C^∞-ring, such that $\operatorname{Spec}^{\rm c}(\mathfrak{D}, \mathfrak{D}_{\rm ex}) \cong \boldsymbol{X}$ and the canonical map $(\mathfrak{D}, \mathfrak{D}_{\rm ex}) \to \Gamma^{\rm c} \circ \operatorname{Spec}^{\rm c}(\mathfrak{D}, \mathfrak{D}_{\rm ex})$ is an isomorphism on $\mathfrak{D}$, and injective on $\mathfrak{D}_{\rm ex}$. If $(\mathfrak{C}, \mathfrak{C}_{\rm ex})$ is firm, or interior, then $(\mathfrak{D}, \mathfrak{D}_{\rm ex})$ is firm, or interior.*

Proof We define $(\mathfrak{D}, \mathfrak{D}_{\rm ex})$ such that $\mathfrak{D} = \Gamma \circ \operatorname{Spec} \mathfrak{C} = \mathcal{O}_X(X)$, and let $\mathfrak{D}_{\rm ex}$ be the submonoid of $\mathcal{O}_X^{\rm ex}(X)$ generated by the invertible elements $\Psi_{\exp}(\mathfrak{D})$ and the image $\phi_{\rm ex}(\mathfrak{C}_{\rm ex})$. One can check that the C^∞-operations from $(\mathcal{O}_X(X), \mathcal{O}_X^{\rm ex}(X))$ restrict to C^∞-operations on $(\mathfrak{D}, \mathfrak{D}_{\rm ex})$, and make $(\mathfrak{D}, \mathfrak{D}_{\rm ex})$ into a C^∞-ring with corners. Let $\boldsymbol{Y} = \operatorname{Spec}^{\rm c}(\mathfrak{D}, \mathfrak{D}_{\rm ex})$. If $(\mathfrak{C}, \mathfrak{C}_{\rm ex})$ is firm, then $(\mathfrak{D}, \mathfrak{D}_{\rm ex})$ is firm, as the sharpening $\mathfrak{D}_{\rm ex}^\sharp$ is the image of $\mathfrak{C}_{\rm ex}^\sharp$ under $\phi_{\rm ex}$, hence the image of the generators generates $\mathfrak{D}_{\rm ex}^\sharp$. If $(\mathfrak{C}, \mathfrak{C}_{\rm ex})$ is interior, then $(\mathfrak{D}, \mathfrak{D}_{\rm ex})$ is interior, as elements in both $\Psi_{\exp}(\mathfrak{D})$ and $\phi_{\rm ex}(\mathfrak{C}_{\rm ex})$ have no zero divisors.

Now the canonical morphism $(\phi, \phi_{\rm ex}) : (\mathfrak{C}, \mathfrak{C}_{\rm ex}) \to \Gamma^{\rm c} \circ \operatorname{Spec}^{\rm c}(\mathfrak{C}, \mathfrak{C}_{\rm ex})$ gives a morphism $(\psi, \psi_{\rm ex}) : (\mathfrak{C}, \mathfrak{C}_{\rm ex}) \to (\mathfrak{D}, \mathfrak{D}_{\rm ex})$, where $\psi = \phi$, and $\psi_{\rm ex} = \phi_{\rm ex}$ with its image restricted to the submonoid $\mathfrak{D}_{\rm ex}$ of $\mathcal{O}_{X,{\rm ex}}(X)$. As $\mathfrak{D}$ is complete, then $\operatorname{Spec}(\mathfrak{D}) \cong \operatorname{Spec}(\mathfrak{C}) \cong (X, \mathcal{O}_X)$, and $\operatorname{Spec}^{\rm c}(\psi, \psi_{\rm ex}) : \boldsymbol{Y} \cong \operatorname{Spec}^{\rm c}(\mathfrak{D}, \mathfrak{D}_{\rm ex}) \to \operatorname{Spec}^{\rm c}(\mathfrak{C}, \mathfrak{C}_{\rm ex}) \cong \boldsymbol{X}$ is an isomorphism on the topological space and the sheaves of C^∞-rings. To show that $\operatorname{Spec}^{\rm c}(\psi, \psi_{\rm ex})$ is an isomorphism on the sheaves of monoids, we show $\psi_{\rm ex}$ induces an isomorphism on the stalks $\mathcal{O}^{\rm ex}_{X,x} \cong \mathfrak{C}_{x,{\rm ex}}$ and $\mathcal{O}^{\rm ex}_{Y,x} \cong \mathfrak{D}_{x,{\rm ex}}$, for all $\mathbb{R}$-points $x \in X$.

The stalk map corresponds to the morphism $\psi_{x,{\rm ex}} : \mathfrak{C}_{x,{\rm ex}} \to \mathfrak{D}_{x,{\rm ex}}$, which is defined by $\psi_{x,{\rm ex}}(\pi_{x,{\rm ex}}(c')) = \hat\pi_{x,{\rm ex}}(\psi_{\rm ex}(c'))$, where $\pi_{x,{\rm ex}} : \mathfrak{C}_{\rm ex} \to \mathfrak{C}_{x,{\rm ex}}$ and $\hat\pi_{x,{\rm ex}} : \mathfrak{D}_{\rm ex} \to \mathfrak{D}_{x,{\rm ex}}$ are the localization morphisms. Now, as $\mathrm{Spec^c}(\psi, \psi_{\rm ex})$ is an isomorphism on the sheaves of C^∞-rings, we know that $\psi_x : \mathfrak{C}_x \to \mathfrak{D}_x$ is an isomorphism, which implies we have an isomorphism $\psi_{x,{\rm ex}}|_{\mathfrak{C}^\times_{x,{\rm ex}}} : \mathfrak{C}^\times_{x,{\rm ex}} \to \mathfrak{D}^\times_{x,{\rm ex}}$ of invertible elements in the monoids. This gives the following commutative diagram of monoids:

$$\begin{array}{ccccccc} & & \mathfrak{C}_{\rm ex} & \xrightarrow{\psi_{\rm ex}} & \mathfrak{D}_{\rm ex} \subset \mathcal{O}^{\rm ex}_X(X) & & \\ & & \downarrow{\scriptstyle \pi_{x,{\rm ex}}} & & \downarrow{\scriptstyle \hat\pi_{x,{\rm ex}}} & & \\ \mathfrak{C}^\times_{x,{\rm ex}} & \hookrightarrow & \mathfrak{C}_{x,{\rm ex}} & \xrightarrow{\psi_{x,{\rm ex}}} & \mathfrak{D}_{x,{\rm ex}} & \hookleftarrow & \mathfrak{D}^\times_{x,{\rm ex}}. \end{array}$$

(with the bottom arrow $\mathfrak{C}^\times_{x,{\rm ex}} \xrightarrow{\cong} \mathfrak{D}^\times_{x,{\rm ex}}$)

To show that $\psi_{x,{\rm ex}}$ is injective, suppose $a'_x, b'_x \in \mathfrak{C}_{x,{\rm ex}}$ with $\psi_{x,{\rm ex}}(a'_x) = \psi_{x,{\rm ex}}(b'_x)$. As $\pi_{x,{\rm ex}}$ is surjective we have $a'_x = \pi_{x,{\rm ex}}(a')$, $b'_x = \pi_{x,{\rm ex}}(b')$ in $\mathfrak{C}_{x,{\rm ex}}$ for $a', b' \in \mathfrak{C}_{\rm ex}$. Then in $\mathfrak{D}_{x,{\rm ex}}$ we have

$$\begin{aligned} \hat\pi_{x,{\rm ex}}(\psi_{\rm ex}(a')) &= \psi_{x,{\rm ex}}(\pi_{x,{\rm ex}}(a')) = \psi_{x,{\rm ex}}(a'_x) \\ &= \psi_{x,{\rm ex}}(b'_x) = \psi_{x,{\rm ex}}(\pi_{x,{\rm ex}}(b')) = \hat\pi_{x,{\rm ex}}(\psi_{\rm ex}(b')), \end{aligned}$$

so Proposition 4.33 gives $e', f' \in \mathfrak{D}_{\rm ex}$ such that $e'\psi_{\rm ex}(a') = f'\psi_{\rm ex}(b')$ in $\mathfrak{D}_{\rm ex}$, with $\Phi_i(e') - \Phi_i(f') \in I$ and $x \circ \Phi_i(e') \neq 0$. Hence

$$\begin{aligned} e'|_x a'_x &= e'|_x \pi_{x,{\rm ex}}(a') = e'|_x \psi_{\rm ex}(a')|_x = (e'\psi_{\rm ex}(a'))|_x \\ &= (f'\psi_{\rm ex}(b'))|_x = f'|_x \psi_{\rm ex}(b')|_x = f'|_x \pi_{x,{\rm ex}}(b') = f'|_x b'_x \end{aligned} \tag{5.6}$$

in $\mathfrak{C}_{x,{\rm ex}}$. Under the maps $\Phi_{i,x} : \mathfrak{C}_{x,{\rm ex}} \to \mathfrak{C}_x$ and $\pi : \mathfrak{C}_x \to \mathbb{R}$ we have $\Phi_{i,x}(e'|_x) = \Phi_{i,x}(f'|_x)$ as $\Phi_i(e') - \Phi_i(f') \in I$, and $\pi \circ \Phi_{i,x}(e'|_x) = x \circ \Phi_i(e') \neq 0$. Hence $e'|_x$ is invertible in $\mathfrak{C}_{x,{\rm ex}}$, so $\Phi_{i,x}(e'|_x) = \Phi_{i,x}(f'|_x)$ and Definition 4.16(i) imply that $e'|_x = f'|_x$, and thus (5.6) shows that $a'_x = b'_x$, so $\psi_{x,{\rm ex}}$ is injective.

To show $\psi_{x,{\rm ex}}$ is surjective, let $d'_x \in \mathfrak{D}_{x,{\rm ex}}$, and write $d'_x = \hat\pi_{x,{\rm ex}}(d')$ for $d' \in \mathfrak{D}_{\rm ex}$. As $\mathfrak{D}_{\rm ex}$ is generated by $\psi_{\rm ex}(\mathfrak{C}_{\rm ex})$ and $\Psi_{\rm exp}(\mathfrak{D})$ we have $d' = \psi_{\rm ex}(c') \cdot e'$ for $c' \in \mathfrak{C}$ and $e' \in \Psi_{\rm exp}(\mathfrak{D})$ invertible. Thus $d'_x = \psi_{x,{\rm ex}} \circ \pi_{x,{\rm ex}}(c') \cdot \hat\pi_{x,{\rm ex}}(e')$, where $\hat\pi_{x,{\rm ex}}(e') \in \mathfrak{D}^\times_{x,{\rm ex}}$. As $\psi_{x,{\rm ex}}|_{\mathfrak{C}^\times_{x,{\rm ex}}} : \mathfrak{C}^\times_{x,{\rm ex}} \to \mathfrak{D}^\times_{x,{\rm ex}}$ is an isomorphism and $\pi_{x,{\rm ex}}$ is surjective, there exists $c'' \in \mathfrak{C}_{\rm ex}$ with $\psi_{x,{\rm ex}} \circ \pi_{x,{\rm ex}}(c'') = \hat\pi_{x,{\rm ex}}(e')$. Then $d'_x = \psi_{x,{\rm ex}}[\pi_{x,{\rm ex}}(c'c'')]$, so $\psi_{x,{\rm ex}}$ is surjective. Hence $\psi_{x,{\rm ex}}$ is an isomorphism, and $\mathrm{Spec^c}(\psi, \psi_{\rm ex}) : \mathrm{Spec^c}(\mathfrak{D}, \mathfrak{D}_{\rm ex}) \to \mathrm{Spec^c}(\mathfrak{C}, \mathfrak{C}_{\rm ex}) = \boldsymbol{X}$ is an isomorphism.

Thus there is an isomorphism $\Gamma^{\rm c} \circ \mathrm{Spec^c}(\mathfrak{D}, \mathfrak{D}_{\rm ex}) \cong \Gamma^{\rm c} \circ \mathrm{Spec^c}(\mathfrak{C}, \mathfrak{C}_{\rm ex})$, which identifies the canonical map $(\mathfrak{D},\mathfrak{D}_{\rm ex})\to\Gamma^{\rm c}\circ\operatorname{Spec^c}(\mathfrak{D},\mathfrak{D}_{\rm ex})$ with the inclusion $(\mathfrak{D},\mathfrak{D}_{\rm ex})\hookrightarrow(\mathcal{O}_X(X),\mathcal{O}_X^{\rm ex}(X))=\Gamma^{\rm c}\circ\operatorname{Spec^c}(\mathfrak{C},\mathfrak{C}_{\rm ex})$. By definition this inclusion is an isomorphism on $\mathfrak{D}$, and injective on $\mathfrak{D}_{\rm ex}$. This completes the proof. □

Definition 5.26 We call a C^∞-ring with corners $\boldsymbol{\mathfrak{D}}=(\mathfrak{D},\mathfrak{D}_{\rm ex})$ *semi-complete* if $\mathfrak{D}$ is complete, and $\Xi_{\boldsymbol{\mathfrak{D}}}:(\mathfrak{D},\mathfrak{D}_{\rm ex})\to\Gamma^{\rm c}\circ\operatorname{Spec^c}(\mathfrak{D},\mathfrak{D}_{\rm ex})$ is an isomorphism on $\mathfrak{D}$ (this is equivalent to $\mathfrak{D}$ complete) and injective on $\mathfrak{D}_{\rm ex}$. Write $\mathbf{C^\infty Rings^c_{sc}}\subset\mathbf{C^\infty Rings^c}$ and $\mathbf{C^\infty Rings^c_{sc,in}}\subset\mathbf{C^\infty Rings^c_{in}}$ for the full subcategories of (interior) semi-complete C^∞-rings with corners.

Given a C^∞-ring with corners $\boldsymbol{\mathfrak{C}}$, Proposition 5.25 constructs a semi-complete C^∞-ring with corners $\boldsymbol{\mathfrak{D}}$ with a morphism $\boldsymbol{\psi}:\boldsymbol{\mathfrak{C}}\to\boldsymbol{\mathfrak{D}}$ such that $\operatorname{Spec^c}\boldsymbol{\psi}:\operatorname{Spec^c}\boldsymbol{\mathfrak{D}}\to\operatorname{Spec^c}\boldsymbol{\mathfrak{C}}$ is an isomorphism. It is easy to show that this map $\boldsymbol{\mathfrak{C}}\mapsto\boldsymbol{\mathfrak{D}}$ is functorial, yielding a functor $\Pi_{\rm all}^{\rm sc}:\mathbf{C^\infty Rings^c}\to\mathbf{C^\infty Rings^c_{sc}}$ mapping $\boldsymbol{\mathfrak{C}}\mapsto\boldsymbol{\mathfrak{D}}$ on objects, with $\operatorname{Spec^c}\cong\operatorname{Spec^c}\circ\Pi_{\rm all}^{\rm sc}$. Similarly we get $\Pi_{\rm in}^{\rm sc,in}:\mathbf{C^\infty Rings^c_{in}}\to\mathbf{C^\infty Rings^c_{sc,in}}$ with $\operatorname{Spec^c_{in}}\cong\operatorname{Spec^c_{in}}\circ\Pi_{\rm in}^{\rm sc,in}$.

Here is a partial analogue of Theorem 2.53.

Theorem 5.27 **(a)** $\operatorname{Spec^c}|_{\cdots}:(\mathbf{C^\infty Rings^c_{sc}})^{\rm op}\to\mathbf{AC^\infty Sch^c}$ *is faithful and essentially surjective, but not full.*

(b) *Let $\boldsymbol{X}$ be an affine C^∞-scheme with corners. Then $\boldsymbol{X}\cong\operatorname{Spec^c}\boldsymbol{\mathfrak{D}}$, where $\boldsymbol{\mathfrak{D}}$ is a semi-complete C^∞-ring with corners.*

(c) *The functor* $\Pi_{\rm all}^{\rm sc}:\mathbf{C^\infty Rings^c}\to\mathbf{C^\infty Rings^c_{sc}}$ *is left adjoint to the inclusion* $\operatorname{inc}:\mathbf{C^\infty Rings^c_{sc}}\hookrightarrow\mathbf{C^\infty Rings^c}$. *That is,* $\Pi_{\rm all}^{\rm sc}$ *is a* ***reflection functor****.*

(d) *All small colimits exist in* $\mathbf{C^\infty Rings^c_{sc}}$, *although they may not coincide with the corresponding small colimits in* $\mathbf{C^\infty Rings^c}$.

(e) *The functor* $\operatorname{Spec^c}|_{\cdots}:(\mathbf{C^\infty Rings^c_{sc}})^{\rm op}\to\mathbf{LC^\infty RS^c}$ *is right adjoint to* $\Pi_{\rm all}^{\rm sc}\circ\Gamma^{\rm c}:\mathbf{LC^\infty RS^c}\to(\mathbf{C^\infty Rings^c_{sc}})^{\rm op}$. *Thus* $\operatorname{Spec}|_{\cdots}$ *takes limits in* $(\mathbf{C^\infty Rings^c_{sc}})^{\rm op}$ *(or colimits in* $\mathbf{C^\infty Rings^c_{sc}}$*) to limits in* $\mathbf{LC^\infty RS^c}$.

The analogues of **(a)–(e)** *hold in the interior case, replacing* $\mathbf{C^\infty Rings^c_{sc}},\ldots,\mathbf{LC^\infty RS^c}$ *by* $\mathbf{C^\infty Rings^c_{sc,in}},\ldots,\mathbf{LC^\infty RS^c_{in}}$.

Proof For (a), if $\boldsymbol{\phi}:\boldsymbol{\mathfrak{D}}\to\boldsymbol{\mathfrak{C}}$ is a morphism in $\mathbf{C^\infty Rings^c_{sc}}$, we have a commutative diagram

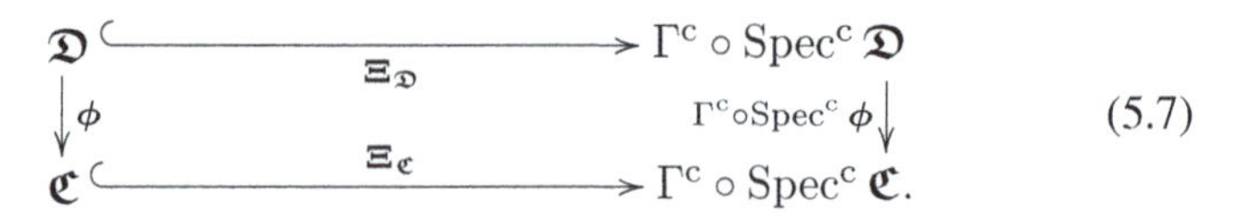

$$\begin{array}{ccc}\boldsymbol{\mathfrak{D}} & \xrightarrow{\Xi_{\boldsymbol{\mathfrak{D}}}} & \Gamma^{\rm c}\circ\operatorname{Spec^c}\boldsymbol{\mathfrak{D}} \\ \downarrow\boldsymbol{\phi} & & \downarrow\Gamma^{\rm c}\circ\operatorname{Spec^c}\boldsymbol{\phi} \\ \boldsymbol{\mathfrak{C}} & \xrightarrow{\Xi_{\boldsymbol{\mathfrak{C}}}} & \Gamma^{\rm c}\circ\operatorname{Spec^c}\boldsymbol{\mathfrak{C}}.\end{array} \tag{5.7}$$

As the rows are injective by semi-completeness, $\Gamma^{\rm c}\circ\operatorname{Spec^c}\boldsymbol{\phi}$ determines $\boldsymbol{\phi}$, so

$\mathrm{Spec^c}\,\phi$ determines ϕ. Thus $\mathrm{Spec^c}|_{\dots}$ is injective on morphisms, that is, it is faithful. Essential surjectivity follows by Proposition 5.25. To see that $\mathrm{Spec^c}|_{\dots}$ is not full, let $\mathfrak{C}$ be as in Example 5.24, and set $\mathfrak{D} = C^\infty([0,\infty))$. Then

$$\mathrm{Spec}^{\mathrm{c}}_{\mathfrak{C},\mathfrak{D}} : \mathrm{Hom}_{\mathbf{C^\infty Rings^c_{sc}}}(\mathfrak{D},\mathfrak{C}) \longrightarrow \mathrm{Hom}_{\mathbf{AC^\infty Sch^c}}(\mathrm{Spec^c}\,\mathfrak{C}, \mathrm{Spec^c}\,\mathfrak{D})$$

may be identified with the non-surjective map

$$\Xi_{\mathrm{ex}} : \mathfrak{C}_{\mathrm{ex}} \longrightarrow \mathcal{O}^{\mathrm{ex}}_X(X).$$

Part (b) is immediate from Proposition 5.25. Part (c) is easy to check, where the unit of the adjunction is $\psi : \mathfrak{C} \to \mathfrak{D} = \mathrm{inc}\circ\Pi^{\mathrm{sc}}_{\mathrm{all}}(\mathfrak{C})$, and the counit is $\mathbf{id}_{\mathfrak{D}} : \Pi^{\mathrm{sc}}_{\mathrm{all}}\circ\mathrm{inc}(\mathfrak{D}) = \mathfrak{D} \to \mathfrak{D}$. For (d), given a small colimit in $\mathbf{C^\infty Rings^c_{sc}}$, the colimit exists in $\mathbf{C^\infty Rings^c}$ by Theorem 4.20(b), and applying the reflection functor $\Pi^{\mathrm{sc}}_{\mathrm{all}}$ gives the (possibly different) colimit in $\mathbf{C^\infty Rings^c_{sc}}$. Part (e) holds as $\Pi^{\mathrm{sc}}_{\mathrm{all}}, \Gamma^{\mathrm{c}}$ are left adjoint to $\mathrm{inc}, \mathrm{Spec^c}$ by (c) and Theorem 5.16. The extension to the interior case is immediate. □

Theorem 5.28 **(a)** *Suppose* $\mathfrak{C} \in \mathbf{C^\infty Rings^c}$ *and* $\boldsymbol{X} = \mathrm{Spec^c}\,\mathfrak{C}$ *lies in* $\mathbf{LC^\infty RS^c_{in}} \subset \mathbf{LC^\infty RS^c}$. *Then* $\boldsymbol{X} \cong \mathrm{Spec^c_{in}}\,\mathfrak{D}$ *for some* $\mathfrak{D} \in \mathbf{C^\infty Rings^c_{in}}$.
(b) *An* $\boldsymbol{X} \in \mathbf{C^\infty Sch^c}$ *lies in* $\mathbf{C^\infty Sch^c_{in}}$ *if and only if it lies in* $\mathbf{LC^\infty RS^c_{in}}$.

Proof For (a), by Theorem 5.27(b) we may take $\mathfrak{C}$ to be semi-complete, so that $\Xi_{\mathfrak{C}} : (\mathfrak{C}, \mathfrak{C}_{\mathrm{ex}}) \to (\mathcal{O}_X(X), \mathcal{O}^{\mathrm{ex}}_X(X))$ is an isomorphism on $\mathfrak{C}$ and injective on $\mathfrak{C}_{\mathrm{ex}}$. As $\boldsymbol{X}$ lies in $\mathbf{LC^\infty RS^c_{in}}$, the stalks $\boldsymbol{\mathcal{O}}_{X,x}$ are interior for $x \in X$, and writing

$$\mathcal{O}^{\mathrm{in}}_X(X) = \{f' \in \mathcal{O}^{\mathrm{ex}}_X(X) : f'|_x \neq 0 \text{ in } \mathcal{O}^{\mathrm{ex}}_{X,x} \text{ for all } x \in X\},$$

then $\boldsymbol{\mathcal{O}}^{\mathrm{in}}_X(X) = (\mathcal{O}_X(X), \mathcal{O}^{\mathrm{in}}_X(X) \amalg \{0\})$ is an interior C^∞-subring with corners of $\boldsymbol{\mathcal{O}}_X(X) = (\mathcal{O}_X(X), \mathcal{O}^{\mathrm{ex}}_X(X))$. Define $\boldsymbol{\mathfrak{E}} = (\mathfrak{E}, \mathfrak{E}_{\mathrm{ex}})$, where $\mathfrak{E} = \mathcal{O}_X(X)$ and

$$\begin{aligned}\mathfrak{E}_{\mathrm{ex}} = \bigl\{e' \in \mathcal{O}^{\mathrm{ex}}_X(X) :\ & \text{there exists a decomposition } X = \textstyle\coprod_{i\in I} X_i \text{ with} \\ & X_i \text{ open and closed in } X \text{ and elements } c'_i \in \mathfrak{C}_{\mathrm{ex}} \text{ for } i \in I \\ & \text{with } \Xi_{\mathfrak{C},\mathrm{ex}}(c'_i)|_{X_i} = e'|_{X_i} \text{ for all } i \in I\bigr\}.\end{aligned} \tag{5.8}$$

We claim $\mathfrak{E}, \mathfrak{E}_{\mathrm{ex}}$ are closed under the C^∞-operations on $\mathcal{O}_X(X), \mathcal{O}^{\mathrm{ex}}_X(X)$, so that $\boldsymbol{\mathfrak{E}}$ is a C^∞-subring with corners of $\boldsymbol{\mathcal{O}}_X(X)$. To see this, let $g : \mathbb{R}^m \times [0,\infty)^n \to [0,\infty)$ be exterior and $e_1, \dots, e_m \in \mathfrak{E}$, $e'_1, \dots, e'_n \in \mathfrak{E}_{\mathrm{ex}}$. Then $e_j = \Xi_{\mathfrak{C}}(c_j)$ for unique $c_j \in \mathfrak{C}$, $j = 1, \dots, m$ as $\Xi_{\mathfrak{C}}$ is an isomorphism. Also for each $k = 1, \dots, n$ we get a decomposition $X = \coprod_{i\in I} X_i$ in (5.8) for e'_k. By intersecting the subsets of the decompositions we can take a common decomposition $X = \coprod_{i\in I} X_i$ for all $k = 1, \dots, n$, and elements $c'_{i,k} \in \mathfrak{C}_{\mathrm{ex}}$

with $\Xi_{\mathfrak{C},\mathrm{ex}}(c'_{i,k})|_{X_i} = e'_k|_{X_i}$ for all $i \in I$ and $k = 1,\ldots,n$. Then $e' = \Psi_g(e_1,\ldots,e_m,e'_1,\ldots,e'_n) \in \mathcal{O}_X^{\mathrm{ex}}(X)$ satisfies the conditions of (5.8) with the same decomposition $X = \coprod_{i\in I} X_i$, and elements $c'_i = \Psi_g(c_1,\ldots,c_m, c'_{i,1},\ldots,c'_{i,n})$, so $e' \in \mathfrak{C}_{\mathrm{ex}}$, and $\boldsymbol{\mathfrak{C}} \in \mathbf{C^\infty Rings^c}$.

Observe that $\Xi_{\mathfrak{C}} : \boldsymbol{\mathfrak{C}} \to \boldsymbol{\mathcal{O}}_X(X)$ factors via the inclusion $\boldsymbol{\mathfrak{C}} \hookrightarrow \boldsymbol{\mathcal{O}}_X(X)$, by taking the decomposition in (5.8) to be into one set $X_1 = X$.

Next define $\mathfrak{D} = \mathfrak{C}$ and $\mathfrak{D}_{\mathrm{in}} = \mathcal{O}_X^{\mathrm{in}}(X) \cap \mathfrak{C}_{\mathrm{ex}}$, so that $\boldsymbol{\mathfrak{D}} = (\mathfrak{D}, \mathfrak{D}_{\mathrm{in}} \amalg \{0\})$ is the intersection of the C^∞-subrings $\boldsymbol{\mathcal{O}}_X^{\mathrm{in}}(X)$ and $\boldsymbol{\mathfrak{C}}$ in $\boldsymbol{\mathcal{O}}_X(X)$, and thus lies in $\mathbf{C^\infty Rings^c_{in}}$, as $\boldsymbol{\mathcal{O}}_X^{\mathrm{in}}(X)$ does. We now have morphisms in $\mathbf{C^\infty Rings^c}$

$$\boldsymbol{\mathfrak{C}} \xrightarrow{\quad \Xi_{\mathfrak{C}} \quad} \boldsymbol{\mathfrak{C}} \xleftarrow{\quad \mathrm{inc} \quad} \boldsymbol{\mathfrak{D}},$$

giving morphisms in $\mathbf{LC^\infty RS^c}$

$$\boldsymbol{X} = \mathrm{Spec^c}\,\boldsymbol{\mathfrak{C}} \xleftarrow{\mathrm{Spec^c}\,\Xi_{\mathfrak{C}}} \mathrm{Spec^c}\,\boldsymbol{\mathfrak{C}} \xrightarrow{\mathrm{Spec^c}\,\mathrm{inc}} \mathrm{Spec^c}\,\boldsymbol{\mathfrak{D}} = \mathrm{Spec^c_{in}}\,\boldsymbol{\mathfrak{D}}. \tag{5.9}$$

We will show that both morphisms in (5.9) are isomorphisms, so that $\boldsymbol{X} \cong \mathrm{Spec^c_{in}}\,\boldsymbol{\mathfrak{D}}$, as we have to prove. As $\mathfrak{C} \cong \mathfrak{D} = \mathfrak{C}$, equation (5.9) is isomorphic on the level of C^∞-schemes. Thus it is sufficient to prove (5.9) induces isomorphisms on the stalks of the sheaves of monoids. That is, for each $x \in X$ we must show that the following maps are isomorphisms:

$$\mathfrak{C}_{x,\mathrm{ex}} \xrightarrow{\quad \Xi_{\mathfrak{C},x,\mathrm{ex}} \quad} \mathfrak{C}_{x,\mathrm{ex}} \xleftarrow{\quad \mathrm{inc}_{x,\mathrm{ex}} \quad} \mathfrak{D}_{x,\mathrm{ex}}.$$

Suppose $e'' \in \mathfrak{C}_{x,\mathrm{ex}}$. Then $e'' = \pi_{\mathfrak{C}_{x,\mathrm{ex}}}(e')$ for $e' \in \mathfrak{C}_{\mathrm{ex}}$. Let $X_i, c'_i, i \in I$ be as in (5.8) for e'. Then $x \in X_i$ for unique $i \in I$. Define $a', b' \in \mathfrak{C}_{\mathrm{ex}}$ by $a'|_{X_i} = 1$, $a'|_{X\setminus X_i} = 0$ and $b' = a'$. Then $a'e' = b'\Xi_{\mathfrak{C},\mathrm{ex}}(c'_i)$, so Proposition 4.33 gives $e'' = \pi_{\mathfrak{C}_{x,\mathrm{ex}}}(e') = \pi_{\mathfrak{C}_{x,\mathrm{ex}}} \circ \Xi_{\mathfrak{C},\mathrm{ex}}(c'_i) = \Xi_{\mathfrak{C},x,\mathrm{ex}} \circ \pi_{\mathfrak{C}_{x,\mathrm{ex}}}(c'_i)$. Thus $\Xi_{\mathfrak{C},x,\mathrm{ex}}$ is surjective.

As the composition $\mathfrak{C}_{x,\mathrm{ex}} \xrightarrow{\Xi_{\mathfrak{C},x,\mathrm{ex}}} \mathfrak{C}_{x,\mathrm{ex}} \to \mathcal{O}_X^{\mathrm{ex}}(X)_x \to \mathcal{O}_{X,x}^{\mathrm{ex}}$ is an isomorphism since $\boldsymbol{X} = \mathrm{Spec^c}\,\boldsymbol{\mathfrak{C}}$, we see that $\Xi_{\mathfrak{C},x,\mathrm{ex}}$ is injective, and thus an isomorphism.

Suppose $0 \neq e'' \in \mathfrak{C}_{x,\mathrm{ex}}$. Then $e'' = \pi_{\mathfrak{C}_{x,\mathrm{ex}}}(e')$ for $e' \in \mathfrak{C}_{\mathrm{ex}} \subseteq \mathcal{O}_X^{\mathrm{ex}}(X)$. Define $Y = \{y \in X : e'|_y \neq 0 \text{ in } \mathcal{O}_{X,y}^{\mathrm{ex}}\}$. Then Y is open and closed in X by Definition 5.2(a), as $\boldsymbol{X}$ is interior, and $x \in Y$ as $e'' \neq 0$. Define $d' \in \mathfrak{D}_{\mathrm{in}}$ by $d'|_Y = e'|_Y$ and $d'|_{X\setminus Y} = 1$. Define $a', b' \in \mathfrak{C}_{\mathrm{ex}}$ by $a'|_Y = 1$, $a'|_{X\setminus Y} = 0$ and $b' = a'$. Then $a'd' = b'e'$ with $x \circ \Phi_i(a') = 1$, so Proposition 4.33 gives $e'' = \pi_{\mathfrak{C}_{x,\mathrm{ex}}}(e') = \pi_{\mathfrak{C}_{x,\mathrm{ex}}} \circ \mathrm{inc}(d') = \mathrm{inc}_{x,\mathrm{ex}} \circ \pi_{\mathfrak{D}_{x,\mathrm{ex}}}(d')$, and $\mathrm{inc}_{x,\mathrm{ex}}$ is surjective.

Let $d''_1, d''_2 \in \mathfrak{D}_{x,\mathrm{ex}}$ with $\mathrm{inc}_{x,\mathrm{ex}}(d''_1) = \mathrm{inc}_{x,\mathrm{ex}}(d''_2)$. Then $d''_i = \pi_{\mathfrak{D}_{x,\mathrm{ex}}}(d'_i)$ for $d'_i \in \mathfrak{D}_{\mathrm{in}} \amalg \{0\}$, $i = 1, 2$. By Proposition 4.33, $\mathrm{inc}_{x,\mathrm{ex}}(d''_1) = \mathrm{inc}_{x,\mathrm{ex}}(d''_2)$

means that there exist $a',b'\in\mathfrak{C}_{\rm ex}$ with $\Phi_i(a')=\Phi_i(b')$ near x, and $x\circ\Phi_i(a')\neq 0$, and $a'd_1'=b'd_2'$ in $\mathfrak{C}_{\rm ex}$. If $a'd_1'|_x=b'd_2'|_x=0$ in $\mathcal{O}^{\rm in}_{X,x}$ then $d_1'=d_2'=0$, so $d_1''=d_2''=0$, as we want. Otherwise $d_1',d_2'\neq 0$, so $d_1',d_2'\in\mathfrak{D}_{\rm in}$.

Define $Y=\{y\in X:a'|_y\neq 0$ in $\mathcal{O}^{\rm ex}_{X,y}\}$. Then Y is open and closed in X as above, and $Y=\{y\in X:b'|_y\neq 0$ in $\mathcal{O}^{\rm ex}_{X,y}\}$, since $a'd_1'=b'd_2'$ with $d_1',d_2'\in\mathfrak{D}_{\rm in}$. Define $a'',b''\in\mathfrak{D}_{\rm in}$ by $a''|_Y=a'|_Y$, $a''|_{X\setminus Y}=d_2'|_{X\setminus Y}$, $b''|_Y=b'|_Y$, $b''|_{X\setminus Y}=d_1'|_{X\setminus Y}$. Then $\Phi_i(a'')=\Phi_i(b'')$ near x as $x\in Y$ and $\Phi_i(a')=\Phi_i(b')$ near x, and $x\circ\Phi_i(a'')\neq 0$, and $a''d_1'=b''d_2'$, so Proposition 4.33 says that $d_1''=\pi_{\mathfrak{D}_{x,\rm ex}}(d_1')=\pi_{\mathfrak{D}_{x,\rm ex}}(d_2')=d_2''$. Hence $\mathrm{inc}_{x,\rm ex}$ is injective, and thus an isomorphism. This proves (a). Part (b) follows from (a) by covering $\boldsymbol{X}$ with open affine C^∞-subschemes. □

5.6 Special classes of C^∞-schemes with corners

Definition 5.29 Let $\boldsymbol{X}$ be a C^∞-scheme with corners. We call $\boldsymbol{X}$ *finitely presented*, or *finitely generated*, or *firm*, or *integral*, or *torsion-free*, or *saturated*, or *toric*, or *simple*, if we can cover $\boldsymbol{X}$ by open $\boldsymbol{U}\subseteq\boldsymbol{X}$ with $\boldsymbol{U}\cong\mathrm{Spec^c}\,\boldsymbol{\mathfrak{C}}$ for $\boldsymbol{\mathfrak{C}}$ a C^∞-ring with corners which is firm, integral, torsion-free, saturated, toric, or simple, respectively, in the sense of Definitions 4.40, 4.43, and 4.45. We write

$$\mathbf{C^\infty Sch^c_{fp}}\subset\mathbf{C^\infty Sch^c_{fg}}\subset\mathbf{C^\infty Sch^c_{fi}}\subset\mathbf{C^\infty Sch^c}$$

for the full subcategories of finitely presented, finitely generated, and firm C^∞-schemes with corners. We write

$$\mathbf{C^\infty Sch^c_{si}}\subset\mathbf{C^\infty Sch^c_{to}}\subset\mathbf{C^\infty Sch^c_{sa}}\subset\mathbf{C^\infty Sch^c_{tf}}\subset\mathbf{C^\infty Sch^c_{\mathbb{Z}}}\subset\mathbf{C^\infty Sch^c_{in}}$$

for the full subcategories of simple, toric, saturated, torsion-free, and integral objects in $\mathbf{C^\infty Sch^c_{in}}$, respectively. We write

$$\begin{aligned}&\mathbf{C^\infty Sch^c_{si,ex}}\subset\mathbf{C^\infty Sch^c_{to,ex}}\subset\mathbf{C^\infty Sch^c_{sa,ex}}\subset\\&\mathbf{C^\infty Sch^c_{tf,ex}}\subset\mathbf{C^\infty Sch^c_{\mathbb{Z},ex}}\subset\mathbf{C^\infty Sch^c_{in,ex}}\subset\mathbf{C^\infty Sch^c}\end{aligned}\tag{5.10}$$

for the full subcategories of simple, ..., interior objects in $\mathbf{C^\infty Sch^c}$, but with exterior (i.e. all) rather than interior morphisms. We write

$$\begin{aligned}&\mathbf{C^\infty Sch^c_{fi,in}}\subset\mathbf{C^\infty Sch^c_{in}}, &&\mathbf{C^\infty Sch^c_{fi,\mathbb{Z}}}\subset\mathbf{C^\infty Sch^c_{\mathbb{Z}}},\\&\mathbf{C^\infty Sch^c_{fi,tf}}\subset\mathbf{C^\infty Sch^c_{tf}}, &&\mathbf{C^\infty Sch^c_{fi,in,ex}}\subset\mathbf{C^\infty Sch^c_{in,ex}},\\&\mathbf{C^\infty Sch^c_{fi,\mathbb{Z},ex}}\subset\mathbf{C^\infty Sch^c_{\mathbb{Z},ex}}, &&\mathbf{C^\infty Sch^c_{fi,tf,ex}}\subset\mathbf{C^\infty Sch^c_{tf,ex}},\end{aligned}$$

for the full subcategories of objects which are also firm. These live in a diagram of subcategories shown in Figure 5.1.

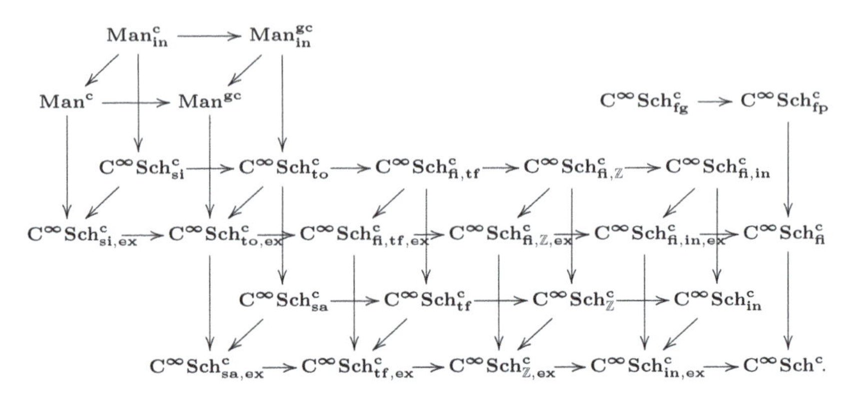

Figure 5.1 Subcategories of C^∞-schemes with corners $\mathbf{C^\infty Sch^c}$

Remark 5.30 **(i)** General C^∞-schemes with corners $\boldsymbol{X} \in \mathbf{C^\infty Sch^c}$ may have 'corner behaviour' which is very complicated and pathological. The subcategories of $\mathbf{C^\infty Sch^c}$ in Figure 5.1 have 'corner behaviour' which is progressively nicer, and more like ordinary manifolds with corners, as we impose more conditions.

(ii) It is easy to see that if X is a manifold with corners then $\boldsymbol{X} = F^{\mathbf{C^\infty Sch^c}}_{\mathbf{Man^c}}(X)$ in Definition 5.21 is an object of $\mathbf{C^\infty Sch^c_{si}}$, since $\boldsymbol{C}^\infty(\mathbb{R}^m_k)$ is a simple C^∞-ring with corners, and similarly if X is a manifold with g-corners then $\boldsymbol{X} = F^{\mathbf{C^\infty Sch^c}}_{\mathbf{Man^{gc}}}(X)$ is an object of $\mathbf{C^\infty Sch^c_{to}}$.

We can think of $\mathbf{C^\infty Sch^c_{si}}, \mathbf{C^\infty Sch^c_{si,ex}}, \mathbf{C^\infty Sch^c_{to}}, \mathbf{C^\infty Sch^c_{to,ex}}$ as good analogues in the world of C^∞-schemes of $\mathbf{Man^c_{in}}, \mathbf{Man^c}, \mathbf{Man^{gc}_{in}}, \mathbf{Man^{gc}}$.

(iii) For a C^∞-scheme with corners $\boldsymbol{X}$ to be *firm* basically means that locally $\boldsymbol{X}$ has only finitely many boundary faces. This is a mild condition, which will hold automatically in most interesting applications. An example of a non-firm C^∞-scheme with corners is $[\mathbf{0},\boldsymbol{\infty})^I := \prod_{i\in I}[\mathbf{0},\boldsymbol{\infty})$ for I an infinite set.

We will need to restrict to firm C^∞-schemes with corners to define fibre products in §5.7, and for parts of the theory of corner functors in Chapter 6.

Parts (b) and (c) of the next theorem are similar to Theorem 5.28(b).

Theorem 5.31 **(a)** *The functor* $\Pi^{\mathbb{Z}}_{\mathrm{in}} : \mathbf{LC^\infty RS^c_{in}} \to \mathbf{LC^\infty RS^c_{\mathbb{Z}}}$ *of Theorem* 5.8 *maps* $\mathbf{C^\infty Sch^c_{in}} \to \mathbf{C^\infty Sch^c_{\mathbb{Z}}}$. *Restricting gives a functor* $\Pi^{\mathbb{Z}}_{\mathrm{in}} : \mathbf{C^\infty Sch^c_{in}} \to \mathbf{C^\infty Sch^c_{\mathbb{Z}}}$ *right adjoint to the inclusion* $\mathrm{inc} : \mathbf{C^\infty Sch^c_{\mathbb{Z}}} \hookrightarrow$

$\mathbf{C^\infty Sch^c_{in}}$. *The analogues also hold for* $\mathrm{inc}:\mathbf{C^\infty Sch^c_{sa}}\hookrightarrow\mathbf{LC^\infty RS^c_{sa}}$ *and* $\mathrm{inc}:\mathbf{C^\infty Sch^c_{tf}}\hookrightarrow\mathbf{LC^\infty RS^c_{tf}}$ *and functors* $\Pi^{\rm tf}_{\mathbb{Z}},\Pi^{\rm sa}_{\rm tf},\Pi^{\rm sa}_{\rm in}$, *giving a diagram*

$$\mathbf{C^\infty Sch^c_{sa}} \underset{\mathrm{inc}}{\overset{\Pi^{\rm sa}_{\rm tf}}{\leftrightarrows}} \mathbf{C^\infty Sch^c_{tf}} \underset{\mathrm{inc}}{\overset{\Pi^{\rm tf}_{\mathbb{Z}}}{\leftrightarrows}} \mathbf{C^\infty Sch^c_{\mathbb{Z}}} \underset{\mathrm{inc}}{\overset{\Pi^{\mathbb{Z}}_{\rm in}}{\leftrightarrows}} \mathbf{C^\infty Sch^c_{in}}, \qquad \Pi^{\rm sa}_{\rm in}:\mathbf{C^\infty Sch^c_{in}}\to\mathbf{C^\infty Sch^c_{sa}}.$$

(b) *An object* $\boldsymbol{X}$ *in* $\mathbf{C^\infty Sch^c_{in}}$ *lies in* $\mathbf{C^\infty Sch^c_{sa}},\mathbf{C^\infty Sch^c_{tf}}$, *or* $\mathbf{C^\infty Sch^c_{\mathbb{Z}}}$ *if and only if it lies in* $\mathbf{LC^\infty RS^c_{sa}},\mathbf{LC^\infty RS^c_{tf}}$, *or* $\mathbf{LC^\infty RS^c_{\mathbb{Z}}}$, *respectively.*

(c) *Let* $\boldsymbol{X}$ *be an object in* $\mathbf{C^\infty Sch^c_{fi}}$. *Then the following are equivalent.*

- **(i)** $\boldsymbol{X}$ *lies in* $\mathbf{C^\infty Sch^c_{fi,in}}$.
- **(ii)** $\boldsymbol{X}$ *lies in* $\mathbf{LC^\infty RS^c_{in}}$.
- **(iii)** $\boldsymbol{X}$ *may be covered by open* $\boldsymbol{U}\subseteq\boldsymbol{X}$ *with* $\boldsymbol{U}\cong\mathrm{Spec}^{\rm c}_{\rm in}\,\mathfrak{C}$ *for* $\mathfrak{C}$ *a firm, interior* C^∞*-ring with corners.*

Also $\boldsymbol{X}$ *lies in* $\mathbf{C^\infty Sch^c_{to}}$ *if and only if it lies in* $\mathbf{LC^\infty RS^c_{sa}}$.

Proof For (a), consider the diagram of functors

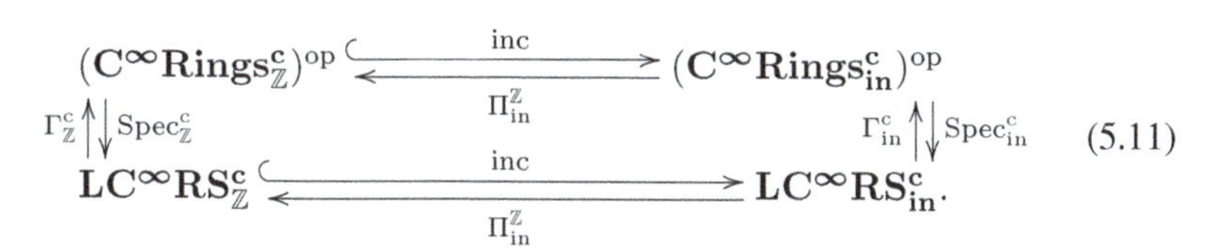

Here $\mathrm{Spec}^{\rm c}_{\mathbb{Z}},\Gamma^{\rm c}_{\mathbb{Z}}$ are the restrictions of $\mathrm{Spec}^{\rm c}_{\rm in},\Gamma^{\rm c}_{\rm in}$ to $(\mathbf{C^\infty Rings^c_{\mathbb{Z}}})^{\rm op}$ and $\mathbf{LC^\infty RS^c_{\mathbb{Z}}}$, and do map to $\mathbf{LC^\infty RS^c_{\mathbb{Z}}}$, $(\mathbf{C^\infty Rings^c_{\mathbb{Z}}})^{\rm op}$. The functors $\Pi^{\mathbb{Z}}_{\rm in}$ are from Theorems 4.48 and 5.8, and are right adjoint to the inclusions inc, noting that taking opposite categories converts left to right adjoints. Also $\mathrm{Spec}^{\rm c}_{\rm in}$ is right adjoint to $\Gamma^{\rm c}_{\rm in}$ by Theorem 5.18, which implies that $\mathrm{Spec}^{\rm c}_{\mathbb{Z}}$ is right adjoint to $\Gamma^{\rm c}_{\mathbb{Z}}$.

Clearly $\mathrm{inc}\circ\Gamma^{\rm c}_{\mathbb{Z}}=\Gamma^{\rm c}_{\rm in}\circ\mathrm{inc}$. As $\mathrm{Spec}^{\rm c}_{\mathbb{Z}}\circ\Pi^{\mathbb{Z}}_{\rm in}$, $\Pi^{\mathbb{Z}}_{\rm in}\circ\mathrm{Spec}^{\rm c}_{\rm in}$ are the right adjoints of these we have a natural isomorphism $\mathrm{Spec}^{\rm c}_{\mathbb{Z}}\circ\Pi^{\mathbb{Z}}_{\rm in}\cong\Pi^{\mathbb{Z}}_{\rm in}\circ\mathrm{Spec}^{\rm c}_{\rm in}$. Thus, for any $\mathfrak{C}\in\mathbf{C^\infty Rings^c_{in}}$ we have $\mathrm{Spec}^{\rm c}_{\mathbb{Z}}\circ\Pi^{\mathbb{Z}}_{\rm in}(\mathfrak{C})\cong\Pi^{\mathbb{Z}}_{\rm in}\circ\mathrm{Spec}^{\rm c}_{\rm in}\,\mathfrak{C}$ in $\mathbf{LC^\infty RS^c_{\mathbb{Z}}}$.

Let $\boldsymbol{X}\in\mathbf{C^\infty Sch^c_{in}}$. Then $\boldsymbol{X}$ is covered by open $\boldsymbol{U}\subset\boldsymbol{X}$ with $\boldsymbol{U}\cong\mathrm{Spec}^{\rm c}_{\rm in}\,\mathfrak{C}$ for $\mathfrak{C}\in\mathbf{C^\infty Rings^c_{in}}$. Hence $\Pi^{\mathbb{Z}}_{\rm in}(\boldsymbol{X})$ is covered by open $\Pi^{\mathbb{Z}}_{\rm in}(\boldsymbol{U})\subset\Pi^{\mathbb{Z}}_{\rm in}(\boldsymbol{X})$ with $\Pi^{\mathbb{Z}}_{\rm in}(\boldsymbol{U})\cong\Pi^{\mathbb{Z}}_{\rm in}\circ\mathrm{Spec}^{\rm c}_{\rm in}\,\mathfrak{C}\cong\mathrm{Spec}^{\rm c}_{\mathbb{Z}}\circ\Pi^{\mathbb{Z}}_{\rm in}(\mathfrak{C})$ for $\Pi^{\mathbb{Z}}_{\rm in}(\mathfrak{C})\in\mathbf{C^\infty Rings^c_{\mathbb{Z}}}$. Thus $\Pi^{\mathbb{Z}}_{\rm in}(\boldsymbol{X})\in\mathbf{C^\infty Sch^c_{\mathbb{Z}}}$, and $\Pi^{\mathbb{Z}}_{\rm in}:\mathbf{C^\infty Sch^c_{in}}\to\mathbf{C^\infty Sch^c_{\mathbb{Z}}}$. That the restriction $\Pi^{\mathbb{Z}}_{\rm in}:\mathbf{C^\infty Sch^c_{in}}\to\mathbf{C^\infty Sch^c_{\mathbb{Z}}}$ is right adjoint to inc follows from Theorem 5.8. This proves the first part of (a). The rest of (a) follows by the same argument.

For (b), the 'only if' part is obvious. For the 'if' part, suppose $\boldsymbol{X}$ lies in $\mathbf{C^\infty Sch^c_{in}}$ and $\mathbf{LC^\infty RS^c_{sa}}$. Then $\Pi^{\rm sa}_{\rm in}(\boldsymbol{X})$ lies in $\mathbf{C^\infty Sch^c_{sa}}$ by (a). But $\Pi^{\rm sa}_{\rm in}$ is isomorphic to the identity on $\mathbf{LC^\infty RS^c_{sa}} \subset \mathbf{LC^\infty RS^c_{in}}$, so $\Pi^{\rm sa}_{\rm in}(\boldsymbol{X}) \cong \boldsymbol{X}$, and $\boldsymbol{X}$ lies in $\mathbf{C^\infty Sch^c_{sa}}$. The arguments for $\mathbf{C^\infty Sch^c_{tf}}, \mathbf{C^\infty Sch^c_{\mathbb{Z}}}$ are the same.

For (c), clearly (iii) $\Rightarrow$ (i) $\Rightarrow$ (ii), so it is enough to show that (ii) $\Rightarrow$ (iii). Suppose $\boldsymbol{X}$ lies in $\mathbf{C^\infty Sch^c_{fi}}$ and $\mathbf{LC^\infty RS^c_{in}}$, and $x \in \boldsymbol{X}$. Then x has an open neighbourhood $\boldsymbol{V} \subset \boldsymbol{X}$ with $\boldsymbol{V} \cong \operatorname{Spec^c}\boldsymbol{\mathfrak{D}}$ for $\boldsymbol{\mathfrak{D}}$ firm. Let $d_1', \ldots, d_n' \in \mathfrak{D}_{\rm ex}$ such that $[d_1'], \ldots, [d_n']$ are generators for $\mathfrak{D}^\sharp_{\rm ex}$. Choose an open neighbourhood U of x in $\boldsymbol{V}$ such that for each $i = 1, \ldots, n$, either $\pi_{u,\rm ex}(d_i') = 0$ in $\mathfrak{D}_{u,\rm ex} \cong \mathcal{O}^{\rm ex}_{X,x}$ for all $u \in U$, or $\pi_{e,\rm ex}(d_i') \neq 0$ in $\mathfrak{D}_{u,\rm ex} \cong \mathcal{O}^{\rm ex}_{X,x}$ for all $u \in U$. As the topology of $\boldsymbol{V} \cong \operatorname{Spec^c}\boldsymbol{\mathfrak{D}}$ is generated by open sets U_d as in Definition 2.42, making U smaller if necessary we can suppose that $U = U_d$ for $d \in \mathfrak{D}$. Then Lemma 5.15 says that $\boldsymbol{U} \cong \operatorname{Spec^c}(\boldsymbol{\mathfrak{D}}(d^{-1}))$.

Define $\boldsymbol{\mathfrak{C}}$ to be the semi-complete C^∞-ring with corners constructed from $\boldsymbol{\mathfrak{D}}(d^{-1})$ in Proposition 5.25, so we have a morphism $\boldsymbol{\psi} : \boldsymbol{\mathfrak{D}}(d^{-1}) \to \boldsymbol{\mathfrak{C}}$ with $\operatorname{Spec^c}\boldsymbol{\psi}$ an isomorphism, implying that $\boldsymbol{U} \cong \operatorname{Spec^c}\boldsymbol{\mathfrak{C}}$. We will show $\boldsymbol{\mathfrak{C}}$ is firm and interior. Write $d_1'', \ldots, d_n''$ for the images of $d_1', \ldots, d_n'$ in $\mathfrak{D}(d^{-1})_{\rm ex}$. Since $[d_1'], \ldots, [d_n']$ generate $\mathfrak{D}^\sharp_{\rm ex}$, $[d_1''], \ldots, [d_n'']$ generate $\mathfrak{D}(d^{-1})^\sharp_{\rm ex}$. As by construction $\mathfrak{C}_{\rm ex}$ is generated by $\psi_{\rm ex}(\mathfrak{D}(d^{-1})_{\rm ex})$ and invertibles $\Psi_{\rm exp}(\mathfrak{C})$, so $[\psi_{\rm ex}(d_1'')], \ldots, [\psi_{\rm ex}(d_n'')]$ generate $\mathfrak{C}^\sharp_{\rm ex}$, and $\boldsymbol{\mathfrak{C}}$ is firm.

By definition the natural map $\Xi_{\rm ex} : \mathfrak{C}_{\rm ex} \to \mathcal{O}^{\rm ex}_X(U)$ is injective. For each $i = 1, \ldots, n$, either $\Xi_{\rm ex} \circ \psi_{\rm ex}(d_i'')|_u = 0$ in $\mathcal{O}^{\rm ex}_{X,u}$ for all $u \in U$, so that $\psi_{\rm ex}(d_i'') = 0$, or $\Xi_{\rm ex} \circ \psi_{\rm ex}(d_i'')|_u = 0$ in $\mathcal{O}^{\rm ex}_{X,u}$ for all $u \in U$, so that $\Xi_{\rm ex} \circ \psi_{\rm ex}(d_i'')$ is not a zero divisor in $\mathcal{O}^{\rm ex}_X(U)$, as the $\mathcal{O}^{\rm ex}_{X,u}$ are interior, and thus $\psi_{\rm ex}(d_i'')$ is not a zero divisor in $\mathfrak{C}_{\rm ex}$. Since $[\psi_{\rm ex}(d_1'')], \ldots, [\psi_{\rm ex}(d_n'')]$ generate $\mathfrak{C}^\sharp_{\rm ex}$, and each $\psi_{\rm ex}(d_i'')$ is either zero or not a zero divisor, we see that $\mathfrak{C}_{\rm ex}$ has no zero divisors. Also $0_{\mathfrak{C}_{\rm ex}} \neq 1_{\mathfrak{C}_{\rm ex}}$ as $U \neq \emptyset$. So $\boldsymbol{\mathfrak{C}}$ is interior. Thus (ii) implies (iii), and the proof is complete. □

Let $\boldsymbol{f} : \boldsymbol{X} \to \boldsymbol{Y}$ be a morphism in $\mathbf{C^\infty Sch^c}$, and $x \in \boldsymbol{X}$ with $\boldsymbol{f}(x) = y$ in $\boldsymbol{Y}$. Then by definition, $\boldsymbol{X}, \boldsymbol{Y}$ are locally isomorphic to $\operatorname{Spec^c}\boldsymbol{\mathfrak{C}}$, $\operatorname{Spec^c}\boldsymbol{\mathfrak{D}}$ near x, y for C^∞-rings with corners $\boldsymbol{\mathfrak{C}}, \boldsymbol{\mathfrak{D}}$. However, it is *not* clear that $\boldsymbol{f}$ must be locally isomorphic to $\operatorname{Spec^c}\boldsymbol{\phi}$ for some morphism $\boldsymbol{\phi} : \boldsymbol{\mathfrak{D}} \to \boldsymbol{\mathfrak{C}}$ in $\mathbf{C^\infty Rings^c}$, and the authors expect this is false in some cases. This is because $\operatorname{Spec^c} : (\mathbf{C^\infty Rings^c})^{\rm op} \to \mathbf{AC^\infty Sch^c}$ is not full, as in Theorem 5.27(a), so morphisms $\operatorname{Spec^c}\boldsymbol{\mathfrak{C}} \to \operatorname{Spec^c}\boldsymbol{\mathfrak{D}}$ in $\mathbf{C^\infty Sch^c}$ need not be of the form $\operatorname{Spec^c}\boldsymbol{\phi}$. We now show that if $\boldsymbol{Y}$ is *firm*, then $\boldsymbol{f}$ is locally isomorphic to $\operatorname{Spec^c}\boldsymbol{\phi}$. We use this in Theorem 5.33 to give a criterion for existence of fibre products in $\mathbf{C^\infty Sch^c}$.

Proposition 5.32 *Suppose $\boldsymbol{f}:\boldsymbol{X}\to\boldsymbol{Y}$ is a morphism in* $\mathbf{C^\infty Sch^c}$ *with* $\boldsymbol{Y}$ *firm. Then for all $x\in\boldsymbol{X}$ with $\boldsymbol{f}(x)=y\in\boldsymbol{Y}$, there exists a commutative diagram in* $\mathbf{C^\infty Sch^c}$:

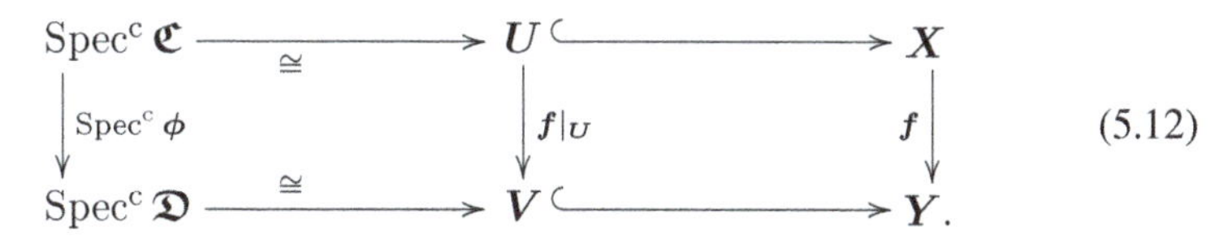

Here $\boldsymbol{U}\subseteq\boldsymbol{X}$, $\boldsymbol{V}\subseteq\boldsymbol{Y}$ are affine open neighbourhoods of x,y with $\boldsymbol{U}\subseteq\boldsymbol{f}^{-1}(\boldsymbol{V})$, and $\boldsymbol{\phi}:\mathfrak{D}\to\mathfrak{C}$ is a morphism in $\mathbf{C^\infty Rings^c_{sc}}$ *with $\mathfrak{D}$ firm. The analogue also holds with* $\mathbf{C^\infty Sch^c}, \mathbf{C^\infty Rings^c_{sc}}$ *replaced by* $\mathbf{C^\infty Sch^c_{in}}$ *and* $\mathbf{C^\infty Rings^c_{sc,in}}$.

Proof As $\boldsymbol{Y}$ is a firm C^∞-scheme with corners there exists an open neighbourhood $\boldsymbol{V}$ of y in $\boldsymbol{Y}$ and an isomorphism $\boldsymbol{V}\cong\operatorname{Spec^c}\mathfrak{D}$ for $\mathfrak{D}$ a firm C^∞-ring with corners. By Proposition 5.25 we may take $\mathfrak{D}$ to be semi-complete.

As $\boldsymbol{X}$ is a C^∞-scheme with corners there exists an affine open neighbourhood $\dot{\boldsymbol{U}}$ of x in $\boldsymbol{X}$ with $\dot{\boldsymbol{U}}\cong\operatorname{Spec^c}\dot{\mathfrak{C}}$. Since subsets $\dot{U}_c=\{x\in\dot{U}:x(c)\neq 0\}$ for $c\in\dot{\mathfrak{C}}$ are a basis for the topology of $\dot{U}$ by Definition 2.43, there exists $c\in\dot{\mathfrak{C}}$ with $\ddot{U}:=\dot{U}_c$ an open neighbourhood of x in $\dot{U}\cap f^{-1}(V)$. Set $\ddot{\mathfrak{C}}=\dot{\mathfrak{C}}(c^{-1})$. Then $\ddot{\boldsymbol{U}}\cong\operatorname{Spec^c}\ddot{\mathfrak{C}}$ is an open neighbourhood of x in $\boldsymbol{f}^{-1}(\boldsymbol{V})\subseteq\boldsymbol{X}$ by Lemma 5.15.

As $\mathfrak{D}$ is firm, the sharpening $\mathfrak{D}^\sharp_{\rm ex}$ is finitely generated. Choose $d'_1,\dots,d'_n$ in $\mathfrak{D}_{\rm ex}$ whose images generate $\mathfrak{D}^\sharp_{\rm ex}$. Then $\Xi^{\rm ex}_{\mathfrak{D}}(d'_i)$ lies in $(\Gamma^{\rm c}\circ\operatorname{Spec^c}\mathfrak{D})_{\rm ex}\cong\mathcal{O}^{\rm ex}_Y(V)$ for $i=1,\dots,n$, so $f^\sharp_{\rm ex}(\ddot{U})\circ\Xi^{\rm ex}_{\mathfrak{D}}(d'_i)$ lies in $(\Gamma^{\rm c}\circ\operatorname{Spec^c}\ddot{\mathfrak{C}})_{\rm ex}\cong\mathcal{O}^{\rm ex}_X(\ddot{U})$ for $i=1,\dots,n$. By definition of the sheaf $\mathcal{O}^{\rm ex}_X$, any section of $\mathcal{O}^{\rm ex}_X$ is locally modelled on $\Xi^{\rm ex}_{\ddot{\mathfrak{C}}}(c')$ for some $c'\in\ddot{\mathfrak{C}}_{\rm ex}$. Thus, we may choose an open neighbourhood U of x in $\ddot{U}$ and elements $c'_1,\dots,c'_n\in\ddot{\mathfrak{C}}_{\rm ex}$ such that $f^\sharp_{\rm ex}(\ddot{U})\circ\Xi^{\rm ex}_{\mathfrak{D}}(d'_i)|_U=\Xi^{\rm ex}_{\ddot{\mathfrak{C}}}(c'_i)|_U$ for $i=1,\dots,n$. Making U smaller if necessary, as above we may take U of the form $\ddot{U}_c$ for $c\in\ddot{\mathfrak{C}}$. Set $\mathfrak{C}=\Pi^{\rm sc}_{\rm all}(\ddot{\mathfrak{C}}(c^{-1}))$. Then $\boldsymbol{U}\cong\operatorname{Spec^c}\mathfrak{C}$ by Lemma 5.15 and Definition 5.26, for $\boldsymbol{U}$ an affine open neighbourhood of x in $\boldsymbol{X}$ with $\boldsymbol{U}\subseteq\boldsymbol{f}^{-1}(\boldsymbol{V})$, and $\mathfrak{C}$ a semi-complete C^∞-ring.

We will show there is a unique morphism $\boldsymbol{\phi}=(\phi,\phi_{\rm ex}):\mathfrak{D}\to\mathfrak{C}$ such that (5.12) commutes. To see this, note that there is a unique C^∞-ring morphism $\phi:\mathfrak{C}\to\mathfrak{D}$ making the C^∞-scheme part of (5.12) commute by Theorem 2.53(a), as $\mathfrak{C},\mathfrak{D}$ are complete. Let $d'\in\mathfrak{D}_{\rm ex}$. Since $d'_1,\dots,d'_n$ generate $\mathfrak{D}^\sharp_{\rm ex}$ we may write $d'=\Psi_{\rm exp}(d)(d'_1)^{a_1}\cdots(d'_n)^{a_n}$ for $d\in\mathfrak{D}$ and $a_1,\dots,a_n>0$.

We claim that defining $\phi_{\rm ex}(d')=\Psi_{\rm exp}(\phi(d))(c'_1)^{a_1}\cdots(c'_n)^{a_n}$ is independent of the presentation $d'=\Psi_{\rm exp}(d)(d'_1)^{a_1}\cdots(d'_n)^{a_n}$, and yields a monoid morphism $\phi_{\rm ex}:\mathfrak{D}_{\rm ex}\to\mathfrak{C}_{\rm ex}$ with all the properties we need. To see this, note that

$$\begin{aligned}\Xi^{\rm ex}_{\mathfrak{C}}(\phi_{\rm ex}(d'))&=\Xi^{\rm ex}_{\mathfrak{C}}\bigl(\Psi_{\rm exp}(\phi(d))(c_1')^{a_1}\cdots(c_n')^{a_n}\bigr)\\&=\bigl(\Xi^{\rm ex}_{\mathfrak{C}}\circ\Psi_{\rm exp}\circ\phi(d)\bigr)\bigl(\Xi^{\rm ex}_{\mathfrak{C}}(c_1')\bigr)^{a_1}\cdots(\Xi^{\rm ex}_{\mathfrak{C}}(c_n'))^{a_n}\\&=\Psi_{\rm exp}\bigl(\Xi_{\mathfrak{C}}\circ\phi(d)\bigr)\bigl(f^\sharp_{\rm ex}(U)\circ\Xi^{\rm ex}_{\mathfrak{D}}(d_1')\bigr)^{a_1}\cdots(f^\sharp_{\rm ex}(U)\circ\Xi^{\rm ex}_{\mathfrak{D}}(d_n'))^{a_n}\\&=\Psi_{\rm exp}\bigl(f^\sharp(U)\circ\Xi_{\mathfrak{D}}(d)\bigr)\bigl(f^\sharp_{\rm ex}(U)\circ\Xi^{\rm ex}_{\mathfrak{D}}(d_1')\bigr)^{a_1}\cdots(f^\sharp_{\rm ex}(U)\circ\Xi^{\rm ex}_{\mathfrak{D}}(d_n'))^{a_n}\\&=f^\sharp_{\rm ex}(U)\circ\Xi^{\rm ex}_{\mathfrak{D}}\bigl(\Psi_{\rm exp}(d)(d_1')^{a_1}\cdots(d_n')^{a_n}\bigr)=f^\sharp_{\rm ex}(U)\circ\Xi^{\rm ex}_{\mathfrak{D}}(d').\end{aligned}$$

Thus $\Xi^{\rm ex}_{\mathfrak{C}}(\phi_{\rm ex}(d'))$ is independent of the presentation, and $\Xi^{\rm ex}_{\mathfrak{C}}$ is injective as $\mathfrak{C}$ is semi-complete, so $\phi_{\rm ex}(d')$ is well defined. By identifying $\mathfrak{C},\mathfrak{D}$ with C^∞-subrings with corners of $\mathcal{O}_X(U),\mathcal{O}_Y(V)$, and $\boldsymbol{\phi}=(\phi,\phi_{\rm ex})$ with the restriction of $\boldsymbol{f}_\sharp(U):\mathcal{O}_Y(V)\to\mathcal{O}_X(U)$ to these C^∞-subrings, in a similar way to (5.7), we see that $\boldsymbol{\phi}$ is a morphism in $\mathbf{C^\infty Rings^c}$, proving the first part. The same arguments work in the interior case. □

5.7 Fibre products of C^∞-schemes with corners

We use Proposition 5.32 to construct fibre products of C^∞-schemes with corners.

Theorem 5.33 *Suppose $\boldsymbol{g}:\boldsymbol{X}\to\boldsymbol{Z}$, $\boldsymbol{h}:\boldsymbol{Y}\to\boldsymbol{Z}$ are morphisms in $\mathbf{C^\infty Sch^c}$, and $\boldsymbol{Z}$ is firm. Then $\boldsymbol{X}\times_{\boldsymbol{g},\boldsymbol{Z},\boldsymbol{h}}\boldsymbol{Y}$ exists in $\mathbf{C^\infty Sch^c}$, and is equal to the fibre product in $\mathbf{LC^\infty RS^c}$. Similarly, if $\boldsymbol{g}:\boldsymbol{X}\to\boldsymbol{Z}$, $\boldsymbol{h}:\boldsymbol{Y}\to\boldsymbol{Z}$ are morphisms in $\mathbf{C^\infty Sch^c_{in}}$, and $\boldsymbol{Z}$ is firm, then the fibre product $\boldsymbol{X}\times_{\boldsymbol{g},\boldsymbol{Z},\boldsymbol{h}}\boldsymbol{Y}$ exists in $\mathbf{C^\infty Sch^c_{in}}$, and is equal to the fibre product in $\mathbf{LC^\infty RS^c_{in}}$.*

The inclusion $\mathbf{C^\infty Sch^c_{in}}\hookrightarrow\mathbf{C^\infty Sch^c}$ preserves fibre products.

Proof Let $\boldsymbol{g}:\boldsymbol{X}\to\boldsymbol{Z}$, $\boldsymbol{h}:\boldsymbol{Y}\to\boldsymbol{Z}$ be morphisms in $\mathbf{C^\infty Sch^c}\subset\mathbf{LC^\infty RS^c}$. By Theorem 5.5 there exists a fibre product $\boldsymbol{W}=\boldsymbol{X}\times_{\boldsymbol{g},\boldsymbol{Z},\boldsymbol{h}}\boldsymbol{Y}$ in $\mathbf{LC^\infty RS^c}$, in a Cartesian square in $\mathbf{LC^\infty RS^c}$:

$$\begin{array}{ccc}\boldsymbol{W}&\xrightarrow{\ \boldsymbol{f}\ }&\boldsymbol{Y}\\ \downarrow{\scriptstyle\boldsymbol{e}}&&\downarrow{\scriptstyle\boldsymbol{h}}\\ \boldsymbol{X}&\xrightarrow{\ \boldsymbol{g}\ }&\boldsymbol{Z}.\end{array}\tag{5.13}$$

We will prove that $\boldsymbol{W}$ is a C^∞-scheme with corners. Then (5.13) is Cartesian in $\mathbf{C^\infty Sch^c}\subset\mathbf{LC^\infty RS^c}$, so $\boldsymbol{W}$ is a fibre product $\boldsymbol{X}\times_{\boldsymbol{g},\boldsymbol{Z},\boldsymbol{h}}\boldsymbol{Y}$ in $\mathbf{C^\infty Sch^c}$.

Let $w\in\boldsymbol{W}$ with $\boldsymbol{e}(w)=x\in\boldsymbol{X}$, $\boldsymbol{f}(w)=y\in\boldsymbol{Y}$ and $\boldsymbol{g}(x)=\boldsymbol{h}(y)=z\in\boldsymbol{Z}$. As $\boldsymbol{Z}$ is firm we may pick an affine open neighbourhood $\boldsymbol{V}$ of z in $\boldsymbol{Z}$ with

$\boldsymbol{V} \cong \operatorname{Spec^c} \mathfrak{F}$ for a firm C^∞-ring with corners $\mathfrak{F}$. By Proposition 5.25 we may take $\mathfrak{F}$ to be semi-complete.

By Proposition 5.32 we may choose affine open neighbourhoods $\boldsymbol{T}$ of x in $\boldsymbol{g}^{-1}(\boldsymbol{V}) \subset \boldsymbol{X}$ and $\boldsymbol{U}$ of y in $\boldsymbol{h}^{-1}(\boldsymbol{V}) \subset \boldsymbol{Y}$, and semi-complete C^∞-rings with corners $\mathfrak{D}, \mathfrak{E}$ with isomorphisms $\boldsymbol{T} \cong \operatorname{Spec^c} \mathfrak{D}$, $\boldsymbol{U} \cong \operatorname{Spec^c} \mathfrak{E}$, and morphisms $\boldsymbol{\phi} : \mathfrak{F} \to \mathfrak{D}$, $\boldsymbol{\psi} : \mathfrak{F} \to \mathfrak{E}$ in $\mathbf{C^\infty Rings^c_{sc}}$ such that $\operatorname{Spec^c} \boldsymbol{\phi}$, $\operatorname{Spec^c} \boldsymbol{\psi}$ are identified with $\boldsymbol{g}|_{\boldsymbol{T}} : \boldsymbol{T} \to \boldsymbol{V}$, $\boldsymbol{h}|_{\boldsymbol{U}} : \boldsymbol{U} \to \boldsymbol{V}$ by the isomorphisms $\boldsymbol{T} \cong \operatorname{Spec^c} \mathfrak{D}$, $\boldsymbol{U} \cong \operatorname{Spec^c} \mathfrak{E}$ and $\boldsymbol{V} \cong \operatorname{Spec^c} \mathfrak{F}$.

Set $\boldsymbol{S} = \boldsymbol{e}^{-1}(\boldsymbol{T}) \cap \boldsymbol{f}^{-1}(\boldsymbol{U})$. Then $\boldsymbol{S}$ is an open neighbourhood of w in $\boldsymbol{W}$. Write $\mathfrak{C} = \mathfrak{D} \amalg_{\boldsymbol{\phi}, \mathfrak{F}, \boldsymbol{\psi}} \mathfrak{E}$ for the pushout in $\mathbf{C^\infty Rings^c_{sc}}$, which exists by Theorem 5.27(d), with projections $\boldsymbol{\eta} : \mathfrak{D} \to \mathfrak{C}$, $\boldsymbol{\theta} : \mathfrak{E} \to \mathfrak{C}$. Consider the diagrams

$$\begin{array}{ccc} \boldsymbol{S} & \xrightarrow{\boldsymbol{f}|_{\boldsymbol{S}}} & \boldsymbol{U} \\ \big\downarrow{\scriptstyle \boldsymbol{e}|_{\boldsymbol{S}}} & & \big\downarrow{\scriptstyle \boldsymbol{h}|_{\boldsymbol{U}}} \\ \boldsymbol{T} & \xrightarrow{\boldsymbol{g}|_{\boldsymbol{T}}} & \boldsymbol{V}, \end{array} \qquad \begin{array}{ccc} \operatorname{Spec^c} \mathfrak{C} & \xrightarrow{\operatorname{Spec^c} \boldsymbol{\theta}} & \operatorname{Spec^c} \mathfrak{E} \\ \big\downarrow{\scriptstyle \operatorname{Spec^c} \boldsymbol{\eta}} & & \big\downarrow{\scriptstyle \operatorname{Spec^c} \boldsymbol{\psi}} \\ \operatorname{Spec^c} \mathfrak{D} & \xrightarrow{\operatorname{Spec^c} \boldsymbol{\phi}} & \operatorname{Spec^c} \mathfrak{F}, \end{array}$$

in $\mathbf{LC^\infty RS^c}$. The left-hand square is Cartesian as (5.13) is, by local properties of fibre products. The right-hand square is Cartesian by Theorem 5.27(e), as $\mathfrak{C} = \mathfrak{D} \amalg_{\boldsymbol{\phi}, \mathfrak{F}, \boldsymbol{\psi}} \mathfrak{E}$. The isomorphisms $\boldsymbol{T} \cong \operatorname{Spec^c} \mathfrak{D}$, $\boldsymbol{U} \cong \operatorname{Spec^c} \mathfrak{E}$, $\boldsymbol{V} \cong \operatorname{Spec^c} \mathfrak{F}$ identify the right and bottom sides of both squares. Hence the two squares are isomorphic. Thus $\boldsymbol{S} \cong \operatorname{Spec^c} \mathfrak{C}$. So each point w in $\boldsymbol{W}$ has an affine open neighbourhood, and $\boldsymbol{W}$ is a C^∞-scheme with corners, as we have to prove.

The proof for the interior case is essentially the same. Theorem 5.31(c) allows us to write $\boldsymbol{V} \cong \operatorname{Spec^c_{in}} \mathfrak{F}$ for $\mathfrak{F} \in \mathbf{C^\infty Sch^c_{fi,in}}$, and Theorem 5.27 and Propositions 5.25 and 5.32 also work in the interior case. The last part then follows from Theorem 5.5. □

Theorem 5.34 *Consider the family of full subcategories:*

$$\begin{aligned} &\mathbf{C^\infty Sch^c_{fp}} \subset \mathbf{LC^\infty RS^c}, && \mathbf{C^\infty Sch^c_{fg}} \subset \mathbf{LC^\infty RS^c}, \\ &\mathbf{C^\infty Sch^c_{fi}} \subset \mathbf{LC^\infty RS^c}, && \mathbf{C^\infty Sch^c_{fi,in}} \subset \mathbf{LC^\infty RS^c_{in}}, \\ &\mathbf{C^\infty Sch^c_{fi,\mathbb{Z}}} \subset \mathbf{LC^\infty RS^c_{\mathbb{Z}}}, && \mathbf{C^\infty Sch^c_{fi,tf}} \subset \mathbf{LC^\infty RS^c_{tf}}, \\ &\mathbf{C^\infty Sch^c_{to}} \subset \mathbf{LC^\infty RS^c_{sa}}. \end{aligned} \tag{5.14}$$

In each case, the subcategory is closed under finite limits in the larger category. Hence, fibre products and all finite limits exist in $\mathbf{C^\infty Sch^c_{fp}}, \ldots, \mathbf{C^\infty Sch^c_{to}}$.

Proof In the proof of Theorem 5.33, if $\boldsymbol{X}, \boldsymbol{Y}, \boldsymbol{Z}$ are finitely presented we may take $\mathfrak{D}, \mathfrak{E}, \mathfrak{F}$ to be finitely presented, and then $\mathfrak{C}$ is finitely presented by

Proposition 4.42, so $\boldsymbol{W} = \boldsymbol{X} \times_{\boldsymbol{g},\boldsymbol{Z},\boldsymbol{h}} \boldsymbol{Y}$ is finitely presented. Hence all fibre products exist in $\mathbf{C^\infty Sch^c_{fp}}$, and agree with fibre products in $\mathbf{LC^\infty RS^c}$.

As $\mathbf{C^\infty Sch^c_{fp}}$ has a final object $\operatorname{Spec^c}(\mathbb{R},[0,\infty))$, and all fibre products in a category with a final object are (iterated) fibre products, then $\mathbf{C^\infty Sch^c_{fp}}$ is closed under finite limits in $\mathbf{LC^\infty RS^c}$, and all such finite limits exist in $\mathbf{C^\infty Sch^c_{fp}}$ by Theorem 5.5. The same argument works for $\mathbf{C^\infty Sch^c_{fg}}$, $\mathbf{C^\infty Sch^c_{fi}}$, $\mathbf{C^\infty Sch^c_{fi,in}}$, using Proposition 4.44 for the latter two.

For $\mathbf{C^\infty Sch^c_{fi,\mathbb{Z}}}$, $\mathbf{C^\infty Sch^c_{fi,tf}}$, $\mathbf{C^\infty Sch^c_{to}}$ we apply the coreflection functors $\Pi^{\mathbb{Z}}_{\rm in}, \Pi^{\rm tf}_{\mathbb{Z}}, \Pi^{\rm sa}_{\rm tf}$ of Theorem 5.8, noting that the restrictions of these map

$$\mathbf{C^\infty Sch^c_{fi,in}} \xrightarrow{\Pi^{\mathbb{Z}}_{\rm in}} \mathbf{C^\infty Sch^c_{fi,\mathbb{Z}}} \xrightarrow{\Pi^{\rm tf}_{\mathbb{Z}}} \mathbf{C^\infty Sch^c_{fi,tf}} \xrightarrow{\Pi^{\rm sa}_{\rm tf}} \mathbf{C^\infty Sch^c_{to}},$$

by Theorem 5.31 and the fact (clear from the proof of Theorem 5.31) that $\Pi^{\mathbb{Z}}_{\rm in}, \Pi^{\rm tf}_{\mathbb{Z}}, \Pi^{\rm sa}_{\rm tf}$ preserve firmness.

Let J be a finite category and let $\boldsymbol{F} : J \to \mathbf{C^\infty Sch^c_{fi,\mathbb{Z}}}$ be a functor. Then a limit $\boldsymbol{X} = \varprojlim \boldsymbol{F}$ exists in $\mathbf{C^\infty Sch^c_{fi,in}}$ from above, which is also a limit in $\mathbf{LC^\infty RS^c_{in}}$. Hence $\Pi^{\mathbb{Z}}_{\rm in}(\boldsymbol{X})$ is a limit $\varprojlim \Pi^{\mathbb{Z}}_{\rm in} \circ \boldsymbol{F}$ in $\mathbf{LC^\infty RS^c_{\mathbb{Z}}}$, since $\Pi^{\mathbb{Z}}_{\rm in} : \mathbf{LC^\infty RS^c_{in}} \to \mathbf{LC^\infty RS^c_{\mathbb{Z}}}$ preserves limits as it is a right adjoint. But $\Pi^{\mathbb{Z}}_{\rm in}$ acts as the identity on $\mathbf{LC^\infty RS^c_{\mathbb{Z}}} \subset \mathbf{LC^\infty RS^c_{in}}$, so $\Pi^{\mathbb{Z}}_{\rm in}(\boldsymbol{X})$ is a limit $\varprojlim \boldsymbol{F}$ in $\mathbf{LC^\infty RS^c_{\mathbb{Z}}}$. As $\boldsymbol{X} \in \mathbf{C^\infty Sch^c_{fi,in}}$ we have $\Pi^{\mathbb{Z}}_{\rm in}(\boldsymbol{X}) \in \mathbf{C^\infty Sch^c_{fi,\mathbb{Z}}}$, so $\mathbf{C^\infty Sch^c_{fi,\mathbb{Z}}}$ is closed under finite limits in $\mathbf{LC^\infty RS^c_{\mathbb{Z}}}$, and all finite limits exist in $\mathbf{C^\infty Sch^c_{fi,\mathbb{Z}}}$ by Corollary 5.10. The arguments for $\mathbf{C^\infty Sch^c_{fi,tf}}$, $\mathbf{C^\infty Sch^c_{to}}$ are the same. □

We will return to the subject of fibre products in §6.6–§6.7, using 'corner functors' from Chapter 6 as a tool to study them.

6

Boundaries, corners, and the corner functor

In §3.4, for manifolds with (g-)corners of mixed dimension, we defined *corner functors* $C : \check{\mathbf{M}}\mathbf{an^c} \to \check{\mathbf{M}}\mathbf{an^c_{in}}$ and $C : \check{\mathbf{M}}\mathbf{an^{gc}} \to \check{\mathbf{M}}\mathbf{an^{gc}_{in}}$, which are right adjoint to the inclusions $\mathrm{inc} : \check{\mathbf{M}}\mathbf{an^c_{in}} \hookrightarrow \check{\mathbf{M}}\mathbf{an^c}$ and $\mathrm{inc} : \check{\mathbf{M}}\mathbf{an^{gc}_{in}} \hookrightarrow \check{\mathbf{M}}\mathbf{an^{gc}}$. These encode the boundary ∂X and k-corners $C_k(X)$ of a manifold with (g-)corners X in a functorial way.

This chapter will show that analogous corner functors $C : \mathbf{LC^\infty RS^c} \to \mathbf{LC^\infty RS^c_{in}}$ and $C : \mathbf{C^\infty Sch^c} \to \mathbf{C^\infty Sch^c_{in}}$ may be defined for local C^∞-ringed spaces and C^∞-schemes with corners $\boldsymbol{X}$, which are right adjoint to the inclusions $\mathrm{inc} : \mathbf{LC^\infty RS^c_{in}} \hookrightarrow \mathbf{LC^\infty RS^c}$ and $\mathrm{inc} : \mathbf{C^\infty Sch^c_{in}} \hookrightarrow \mathbf{C^\infty Sch^c}$. This allows us to define the *boundary* $\partial\boldsymbol{X}$ and *k-corners* $C_k(\boldsymbol{X})$ for firm C^∞-schemes with corners $\boldsymbol{X}$. We use corner functors to study existence of fibre products in categories such as $\mathbf{C^\infty Sch^c_{to,ex}}$.

6.1 The corner functor for C^∞-ringed spaces with corners

Definition 6.1 Let $\boldsymbol{X} = (X, \boldsymbol{\mathcal{O}}_X)$ be a local C^∞-ringed space with corners. We will define the *corners* $C(\boldsymbol{X})$ in $\mathbf{LC^\infty RS^c_{in}}$. As a set, we define

$$C(X) = \{(x, P) : x \in X,\, P \text{ is a prime ideal in } \mathcal{O}^{\mathrm{ex}}_{X,x}\}, \qquad (6.1)$$

where prime ideals in monoids are as in Definition 3.5. Define $\Pi_X : C(X) \to X$ as a map of sets by $\Pi_X : (x, P) \mapsto x$.

We define a topology on $C(X)$ to be the weakest topology such that Π_X is continuous, so $\Pi_X^{-1}(U)$ is open for all $U \subset X$, and such that for all open $U \subset X$, for all elements $s' \in \mathcal{O}^{\mathrm{ex}}_X(U)$, then

$$\dot{U}_{s'} = \{(x, P) : x \in U,\ s'_x \notin P\} \subseteq \Pi_X^{-1}(U) \ \text{ and } \ \ddot{U}_{s'} = \Pi_X^{-1}(U) \setminus \dot{U}_{s'} \qquad (6.2)$$

are both open and closed in $\Pi_X^{-1}(U)$. Then $\Pi_X^{-1}(U) = \dot{U}_1 = \ddot{U}_0$, and $\emptyset = \ddot{U}_1 =$

$\dot{U}_0$ for $0, 1 \in \mathcal{O}_X^{\rm ex}(U)$. The collection $\{\dot{U}_{s'}, \ddot{U}_{s'} : \text{open } U \subset X,\ s' \in \mathcal{O}_X^{\rm ex}(U)\}$ is a subbase for the topology.

Pull back the sheaves $\mathcal{O}_X, \mathcal{O}_X^{\rm ex}$ by Π_X to get sheaves $\Pi_X^{-1}(\mathcal{O}_X), \Pi_X^{-1}(\mathcal{O}_X^{\rm ex})$ on $C(X)$. For each $x \in X$ and each prime P in $\mathcal{O}_{X,x}^{\rm ex}$, we have ideals

$$P \subset \mathcal{O}_{X,x}^{\rm ex} \cong (\Pi_X^{-1}(\mathcal{O}_X^{\rm ex}))_{(x,P)}, \quad \langle \Phi_i(P) \rangle \subset \mathcal{O}_{X,x} \cong (\Pi_X^{-1}(\mathcal{O}_X))_{(x,P)}.$$

Here $\langle \Phi_i(P) \rangle$ is the ideal generated by the image of P under $\Phi_i : \mathcal{O}_{X,x}^{\rm ex} \to \mathcal{O}_{X,x}$.

We define the sheaf of C^∞-rings $\mathcal{O}_{C(X)}$ on $C(X)$ to be the sheafification of the presheaf of C^∞-rings $\Pi_X^{-1}(\mathcal{O}_X)/I$ on $C(X)$, where for open $U \subset C(X)$

$$I(U) = \{ s \in \Pi_X^{-1}(\mathcal{O}_X)(U) : s_{(x,P)} \in \langle \Phi_i(P) \rangle \text{ for all } (x,P) \in U \}.$$

Let $\mathcal{O}_{C(X)}^{\rm ex}$ be the sheafification of the presheaf $U \mapsto \Pi_X^{-1}(\mathcal{O}_X^{\rm ex})(U)/\!\sim$ where the equivalence relation $\sim$ is generated by the relations that $s_1' \sim s_2'$ if for each $(x,P) \in U$ either $s_{1,x}', s_{2,x}' \in P$, or there is $p \in \langle \Phi_i(P) \rangle$ such that $s_{1,x}' = \Psi_{\exp}(p) s_{2,x}'$, for $s_1', s_2' \in \Pi_X^{-1}(\mathcal{O}_X^{\rm ex})(U)$. This is similar to quotienting the C^∞-ring with corners $(\Pi_X^{-1}(\mathcal{O}_X)(U), \Pi_X^{-1}(\mathcal{O}_X^{\rm ex})(U))$ by a prime ideal in $\Pi_X^{-1}(\mathcal{O}_X^{\rm ex})(U)$, which we described in Example 4.25(b), and creates a sheaf of C^∞-rings with corners $\boldsymbol{\mathcal{O}}_{C(X)} = (\mathcal{O}_{C(X)}, \mathcal{O}_{C(X)}^{\rm ex})$ on $C(X)$.

Lemma 6.2 shows $C(\boldsymbol{X}) = (C(X), \boldsymbol{\mathcal{O}}_{C(X)})$ is an interior local C^∞-ringed space with corners, and the stalk of $\boldsymbol{\mathcal{O}}_{C(X)}$ at (x,P) is an interior local C^∞-ring with corners isomorphic to $\boldsymbol{\mathcal{O}}_{X,x}/\!\sim_P$, in the notation of Example 4.25(b).

Define sheaf morphisms $\Pi_X^\sharp : \Pi_X^{-1}(\mathcal{O}_X) \to \mathcal{O}_{C(X)}$ and $\Pi_{X,\rm ex}^\sharp : \Pi_X^{-1}(\mathcal{O}_X^{\rm ex}) \to \mathcal{O}_{C(X)}^{\rm ex}$ as the sheafifications of the projections $\Pi_X^{-1}(\mathcal{O}_X) \to \Pi_X^{-1}(\mathcal{O}_X)/I$ and $\Pi_X^{-1}(\mathcal{O}_X^{\rm ex}) \to \Pi_X^{-1}(\mathcal{O}_X^{\rm ex})/\!\sim$. Then $\boldsymbol{\Pi}_{\boldsymbol{X}}^\sharp = (\Pi_X^\sharp, \Pi_{X,\rm ex}^\sharp) : \Pi_X^{-1}(\boldsymbol{\mathcal{O}}_X) \to \boldsymbol{\mathcal{O}}_{C(X)}$ is a morphism of sheaves of C^∞-rings with corners, and $\boldsymbol{\Pi}_{\boldsymbol{X}} = (\Pi_X, \boldsymbol{\Pi}_{\boldsymbol{X}}^\sharp)$ is a morphism $\boldsymbol{\Pi}_{\boldsymbol{X}} : C(\boldsymbol{X}) \to \boldsymbol{X}$ in $\mathbf{LC^\infty RS^c}$.

Let $\boldsymbol{f} = (f, f^\sharp, f_{\rm ex}^\sharp) : \boldsymbol{X} \to \boldsymbol{Y}$ be a morphism in $\mathbf{LC^\infty RS^c}$. We will define a morphism $C(\boldsymbol{f}) : C(\boldsymbol{X}) \to C(\boldsymbol{Y})$. On topological spaces, we define $C(f) : C(X) \to C(Y)$ by $(x,P) \mapsto (f(x), (f_{{\rm ex},x}^\sharp)^{-1}(P))$ where $f_{{\rm ex},x}^\sharp : \mathcal{O}_{Y,f(x)}^{\rm ex} \to \mathcal{O}_{X,x}^{\rm ex}$ is the stalk map of $f_{\rm ex}^\sharp$. This is continuous as if $t \in \mathcal{O}_Y^{\rm ex}(U)$ for some open $U \subset Y$ then $C(f)^{-1}(\dot{U}_t) = (f^{-1}(U))^{\cdot}_{f_\sharp^{\rm ex}(U)(t)}$ is open, and similarly $C(f)^{-1}(\ddot{U}_t) = (f^{-1}(U))^{\cdot\cdot}_{f_\sharp^{\rm ex}(U)(t)}$ is open. Also $\Pi_Y \circ C(f) = f \circ \Pi_X$.

To define the sheaf morphism $C(\boldsymbol{f})^\sharp : C(f)^{-1}(\boldsymbol{\mathcal{O}}_{C(Y)}) \to \boldsymbol{\mathcal{O}}_{C(X)}$, consider the diagram of presheaves on $C(X)$:

$$\begin{array}{ccc}
C(f)^{-1} \circ \Pi_Y^{-1}(\mathcal{O}_Y) = \Pi_X^{-1} \circ f^{-1}(\mathcal{O}_Y) & \xrightarrow[\Pi_X^{-1}(f^\sharp)]{} & \Pi_X^{-1}(\mathcal{O}_X) \\
\downarrow & & \downarrow \\
C(f)^{-1}(\Pi_Y^{-1}(\mathcal{O}_Y)/I_Y) & \xrightarrow{\Pi_X^{-1}(f^\sharp)_*} & \Pi_X^{-1}(\mathcal{O}_X)/I_X \\
\downarrow{\scriptstyle \text{sheafify}} & & {\scriptstyle \text{sheafify}}\downarrow \\
C(f)^{-1}(\mathcal{O}_{C(Y)}) & \xrightarrow{C(f)^\sharp} & \mathcal{O}_{C(X)}.
\end{array}$$

Here there is a unique morphism $\Pi_X^{-1}(f^\sharp)_*$ making the upper rectangle commute as $\Pi_X^{-1}(f^\sharp)$ maps $C(f)^{-1}(I_Y) \to I_X$ by definition of $I_Y, I_X, C(f)$. Hence there is a unique morphism $C(f)^\sharp$ making the lower rectangle commute by properties of sheafification.

Similarly we define $C(f)^\sharp_{\rm ex} : C(f)^{-1}(\mathcal{O}^{\rm ex}_{C(Y)}) \to \mathcal{O}_{C(X)}$, and $C(\boldsymbol{f})^\sharp = (C(f)^\sharp, C(f)^\sharp_{\rm ex}) : C(f)^{-1}(\boldsymbol{\mathcal{O}}_{C(Y)}) \to \boldsymbol{\mathcal{O}}_{C(X)}$ is a morphism of sheaves of C^∞-rings with corners on $C(X)$, so $C(\boldsymbol{f}) = (C(f), C(\boldsymbol{f})^\sharp) : C(\boldsymbol{X}) \to C(\boldsymbol{Y})$ is a morphism in $\mathbf{LC^\infty RS^c}$. We see that $\boldsymbol{\Pi_Y} \circ C(\boldsymbol{f}) = \boldsymbol{f} \circ \boldsymbol{\Pi_X}$.

On the stalks at (x, P) in $C(X)$, we have

$$C(\boldsymbol{f})^\sharp_{(x,P)} = (\boldsymbol{f}^\sharp_x)_* : \boldsymbol{\mathcal{O}}_{Y,f(x)}/\sim_{(f^\sharp_{{\rm ex},x})^{-1}(P)} \longrightarrow \boldsymbol{\mathcal{O}}_{X,x}/\sim_P .$$

For the monoid sheaf, if $s' \mapsto 0$ in the stalk, then $s' = [s'']$ for $s'' \in \mathcal{O}^{\rm ex}_{Y,f(x)}$, and $f^\sharp_{{\rm ex},(x,P)}(s'')_x \in P$. Then $s''_{f(x)} \in (f^\sharp_{{\rm ex},x})^{-1}(P)$, so $s'' \sim_{(f^\sharp_{{\rm ex},x})^{-1}(P)} 0$, giving $s' = 0$. Therefore $C(\boldsymbol{f})$ is an interior morphism. One can check that $C(\boldsymbol{f} \circ \boldsymbol{g}) = C(\boldsymbol{f}) \circ C(\boldsymbol{g})$ and $C(\mathbf{id}_{\boldsymbol{X}}) = \mathbf{id}_{C(\boldsymbol{X})}$, and thus $C : \mathbf{LC^\infty RS^c} \to \mathbf{LC^\infty RS^c_{in}}$ is a well defined functor.

If we now assume $\boldsymbol{X}$ is interior, then $\{(x, \{0\}) : x \in X\}$ is contained in $C(X)$, and there is an inclusion of sets $\iota_X : X \hookrightarrow C(X)$, $\iota_X : x \mapsto (x, \{0\})$. The image of X under $\iota_X : X \hookrightarrow C(X)$ is closed in $C(X)$, since for all open $U \subset X$,

$$\iota_X(U) = \bigcap\nolimits_{s' \in \mathcal{O}^{\rm in}_X(U)} \dot{U}_{s'}$$

is closed in $\Pi_X^{-1}(U)$. Also, ι_X is continuous, as the definition of interior implies $\iota_X^{-1}(\dot{U}_{s'}) = \{x \in U : s'_x \neq 0 \in \mathcal{O}^{\rm ex}_{X,x}\}$ is open, and $\iota_X^{-1}(\ddot{U}_{s'}) = \{x \in U : s'_x = 0 \text{ in } \mathcal{O}^{\rm ex}_{X,x}\}$ is open for all $s' \in \mathcal{O}^{\rm ex}_X(U)$.

Now any $s \in \mathcal{O}_X(U)$ gives equivalence classes in $\Pi_X^{-1}(\mathcal{O}_X)(\Pi_X^{-1}(U))$, and $\Pi_X^{-1}(\mathcal{O}_X)(\Pi_X^{-1}(U))/\sim$, and $\mathcal{O}_{C(X)}(\Pi^{-1}(U))$, and $\iota_X^{-1}(\mathcal{O}_{C(X)})(U)$. So there is a map $\mathcal{O}_X(U) \to \iota_X^{-1}(\mathcal{O}_{C(X)})(U)$, and a similar map $\mathcal{O}^{\rm ex}_X(U) \to \iota_X^{-1}(\mathcal{O}^{\rm ex}_{C(X)})(U)$. These respect restriction and give morphisms of sheaves of C^∞-rings with corners. The stalks of $\iota_X^{-1}(\boldsymbol{\mathcal{O}}_{C(X)})$ are isomorphic to the stalks of $\boldsymbol{\mathcal{O}}_{C(X)}$, which are isomorphic to $\boldsymbol{\mathcal{O}}_{X,x}/\sim_{\{0\}} \cong \boldsymbol{\mathcal{O}}_{X,x}$ by Lemma 6.2.

So there is a canonical isomorphism $\iota_{\boldsymbol{X}}^\sharp : \iota_X^{-1}(\mathcal{O}_{C(X)}) \to \mathcal{O}_X$, and $\iota_{\boldsymbol{X}} = (\iota_X, \iota_{\boldsymbol{X}}^\sharp) : \boldsymbol{X} \to C(\boldsymbol{X})$ is a morphism, which is interior, as the stalk maps are isomorphisms.

Thus, for each interior $\boldsymbol{X}$ we have defined a morphism $\iota_{\boldsymbol{X}} : \boldsymbol{X} \to C(\boldsymbol{X})$ in $\mathbf{LC^\infty RS^c_{in}}$, which is an inclusion of $\boldsymbol{X}$ as an open and closed C^∞-subscheme with corners in $C(\boldsymbol{X})$. One can show that $\boldsymbol{\Pi}_{\boldsymbol{X}} \circ \iota_{\boldsymbol{X}} = \mathbf{id}_{\boldsymbol{X}}$. Also, if $\boldsymbol{f} : \boldsymbol{X} \to \boldsymbol{Y}$ is a morphism in $\mathbf{LC^\infty RS^c_{in}}$ one can show that $\iota_{\boldsymbol{Y}} \circ \boldsymbol{f} = C(\boldsymbol{f}) \circ \iota_{\boldsymbol{X}}$.

Lemma 6.2 *In Definition* 6.1, $C(\boldsymbol{X}) = (C(X), \mathcal{O}_{C(X)})$ *is an interior local C^∞-ringed space with corners in the sense of Definition* 5.2. *The stalk $\mathcal{O}_{C(X),(x,P)}$ of $\mathcal{O}_{C(X)}$ at (x,P) in $C(X)$ is an interior local C^∞-ring with corners isomorphic to $\mathcal{O}_{X,x}/\sim_P$, in the notation of Example* 4.25(b).

Proof We first verify the openness conditions for $C(\boldsymbol{X})$ to be an *interior C^∞-ringed space with corners* in Definition 5.2(a),(b). Let $U \subset C(X)$ be open and $s' \in \mathcal{O}^{\rm ex}_{C(X)}(U)$, and as in Definition 5.2(a),(b) write

$$U_{s'} = \{(x,P) \in U : s'_{(x,P)} \neq 0 \in \mathcal{O}^{\rm ex}_{C(X),(x,P)}\} \quad \text{and} \quad \hat{U}_{s'} = U \setminus U_{s'}.$$

Then there is an open cover $\{U_i\}_{i\in I}$ of U such that $s'|_{U_i} = [s'_i]$ for $s'_i \in \Pi^{-1}(\mathcal{O}^{\rm ex}_X)(U_i)$, where $s'_i|_{(x,P)} \notin P$ if $(x,P) \in U_{s'} \cap U_i$ and $s'_i|_{(x,P)} \in P$ if $(x,P) \in U_{s'} \cap \hat{U}_i$. The definition of the inverse image sheaf implies there is an open cover $\{W^i_j\}_{j\in J_i}$ of U_i such that $s'_i|_{W_{i,j}} = [s'_{i,j}]$ for some $s'_{i,j} \in \mathcal{O}^{\rm ex}_X(V_{i,j})$ for open $V_{i,j} \subset X$ such that $V_{i,j} \supset \Pi(W_{i,j})$. Then $s'_{i,j}|_x \notin P$ for all $(x,P) \in U_{s'} \cap W_{i,j}$, and $s'_{i,j}|_x \in P$ for all $(x,P) \in \hat{U}_{s'} \cap W_{i,j}$. Thus

$$U_{s'} \cap W_{i,j} = W_{i,j} \cap (\dot{V_{i,j}})_{s'_{i,j}} \qquad \hat{U}_{s'} \cap W_{i,j} = W_{i,j} \cap (\ddot{V_{i,j}})_{s'_{i,j}}$$

are both open in $C(X)$ by the definition of the topology on $C(X)$ involving $\dot{U}_{s'}, \ddot{U}_{s'}$ in (6.2). Taking the union over $i \in I$, $j \in J_i$, and using that $\bigcup_{i,j} W_{i,j} = U$, we see that $U_{s'}, \hat{U}_{s'}$ are open in U, as we have to prove.

The stalks of $\mathcal{O}_{C(X)}, \mathcal{O}^{\rm ex}_{C(X)}$ at $(x,P) \in C(X)$ are $\mathcal{O}_{X,x}/\langle \Phi_i(P)\rangle$ and $\mathcal{O}^{\rm ex}_{X,x}/\sim_P$, where $s'_1 \sim_P s'_2$ in $\mathcal{O}^{\rm ex}_{X,x}$ if $s'_1, s'_2 \in P$ or there is a $p \in \langle \Phi_i(P)\rangle$ such that $\Psi_{\exp}(p)s'_1 = s'_2$, as in Example 4.25(b). To see this, consider that the definitions give us the following diagram, where we will show the arrow $\boldsymbol{t}$ exists and is an isomorphism

$$\begin{array}{ccccc} \mathcal{O}_{X,x} & \xrightarrow[\boldsymbol{\Pi}^\sharp_x]{\sim} & \Pi^{-1}(\mathcal{O}_X)_{(x,P)} & \longrightarrow\!\!\!\!\rightarrow & \mathcal{O}_{C(X),(x,P)}. \\ \big\downarrow & & & \nearrow_{\boldsymbol{t}}^{\cong} & \\ \mathcal{O}_{X,x}/\sim_P & & & & \end{array} \tag{6.3}$$

To see that $\boldsymbol{t}$ exists, we use the universal property of $\mathcal{O}_{X,x}/\sim_P$ as in Example

4.25(b). Let U be an open set in X and take $s' \in \mathcal{O}_X^{\mathrm{ex}}(U)$ such that $s'_x \in P$. Then s' maps to an equivalence class in $\mathcal{O}_{C(X)}^{\mathrm{ex}}(\Pi^{-1}(U))$. Consider the open set $\ddot{U}_{s'} = \{(x,P) \in C(X) : x \in U,\ s'_x \in P\}$, then restricting to this open set, we see that our (x,P) is in $\ddot{U}_{s'} \subset \Pi^{-1}(U)$ and so we can restrict the equivalence class of s' to $\ddot{U}_{s'}$. In this open set, however, $s'_x \in P$ for all $(x,P) \in \ddot{U}_{s'}$ so $s' \sim 0$, and s' lies in the kernel of the composition of the top row of (6.3). Then the universal property of $\boldsymbol{\mathcal{O}}_{X,x}/\sim_P$ says that $\boldsymbol{t}$ must exist and commute with the diagram. Also, $\boldsymbol{t}$ must be surjective as the top line is surjective.

To see that $\boldsymbol{t}$ is injective is straightforward. For example, in the monoid case, if $[s'_{1,x}], [s'_{2,x}] \in \mathcal{O}_{X,x}^{\mathrm{ex}}/\sim_P$ with representatives $s'_1, s'_2 \in \mathcal{O}_X^{\mathrm{ex}}(U)$, and if $t_{\mathrm{ex}}([s'_{1,x}]) = t_{\mathrm{ex}}([s'_{2,x}])$, then $s'_{1,x} \sim s'_{2,x}$ for all $(x,P) \in V$ for some open $V \subset \Pi^{-1}(U)$. This means at every $(x,P) \in V$ then $s'_{1,x} \sim_P s'_{2,x}$, so this must be true at our (x,P), so that $[s'_{1,x}] = [s'_{2,x}] \in \mathcal{O}_{X,x}^{\mathrm{ex}}/\sim_P$ as required.

Now $\boldsymbol{\mathcal{O}}_{X,x}/\sim_P$ is interior as the complement of $P \in \mathcal{O}_{X,x}^{\mathrm{ex}}$ has no zero divisors. It is also local, as the unique morphism $\mathcal{O}_{X,x} \to \mathbb{R}$ must have P in its kernel so it factors through the morphism $\mathcal{O}_{X,x} \to \mathcal{O}_{X,x}^{\mathrm{ex}}/\sim_P$ giving a unique morphism $\mathcal{O}_{X,x}^{\mathrm{ex}}/\sim_P \to \mathbb{R}$ with the correct properties to be local. Thus $\boldsymbol{\mathcal{O}}_{C(X),(x,P)}$ is an interior local C^∞-ring with corners for all $(x,P) \in C(X)$, so $C(\boldsymbol{X})$ is an interior local C^∞-ringed space with corners by Definition 5.2. □

Theorem 6.3 *The corner functor* $C : \mathbf{LC^\infty RS^c} \to \mathbf{LC^\infty RS^c_{in}}$ *is right adjoint to the inclusion* $\mathrm{inc} : \mathbf{LC^\infty RS^c_{in}} \hookrightarrow \mathbf{LC^\infty RS^c}$. *Thus we have functorial isomorphisms* $\mathrm{Hom}_{\mathbf{LC^\infty RS^c_{in}}}(C(\boldsymbol{X}), \boldsymbol{Y}) \cong \mathrm{Hom}_{\mathbf{LC^\infty RS^c}}(\boldsymbol{X}, \mathrm{inc}(\boldsymbol{Y}))$.

As C *is a right adjoint, it preserves limits in* $\mathbf{LC^\infty RS^c}, \mathbf{LC^\infty RS^c_{in}}$.

Proof We describe the unit $\eta : \mathrm{Id} \Rightarrow C \circ \mathrm{inc}$ and counit $\epsilon : \mathrm{inc} \circ C \Rightarrow \mathrm{Id}$ of the adjunction. Here $\epsilon_{\boldsymbol{X}} = \boldsymbol{\Pi}_{\boldsymbol{X}}$ for $\boldsymbol{X}$ a C^∞-scheme with corners, and $\eta_{\boldsymbol{X}} = \iota_{\boldsymbol{X}}$ for $\boldsymbol{X}$ an interior C^∞-scheme with corners. That ϵ is a natural transformation follows from $\boldsymbol{\Pi}_{\boldsymbol{Y}} \circ C(\boldsymbol{f}) = \boldsymbol{f} \circ \boldsymbol{\Pi}_{\boldsymbol{X}}$ for morphisms $\boldsymbol{f} : \boldsymbol{X} \to \boldsymbol{Y}$ in $\mathbf{LC^\infty RS^c}$, and that η is a natural transformation follows from $\iota_{\boldsymbol{Y}} \circ \boldsymbol{f} = C(\boldsymbol{f}) \circ \iota_{\boldsymbol{X}}$ for morphisms $\boldsymbol{f} : \boldsymbol{X} \to \boldsymbol{Y}$ in $\mathbf{LC^\infty RS^c_{in}}$. Finally, to prove the adjunction, we must show the following compositions are the identity natural transformations:

$$C \xRightarrow{\eta * \mathrm{id}_C} C \circ \mathrm{inc} \circ C \xRightarrow{\mathrm{id}_C * \epsilon} C, \quad \mathrm{inc} \xRightarrow{\mathrm{id}_{\mathrm{inc}} * \eta} \mathrm{inc} \circ C \circ \mathrm{inc} \xRightarrow{\eta * \mathrm{id}_{\mathrm{inc}}} \mathrm{inc}.$$

Both of these follow as $\boldsymbol{\Pi}_{\boldsymbol{X}} \circ \iota_{\boldsymbol{X}} = \mathbf{id}_{\boldsymbol{X}}$. □

Definition 5.7 defined subcategories $\mathbf{LC^\infty RS^c_{sa}} \subset \cdots \subset \mathbf{LC^\infty RS^c_{in}}$ and $\mathbf{LC^\infty RS^c_{sa,ex}} \subset \cdots \subset \mathbf{LC^\infty RS^c_{in,ex}} \subset \mathbf{LC^\infty RS^c}$, where the objects $\boldsymbol{X}$

have stalks $\mathcal{O}_{X,x}$ which are saturated, torsion-free, integral, or interior. From the definition of $C(\boldsymbol{X})$ in Definition 6.1, we see that if the stalks $\mathcal{O}_{X,x}$ are saturated, ..., interior then the stalks $\mathcal{O}_{C(\boldsymbol{X}),(x,P)} \cong \mathcal{O}_{X,x}/\sim_P$ are also saturated, ..., interior, as these properties are preserved by quotients by prime ideals, so $C(\boldsymbol{X})$ is saturated, ..., interior, respectively. Hence from Theorem 6.3 we deduce the following.

Theorem 6.4 *Restricting C in Theorem 6.3 gives corner functors*

$$C : \mathbf{LC^\infty RS^c_{sa,ex}} \longrightarrow \mathbf{LC^\infty RS^c_{sa}}, \quad C : \mathbf{LC^\infty RS^c_{tf,ex}} \longrightarrow \mathbf{LC^\infty RS^c_{tf}},$$
$$C : \mathbf{LC^\infty RS^c_{\mathbb{Z},ex}} \longrightarrow \mathbf{LC^\infty RS^c_{\mathbb{Z}}}, \quad C : \mathbf{LC^\infty RS^c_{in,ex}} \longrightarrow \mathbf{LC^\infty RS^c_{in}},$$

which are right adjoint to the corresponding inclusions.

6.2 The corner functor for C^∞-schemes with corners

Motivated by Theorem 6.3, naïvely we might expect that there exists $\tilde{C}$: $\mathbf{C^\infty Rings^c} \to \mathbf{C^\infty Rings^c_{in}}$ left adjoint to $\mathbf{C^\infty Rings^c_{in}} \hookrightarrow \mathbf{C^\infty Rings^c}$, fitting into a diagram of adjoint functors similar to (5.11):

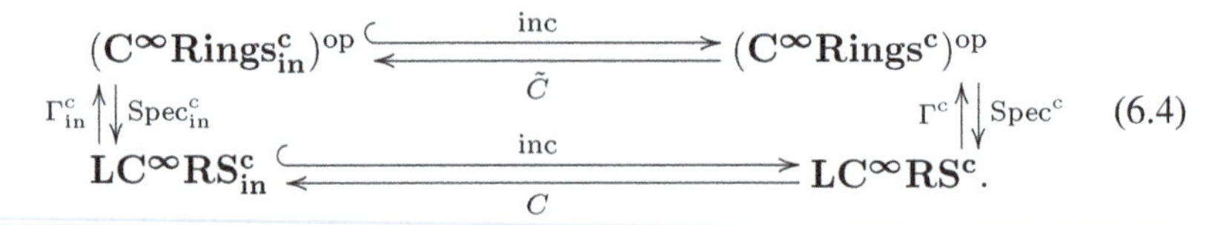

As in the proof of Theorem 5.31(a) this would imply $\mathrm{Spec}^{\mathrm{c}}_{\mathrm{in}} \circ \tilde{C} \cong C \circ \mathrm{Spec}^{\mathrm{c}}$, allowing us to deduce that C maps $\mathbf{C^\infty Sch^c} \to \mathbf{C^\infty Sch^c_{in}}$. However, no such left adjoint $\tilde{C}$ for inc exists by Theorem 4.20(d).

Surprisingly, there is a 'corner functor' $\tilde{C} : \mathbf{C^\infty Rings^c} \to \mathbf{C^\infty Rings^c_{sc}}$ such that $\mathrm{Spec}^{\mathrm{c}} \circ \tilde{C}$ maps to $\mathbf{LC^\infty RS^c_{in}} \subset \mathbf{LC^\infty RS^c}$, with $\mathrm{Spec}^{\mathrm{c}} \circ \tilde{C} \cong C \circ \mathrm{Spec}^{\mathrm{c}}$, as we prove in the next definition and theorem.

Definition 6.5 Let $\mathfrak{C} = (\mathfrak{C}, \mathfrak{C}_{\mathrm{ex}})$ be a C^∞-ring with corners. Using the notation of Definition 4.24, define a C^∞-ring with corners $\tilde{\mathfrak{C}}$ by

$$\tilde{\mathfrak{C}} = \mathfrak{C}\big[\alpha_{c'}, \beta_{c'} : c' \in \mathfrak{C}_{\mathrm{ex}}\big] \big/ \big(\Phi_i(\alpha_{c'}) + \Phi_i(\beta_{c'}) = 1 \;\; \forall c' \in \mathfrak{C}_{\mathrm{ex}}\big) \qquad (6.5)$$
$$\big[\alpha_{1_{\mathfrak{C}_{\mathrm{ex}}}} = 1_{\mathfrak{C}_{\mathrm{ex}}},\; \alpha_{c'}\alpha_{c''} = \alpha_{c'c''} \;\forall c', c'' \in \mathfrak{C}_{\mathrm{ex}},\; c'\beta_{c'} = \alpha_{c'}\beta_{c'} = 0 \;\forall c' \in \mathfrak{C}_{\mathrm{ex}}\big].$$

That is, we add two $[0,\infty)$-type generators $\alpha_{c'}, \beta_{c'}$ to $\mathfrak{C}$ for each $c' \in \mathfrak{C}_{\mathrm{ex}}$, and impose $\mathbb{R}$-type relations $(\cdots)$ and $[0,\infty)$-type relations $[\cdots]$ as shown.

Let $\phi : \mathfrak{C} \to \mathfrak{D}$ be a morphism in $\mathbf{C^\infty Rings^c}$, and define $\tilde{\mathfrak{C}}, \tilde{\mathfrak{D}}$ from

$\mathfrak{C}, \mathfrak{D}$ as above, writing the additional generators in $\mathfrak{D}_{\rm ex}$ as $\bar\alpha_{d'}, \bar\beta_{d'}$. Define a morphism $\tilde\phi : \tilde{\mathfrak{C}} \to \tilde{\mathfrak{D}}$ by the commutative diagram

$$\begin{array}{ccc} \mathfrak{C}\bigl[\alpha_{c'}, \beta_{c'} : c' \in \mathfrak{C}_{\rm ex}\bigr] & \xrightarrow{\text{project}} & \tilde{\mathfrak{C}} \\ \big\downarrow \phi[\alpha_{c'} \mapsto \bar\alpha_{\phi_{\rm ex}(c')},\, \beta_{c'} \mapsto \bar\beta_{\phi_{\rm ex}(c')},\, c' \in \mathfrak{C}_{\rm ex}] & & \big\downarrow \tilde\phi \\ \mathfrak{D}\bigl[\bar\alpha_{d'}, \bar\beta_{d'} : d' \in \mathfrak{D}_{\rm ex}\bigr] & \xrightarrow{\text{project}} & \tilde{\mathfrak{D}}. \end{array}$$

That is, we begin with $\phi : \mathfrak{C} \to \mathfrak{D}$, and act on the extra generators by mapping $\alpha_{c'} \mapsto \bar\alpha_{\phi_{\rm ex}(c')}$, $\beta_{c'} \mapsto \bar\beta_{\phi_{\rm ex}(c')}$ for all $c' \in \mathfrak{C}_{\rm ex}$. These map the relations in $\tilde{\mathfrak{C}}$ to the relations in $\tilde{\mathfrak{D}}$, and so descend to a unique morphism $\tilde\phi$ as shown.

Clearly mapping $\phi \mapsto \tilde\phi$ takes compositions and identities to compositions and identities, so mapping $\mathfrak{C} \mapsto \tilde{\mathfrak{C}}$ on objects and $\phi \mapsto \tilde\phi$ on morphisms defines a functor $\mathbf{C^\infty Rings^c} \to \mathbf{C^\infty Rings^c}$. For reasons explained in §6.3, we define $\tilde C : \mathbf{C^\infty Rings^c} \to \mathbf{C^\infty Rings^c_{sc}}$ to be the composition of this functor with $\Pi_{\rm all}^{\rm sc} : \mathbf{C^\infty Rings^c} \to \mathbf{C^\infty Rings^c_{sc}}$ in Definition 5.26. That is, we define the *corner functor* $\tilde C : \mathbf{C^\infty Rings^c} \to \mathbf{C^\infty Rings^c_{sc}}$ to map $\mathfrak{C} \mapsto \Pi_{\rm all}^{\rm sc}(\tilde{\mathfrak{C}})$ on objects, and $\phi \mapsto \Pi_{\rm all}^{\rm sc}(\tilde\phi)$ on morphisms, for $\mathfrak{C}, \mathfrak{D}, \phi, \tilde{\mathfrak{C}}, \tilde{\mathfrak{D}}, \tilde\phi$ as above.

For each $\mathfrak{C} \in \mathbf{C^\infty Rings^c}$, define $\tilde{\boldsymbol{\Pi}}_{\mathfrak{C}} : \mathfrak{C} \to \tilde C(\mathfrak{C})$ to be the composition

$$\mathfrak{C} \xhookrightarrow{\text{inc}} \mathfrak{C}\bigl[\alpha_{c'}, \beta_{c'} : c' \in \mathfrak{C}_{\rm ex}\bigr] \xrightarrow{\text{project}} \tilde{\mathfrak{C}} \xrightarrow{\psi_{\tilde{\mathfrak{C}}}} \Pi_{\rm all}^{\rm sc}(\tilde{\mathfrak{C}}) = \tilde C(\mathfrak{C}), \qquad (6.6)$$

for $\psi_{\tilde{\mathfrak{C}}}$ as in Definition 5.26. This is functorial in $\mathfrak{C}$, and so defines a natural transformation $\tilde{\boldsymbol{\Pi}} : \mathrm{Id} \Rightarrow \tilde C$ of functors $\mathbf{C^\infty Rings^c} \to \mathbf{C^\infty Rings^c}$.

We will also write $\alpha_{c'}, \beta_{c'}$ for the images of $\alpha_{c'}, \beta_{c'}$ in $\tilde C(\mathfrak{C})_{\rm ex}$, and write

$$\mathrm{Pr}_{\mathfrak{C}} = \{P \subset \mathfrak{C}_{\rm ex} : P \text{ is a prime ideal}\}. \qquad (6.7)$$

Theorem 6.6 *In Definition* 6.5, *with* $C : \mathbf{LC^\infty RS^c} \to \mathbf{LC^\infty RS^c_{in}}$ *as in Definition* 6.1, *for each* $\mathfrak{C} \in \mathbf{C^\infty Rings^c}$ *there is a natural isomorphism*

$$\eta_{\mathfrak{C}} : \mathrm{Spec^c}\, \tilde C(\mathfrak{C}) \longrightarrow C(\mathrm{Spec^c}\, \mathfrak{C}) \qquad (6.8)$$

in $\mathbf{LC^\infty RS^c}$. *So* $\mathrm{Spec^c}\, \tilde C(\mathfrak{C})$ *lies in* $\mathbf{LC^\infty RS^c_{in}} \subset \mathbf{LC^\infty RS^c}$. *The following diagram commutes in* $\mathbf{LC^\infty RS^c}$*:*

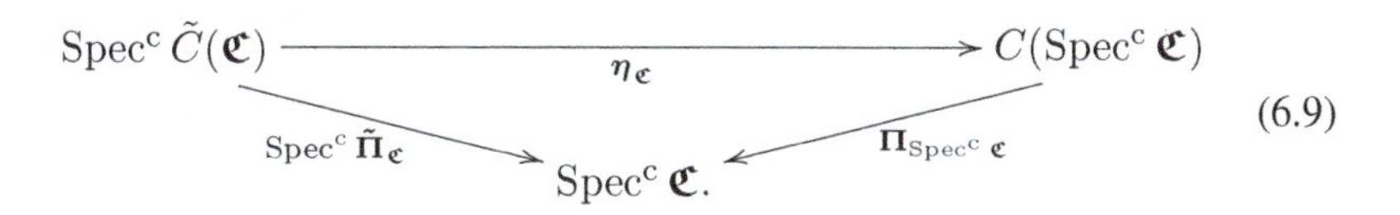

(6.9)

For each morphism $\phi : \mathfrak{C} \to \mathfrak{D}$ *in* $\mathbf{C^\infty Rings^c}$ *the following commutes:*

$$\begin{array}{ccc} \mathrm{Spec^c}\,\tilde{C}(\mathfrak{D}) & \xrightarrow[\cong]{\eta_{\mathfrak{D}}} & C(\mathrm{Spec^c}\,\mathfrak{D}) \\ \downarrow{\scriptstyle \mathrm{Spec^c}\,\tilde{C}(\phi)} & & \downarrow{\scriptstyle C(\mathrm{Spec^c}\,\phi)} \\ \mathrm{Spec^c}\,\tilde{C}(\mathfrak{C}) & \xrightarrow[\cong]{\eta_{\mathfrak{C}}} & C(\mathrm{Spec^c}\,\mathfrak{C}), \end{array} \tag{6.10}$$

so $\mathrm{Spec^c}\,\tilde{C}(\phi)$ *is a morphism in* $\mathbf{LC^\infty RS^c_{in}}$. *Thus* $\boldsymbol{\eta} : \mathrm{Spec^c} \circ \tilde{C} \Rightarrow C \circ \mathrm{Spec^c}$ *is a natural isomorphism of functors* $(\mathbf{C^\infty Rings^c})^{\mathrm{op}} \to \mathbf{LC^\infty RS^c_{in}}$.

Proof Let $\mathfrak{C} \in \mathbf{C^\infty Rings^c}$, and define $\tilde{\mathfrak{C}}$ as in (6.4), so that $\tilde{C}(\mathfrak{C}) = \Pi^{\mathrm{sc}}_{\mathrm{all}}(\tilde{\mathfrak{C}})$. Write $\boldsymbol{X} = \mathrm{Spec^c}\,\mathfrak{C}$ and $\tilde{\boldsymbol{X}} = \mathrm{Spec^c}\,\tilde{\mathfrak{C}}$. As in (6.6), there is a functorial morphism $\psi_{\tilde{\mathfrak{C}}} : \tilde{\mathfrak{C}} \to \tilde{C}(\mathfrak{C})$ such that $\mathrm{Spec^c}\,\psi_{\tilde{\mathfrak{C}}} : \mathrm{Spec^c}\,\tilde{C}(\mathfrak{C}) \to \mathrm{Spec^c}\,\tilde{\mathfrak{C}}$ is an isomorphism. Write $\check{\boldsymbol{\Pi}}_{\mathfrak{C}} : \mathfrak{C} \to \tilde{\mathfrak{C}}$ for the composition of the first two morphisms in (6.6), so that $\tilde{\boldsymbol{\Pi}}_{\mathfrak{C}} = \psi_{\tilde{\mathfrak{C}}} \circ \check{\boldsymbol{\Pi}}_{\mathfrak{C}}$, and set $\check{\boldsymbol{\Pi}}_{\boldsymbol{X}} = \mathrm{Spec^c}\,\check{\boldsymbol{\Pi}}_{\mathfrak{C}} : \tilde{\boldsymbol{X}} \to \boldsymbol{X}$. We will construct an isomorphism

$$\boldsymbol{\zeta}_{\mathfrak{C}} : \tilde{\boldsymbol{X}} = \mathrm{Spec^c}\,\tilde{\mathfrak{C}} \longrightarrow C(\mathrm{Spec^c}\,\mathfrak{C}) = C(\boldsymbol{X}) \tag{6.11}$$

in $\mathbf{LC^\infty RS^c}$, and define $\boldsymbol{\eta}_{\mathfrak{C}}$ in (6.8) by $\boldsymbol{\eta}_{\mathfrak{C}} = \boldsymbol{\zeta}_{\mathfrak{C}} \circ \mathrm{Spec^c}\,\psi_{\tilde{\mathfrak{C}}}$.

First we define the isomorphism (6.11) at the level of points. A point of $\tilde{\boldsymbol{X}}$ is an $\mathbb{R}$-algebra morphism $\tilde{x} : \tilde{\mathfrak{C}} \to \mathbb{R}$. Composing with the projection $\mathfrak{C}[\alpha_{c'}, \beta_{c'} : c' \in \mathfrak{C}_{\mathrm{ex}}] \to \tilde{\mathfrak{C}}$ gives a morphism $\hat{x} : \mathfrak{C}[\alpha_{c'}, \beta_{c'} : c' \in \mathfrak{C}_{\mathrm{ex}}] \to \mathbb{R}$. Such morphisms are in 1-1 correspondence with data $(x, a_{c'}, b_{c'} : c' \in \mathfrak{C}_{\mathrm{ex}})$, where $x : \mathfrak{C} \to \mathbb{R}$ is an $\mathbb{R}$-algebra morphism (i.e. $x \in X$) and $a_{c'} = \hat{x}(\alpha_{c'})$, $b_{c'} = \hat{x}(\beta_{c'})$ lie in $[0, \infty) \subseteq \mathbb{R}$ for all $c' \in \mathfrak{C}_{\mathrm{ex}}$. A morphism $\hat{x} : \mathfrak{C}[\alpha_{c'}, \beta_{c'} : c' \in \mathfrak{C}_{\mathrm{ex}}] \to \mathbb{R}$ descends to a (unique) morphism $\tilde{x} : \tilde{\mathfrak{C}} \to \mathbb{R}$ if and only if $\hat{x}$ maps the relations in (6.5) to relations in $\mathbb{R}$. Hence we may identify

$$\begin{aligned} \tilde{X} \cong \bigl\{ &(x, a_{c'}, b_{c'} : c' \in \mathfrak{C}_{\mathrm{ex}}) : x \in X, \;\; a_{c'}, b_{c'} \in [0, \infty) \;\; \forall c' \in \mathfrak{C}_{\mathrm{ex}}, \\ &a_{c'} + b_{c'} = 1 \;\; \forall c' \in \mathfrak{C}_{\mathrm{ex}}, \;\; a_{1_{\mathfrak{C}_{\mathrm{ex}}}} = 1, \;\; a_{c'} a_{c''} = a_{c'c''} \;\; \forall c', c'' \in \mathfrak{C}_{\mathrm{ex}}, \\ &x \circ \Phi_i(c') b_{c'} = a_{c'} b_{c'} = 0 \;\; \forall c' \in \mathfrak{C}_{\mathrm{ex}} \bigr\}. \end{aligned} \tag{6.12}$$

We can simplify (6.12). Let $(x, a_{c'}, b_{c'} : c' \in \mathfrak{C}_{\mathrm{ex}})$ be a point in the right-hand side. The equations $a_{c'} + b_{c'} = 1$ and $a_{c'} b_{c'} = 0$ imply that $(a_{c'}, b_{c'})$ is $(1, 0)$ or $(0, 1)$ for each c'. Define $P = \{c' \in \mathfrak{C}_{\mathrm{ex}} : a_{c'} = 0\}$. Then $a_{1_{\mathfrak{C}_{\mathrm{ex}}}} = 1$ implies that $1_{\mathfrak{C}_{\mathrm{ex}}} \notin P$, and $a_{c'} a_{c''} = a_{c'c''}$ implies that if $c' \in \mathfrak{C}_{\mathrm{ex}}$ and $c'' \in P$ then $c'c'' \in P$, so P is an ideal. Also $a_{c'} a_{c''} = a_{c'c''}$ implies that if $c'c'' \in P$ then $c' \in P$ or $c'' \in P$, so P is a prime ideal. The condition $x \circ \Phi_i(c') b_{c'} = 0$ then becomes $x \circ \Phi_i(c') = 0$ if $c' \in P$.

Conversely, if P is a prime ideal and $x \circ \Phi_i(c') = 0$ for $c' \in P$, then setting $a_{c'} = 0$, $b_{c'} = 1$ if $c' \in P$ and $a_{c'} = 1$, $b_{c'} = 0$ if $c' \in \mathfrak{C}_{\mathrm{ex}} \setminus P$ then all the equations of (6.12) hold. Thus, mapping $\bigl(x, a_{c'}, b_{c'} : c' \in \mathfrak{C}_{\rm ex}\bigr) \mapsto \bigl(x, P = \{c' \in \mathfrak{C}_{\rm ex} : a_{c'} = 0\}\bigr)$ turns (6.12) into a bijection

$$\tilde X \cong \bigl\{(x,P) : x \in X,\, P \in \mathrm{Pr}_{\mathfrak{C}},\, x \circ \Phi_i(c') = 0 \text{ if } c' \in P\bigr\}. \tag{6.13}$$

Lemma 6.7 *Let $x \in \boldsymbol{X}$, so that $x : \mathfrak{C} \to \mathbb{R}$ is an $\mathbb{R}$-point, with localization $\pi_x : \boldsymbol{\mathfrak{C}} \to \boldsymbol{\mathfrak{C}}_x \cong \boldsymbol{\mathcal{O}}_{X,x}$ giving a morphism $\pi_{x,\rm ex} : \mathfrak{C}_{\rm ex} \to \mathfrak{C}_{x,\rm ex} \cong \mathcal{O}^{\rm ex}_{X,x}$. Then we have inverse bijective maps*

$$\bigl\{P \in \mathrm{Pr}_{\mathfrak{C}} : x\circ\Phi_i(c') = 0\ \forall c' \in P\bigr\} \underset{P_x \longmapsto P = \pi_{x,\rm ex}^{-1}(P_x)}{\overset{P \longmapsto P_x = \pi_{x,\rm ex}(P)}{\rightleftarrows}} \mathrm{Pr}_{\mathfrak{C}_x} \cong \mathrm{Pr}_{\boldsymbol{\mathcal{O}}_{X,x}}. \tag{6.14}$$

Proof Let P lie in the left-hand side of (6.14), and set $P_x = \pi_{x,\rm ex}(P)$. Then $1 \notin P_x$ as $x \circ \Phi_i(c') = 0$ for all $c' \in P$. If $c'_x \in \mathfrak{C}_{x,\rm ex}$ and $c''_x \in P_x$ then $c'_x c''_x \in P_x$ as P is an ideal and $\pi_{x,\rm ex}$ is surjective. Thus P_x is an ideal. To show P_x is prime, suppose $c'_x, d'_x \in \mathfrak{C}_{{\rm ex},x}$ with $c'_x d'_x \in P_x$. As $\pi_{x,\rm ex}$ is surjective we can choose $c', d' \in \mathfrak{C}_{\rm ex}$ and $p' \in P$ with $\pi_{x,\rm ex}(c') = c'_x$, $\pi_{x,\rm ex}(d') = d'_x$ and $\pi_{x,\rm ex}(c'd') = \pi_{x,\rm ex}(p)$. Hence Proposition 4.33 gives $a', b' \in \mathfrak{C}_{\rm ex}$ with $a'c'd' = b'p'$ and $x \circ \Phi_i(a') \neq 0$. But $b'p' \in P$ as P is an ideal, so $a'c'd' \in P$. Since P is prime, one of a', c', d' must lie in P, and $a' \notin P$ as $x \circ \Phi_i(a') \neq 0$, so one of c', d' must lie in P. Thus one of c'_x, d'_x lie in P_x, and P_x is prime. Hence the top morphism in (6.14) is well defined.

Next let $P_x \in \mathrm{Pr}_{\mathfrak{C}_x}$, and set $P = \pi_{x,\rm ex}^{-1}(P_x)$. It is easy to check P lies in the left-hand side of (6.14). So the bottom morphism in (6.14) is well defined.

Suppose P lies in the left-hand side of (6.14). Clearly $P \subseteq \pi_{x,\rm ex}^{-1}(\pi_{x,\rm ex}(P))$. If $c' \in \pi_{x,\rm ex}^{-1}(\pi_{x,\rm ex}(P))$ then there exists $p' \in P$ with $\pi_{x,\rm ex}(c') = \pi_{x,\rm ex}(p')$. Proposition 4.33 gives $a', b' \in \mathfrak{C}_{\rm ex}$ with $a'c' = b'p'$ and $x \circ \Phi_i(a') \neq 0$, and $p' \in P$ implies $a'c' = b'p' \in P$, which implies $c' \in P$ as $\pi_{x,\rm ex}(a') \neq 0$ so $a' \notin P$. Hence $P = \pi_{x,\rm ex}^{-1}(\pi_{x,\rm ex}(P))$. If $P_x \in \mathrm{Pr}_{\mathfrak{C}_x}$ then $P_x = \pi_{x,\rm ex}(\pi_{x,\rm ex}^{-1}(P_x))$ as $\pi_{x,\rm ex}$ is surjective. Thus the maps in (6.14) are inverse. □

Combining equations (6.1) and (6.13) and Lemma 6.7 gives a bijection of sets $\zeta_{\mathfrak{C}} : \tilde X \to C(X)$, defined explicitly by, if $\tilde x : \tilde{\mathfrak{C}} \to \mathbb{R}$ corresponds to $(x, a_{c'}, b_{c'} : c' \in \mathfrak{C}_{\rm ex})$ in (6.25), and $P = \{c' \in \mathfrak{C}_{\rm ex} : a_{c'} = 0\}$ is the associated prime ideal, and $P_x = \pi_{x,\rm ex}(P)$ in $\mathfrak{C}_{x,\rm ex} \cong \mathcal{O}^{\rm ex}_{X,x}$, then $\zeta_{\mathfrak{C}}(\tilde x) = (x, P_x)$ in $C(X)$ in (6.1). Then in an analogue of (6.9), on the level of sets we have

$$\check\Pi_X = \Pi_X \circ \zeta_{\mathfrak{C}} : \tilde X \longrightarrow X, \tag{6.15}$$

as both sides map $\tilde x \mapsto x$.

Lemma 6.8 $\zeta_{\mathfrak{C}} : \tilde X \to C(X)$ *is a homeomorphism of topological spaces.*

Proof By definition of the topology on Spec $\tilde{\mathfrak{C}}$ in Definition 2.42, the topology on $\tilde{X}$ is generated by open sets $\tilde{U}_{\tilde{c}} = \{\tilde{x} \in \tilde{X} : \tilde{x}(\tilde{c}) \neq 0\}$ for $\tilde{c} \in \tilde{\mathfrak{C}}$. As $\tilde{\mathfrak{C}}$ is a quotient of $\mathfrak{C}[\alpha_{c'}, \beta_{c'} : c' \in \mathfrak{C}_{\text{ex}}]$, the topology is generated by $\tilde{U}_c$ for $c \in \mathfrak{C}$, and $\tilde{U}_{\Phi_f(\alpha_{c'})}, \tilde{U}_{\Phi_f(\beta_{c'})}$ for all $c' \in \mathfrak{C}_{\text{ex}}$ and smooth $f : [0, \infty) \to \mathbb{R}$. But as $\alpha_{c'}, \beta_{c'}$ take only values 0,1 on $\tilde{X}$ with $\alpha_{c'} + \beta_{c'} = 1$, the only possibilities for $\tilde{U}_{\Phi_f(\alpha_{c'})}, \tilde{U}_{\Phi_f(\beta_{c'})}$ are $\dot{V}_{c'} = \{\tilde{x} \in \tilde{X} : \alpha_{c'}|_{\tilde{x}} \neq 0\}$ and $\ddot{V}_{c'} = \{\tilde{x} \in \tilde{X} : \alpha_{c'}|_{\tilde{x}} \neq 1\}$. So the topology on $\tilde{X}$ is generated by $\tilde{U}_c$ for all $c \in \mathfrak{C}$ and $\dot{V}_{c'}, \ddot{V}_{c'}$ for all $c' \in \mathfrak{C}_{\text{ex}}$.

For $C(X)$ in Definition 6.1, the topology is generated by open sets $\Pi_X^{-1}(U_c)$ for $c \in \mathfrak{C}$ and subsets $\dot{U}_{s'}, \ddot{U}_{s'}$ for open $U \subset X$ and $s' \in \mathcal{O}_X^{\text{ex}}(U)$. As $\boldsymbol{X} = \text{Spec}^{\text{c}}\, \boldsymbol{\mathfrak{C}}$, rather than taking all such $\dot{U}_{s'}, \ddot{U}_{s'}$, it is enough to take $U = X$ and $s' = \Xi_{\mathfrak{C},\text{ex}}(c')$ for all $c' \in \mathfrak{C}_{\text{ex}}$, as all other $\dot{U}_{s'}, \ddot{U}_{s'}$ can be generated from $\Pi_X^{-1}(U_c)$ and $\dot{X}_{\Xi_{\mathfrak{C},\text{ex}}(c')}, \ddot{X}_{\Xi_{\mathfrak{C},\text{ex}}(c')}$. Thus, the topology on $C(X)$ is generated by $\Pi_X^{-1}(U_c)$ for all $c \in \mathfrak{C}$ and $\dot{X}_{\Xi_{\mathfrak{C},\text{ex}}(c')}, \ddot{X}_{\Xi_{\mathfrak{C},\text{ex}}(c')}$ for all $c' \in \mathfrak{C}_{\text{ex}}$.

It is easy to check that $\zeta_{\mathfrak{C}}$ maps $\tilde{U}_c \mapsto \Pi_X^{-1}(U_c)$, $\dot{V}_{c'} \mapsto \dot{X}_{\Xi_{\mathfrak{C},\text{ex}}(c')}$ and $\ddot{V}_{c'} \mapsto \ddot{X}_{\Xi_{\mathfrak{C},\text{ex}}(c')}$. Therefore $\zeta_{\mathfrak{C}}$ is a bijection and maps a basis for the topology of $\tilde{X}$ to a basis for the topology of $C(X)$, so it is a homeomorphism. □

Lemma 6.9 *In sheaves of C^∞-rings with corners on $\tilde{X}$, there is a unique morphism $\boldsymbol{\zeta}_{\mathfrak{C}}^\sharp$ making the following diagram commute:*

$$\begin{array}{ccc} \zeta_{\mathfrak{C}}^{-1} \circ \Pi_X^{-1}(\mathcal{O}_X) & \xrightarrow[\text{by (6.15)}]{=\!=\!=} & \check{\Pi}_X^{-1}(\mathcal{O}_X) \\ \downarrow \zeta_{\mathfrak{C}}^{-1}(\boldsymbol{\Pi}_{\boldsymbol{X}}^\sharp) & & \check{\boldsymbol{\Pi}}_{\boldsymbol{X}}^\sharp \downarrow \\ \zeta_{\mathfrak{C}}^{-1}(\mathcal{O}_{C(X)}) & \xrightarrow{\boldsymbol{\zeta}_{\mathfrak{C}}^\sharp} & \mathcal{O}_{\tilde{X}}. \end{array} \tag{6.16}$$

Furthermore, $\boldsymbol{\zeta}_{\mathfrak{C}}^\sharp$ is an isomorphism.

Proof By Definition 6.1, $\boldsymbol{\Pi}_{\boldsymbol{X}}^\sharp : \Pi_X^{-1}(\mathcal{O}_X) \to \mathcal{O}_{C(X)}$ is the universal surjective quotient of $\Pi_X^{-1}(\mathcal{O}_X)$ in sheaves of C^∞-rings with corners, such that if $U \subseteq C(X)$ is open and $s' \in \Pi_X^{-1}(\mathcal{O}_X^{\text{ex}})(U)$ with $s'|_{(x,P_x)} \in P_x$ for each $(x, P_x) \in U$ then $\Pi_{X,\text{ex}}^\sharp(U)(s') = 0$ in $\mathcal{O}_{C(X)}^{\text{ex}}(U)$. As $\boldsymbol{X} = \text{Spec}^{\text{c}}\, \boldsymbol{\mathfrak{C}}$, locally on X any section of $\mathcal{O}_{\boldsymbol{X}}^{\text{ex}}$ is of the form $\Xi_{\mathfrak{C},\text{ex}}(c')$ for $c' \in \mathfrak{C}_{\text{ex}}$, so locally on $C(X)$ any section s' of $\Pi_X^{-1}(\mathcal{O}_X)$ is of the form $\Pi_{X,\text{ex}}^{-1} \circ \Xi_{\mathfrak{C},\text{ex}}(c')$. Thus it is enough to verify the universal property above when $s' = \Pi_{X,\text{ex}}^{-1} \circ \Xi_{\mathfrak{C},\text{ex}}(c')|_U$ for $c' \in \mathfrak{C}_{\text{ex}}$, as it is local in s' anyway. That is, $\boldsymbol{\Pi}_{\boldsymbol{X}}^\sharp : \Pi_X^{-1}(\mathcal{O}_X) \to \mathcal{O}_{C(X)}$ is characterized by the property that if $c' \in \mathfrak{C}_{\text{ex}}$ and $U \subset C(X)$ is open with $\pi_{x,\text{ex}}(c') \in P_x \subset \mathfrak{C}_{x,\text{ex}} \cong \mathcal{O}_{X,x}^{\text{ex}}$ for all $(x, P_x) \in U$ then $\Pi_{X,\text{ex}}^\sharp(C(X)) \circ \Xi_{\mathfrak{C},\text{ex}}(c')|_U = 0$.

Since $\zeta_{\mathfrak{C}}$ is a homeomorphism by Lemma 6.8, the pullback $\zeta_{\mathfrak{C}}^{-1}(\boldsymbol{\Pi}_{\boldsymbol{X}}^\sharp)$ in (6.16) has the analogous universal property. Let $U \subseteq C(X)$ and $c' \in \mathfrak{C}_{\text{ex}}$ have the properties above, and set $\tilde{U} = \zeta_{\mathfrak{C}}^{-1}(U)$. Suppose $\tilde{x} \in \tilde{U}$ with $\zeta_{\mathfrak{C}}(\tilde{x}) = (x, P_x)$ in $U \subseteq C(X)$, and let $\tilde{x}$ correspond to $(x, a_{c''}, b_{c''} : c'' \in \mathfrak{C}_{\rm ex})$ in (6.12). By construction of $\zeta_{\mathfrak{C}}$ we have $a_{c'} = 0, b_{c'} = 1$ as $\pi_{x,{\rm ex}}(c') \in P_x$. Thus the relation $c'\beta_{c'} = 0$ in $\tilde{\mathfrak{C}}_{\rm ex}$ in (6.5) implies that $\tilde{\pi}_{\tilde{x},{\rm ex}}(c') = 0$ in $\tilde{\mathfrak{C}}_{\tilde{x},{\rm ex}} \cong \mathcal{O}^{\rm ex}_{\tilde{X},\tilde{x}}$. As this holds for all $\tilde{x} \in \tilde{U}$ we have $\tilde{\Pi}^\sharp_{X,{\rm ex}}(\tilde{X}) \circ \Xi_{\mathfrak{C},{\rm ex}}(c')|_{\tilde{U}} = 0$. Hence the universal property of $\zeta_{\mathfrak{C}}^{-1}(\boldsymbol{\Pi}^\sharp_{\boldsymbol{X}})$ gives a unique morphism $\boldsymbol{\zeta}^\sharp_{\mathfrak{C}}$ making (6.16) commute.

To show $\boldsymbol{\zeta}^\sharp_{\mathfrak{C}}$ is an isomorphism, it is enough to show it is an isomorphism on stalks at each $\tilde{x} \in \tilde{X}$. Let $\tilde{x}$ correspond to $(x, a_{c'}, b_{c'} : c' \in \mathfrak{C}_{\rm ex})$ in (6.12), with $\zeta_{\mathfrak{C}}(\tilde{x}) = (x, P_x)$, so that $a_{c'} = 0$, $b_{c'} = 1$ if $\pi_{x,{\rm ex}}(c') \in P_x$ and $a_{c'} = 1$, $b_{c'} = 0$ otherwise. The stalk $\zeta_{\mathfrak{C}}^{-1}(\boldsymbol{\mathcal{O}}_{C(X)})_{\tilde{x}}$ is $\boldsymbol{\mathcal{O}}_{C(X),(x,P_x)}$, which is $\mathfrak{C}_x/\sim_{P_x}$. The stalk $\boldsymbol{\mathcal{O}}_{\tilde{X},\tilde{x}}$ is $\tilde{\mathfrak{C}}_{\tilde{x}}$. By the values of $a_{c'}, b_{c'}$, in (6.5) we know that $\alpha_{c'} = 0$, $\beta_{c'} = 1$ near $\tilde{x}$ if $\pi_{x,{\rm ex}}(c') \in P_x$, and $\alpha_{c'} = 1$, $\beta_{c'} = 0$ near $\tilde{x}$ otherwise. Hence the generators $\alpha_{c'}, \beta_{c'}$ add no new generators to $\tilde{\mathfrak{C}}_{\tilde{x}}$, and $\tilde{\mathfrak{C}}_{\tilde{x}}$ is a quotient of $\mathfrak{C}_x$. The relations in (6.5) turn into relations in $\tilde{\mathfrak{C}}_{\tilde{x}}$ according to the local values 0 or 1 of $\alpha_{c'}, \beta_{c'}$. Therefore we see that

$$\tilde{\mathfrak{C}}_{\tilde{x}} \cong \mathfrak{C}_x/\big[\pi_{x,{\rm ex}}(c') = 0 \text{ if } c' \in \mathfrak{C}_{\rm ex} \text{ with } \pi_{x,{\rm ex}}(c') \in P_x\big] = \mathfrak{C}_x/\sim_{P_x},$$

where the second step holds as $\pi_{x,{\rm ex}}$ is surjective. So we have an isomorphism of stalks $\zeta_{\mathfrak{C}}^{-1}(\boldsymbol{\mathcal{O}}_{C(X)})_{\tilde{x}} \cong \boldsymbol{\mathcal{O}}_{\tilde{X},\tilde{x}}$, which is the natural map $\boldsymbol{\zeta}^\sharp_{\mathfrak{C},\tilde{x}}$. □

Lemmas 6.8 and 6.9 show $\boldsymbol{\zeta}_{\mathfrak{C}} = (\zeta_{\mathfrak{C}}, \zeta^\sharp_{\mathfrak{C}})$ is an isomorphism $\tilde{\boldsymbol{X}} \to C(\boldsymbol{X})$, as in (6.11). So $\boldsymbol{\eta}_{\mathfrak{C}} = \boldsymbol{\zeta}_{\mathfrak{C}} \circ \operatorname{Spec^c} \boldsymbol{\psi}_{\tilde{\mathfrak{C}}}$ is an isomorphism in (6.8). Equations (6.15)–(6.16) imply that (6.9) commutes. This proves the first part of Theorem 6.6.

For the second part, consider the diagram

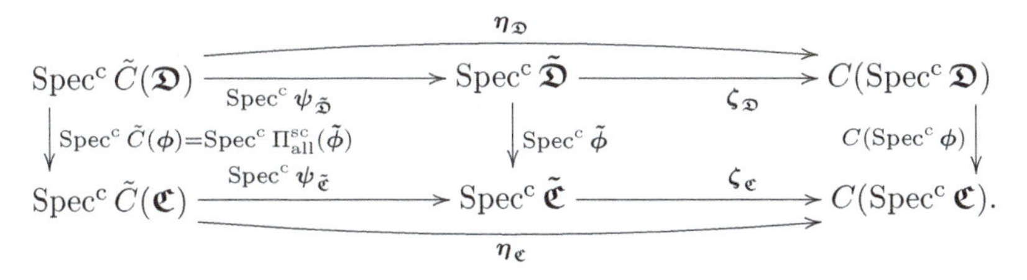

The left-hand square commutes as $\boldsymbol{\psi} : \mathrm{Id} \Rightarrow \Pi^{\rm sc}_{\rm all}$ is a natural transformation in Definition 5.26. The right-hand square commutes by functoriality of $\boldsymbol{\zeta}_{\mathfrak{C}}$ in (6.11). The semi-circles commute by definition of $\boldsymbol{\eta}_{\mathfrak{C}}, \boldsymbol{\eta}_{\mathfrak{D}}$. Hence (6.10) commutes, and Theorem 6.6 follows. □

The next proposition follows from the proof of Theorem 6.6. Here $C(X)^P$ is closed as it may be defined by the equations $\alpha_{c'} = 0$ for $c' \in P$, $\alpha_{c'} = 1$ for $c' \in \mathfrak{C}_{\rm ex} \setminus P$. Example 6.18 below shows the $C(X)^P$ need not be open.

Proposition 6.10 *Let* $\mathfrak{C}=(\mathfrak{C},\mathfrak{C}_{\rm ex})\in\mathbf{C^\infty Rings^c}$, *and write* $\boldsymbol{X}=\operatorname{Spec}^{\rm c}\mathfrak{C}$. *With* $C(X)$ *as in* (6.1), *for each* $P\in\Pr_{\mathfrak{C}}$, *define*

$$X^P=\bigl\{x\in X: x\circ\Phi_i(c')=0\textit{ for all } c'\in P\bigr\}, \tag{6.17}$$

$$\begin{aligned} C(X)^P&=\bigl\{(x,P_x)\in C(X):\pi_{x,\rm ex}(P)=P_x \textit{ in } \mathfrak{C}_{x,\rm ex}\cong \mathcal{O}^{\rm ex}_{X,x}\bigr\}\\ &=\bigl\{(x,P_x)\in C(X): P=\pi^{-1}_{x,\rm ex}(P_x)\bigr\}\\ &=\bigl\{(x,P_x)\in C(X): \textit{if } c'\in\mathfrak{C}_{\rm ex} \textit{ then } \pi_{x,\rm ex}(c')\in P_x\\ &\qquad\qquad \textit{in } \mathfrak{C}_{x,\rm ex}\cong\mathcal{O}^{\rm ex}_{X,x} \textit{ if and only if } c'\in P\bigr\}\subseteq C(X), \end{aligned} \tag{6.18}$$

where the three expressions in (6.18) *are equivalent. Then* $\Pi_X: C(X)\to X$ *restricts to a bijection* $C(X)^P\to X^P$. *Also* $C(X)^P$ *is closed in* $C(X)$, *but not necessarily open, and just as sets we have*

$$C(X)=\coprod\nolimits_{P\in\Pr_{\mathfrak{C}}} C(X)^P. \tag{6.19}$$

As an easy consequence of Theorem 6.6, we deduce the following.

Theorem 6.11 *The corner functor* $C:\mathbf{LC^\infty RS^c}\to\mathbf{LC^\infty RS^c_{in}}$ *of* §6.1 *maps* $\mathbf{C^\infty Sch^c}\to\mathbf{C^\infty Sch^c_{in}}$. *We write* $C:\mathbf{C^\infty Sch^c}\to\mathbf{C^\infty Sch^c_{in}}$ *for the restriction, and call it a* ***corner functor***. *It is right adjoint to* inc : $\mathbf{C^\infty Sch^c_{in}}\hookrightarrow\mathbf{C^\infty Sch^c}$. *Thus it preserves limits in* $\mathbf{C^\infty Sch^c},\mathbf{C^\infty Sch^c_{in}}$.

Proof Let $\boldsymbol{X}$ lie in $\mathbf{C^\infty Sch^c}$. Then we can cover X by open $\boldsymbol{U}\subseteq\boldsymbol{X}$ with $\boldsymbol{U}\cong\operatorname{Spec}^{\rm c}\mathfrak{C}$ for $\mathfrak{C}\in\mathbf{C^\infty Rings^c}$. Since the definition of $\boldsymbol{\Pi_X}: C(\boldsymbol{X})\to\boldsymbol{X}$ is local over $\boldsymbol{X}$, we have $\boldsymbol{\Pi}_{\boldsymbol{X}}^{-1}(\boldsymbol{U})\cong C(\boldsymbol{U})\cong C(\operatorname{Spec}^{\rm c}\mathfrak{C})$. Theorem 6.6 gives $C(\operatorname{Spec}^{\rm c}\mathfrak{C})\cong\operatorname{Spec}^{\rm c}\tilde{C}(\mathfrak{C})$, with $\tilde{C}(\mathfrak{C})\in\mathbf{C^\infty Rings^c}$.

Since $\operatorname{Spec}^{\rm c}\tilde{C}(\mathfrak{C})$ lies in $\mathbf{LC^\infty RS^c_{in}}\subset\mathbf{LC^\infty RS^c}$, Theorem 5.28 shows that $\operatorname{Spec}^{\rm c}\tilde{C}(\mathfrak{C})\cong\operatorname{Spec}^{\rm c}_{\rm in}\mathfrak{D}$ for $\mathfrak{D}\in\mathbf{C^\infty Rings^c_{in}}$. Thus we can cover $C(\boldsymbol{X})$ by open $\boldsymbol{\Pi}_{\boldsymbol{X}}^{-1}(\boldsymbol{U})\subseteq C(\boldsymbol{X})$ with $\boldsymbol{\Pi}_{\boldsymbol{X}}^{-1}(\boldsymbol{U})\cong\operatorname{Spec}^{\rm c}_{\rm in}\mathfrak{D}$ for $\mathfrak{D}\in\mathbf{C^\infty Rings^c_{in}}$, so $C(\boldsymbol{X})$ is an interior C^∞-scheme with corners. Therefore $C:\mathbf{LC^\infty RS^c}\to\mathbf{LC^\infty RS^c_{in}}$ maps $\mathbf{C^\infty Sch^c}\to\mathbf{C^\infty Sch^c_{in}}$. The rest is immediate from Theorem 6.3. □

The next proposition is easy to prove by considering local models.

Proposition 6.12 *The corner functors* C *for* $\mathbf{Man^c},\mathbf{Man^{gc}},\check{\mathbf{M}}\mathbf{an^c},\check{\mathbf{M}}\mathbf{an^{gc}}$ *in Definition* 3.16, *and for* $\mathbf{C^\infty Sch^c}$ *in Theorem* 6.11, *commute with the functors* $F_{\mathbf{Man^c}}^{\mathbf{C^\infty Sch^c}},\ldots,F_{\check{\mathbf{M}}\mathbf{an^{gc}}}^{\mathbf{C^\infty Sch^c}}$ *in Definition* 5.21 *up to natural isomorphism.*

Combining Theorems 5.31(b), 6.4, and 6.11 yields the following.

Theorem 6.13 *Restricting C in Theorem 6.11 gives corner functors*

$$C : \mathbf{C^\infty Sch^c_{sa,ex}} \longrightarrow \mathbf{C^\infty Sch^c_{sa}}, \quad C : \mathbf{C^\infty Sch^c_{tf,ex}} \longrightarrow \mathbf{C^\infty Sch^c_{tf}},$$
$$C : \mathbf{C^\infty Sch^c_{\mathbb{Z},ex}} \longrightarrow \mathbf{C^\infty Sch^c_{\mathbb{Z}}}, \quad C : \mathbf{C^\infty Sch^c_{in,ex}} \longrightarrow \mathbf{C^\infty Sch^c_{in}},$$

in the notation of Definition 5.29, which are right adjoint to the corresponding inclusions. As these corner functors are right adjoints, they preserve limits.

6.3 The corner functor for firm C^∞-schemes with corners

The following proposition is why we included $\Pi^{\rm sc}_{\rm all}$ in defining $\tilde C$ in Definition 6.5. It forces relations on $\alpha_{c'}, \beta_{c'}$ in $\tilde C(\mathfrak{C}) = \Pi^{\rm sc}_{\rm all}(\tilde{\mathfrak{C}})$, which need not hold in $\tilde{\mathfrak{C}}$.

Proposition 6.14 *Let $\mathfrak{C} \in \mathbf{C^\infty Rings^c}$ and write $\mathfrak{D} = \tilde C(\mathfrak{C})$ with morphism $\tilde{\boldsymbol{\Pi}}_{\mathfrak{C}} : \mathfrak{C} \to \mathfrak{D}$, so we have elements $\alpha_{c'}, \beta_{c'}, \tilde\Pi_{\mathfrak{C},\rm ex}(c')$ in $\mathfrak{D}_{\rm ex}$ for all $c' \in \mathfrak{C}_{\rm ex}$. Write $c' \mapsto [c']$, $d' \mapsto [d']$ for the projections $\mathfrak{C}_{\rm ex} \mapsto \mathfrak{C}^\sharp_{\rm ex}$, $\mathfrak{D}_{\rm ex} \mapsto \mathfrak{D}^\sharp_{\rm ex}$. Then we have the following.*

(a) *If $c'_1, c'_2 \in \mathfrak{C}_{\rm ex}$ with $[c'_1] = [c'_2]$ in $\mathfrak{C}^\sharp_{\rm ex}$ then $\alpha_{c'_1} = \alpha_{c'_2}$, $\beta_{c'_1} = \beta_{c'_2}$ in $\mathfrak{D}_{\rm ex}$.*
(b) *If $c'_1, \ldots, c'_n \in \mathfrak{C}_{\rm ex}$ and $i_1, \ldots, i_n, j_1, \ldots, j_n$ are positive integers then $\alpha_{c'^{i_1}_1 \cdots c'^{i_n}_n} = \alpha_{c'^{j_1}_1 \cdots c'^{j_n}_n}$ and $\beta_{c'^{i_1}_1 \cdots c'^{i_n}_n} = \beta_{c'^{j_1}_1 \cdots c'^{j_n}_n}$ in $\mathfrak{D}_{\rm ex}$.*
(c) *$\mathfrak{D}^\sharp_{\rm ex}$ is generated by $[\alpha_{c'}], [\beta_{c'}], [\tilde\Pi_{\mathfrak{C},\rm ex}(c')]$ for all $c' \in \mathfrak{C}_{\rm ex}$.*
(d) *If $\mathfrak{C}$ is firm then $\mathfrak{D}$ is firm. Thus $\tilde C$ maps $\mathbf{C^\infty Rings^c_{fi}} \to \mathbf{C^\infty Rings^c_{fi}}$.*

Proof Let $\tilde{\mathfrak{C}}$ be as in (6.5) and $\tilde{\boldsymbol{X}} = \mathrm{Spec^c}\, \tilde{\mathfrak{C}}$, so that $\mathfrak{D} \subseteq \boldsymbol{\mathcal{O}}_{\tilde X}(\tilde X)$, and elements of $\mathfrak{D}_{\rm ex}$ are global sections of the sheaf $\mathcal{O}^{\rm ex}_{\tilde X}$ on $\tilde X$. Hence $d'_1, d'_2 \in \mathfrak{D}_{\rm ex}$ have $d'_1 = d'_2$ if and only if $d'_1|_{\tilde x} = d'_2|_{\tilde x}$ in $\mathcal{O}^{\rm ex}_{\tilde X, \tilde x} \cong \tilde{\mathfrak{C}}_{\tilde x, \rm ex}$ for all $\tilde x \in \tilde X$.

In the proof of Theorem 6.6 we saw that if $\tilde x \in \tilde X$ with $\zeta_{\mathfrak{C}}(\tilde x) = (x, P_x)$ in $C(X)$ and $c' \in \mathfrak{C}_{\rm ex}$, and P corresponds to P_x under (6.14), then $\alpha_{c'}|_{\tilde x} = 0$, $\beta_{c'}|_{\tilde x} = 1$ in $\mathcal{O}^{\rm ex}_{\tilde X, \tilde x} \cong \tilde{\mathfrak{C}}_{\tilde x, \rm ex}$ if $c' \in P$, and $\alpha_{c'}|_{\tilde x} = 1$, $\beta_{c'}|_{\tilde x} = 0$ otherwise. Thus, if $c'_1, c'_2 \in \mathfrak{C}_{\rm ex}$ satisfy $c'_1 \in P$ if and only if $c'_2 \in P$ for any prime ideal $P \subset \mathfrak{C}_{\rm ex}$, then $\alpha_{c'_1}|_{\tilde x} = \alpha_{c'_2}|_{\tilde x}$, $\beta_{c'_1}|_{\tilde x} = \beta_{c'_2}|_{\tilde x}$ for all $\tilde x \in \tilde X$, so $\alpha_{c'_1} = \alpha_{c'_2}$, $\beta_{c'_1} = \beta_{c'_2}$ in $\mathfrak{D}_{\rm ex}$.

Parts (a) and (b) are consequences of this criterion. For (a), as prime ideals in $\mathfrak{C}_{\rm ex}$ are preimages of prime ideals in $\mathfrak{C}^\sharp_{\rm ex}$, if $[c'_1] = [c'_2]$ in $\mathfrak{C}^\sharp_{\rm ex}$ then $c'_1 \in P$ if and only if $c'_2 \in P$ for any prime $P \subset \mathfrak{C}_{\rm ex}$. For (b), if $i_1, \ldots, j_n > 0$ the prime property of P implies that $c'^{i_1}_1 \cdots c'^{i_n}_n \in P$ if and only if $c'^{j_1}_1 \cdots c'^{j_n}_n \in P$.

For (c), $\tilde{\mathfrak{C}}^\sharp_{\rm ex}$ is generated by $[\alpha_{c'}], [\beta_{c'}], [c']$ for $c' \in \mathfrak{C}_{\rm ex}$ by (6.5), and $\psi^\sharp_{\tilde{\mathfrak{C}},\rm ex}$:

$\tilde{\mathfrak{C}}^\sharp_{\rm ex}\to\mathfrak{D}^\sharp_{\rm ex}$ is surjective as in the construction of $\mathfrak{D}=\Pi^{\rm sc}_{\rm all}(\tilde{\mathfrak{C}})$ in Proposition 5.25, $\mathfrak{D}_{\rm ex}$ is generated in $\mathcal{O}^{\rm ex}_{\tilde{X}}(X)$ by invertibles and $\tilde{\mathfrak{C}}_{\rm ex}$, so (c) follows.

For (d), let $\mathfrak{C}$ be firm, so that $\mathfrak{C}^\sharp_{\rm ex}$ is finitely generated, and choose $c'_1,\ldots,c'_n$ in $\mathfrak{C}_{\rm ex}$ such that $[c'_1],\ldots,[c'_n]$ generate $\mathfrak{C}^\sharp_{\rm ex}$. Then (a),(b) imply that there are at most 2^n distinct $\alpha_{c'}$ and 2^n distinct $\beta_{c'}$ in $\mathfrak{D}_{\rm ex}$, as every $\alpha_{c'}$ equals $\alpha_{\prod_{i\in S}c'_i}$ for some subset $S\subseteq\{1,\ldots,n\}$. Hence (c) shows that $\mathfrak{D}^\sharp_{\rm ex}$ is generated by the $2^{n+1}+n$ elements $[\alpha_{\prod_{i\in S}c'_i}]$, $[\beta_{\prod_{i\in S}c'_i}]$ for $S\subseteq\{1,\ldots,n\}$ and $[\tilde{\Pi}_{\mathfrak{C},{\rm ex}}(c'_i)]$ for $i=1,\ldots,m$. Therefore $\mathfrak{D}$ is firm. □

As for Theorem 6.11, Theorems 5.31(c) and 6.11 and Proposition 6.14(d) imply the following.

Theorem 6.15 *The corner functor $C:\mathbf{C^\infty Sch^c}\to\mathbf{C^\infty Sch^c_{in}}$ from §6.2 maps $\mathbf{C^\infty Sch^c_{fi}}\to\mathbf{C^\infty Sch^c_{fi,in}}$. We call the restriction $C:\mathbf{C^\infty Sch^c_{fi}}\to\mathbf{C^\infty Sch^c_{fi,in}}$ a* ***corner functor***. *It is right adjoint to* inc $:\mathbf{C^\infty Sch^c_{fi,in}}\hookrightarrow\mathbf{C^\infty Sch^c_{fi}}$.

As C is a right adjoint, it preserves limits in $\mathbf{C^\infty Sch^c_{fi}},\mathbf{C^\infty Sch^c_{fi,in}}$.

Combining Theorems 5.31(c), 6.13, and 6.15 yields the following.

Theorem 6.16 *Restricting C in Theorem 6.15 gives corner functors*

$$C:\mathbf{C^\infty Sch^c_{to,ex}}\longrightarrow\mathbf{C^\infty Sch^c_{to}},\qquad C:\mathbf{C^\infty Sch^c_{fi,tf,ex}}\longrightarrow\mathbf{C^\infty Sch^c_{fi,tf}},$$
$$C:\mathbf{C^\infty Sch^c_{fi,\mathbb{Z},ex}}\longrightarrow\mathbf{C^\infty Sch^c_{fi,\mathbb{Z}}},\quad C:\mathbf{C^\infty Sch^c_{fi,in,ex}}\longrightarrow\mathbf{C^\infty Sch^c_{fi,in}},$$

in the notation of Definition 5.29, which are right adjoint to the corresponding inclusions. As these corner functors are right adjoints, they preserve limits.

For a firm C^∞-ring with corners $\mathfrak{C}$ with $\boldsymbol{X}={\rm Spec^c}\,\mathfrak{C}$, we can give an alternative description of $C(\boldsymbol{X})$.

Theorem 6.17 *Let $\mathfrak{C}$ be a firm C^∞-ring with corners and $\boldsymbol{X}={\rm Spec^c}\,\mathfrak{C}$. Then ${\rm Pr}_{\mathfrak{C}}$ in (6.7) is finite. Hence in the decomposition (6.19), the subsets $C(X)^P\subseteq C(X)$ are open as well as closed, and (6.19) lifts to a disjoint union*

$$C(\boldsymbol{X})=\coprod\nolimits_{P\in{\rm Pr}_{\mathfrak{C}}}C(\boldsymbol{X})^P\qquad\text{in }\mathbf{C^\infty Sch^c_{fi,in}}.\tag{6.20}$$

For each $P\in{\rm Pr}_{\mathfrak{C}}$, Example 4.25(b) defines a C^∞-ring with corners $\mathfrak{C}/{\sim_P}$ with projection $\boldsymbol{\pi}_P:\mathfrak{C}\to\mathfrak{C}/{\sim_P}$, where $\mathfrak{C}/{\sim_P}$ is firm and interior. Write $\boldsymbol{X}^P={\rm Spec^c_{in}}(\mathfrak{C}/{\sim_P})$ and $\boldsymbol{\Pi}^P={\rm Spec^c}\,\boldsymbol{\pi}_P:\boldsymbol{X}^P\to\boldsymbol{X}$. Then

$$C(\boldsymbol{\Pi}^P)\circ\boldsymbol{\iota}_{\boldsymbol{X}^P}:\boldsymbol{X}^P\longrightarrow C(\boldsymbol{X})\tag{6.21}$$

is an isomorphism with the open and closed C^∞-subscheme $C(\boldsymbol{X})^P\subseteq C(\boldsymbol{X})$. Thus $\coprod_{P\in{\rm Pr}_{\mathfrak{C}}}\boldsymbol{X}^P\cong C(\boldsymbol{X})$.

Proof First note that prime ideals $P \subset \mathfrak{C}_{\rm ex}$ are in 1-1 correspondence with prime ideals $P^\sharp$ in $\mathfrak{C}_{\rm ex}^\sharp$, which is finitely generated as $\mathfrak{C}$ is firm. If $[c'_1],\ldots,[c'_n]$ are generators of $\mathfrak{C}_{\rm ex}^\sharp$ then $P^\sharp = \langle\{[c'_1],\ldots,[c'_n]\}\cap P^\sharp\rangle$, so there are at most 2^n such prime ideals $P^\sharp$. Hence $\mathrm{Pr}_{\mathfrak{C}}$ is finite. Thus in (6.19), $C(X)$ is the disjoint union of *finitely many* closed sets $C(X)^P$, so the $C(X)^P$ are also open. Equation (6.20) follows.

Write $\mathfrak{D} = \tilde{C}(\mathfrak{C})$ as in Proposition 6.14 and $\boldsymbol{Y} = \mathrm{Spec^c}\,\mathfrak{D}$, so Theorem 6.6 gives an isomorphism $\boldsymbol{\eta}_{\mathfrak{C}} : \boldsymbol{Y} \to C(\boldsymbol{X})$. Fix $P \in \mathrm{Pr}_{\mathfrak{C}}$, and write $\boldsymbol{Y}^P = \boldsymbol{\eta}_{\mathfrak{C}}^{-1}(C(\boldsymbol{X})^P)$, so that $\boldsymbol{Y}^P$ is open and closed in $\boldsymbol{Y}$ with $\boldsymbol{\eta}_{\mathfrak{C}}|_{\boldsymbol{Y}^P} : \boldsymbol{Y}^P \to C(\boldsymbol{X})^P$ an isomorphism. The proof of Theorem 6.6 implies that $\boldsymbol{Y}^P$ may be defined as a C^∞-subscheme of $\boldsymbol{Y}$ by the equations $\alpha_{c'} = 0$, $\beta_{c'} = 1$ for $c' \in P$ and $\alpha_{c'} = 1$, $\beta_{c'} = 0$ for $c' \in \mathfrak{C}_{\rm ex} \setminus P$, where this is only finitely many equations, as there are only finitely many distinct $\alpha_{c'}$ in $\mathfrak{D}_{\rm ex}$ by the proof of Proposition 6.14(d). Thus we see that

$$
\begin{aligned}
\boldsymbol{Y}^P &\cong \mathrm{Spec^c}\bigl(\mathfrak{D}/[\alpha_{c'}=0,\ \beta_{c'}=1,\ c'\in P,\ \alpha_{c''}=1,\ \beta_{c''}=0,\ c''\in\mathfrak{C}_{\rm ex}\setminus P]\bigr)\\
&= \mathrm{Spec^c}\bigl(\Pi^{\rm sc}_{\rm all}(\tilde{\mathfrak{C}})/[\alpha_{c'}=0,\ \beta_{c'}=1,\ c'\in P,\ \alpha_{c''}=1,\ \beta_{c''}=0,\ c''\in\mathfrak{C}_{\rm ex}\setminus P]\bigr)\\
&\cong \mathrm{Spec^c}\circ\Pi^{\rm sc}_{\rm all}\bigl(\tilde{\mathfrak{C}}/[\alpha_{c'}=0,\ \beta_{c'}=1,\ c'\in P,\ \alpha_{c''}=1,\ \beta_{c''}=0,\ c''\in\mathfrak{C}_{\rm ex}\setminus P]\bigr)\\
&= \mathrm{Spec^c}\bigl(\mathfrak{C}\bigl[\alpha_{c'},\beta_{c'} : c'\in\mathfrak{C}_{\rm ex}\bigr]/\bigl(\Phi_i(\alpha_{c'})+\Phi_i(\beta_{c'})=1\ \forall c'\in\mathfrak{C}_{\rm ex}\bigr)\\
&\qquad \bigl[\alpha_{1_{\mathfrak{C}_{\rm ex}}}=1_{\mathfrak{C}_{\rm ex}},\ \alpha_{c'}\alpha_{c''}=\alpha_{c'c''}\ \forall c',c''\in\mathfrak{C}_{\rm ex},\ c'\beta_{c'}=\alpha_{c'}\beta_{c'}=0\ \forall c'\in\mathfrak{C}_{\rm ex}\\
&\qquad \alpha_{c'}=0,\ \beta_{c'}=1,\ c'\in P,\ \alpha_{c''}=1,\ \beta_{c''}=0,\ c''\in\mathfrak{C}_{\rm ex}\setminus P\bigr]\bigr)\\
&\cong \mathrm{Spec^c}\bigl(\mathfrak{C}/[c'=0,\ c'\in P]\bigr) = \mathrm{Spec^c}(\mathfrak{C}/{\sim_P}) = \boldsymbol{X}^P,
\end{aligned}
\tag{6.22}
$$

where in the third step we use that $\Pi^{\rm sc}_{\rm all}$ commutes with applying relations after composing with $\mathrm{Spec^c}$, in the fourth that $\mathrm{Spec^c}\circ\Pi^{\rm sc}_{\rm all} \cong \mathrm{Spec^c}$ and (6.5), and in the fifth that adding generators $\alpha_{c'},\beta_{c'}$ and setting them to 0 or 1 is equivalent to adding no generators, and the only remaining effective relations are $c'\beta_{c'} = 0$ when $c' \in P$, so that $\beta_{c'} = 1$, giving $c' = 0$.

Composing the inverse of (6.22) with $\boldsymbol{\eta}_{\mathfrak{C}}|_{\boldsymbol{Y}^P}$ gives a morphism $\boldsymbol{\theta}_P : \boldsymbol{X}^P \to C(\boldsymbol{X})$ which is an isomorphism with $C(\boldsymbol{X})^P \subset C(\boldsymbol{X})$. As $\mathfrak{C}/{\sim_P}$ is interior, $\boldsymbol{X}^P$ is interior, so (6.21) is well defined. One can check, from the actions on points and on stalks $\mathcal{O}_{X^P,x}, \mathcal{O}_{C(X),(x,P_x)}$, that (6.21) is $\boldsymbol{\theta}_P$. The theorem follows. □

We explain by example how Theorem 6.17 goes wrong in the non-firm case.

Example 6.18 Write $[\mathbf{0},\boldsymbol{\infty}) = F^{\mathbf{C^\infty Sch^c}}_{\mathbf{Man^c}}([0,\infty)) = \mathrm{Spec^c}\, \boldsymbol{C^\infty}([0,\infty))$ as an affine C^∞-scheme with corners. Proposition 6.12 gives $C([\mathbf{0},\boldsymbol{\infty})) \cong \{*\} \amalg [\mathbf{0},\boldsymbol{\infty})$, writing $\partial[0,\infty) = \{*\}$, a single point.

Let S be an infinite set. Consider $\boldsymbol{X} = \prod_{s\in S} [\mathbf{0},\boldsymbol{\infty})_s$, the product in

$\mathbf{C^\infty Sch^c}$ of a copy $[\mathbf{0},\boldsymbol{\infty})_s$ of $[\mathbf{0},\boldsymbol{\infty})$ for each s in S. Since $\mathrm{Spec^c}$ preserves limits, we have $\boldsymbol{X} = \mathrm{Spec^c}\,\mathfrak{C}$ for $\mathfrak{C} = \bigotimes_\infty^{s\in S} C^\infty([0,\infty))_s$.

As $C : \mathbf{C^\infty Sch^c} \to \mathbf{C^\infty Sch^c_{in}}$ preserves limits by Theorem 6.11 we have

$$C(\boldsymbol{X}) = C\bigl(\textstyle\prod_{s\in S}[\mathbf{0},\boldsymbol{\infty})_s\bigr) \cong \textstyle\prod_{s\in S}\bigl(\{*\}_s \amalg [\mathbf{0},\boldsymbol{\infty})_s\bigr). \qquad (6.23)$$

Just as a set, we may decompose the right-hand side of (6.23) as

$$C(X) \cong \textstyle\coprod_{T\subseteq S}\bigl(\prod_{t\in T}\{*\}_t \times \prod_{s\in S\setminus T}[0,\infty)_s\bigr) =: \coprod_{T\subseteq S} C(X)^T. \quad (6.24)$$

One can show that subsets $T \subseteq S$ are in natural 1-1 correspondence with prime ideals $P \subset \mathfrak{C}_{\mathrm{ex}}$, and (6.24) is the decomposition (6.19) for $\boldsymbol{X} = \mathrm{Spec^c}\,\mathfrak{C}$.

Since S is an infinite set, the subsets $C(X)^T \subset C(X)$ in (6.24) are closed, but not open, in (6.24). This is because $C(X)$ has the product topology, as the forgetful functor $\mathbf{C^\infty Sch^c} \to \mathbf{Top}$ preserves limits, and an infinite product of topological spaces $\prod_{s\in S} X_s$ has a weaker topology than one might expect: if $U_s \subseteq X_s$ is open for $s \in S$ then $\prod_{s\in S} U_s$ is generally *not* open in $\prod_{s\in S} X_s$, unless $U_s = X_s$ for all but finitely many $s \in S$.

Although Theorem 6.17 makes sense when $\mathfrak{C}$ is not firm and $\mathrm{Pr}_{\mathfrak{C}}$ is infinite, it is false in general, as $C(\boldsymbol{X}) \cong \coprod_P \boldsymbol{X}^P$ fails already for topological spaces.

6.4 The sheaves of monoids $\check M^{\mathrm{ex}}_{C(X)}, \check M^{\mathrm{in}}_{C(X)}$ on $C(\boldsymbol{X})$

As in Remark 3.18(d), for a manifold with g-corners X, in [46, §3.6] we defined the *comonoid bundle* $M^\vee_{C(X)} \to C(X)$, a local system of toric monoids on $C(X)$ with the property that if $(x,\gamma) \in C(X)^\circ$ and $M^\vee_{C(X)}|_{(x,\gamma)} = P$ then X near x is locally diffeomorphic to $X_P \times \mathbb{R}^{\dim X - \mathrm{rank}\, P}$ near $(\delta_0, 0)$. If X is a manifold with corners then $M^\vee_{C(X)}$ has fibre $\mathbb{N}^k$ over $C_k(X)$. We now generalize these ideas to firm C^∞-schemes with corners.

Definition 6.19 Let $\boldsymbol{X} = (X, \mathcal{O}_X)$ be a local C^∞-ringed space with corners, so that Definition 6.1 defines the corners $C(\boldsymbol{X}) = (C(X), \mathcal{O}_{C(X)})$ in $\mathbf{LC^\infty RS^c_{in}}$, with a morphism $\boldsymbol{\Pi_X} : C(\boldsymbol{X}) \to \boldsymbol{X}$. We have a sheaf of monoids $\Pi_X^{-1}(\mathcal{O}_X^{\mathrm{ex}})$ on $C(X)$, with stalks $(\Pi_X^{-1}(\mathcal{O}_X^{\mathrm{ex}}))_{(x,P)} = \mathcal{O}^{\mathrm{ex}}_{X,x}$ at (x,P) in $C(X)$, and we have a natural prime ideal $P \subset (\Pi_X^{-1}(\mathcal{O}_X^{\mathrm{ex}}))_{(x,P)}$ in the stalk at each $(x,P) \in C(X)$.

Define a presheaf of monoids $\mathcal{P}\check M^{\mathrm{ex}}_{C(X)}$ on $C(X)$ by, for open $U \subseteq C(X)$,

$$\mathcal{P}\check M^{\mathrm{ex}}_{C(X)}(U) = \Pi_X^{-1}(\mathcal{O}_X^{\mathrm{ex}})(U)/\bigl[s = 1 : s \in \Pi_X^{-1}(\mathcal{O}_X^{\mathrm{ex}})(U),\ s|_{(x,P)} \notin P \text{ for all } (x,P) \in U\bigr].$$

For open $V \subseteq U \subseteq C(X)$, the restriction map $\rho_{UV} : \mathcal{P}\check{M}^{\rm ex}_{C(X)}(U) \to \mathcal{P}\check{M}^{\rm ex}_{C(X)}(V)$ is induced by $\rho_{UV} : \Pi_X^{-1}(\mathcal{O}_X^{\rm ex})(U) \to \Pi_X^{-1}(\mathcal{O}_X^{\rm ex})(V)$. This gives a well-defined presheaf $\mathcal{P}\check{M}^{\rm ex}_{C(X)}$ valued in $\mathbf{Mon}$. Write $\check{M}^{\rm ex}_{C(X)}$ for its sheafification.

The surjective projections $\Pi_X^{-1}(\mathcal{O}_X^{\rm ex})(U) \to \mathcal{P}\check{M}^{\rm ex}_{C(X)}(U)$ for open $U \subseteq C(X)$ induce a surjective morphism $\Pi_{\check{M}^{\rm ex}_{C(X)}} : \Pi_X^{-1}(\mathcal{O}_X^{\rm ex}) \to \check{M}^{\rm ex}_{C(X)}$. Local sections s of $\Pi_X^{-1}(\mathcal{O}_X^{\rm ex})$ lying in $\mathcal{O}^{\rm ex}_{X,x} \setminus P$ at each (x,P) have $\pi(s) = 1$ in $\check{M}^{\rm ex}_{C(X)}$, and $\check{M}^{\rm ex}_{C(X)}$ is universal with this property. The stalk of $\check{M}^{\rm ex}_{C(X)}$ at (x,P) is

$$(\check{M}^{\rm ex}_{C(X)})_{(x,P)} \cong \mathcal{O}^{\rm ex}_{X,x} / \big[s = 1 : s \in \mathcal{O}^{\rm ex}_{X,x} \setminus P \big]. \tag{6.25}$$

Now suppose that $\boldsymbol{X}$ is interior. Then as $\boldsymbol{\mathcal{O}}_{X,x}$ is interior for $x \in X$, the monoid $\mathcal{O}^{\rm ex}_{X,x}$ has no zero divisors, and $\mathcal{O}^{\rm in}_{X,x} = \mathcal{O}^{\rm ex}_{X,x} \setminus \{0\}$ is also a monoid. Write $\check{M}^{\rm in}_{C(X)} \subset \check{M}^{\rm ex}_{C(X)}$ for the subsheaf of sections of $\check{M}^{\rm ex}_{C(X)}$ which are non-zero at every $(x,P) \in C(X)$. Then $\check{M}^{\rm in}_{C(X)}$ is also a sheaf of monoids on $C(X)$, with stalk $(\check{M}^{\rm in}_{C(X)})_{(x,P)} = (\check{M}^{\rm ex}_{C(X)})_{(x,P)} \setminus \{0\}$ at each $(x,P) \in C(X)$.

Next let $\boldsymbol{f} : \boldsymbol{X} \to \boldsymbol{Y}$ be a morphism in $\mathbf{LC^\infty RS^c}$, so that $C(\boldsymbol{f}) : C(\boldsymbol{X}) \to C(\boldsymbol{Y})$ is a morphism in $\mathbf{LC^\infty RS^c_{in}}$. Consider the diagram:

$$\begin{array}{ccccc}
C(f)^{-1}(\mathcal{O}^{\rm ex}_{C(Y)}) & \xleftarrow{C(f)^{-1}(\Pi^\sharp_{Y,\rm ex})} & \begin{array}{c} C(f)^{-1}\circ\Pi_Y^{-1}(\mathcal{O}_Y^{\rm ex}) = \\ \Pi_X^{-1}\circ f^{-1}\circ(\mathcal{O}_Y^{\rm ex}) \end{array} & \xrightarrow{C(f)^{-1}(\Pi_{\check{M}^{\rm ex}_{C(Y)}})} & C(f)^{-1}(\check{M}^{\rm ex}_{C(Y)}) \\
\Big\downarrow {\scriptstyle C(f)^\sharp_{\rm ex}} & & \Big\downarrow {\scriptstyle \Pi_X^{-1}(f^\sharp_{\rm ex})} & & \Big\downarrow {\scriptstyle \check{M}^{\rm ex}_{C(f)}} \\
\mathcal{O}^{\rm ex}_{C(X)} & \xleftarrow{\Pi^\sharp_{X,\rm ex}} & \Pi_X^{-1}(\mathcal{O}_X^{\rm ex}) & \xrightarrow{\Pi_{\check{M}^{\rm ex}_{C(X)}}} & \check{M}^{\rm ex}_{C(X)}.
\end{array} \tag{6.26}$$

Here $C(f)^{-1}(\Pi_{\check{M}^{\rm ex}_{C(Y)}}) : C(f)^{-1}\circ\Pi_Y^{-1}(\mathcal{O}_Y^{\rm ex}) \twoheadrightarrow C(f)^{-1}(\check{M}^{\rm ex}_{C(Y)})$ is the quotient sheaf of monoids which is universal for morphisms $C(f)^{-1}\circ\Pi_Y^{-1}(\mathcal{O}_Y^{\rm ex}) \to M$, for M a sheaf of monoids on $C(X)$, such that a local section s of $C(f)^{-1}\circ \Pi_Y^{-1}(\mathcal{O}_Y^{\rm ex})$ is mapped to 1 in M if $s|_{(x,P)} \in \mathcal{O}^{\rm ex}_{Y,y} \setminus Q$ at each $(x,P) \in C(X)$ with $C(f)(x,P) = (y,Q)$ in $C(Y)$.

Now taking $M = \check{M}^{\rm ex}_{C(X)}$, the morphism $\Pi_{\check{M}^{\rm ex}_{C(X)}} \circ \Pi_X^{-1}(f^\sharp_{\rm ex}) : C(f)^{-1} \circ \Pi_Y^{-1}(\mathcal{O}_Y^{\rm ex}) \to \check{M}^{\rm ex}_{C(X)}$ has the property required. This is because $C(\boldsymbol{f})$ is interior, so $C(f)^\sharp_{\rm ex}$ maps $\mathcal{O}^{\rm ex}_{C(Y),(y,Q)} \setminus \{0\} \to \mathcal{O}^{\rm ex}_{C(X),(x,P)} \setminus \{0\}$, and thus $\Pi_X^{-1}(f^\sharp_{\rm ex})$ maps $\mathcal{O}^{\rm ex}_{Y,y} \setminus Q \to \mathcal{O}^{\rm ex}_{X,x} \setminus P$, and $\Pi_{\check{M}^{\rm ex}_{C(X)}}$ maps $\mathcal{O}^{\rm ex}_{X,x} \setminus P \to 1$. Hence there is a unique morphism $\check{M}^{\rm ex}_{C(f)} : C(f)^{-1} \circ \Pi_Y^{-1}(\mathcal{O}_Y^{\rm ex}) \to \check{M}^{\rm ex}_{C(X)}$ of sheaves of monoids on $C(X)$ making (6.26) commute.

If $\boldsymbol{g} : \boldsymbol{Y} \to \boldsymbol{Z}$ is another morphism in $\mathbf{LC^\infty RS^c}$, then composing (6.26) for $\boldsymbol{f},\boldsymbol{g}$ vertically we see that $\check{M}^{\rm ex}_{C(g\circ f)} = \check{M}^{\rm ex}_{C(f)} \circ C(f)^{-1}(\check{M}^{\rm ex}_{C(g)})$. So the $\check{M}^{\rm ex}_{C(X)}, \check{M}^{\rm ex}_{C(f)}$ are contravariantly functorial.

Proposition 6.20 *Let $\boldsymbol{X}$ be a firm C^∞-scheme with corners. Then $\check{M}^{\rm ex}_{C(X)}$ in Definition 6.19 is a* **locally constant** *sheaf of monoids on $C(X)$, and the stalks $(\check{M}^{\rm ex}_{C(X)})_{(x,P)}$ for $(x,P) \in C(X)$ are finitely generated sharp monoids, with zero elements not equal to the identity. If $\boldsymbol{X}$ is interior, the same holds for $\check{M}^{\rm in}_{C(X)}$, but without with zero elements.*

Proof As the proposition is local in $\boldsymbol{X}$, it is sufficient to prove it in the case that $\boldsymbol{X} = \operatorname{Spec}^{\rm c}\mathfrak{C}$ for $\mathfrak{C}$ a firm C^∞-ring with corners. Then Theorem 6.17 defines an isomorphism $\coprod_{P\in \Pr_{\mathfrak{C}}} \boldsymbol{X}^P \to C(\boldsymbol{X})$.

We will prove that for $P \in \Pr_{\mathfrak{C}}$, the restriction of $\check{M}^{\rm ex}_{C(X)}$ to the image $C(\boldsymbol{X})^P$ of $\boldsymbol{X}^P$ under (6.21) is the constant sheaf with fibre the monoid

$$\mathfrak{C}_{\rm ex}/\big[c'=1 : c' \in \mathfrak{C}_{\rm ex}\setminus P\big] \cong \mathfrak{C}^\sharp_{\rm ex}/\big[c''=1 : c'' \in \mathfrak{C}^\sharp_{\rm ex}\setminus P^\sharp\big]. \qquad (6.27)$$

Here $\mathfrak{C}^\sharp_{\rm ex} = \mathfrak{C}_{\rm ex}/\mathfrak{C}^\times_{\rm ex}$ and $P^\sharp = P/\mathfrak{C}^\times_{\rm ex}$, and the two sides of (6.27) are isomorphic as $\mathfrak{C}^\times_{\rm ex} \subseteq \mathfrak{C}_{\rm ex}\setminus P$, so quotienting by $\mathfrak{C}^\times_{\rm ex}$ commutes with setting all elements of $\mathfrak{C}_{\rm ex}\setminus P$ or $\mathfrak{C}^\sharp_{\rm ex}\setminus P^\sharp$ equal to 1. Equation (6.27) is finitely generated as $\mathfrak{C}^\sharp_{\rm ex}$ is, since $\mathfrak{C}$ is firm. Also (6.27) is the disjoint union of $\{1\}$, and the image of P, which is an ideal in (6.27), and so contains no units. Hence (6.27) is sharp. The image of $0_{\mathfrak{C}_{\rm ex}}$ is a zero element not equal to the identity.

Suppose $P \in \Pr_{\mathfrak{C}}$, and $x' \in \boldsymbol{X}^P$ maps to $(x,P_x) \in C(\boldsymbol{X})$ under (6.21), where $P_x \subset O^{\rm ex}_{X,x} = \mathfrak{C}_{x,\rm ex}$ is a prime ideal, with $P_x = \pi_{x,\rm ex}(P) \subsetneq \mathfrak{C}_{x,\rm ex}$ and $P = \pi^{-1}_{x,\rm ex}(P_x) \subsetneq \mathfrak{C}_{\rm ex}$ by Lemma 6.7. Consider the diagram of monoids:

$$\begin{array}{ccc}
\mathfrak{C}_{\rm ex} & \xrightarrow[\text{project}]{} & \mathfrak{C}_{\rm ex}/\big[c'=1 : c' \in \mathfrak{C}_{\rm ex}\setminus P\big] = (6.27) \\
\downarrow{\scriptstyle \pi_{x,\rm ex}} & & {\scriptstyle (\pi_{x,\rm ex})_*}\downarrow \\
\mathfrak{C}_{x,\rm ex} & \xrightarrow{\text{project}} & \mathfrak{C}_{x,\rm ex}/\big[c''=1 : c'' \in \mathfrak{C}_{x,\rm ex}\setminus P_x\big] = (\check{M}^{\rm ex}_{C(X)})_{(x,P_x)}.
\end{array} \qquad (6.28)$$

As $P = \pi^{-1}_{x,\rm ex}(P_x)$, if $c' \in \mathfrak{C}_{\rm ex}\setminus P$ then $c'' = \pi_{x,\rm ex}(c') \in \mathfrak{C}_{x,\rm ex}\setminus P_x$, so there is a unique, surjective morphism $(\pi_{x,\rm ex})_*$ making (6.28) commute.

We claim that $(\pi_{x,\rm ex})_*$ is an isomorphism. To see this, note that as in §4.6, $\mathfrak{C}_{x,\rm ex}$ may be obtained from $\mathfrak{C}_{\rm ex}$ by inverting all elements $c' \in \mathfrak{C}_{\rm ex}$ with $x \circ \Phi_i(c') \neq 0$, and setting $c'' = 1$ for all $c'' \in \Psi_{\rm exp}(I)$, where $I \subset \mathfrak{C}$ is the ideal vanishing near x. Here $c' \in \mathfrak{C}_{\rm ex}$ with $x \circ \Phi_i(c') \neq 0$ and $c'' \in \Psi_{\rm exp}(I)$ both lie in the set $\mathfrak{C}_{\rm ex}\setminus P$ of elements set to 1 in the right-hand side of (6.28). Also first inverting c' and then setting $c' = 1$ is equivalent to just setting $c' = 1$.

Thus $(\pi_{x,\rm ex})_*$ is an isomorphism. So the fibre of $(\check{M}^{\rm ex}_{C(X)})_{(x,P_x)}$ at each

(x, P_x) in the image of $\boldsymbol{X}^P$ is naturally isomorphic to (6.27), which is independent of (x, P_x). It easily follows that $\check{M}^{\rm ex}_{C(X)}$ is constant with fibre (6.27) on the image $\boldsymbol{X}^P$. The analogue for $\check{M}^{\rm in}_{C(X)}$ when $\boldsymbol{X}$ is interior is immediate. □

Remark 6.21 If $\boldsymbol{X}$ lies in $\mathbf{LC^\infty RS^c}$ or $\mathbf{C^\infty Sch^c}$, but not in $\mathbf{C^\infty Sch^c_{fi}}$, then $\check{M}^{\rm ex}_{C(X)}$ in Definition 6.19 may not be locally constant.

For example, suppose $\mathfrak{C} \in \mathbf{C^\infty Rings^c}$ is not firm, and $\mathfrak{C}_{\rm ex}$ has infinitely many prime ideals P, and $\boldsymbol{X} = \operatorname{Spec^c} \mathfrak{C}$. Then (6.19) decomposes $C(X)$ into infinitely many disjoint closed subsets $C(X)^P$, which need not be open. The proof of Proposition 6.20 shows that $\check{M}^{\rm ex}_{C(X)}|_{C(X)^P}$ is constant on $C(X)^P$ with value (6.27) for each P. But if the $C(X)^P$ are not open, as happens in Example 6.18, then in general $\check{M}^{\rm ex}_{C(X)}$ is not locally constant.

Proposition 6.20 implies that if $\boldsymbol{X}$ is a firm C^∞-scheme with corners then

$$C(\boldsymbol{X}) = \coprod_{\substack{\text{iso. classes } [M] \text{ of finitely generated}\\ \text{sharp monoids } M \text{ with zero elements}}} C_M(\boldsymbol{X}),$$

where $C_M(\boldsymbol{X}) \subseteq C(\boldsymbol{X})$ is the open and closed C^∞-subscheme of points (x, P_x) in $C(\boldsymbol{X})$ with $(\check{M}^{\rm ex}_{C(X)})_{(x,P_x)} \cong M$.

Example 6.22 In Theorem 6.17, let $\mathfrak{C} = C^\infty(\mathbb{R}^m_k)$, so that $\boldsymbol{X} = \operatorname{Spec^c} \mathfrak{C} = F^{\mathbf{C^\infty Sch^c}}_{\mathbf{Man^c}}(\mathbb{R}^m_k)$. Take P to be the prime ideal $P = \langle x_1, \ldots, x_k \rangle$ in $\mathfrak{C}_{\rm ex}$, so that (6.21) maps $\boldsymbol{X}^P$ to the corner stratum $\{x_1 = \cdots = x_k = 0\}$ of $\mathbb{R}^m_k$ in $C(\boldsymbol{X})$. The proof of Proposition 6.20 shows that $\check{M}^{\rm ex}_{C(X)}$ on $\boldsymbol{X}^P$ is the constant sheaf with fibre (6.27). In this case $\mathfrak{C}$ is interior with

$$\mathfrak{C}_{\rm in} = \{x_1^{a_1} \cdots x_k^{a_k} \exp f : a_i \in \mathbb{N},\ f \in C^\infty(\mathbb{R}^m_k)\},\quad \mathfrak{C}_{\rm ex} = \mathfrak{C}_{\rm in} \amalg \{0\}. \quad (6.29)$$

Since $\mathfrak{C}_{\rm ex} \setminus P = \mathfrak{C}_{\rm ex}^\times = \{\exp f : f \in C^\infty(\mathbb{R}^m_k)\}$, equation (6.27) is $\mathfrak{C}_{\rm ex}/\mathfrak{C}_{\rm ex}^\times = \mathfrak{C}_{\rm ex}^\sharp \cong \mathbb{N}^k \amalg \{0\}$. So $\check{M}^{\rm ex}_{C(X)}$ has fibre $\mathbb{N}^k \amalg \{0\}$, and $\check{M}^{\rm in}_{C(X)}$ has fibre $\mathbb{N}^k$, on $\boldsymbol{X}^P$.

Now let Y be a manifold with corners, and $\boldsymbol{Y} = F^{\mathbf{C^\infty Sch^c}}_{\mathbf{Man^c}}(Y)$. Then Proposition 6.12 says that $C(\boldsymbol{Y}) \cong F^{\mathbf{C^\infty Sch^c}}_{\mathbf{\check{M}an^c}}(C(Y))$. Let $(y, \gamma) \in C_k(Y)^\circ$. Then $y \in S^k(Y)$, and Y near y is locally diffeomorphic to $\mathbb{R}^m_k$ near 0, such that the local k-corner component γ of Y at y is identified with $\{x_1 = \cdots = x_k = 0\}$ in $\mathbb{R}^m_k$. Hence $C(\boldsymbol{Y})$ and $M^{\rm ex}_{C(Y)}, M^{\rm in}_{C(Y)}$ near (y, γ) are locally isomorphic to $C(\boldsymbol{X}), \check{M}^{\rm ex}_{C(X)}, \check{M}^{\rm in}_{C(X)}$ near $(0, \{x_1 = \cdots = x_k = 0\})$. Therefore $\check{M}^{\rm ex}_{C(Y)}, \check{M}^{\rm in}_{C(Y)}$ have fibres $\mathbb{N}^k \amalg \{0\}, \mathbb{N}^k$ on $C_k(Y)^\circ$, and hence on $C_k(Y)$, as $C_k(Y)^\circ$ is dense in $C_k(Y)$ and $\check{M}^{\rm ex}_{C(Y)}, \check{M}^{\rm in}_{C(Y)}$ are locally constant by Proposition 6.20.

As in Remark 3.18(d), in [46, §3.6] we define the *comonoid bundle* $M^\vee_{C(Y)}$

$\to C(Y)$, which has fibre $\mathbb{N}^k$ over $C_k(Y)$. It is easy to check that the sheaf of continuous sections of $M^\vee_{C(Y)}$ is canonically isomorphic to $\check M^{\rm in}_{C(Y)}$.

Example 6.23 We can generalize Example 6.22 to manifolds with g-corners. Let Q be a toric monoid and $l \geqslant 0$, and set $\mathfrak{C} = C^\infty(X_Q \times \mathbb{R}^l)$, so that $\boldsymbol{X} = \operatorname{Spec^c}\mathfrak{C} = F^{\mathbf{C^\infty Sch^c}}_{\mathbf{Man^{gc}}}(X_Q \times \mathbb{R}^l)$. Take P to be the prime ideal of functions in $\mathfrak{C}_{\rm ex}$ vanishing on $\{\delta_0\} \times \mathbb{R}^l$, so that (6.21) maps $\boldsymbol{X}^P$ to the corner stratum $\{\delta_0\} \times \mathbb{R}^l$ in $C(\boldsymbol{X})$. The analogue of (6.29) is

$$\mathfrak{C}_{\rm in} = \bigl\{\lambda_q \exp f : q \in Q,\ f \in C^\infty(X_Q \times \mathbb{R}^l)\bigr\},\quad \mathfrak{C}_{\rm ex} = \mathfrak{C}_{\rm in} \amalg \{0\},$$

for $\lambda_q : X_Q \to [0,\infty)$ as in Definition 3.7, so $\mathfrak{C}^\sharp_{\rm in} \cong Q$ and $\mathfrak{C}^\sharp_{\rm ex} \cong Q \amalg \{0\}$, and $\check M^{\rm ex}_{C(\boldsymbol{X})}, \check M^{\rm in}_{C(\boldsymbol{X})}$ have fibres $Q \amalg \{0\}, Q$ on $\boldsymbol{X}^P$.

Now let Y be a manifold with g-corners, and $\boldsymbol{Y} = F^{\mathbf{C^\infty Sch^c}}_{\mathbf{Man^c}}(Y)$, so that $C(\boldsymbol{Y}) \cong F^{\mathbf{C^\infty Sch^c}}_{\mathbf{Man^c}}(C(Y))$ by Proposition 6.12. Let $(y,\gamma) \in C(Y)^\circ$. Then Y near y is locally diffeomorphic to $X_Q \times \mathbb{R}^l$ near $(\delta_0, 0)$ for some toric monoid Q and $l \geqslant 0$, such that the local corner component γ of Y at y is identified with $\{\delta_0\} \times \mathbb{R}^l$ in $X_Q \times \mathbb{R}^l$. Hence $\check M^{\rm ex}_{C(\boldsymbol{Y})}, \check M^{\rm in}_{C(\boldsymbol{Y})}$ have fibres $Q \amalg \{0\}, Q$ at (y,γ).

As in Remark 3.18(d), the comonoid bundle $M^\vee_{C(Y)} \to C(Y)$ of [46, §3.6] also has fibre Q at (y,γ). It is easy to check that the sheaf of continuous sections of $M^\vee_{C(Y)}$ is canonically isomorphic to $\check M^{\rm in}_{C(\boldsymbol{Y})}$.

6.5 The boundary $\partial \boldsymbol{X}$ and k-corners $C_k(\boldsymbol{X})$

If X is a manifold with (g-)corners then the corners $C(X)$ from §3.4 has a decomposition $C(X) = \coprod_{k=0}^{\dim X} C_k(X)$ with $C_0(X) \cong X$ and $C_1(X) = \partial X$. We generalize this to firm (interior) C^∞-schemes with corners $\boldsymbol{X}$.

Definition 6.24 The *dimension* $\dim M$ in $\mathbb{N} \amalg \{\infty\}$ of a monoid M is the maximum length d (or ∞ if there is no maximum) of a chain of prime ideals $\emptyset \subsetneq P_1 \subsetneq P_2 \subsetneq \cdots \subsetneq P_d \subsetneq M$. If M is finitely generated then $\dim M < \infty$. If M is toric then $\dim M = \dim_{\mathbb{R}}(M \otimes_{\mathbb{N}} \mathbb{R})$.

Let $\boldsymbol{X}$ be a firm C^∞-scheme with corners. For each $k = 0, 1, \ldots,$ define the *k-corners* $C_k(\boldsymbol{X}) \subseteq C(\boldsymbol{X})$ to be the C^∞-subscheme of $(x,P) \in C(\boldsymbol{X})$ with $\dim(\check M^{\rm ex}_{C(X)})_{(x,P)} = k+1$. Here $(x,P) \mapsto \dim(\check M^{\rm ex}_{C(X)})_{(x,P)}$ is a locally constant function $C(X) \to \mathbb{N}$ by Proposition 6.20, so $C_k(\boldsymbol{X})$ is open and closed in $C(\boldsymbol{X})$. Also $\dim(\check M^{\rm ex}_{C(X)})_{(x,P)} \geqslant 1$ as $(\check M^{\rm ex}_{C(X)})_{(x,P)}$ has at least one prime ideal $(\check M^{\rm ex}_{C(X)})_{(x,P)} \setminus \{1\}$, since $(\check M^{\rm ex}_{C(X)})_{(x,P)}$ is sharp and $(\check M^{\rm ex}_{C(X)})_{(x,P)} \setminus \{1\} \neq \emptyset$ as $0 \neq 1$. Hence $C(\boldsymbol{X}) = \coprod_{k \geqslant 0} C_k(\boldsymbol{X})$.

We define the *boundary* $\partial \boldsymbol{X}$ to be $\partial \boldsymbol{X} = C_1(\boldsymbol{X}) \subset C(\boldsymbol{X})$.

If $\boldsymbol{X}$ is interior then $\dim(\check{M}^{\mathrm{ex}}_{C(X)})_{(x,P)} = \dim(\check{M}^{\mathrm{in}}_{C(X)})_{(x,P)} + 1$, as chains of ideals in $(\check{M}^{\mathrm{in}}_{C(X)})_{(x,P)}$ lift to chains in $(\check{M}^{\mathrm{ex}}_{C(X)})_{(x,P)}$, plus $\{0\}$. Hence $C_k(\boldsymbol{X})$ is the C^∞-subscheme of $(x,P) \in C(\boldsymbol{X})$ with $\dim(\check{M}^{\mathrm{in}}_{C(X)})_{(x,P)} = k$.

For $\boldsymbol{X}$ interior, if $(\check{M}^{\mathrm{in}}_{C(X)})_{(x,P)} \neq \{1\}$ then $\dim(\check{M}^{\mathrm{in}}_{C(X)})_{(x,P)} > 0$ as $(\check{M}^{\mathrm{in}}_{C(X)})_{(x,P)} \setminus \{1\}$ is prime. But $(\check{M}^{\mathrm{in}}_{C(X)})_{(x,P)} = \{1\}$ if and only if $P = \{0\}$, as $0 \neq p \in P$ descends to $[p] \neq 1$ in $(\check{M}^{\mathrm{in}}_{C(X)})_{(x,P)}$ since $\mathcal{O}^{\mathrm{ex}}_{X,x}$ has no zero divisors. Hence $C_0(\boldsymbol{X}) = \{(x,\{0\}) : x \in \boldsymbol{X}\} \subseteq C(\boldsymbol{X})$. It is now easy to see that $\iota_{\boldsymbol{X}} : \boldsymbol{X} \to C(\boldsymbol{X})$ in Definition 6.1 is an isomorphism $\iota_{\boldsymbol{X}} : \boldsymbol{X} \to C_0(\boldsymbol{X})$.

Example 6.25 Let X be a manifold with (g-)corners and $\boldsymbol{X} = F^{\mathbf{C^\infty Sch^c}}_{\mathbf{Man^{gc}}}(X)$, which is a firm interior C^∞-scheme with corners. As in Proposition 6.12 there is a natural isomorphism $C(\boldsymbol{X}) \cong F^{\mathbf{C^\infty Sch^c}}_{\mathbf{\check{M}an^{gc}}}(C(X))$. It is easy to check that this identifies $C_k(\boldsymbol{X}) \cong F^{\mathbf{C^\infty Sch^c}}_{\mathbf{Man^{gc}}}(C_k(X))$ for each $k = 0, \ldots, \dim X$.

Example 6.26 If $\boldsymbol{X}$ lies in $\mathbf{C^\infty Sch^c_{fi}}$ but not $\mathbf{C^\infty Sch^c_{fi,in}}$ then $C_0(\boldsymbol{X})$ can contain (x,P) with $P \neq \{0\}$. For instance, if $\boldsymbol{X} = \mathrm{Spec^c}\big(\mathbb{R}[x,y]/[xy=0]\big)$ then $\big((0,0),\langle x\rangle\big)$ and $\big((0,0),\langle y\rangle\big)$ lie in $C_0(\boldsymbol{X})$. Also $\{0\}$ is not prime in $\mathcal{O}^{\mathrm{ex}}_{X,(0,0)}$, so $\big((0,0),\{0\}\big) \notin C(\boldsymbol{X})$. Clearly $\boldsymbol{X} \not\cong C_0(\boldsymbol{X})$, as only one of the two is interior.

6.6 Fibre products and corner functors

The authors' favourite categories of 'nice' C^∞-schemes with corners generalizing $\mathbf{Man^{gc}_{in}}, \mathbf{Man^{gc}}$ are $\mathbf{C^\infty Sch^c_{to}}, \mathbf{C^\infty Sch^c_{to,ex}}$. Theorem 5.34 says that fibre products and finite limits exist in $\mathbf{C^\infty Sch^c_{to}}$. Now we address the question of when fibre products exist in $\mathbf{C^\infty Sch^c_{to,ex}}$, or more generally in $\mathbf{C^\infty Sch^c_{fi,in,ex}}$, or in categories like those in (5.10) in which the objects are interior, but the morphisms need not be interior. We use corner functors as a tool to do this.

Theorem 6.27 *Let $\boldsymbol{g} : \boldsymbol{X} \to \boldsymbol{Z}$, $\boldsymbol{h} : \boldsymbol{Y} \to \boldsymbol{Z}$ be morphisms in $\mathbf{C^\infty Sch^c_{to,ex}}$. Consider the question of whether a fibre product $\boldsymbol{W} = \boldsymbol{X} \times_{\boldsymbol{g},\boldsymbol{Z},\boldsymbol{h}} \boldsymbol{Y}$ exists in $\mathbf{C^\infty Sch^c_{to,ex}}$, with projections $\boldsymbol{e} : \boldsymbol{W} \to \boldsymbol{X}$, $\boldsymbol{f} : \boldsymbol{W} \to \boldsymbol{Y}$.*

(a) *If $\boldsymbol{W}$ exists then the following is a bijection of sets:*

$$(e,f) : W \longrightarrow \big\{(x,y) \in X \times Y : g(x) = h(y)\big\}. \tag{6.30}$$

Note that this need not hold for fibre products in $\mathbf{C^\infty Sch^c_{to}}$.

(b) *If* $\boldsymbol{W}$ *exists then* $\boldsymbol{W}$ *is isomorphic to an open and closed* C^∞*-subscheme* $\boldsymbol{W}'$ *of* $C(\boldsymbol{X}) \times_{C(\boldsymbol{g}),C(\boldsymbol{Z}),C(\boldsymbol{h})} C(\boldsymbol{Y})$*, where the fibre product is taken in* $\mathbf{C^\infty Sch^c_{to}}$*, using* $C(\boldsymbol{g}),C(\boldsymbol{h})$ *interior, and exists by Theorem 5.34.*

(c) *Writing* $C(X) \times_{C(Z)} C(Y)$ *for the topological space of* $C(\boldsymbol{X}) \times_{C(\boldsymbol{Z})} C(\boldsymbol{Y})$*, we have a natural continuous map*

$$\begin{aligned} C(X) \times_{C(Z)} C(Y) &\longrightarrow \{(x,y) \in X \times Y : g(x) = h(y)\}, \\ w' \xmapsto{(\Pi_{C(X)},\Pi_{C(Y)})} ((x,P_x),(y,P_y)) &\xmapsto{\Pi_X \times \Pi_Y} (x,y). \end{aligned} \tag{6.31}$$

If there does not exist an open and closed subset $W' \subset C(X) \times_{C(Z)} C(Y)$ *such that* (6.31) *restricts to a bijection on* W'*, then no fibre product* $\boldsymbol{W} = \boldsymbol{X} \times_{\boldsymbol{g},\boldsymbol{Z},\boldsymbol{h}} \boldsymbol{Y}$ *exists in* $\mathbf{C^\infty Sch^c_{to,ex}}$.

(d) *Suppose* $\boldsymbol{g},\boldsymbol{h}$ *are interior, and a fibre product* $\boldsymbol{W} = \boldsymbol{X} \times_{\boldsymbol{g},\boldsymbol{Z},\boldsymbol{h}} \boldsymbol{Y}, \boldsymbol{e},\boldsymbol{f}$ *exists in* $\mathbf{C^\infty Sch^c_{to}}$*. Then* $\boldsymbol{W}$ *is also a fibre product in* $\mathbf{C^\infty Sch^c_{to,ex}}$ *if and only if the following diagram is Cartesian in* $\mathbf{C^\infty Sch^c_{to}}$*:*

$$\begin{array}{ccc} C(\boldsymbol{W}) & \xrightarrow{\quad C(\boldsymbol{f}) \quad} & C(\boldsymbol{Y}) \\ \downarrow{\scriptstyle C(\boldsymbol{e})} & & \downarrow{\scriptstyle C(\boldsymbol{h})} \\ C(\boldsymbol{X}) & \xrightarrow{\quad C(\boldsymbol{g}) \quad} & C(\boldsymbol{Z}), \end{array} \tag{6.32}$$

that is, if $C(\boldsymbol{W}) \cong C(\boldsymbol{X}) \times_{C(\boldsymbol{g}),C(\boldsymbol{Z}),C(\boldsymbol{h})} C(\boldsymbol{Y})$ *in* $\mathbf{C^\infty Sch^c_{to}}$.

The analogues hold for $\mathbf{C^\infty Sch^c_{fi,in,ex}}$, $\mathbf{C^\infty Sch^c_{fi,\mathbb{Z},ex}}$, $\mathbf{C^\infty Sch^c_{fi,tf,ex}}$*, except that* (6.30) *is a bijection for fibre products in* $\mathbf{C^\infty Sch^c_{fi,in}}$.

Proof For (a), suppose $\boldsymbol{W}$ exists, and apply the universal property of fibre products for morphisms from the point $*$ into $\boldsymbol{W},\boldsymbol{X},\boldsymbol{Y},\boldsymbol{Z}$. There is a 1-1 correspondence between morphisms $w : * \to \boldsymbol{W}$ in $\mathbf{C^\infty Sch^c_{to,ex}}$ and points $w' = w(*) \in \boldsymbol{W}$. Using this, the universal property gives the bijection (6.30).

This argument does not work for fibre products in $\mathbf{C^\infty Sch^c_{to}}$, as (necessarily interior) morphisms $w : * \to \boldsymbol{W}$ need not correspond to points of $\boldsymbol{W}$. For example, morphisms $w : * \to [\boldsymbol{0},\boldsymbol{\infty})$ in $\mathbf{C^\infty Sch^c_{to}}$ correspond to points of $(0,\infty)$ not $[0,\infty)$. Equation (6.30) is not a bijection for the fibre products in $\mathbf{C^\infty Sch^c_{to}}$ in Examples 6.29–6.30. This also holds for $\mathbf{C^\infty Sch^c_{fi,\mathbb{Z},ex}}$ and $\mathbf{C^\infty Sch^c_{fi,tf,ex}}$.

The inclusion $\mathrm{inc} : \mathbf{C^\infty Sch^c_{fi,in}} \hookrightarrow \mathbf{C^\infty Sch^c_{fi,in,ex}}$ preserves fibre products by Theorems 5.33 and 5.34. Thus a fibre product in $\mathbf{C^\infty Sch^c_{fi,in}}$ is also a fibre product in $\mathbf{C^\infty Sch^c_{fi,in,ex}}$, and (6.30) is a bijection.

For (b), suppose $\boldsymbol{W}$ exists. As $C : \mathbf{C^\infty Sch^c_{to,ex}} \to \mathbf{C^\infty Sch^c_{to}}$ preserves limits by Theorem 6.16, we have $C(\boldsymbol{W}) \cong C(\boldsymbol{X}) \times_{C(\boldsymbol{g}),C(\boldsymbol{Z}),C(\boldsymbol{h})} C(\boldsymbol{Y})$.

But $\iota_{\boldsymbol{W}} : \boldsymbol{W} \to C(\boldsymbol{W})$ is an isomorphism with an open and closed C^∞-subscheme $C_0(\boldsymbol{W}) \subseteq C(\boldsymbol{W})$ by Definition 6.24. Part (c) follows from (a) and (b).

For (d), let $\boldsymbol{g}, \boldsymbol{h}$ be interior and $\boldsymbol{W}$ be a fibre product in $\mathbf{C^\infty Sch^c_{to}}$. If $\boldsymbol{W}$ is a fibre product in $\mathbf{C^\infty Sch^c_{to,ex}}$ then (6.32) is Cartesian as $C : \mathbf{C^\infty Sch^c_{to,ex}} \to \mathbf{C^\infty Sch^c_{to}}$ preserves limits by Theorem 6.16, proving the 'only if' part.

Suppose (6.32) is Cartesian. Let $\boldsymbol{c} : \boldsymbol{V} \to \boldsymbol{X}$ and $\boldsymbol{d} : \boldsymbol{V} \to \boldsymbol{Y}$ be morphisms in $\mathbf{C^\infty Sch^c_{to,ex}}$ with $\boldsymbol{g} \circ \boldsymbol{c} = \boldsymbol{h} \circ \boldsymbol{d}$. Consider the diagram:

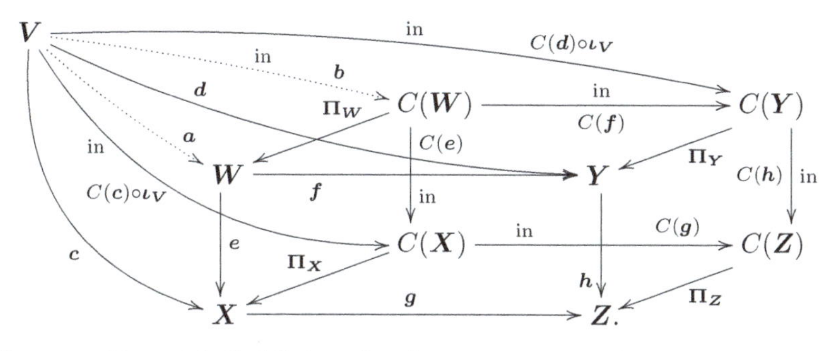

Here morphisms labelled 'in' are interior.

The morphisms $C(\boldsymbol{c}) \circ \iota_{\boldsymbol{V}} : \boldsymbol{V} \to C(\boldsymbol{X})$, $C(\boldsymbol{d}) \circ \iota_{\boldsymbol{V}} : \boldsymbol{V} \to C(\boldsymbol{Y})$ are interior with $C(\boldsymbol{g}) \circ C(\boldsymbol{c}) \circ \iota_{\boldsymbol{V}} = C(\boldsymbol{h}) \circ C(\boldsymbol{d}) \circ \iota_{\boldsymbol{V}}$ as $\boldsymbol{g} \circ \boldsymbol{c} = \boldsymbol{h} \circ \boldsymbol{d}$. So the Cartesian property of (6.32) gives unique interior $\boldsymbol{b} : \boldsymbol{V} \to C(\boldsymbol{W})$ with $C(\boldsymbol{e}) \circ \boldsymbol{b} = C(\boldsymbol{c}) \circ \iota_{\boldsymbol{V}}$ and $C(\boldsymbol{f}) \circ \boldsymbol{b} = C(\boldsymbol{d}) \circ \iota_{\boldsymbol{V}}$. Set $\boldsymbol{a} = \boldsymbol{\Pi}_{\boldsymbol{W}} \circ \boldsymbol{b} : \boldsymbol{V} \to \boldsymbol{W}$. Then

$$\boldsymbol{e} \circ \boldsymbol{a} = \boldsymbol{e} \circ \boldsymbol{\Pi}_{\boldsymbol{W}} \circ \boldsymbol{b} = \boldsymbol{\Pi}_{\boldsymbol{X}} \circ C(\boldsymbol{e}) \circ \boldsymbol{b} = \boldsymbol{\Pi}_{\boldsymbol{X}} \circ C(\boldsymbol{c}) \circ \iota_{\boldsymbol{V}} = \boldsymbol{c} \circ \boldsymbol{\Pi}_{\boldsymbol{V}} \circ \iota_{\boldsymbol{V}} = \boldsymbol{c},$$

and similarly $\boldsymbol{f} \circ \boldsymbol{a} = \boldsymbol{d}$. We claim that $\boldsymbol{a} : \boldsymbol{V} \to \boldsymbol{W}$ is unique with $\boldsymbol{e} \circ \boldsymbol{a} = \boldsymbol{c}$ and $\boldsymbol{f} \circ \boldsymbol{a} = \boldsymbol{d}$, which shows that $\boldsymbol{W} = \boldsymbol{X} \times_{\boldsymbol{g},\boldsymbol{Z},\boldsymbol{h}} \boldsymbol{Y}$ in $\mathbf{C^\infty Sch^c_{to,ex}}$. To see this, note that as C is right adjoint to inc : $\mathbf{C^\infty Sch^c_{to}} \hookrightarrow \mathbf{C^\infty Sch^c_{to,ex}}$, there is a 1-1 correspondence between morphisms $\boldsymbol{a} : \boldsymbol{V} \to \boldsymbol{W}$ in $\mathbf{C^\infty Sch^c_{to,ex}}$ and morphisms $\boldsymbol{b} : \boldsymbol{V} \to C(\boldsymbol{W})$ in $\mathbf{C^\infty Sch^c_{to}}$ with $\boldsymbol{a} = \boldsymbol{\Pi}_{\boldsymbol{W}} \circ \boldsymbol{b}$. But $\boldsymbol{e} \circ \boldsymbol{a} = \boldsymbol{c}$, $\boldsymbol{f} \circ \boldsymbol{a} = \boldsymbol{d}$ imply that $C(\boldsymbol{e}) \circ \boldsymbol{b} = C(\boldsymbol{c}) \circ \iota_{\boldsymbol{V}}$, $C(\boldsymbol{f}) \circ \boldsymbol{b} = C(\boldsymbol{d}) \circ \iota_{\boldsymbol{V}}$, and $\boldsymbol{b}$ is unique under these conditions as above. This proves the 'if' part. □

Here are three examples.

Example 6.28 Define manifolds with corners $X = [0, \infty)^2$, $Y = *$ a point, and $Z = [0, \infty)$, and exterior maps $g : X \to Z$, $h : Y \to Z$ by $g(x_1, x_2) = x_1 x_2$ and $h(*) = 0$. Let $\boldsymbol{X}, \boldsymbol{Y}, \boldsymbol{Z}, \boldsymbol{g}, \boldsymbol{h}$ be the images of $X, \ldots, h$ under $F_{\mathbf{Man^c}}^{\mathbf{C^\infty Sch^c}}$. As $\boldsymbol{X}, \boldsymbol{Y}, \boldsymbol{Z}$ are firm, a fibre product $\tilde{\boldsymbol{W}} = \boldsymbol{X} \times_{\boldsymbol{g},\boldsymbol{Z},\boldsymbol{h}} \boldsymbol{Y}$ exists in $\mathbf{C^\infty Sch^c_{fi}}$, with $\tilde{\boldsymbol{W}} = \operatorname{Spec^c} \mathfrak{C}$, where $\mathfrak{C} = \mathbb{R}[x_1, x_2]/[x_1 x_2 = 0]$.

Here $\mathfrak{C}$, $\tilde{\boldsymbol{W}}$ are not interior, as the relation $x_1x_2 = 0$ is not of interior type. In Theorem 6.27 we have

$$C(X) \times_{C(Z)} C(Y) \cong ([0,\infty)\times\{0\}) \amalg (\{0\}\times[0,\infty)) \amalg \{(0,0)\},$$
$$\{(x,y)\in X\times Y : g(x) = h(y)\} = \{(x_1,x_2)\in[0,\infty)^2 : x_1x_2 = 0\},$$

where the map (6.31) takes $(x_1,x_2)\mapsto(x_1,x_2)$. Considering a neighbourhood of $(0,0)$ we see that no such open and closed subset W' exists in Theorem 6.27(c). Hence no fibre product $\boldsymbol{W} = \boldsymbol{X}\times_{\boldsymbol{Z}}\boldsymbol{Y}$ exists in $\mathbf{C^\infty Sch^c_{fi,in,ex}}$, or $\mathbf{C^\infty Sch^c_{to,ex}}$, or in any category like those in (5.10) in which the objects are interior, but the morphisms need not be interior.

The next two examples are 'b-transverse', but not 'c-transverse', in the sense of Definition 6.31 below.

Example 6.29 Define manifolds with corners $X = Y = [0,\infty)$, $Z = [0,\infty)^2$, and interior maps $g : X\to Z$, $h : Y\to Z$ by $g(x) = (x,x)$ and $h(y) = (y,y^2)$. Let $\boldsymbol{X},\boldsymbol{Y},\boldsymbol{Z},\boldsymbol{g},\boldsymbol{h}$ be the images of $X,\ldots,h$ under $F^{\mathbf{C^\infty Sch^c_{in}}}_{\mathbf{Man^c_{in}}}$. Then $\boldsymbol{g}:\boldsymbol{X}\to\boldsymbol{Z}$, $\boldsymbol{h}:\boldsymbol{Y}\to\boldsymbol{Z}$ are morphisms in $\mathbf{C^\infty Sch^c_{to}}$, so a fibre product $\tilde{\boldsymbol{W}} = \boldsymbol{X}\times_{\boldsymbol{g},\boldsymbol{Z},\boldsymbol{h}}\boldsymbol{Y}$ exists in $\mathbf{C^\infty Sch^c_{to}}$ by Theorem 5.34.

Using the proof of Theorem 5.34 we can show that $\tilde{\boldsymbol{W}} = \mathrm{Spec^c_{in}}\,\mathfrak{C}$, where

$$\mathfrak{C} = \Pi^{\mathrm{sa}}_{\mathrm{in}}(\mathbb{R}[x,y]/[x=y,\ x=y^2]) \cong \mathbb{R}[x,y]/[x=y=1] \cong \boldsymbol{C}^\infty(*), \qquad (6.33)$$

so $\tilde{\boldsymbol{W}}$ is a single ordinary point, which maps to $x = 1$ in $\boldsymbol{X}$ and $y = 1$ in $\boldsymbol{Y}$. This is because on applying $\Pi^{\mathbb{Z}}_{\mathrm{in}}$ in Theorem 4.48, $x = y$, $x = y^2$ force $x = y = 1$.

In contrast, the fibre product $\hat{\boldsymbol{W}} = \boldsymbol{X}\times_{\boldsymbol{Z}}\boldsymbol{Y}$ in $\mathbf{C^\infty Sch^c_{fi,in}}$ is

$$\hat{\boldsymbol{W}} = \mathrm{Spec^c_{in}}(\mathbb{R}[x,y]/[x=y,\ x=y^2]),$$

which is two points, an ordinary point at $x = y = 1$, and another at $x = y = 0$ with a non-toric corner structure. Theorems 5.33–5.34 imply that $\hat{\boldsymbol{W}}$ is also the fibre product in $\mathbf{C^\infty Sch^c_{fi}}$.

For the fibre product $\boldsymbol{W} = \boldsymbol{X}\times_{\boldsymbol{Z}}\boldsymbol{Y}$ in $\mathbf{C^\infty Sch^c_{to,ex}}$, we have

$$C(\boldsymbol{X})\times_{C(\boldsymbol{Z})} C(\boldsymbol{Y}) \cong \{(\mathbf{1},\mathbf{1})\}\amalg\{(\mathbf{0},\mathbf{0})\} \qquad \text{in } \mathbf{C^\infty Sch^c_{to}},$$

which is two ordinary points, $(1,1)$ from $C_0(\boldsymbol{X})\times_{C_0(\boldsymbol{Z})} C_0(\boldsymbol{Y}) \cong \tilde{\boldsymbol{W}}$, and $(0,0)$ from $C_1(\boldsymbol{X})\times_{C_2(\boldsymbol{Z})} C_1(\boldsymbol{Y}) \cong *\times_* *$. (If we had formed the fibre product in $\mathbf{C^\infty Sch^c_{fi,in}}$ there would have been a third, non-toric point over $(0,0)$ in $C_0(\boldsymbol{X})\times_{C_0(\boldsymbol{Z})} C_0(\boldsymbol{Y}) \cong \hat{\boldsymbol{W}}$.) Also

$$\{(x,y)\in X\times Y : g(x) = h(y)\} = \{(1,1),(0,0)\}.$$

Thus $W' = C(X) \times_{C(Z)} C(Y)$ satisfies the condition in Theorem 6.27(c), and in fact $\boldsymbol{W} = C(\boldsymbol{X}) \times_{C(\boldsymbol{Z})} C(\boldsymbol{Y}) \cong \{(\mathbf{1},\mathbf{1})\} \amalg \{(\mathbf{0},\mathbf{0})\}$ is a fibre product in $\mathbf{C^\infty Sch^c_{to,ex}}$. Hence in this case fibre products $\boldsymbol{X} \times_{\boldsymbol{Z}} \boldsymbol{Y}$ exist in $\mathbf{C^\infty Sch^c_{to}}$ and $\mathbf{C^\infty Sch^c_{to,ex}}$, but are different, so inc : $\mathbf{C^\infty Sch^c_{to}} \hookrightarrow \mathbf{C^\infty Sch^c_{to,ex}}$ does not preserve limits, in contrast to the last part of Theorem 5.33. The same holds for inc : $\mathbf{C^\infty Sch^c_{fi,\mathbb{Z}}} \hookrightarrow \mathbf{C^\infty Sch^c_{fi,\mathbb{Z},ex}}$ and inc : $\mathbf{C^\infty Sch^c_{fi,tf}} \hookrightarrow \mathbf{C^\infty Sch^c_{fi,tf,ex}}$, as the fibre products $\tilde{\boldsymbol{W}}, \boldsymbol{W}$ are the same in these categories.

Example 6.30 Define manifolds with corners $X = [0,\infty) \times \mathbb{R}$, $Y = [0,\infty)$ and $Z = [0,\infty)^2$, and interior maps $g : X \to Z$, $h : Y \to Z$ by $g(x_1,x_2) = (x_1, x_1 e^{x_2})$ and $h(y) = (y,y)$. Let $\boldsymbol{X},\boldsymbol{Y},\boldsymbol{Z},\boldsymbol{g},\boldsymbol{h}$ be the images of $X,\ldots,h$ under $F_{\mathbf{Man^c}}^{\mathbf{C^\infty Sch^c}}$. Then $\boldsymbol{g} : \boldsymbol{X} \to \boldsymbol{Z}, \boldsymbol{h} : \boldsymbol{Y} \to \boldsymbol{Z}$ are morphisms in $\mathbf{C^\infty Sch^c_{to}}$, so a fibre product $\tilde{\boldsymbol{W}} = \boldsymbol{X} \times_{\boldsymbol{g},\boldsymbol{Z},\boldsymbol{h}} \boldsymbol{Y}$ exists in $\mathbf{C^\infty Sch^c_{to}}$ by Theorem 5.34. As for (6.33) we find that $\tilde{\boldsymbol{W}} = \mathrm{Spec}^c_{\mathrm{in}}\, \mathfrak{C}$, where

$$\begin{aligned}\mathfrak{C} &= \Pi^{\mathrm{sa}}_{\mathrm{in}}\big(\mathbb{R}[x_1,x_2,y]/[x_1 = y,\ x_1 e^{x_2} = y]\big) \cong \mathbb{R}[x_1,x_2,y]/[x_1 = y,\ x_2 = 0] \\ &\cong \mathbb{R}[y] \cong \boldsymbol{C^\infty}([0,\infty)),\end{aligned}$$

so $\tilde{\boldsymbol{W}} \cong [\mathbf{0},\boldsymbol{\infty})$. Similarly, we have a fibre product in $\mathbf{C^\infty Sch^c_{to}}$

$$C(\boldsymbol{X}) \times_{C(\boldsymbol{Z})} C(\boldsymbol{Y}) \cong \big\{(\boldsymbol{y},\mathbf{0},\boldsymbol{y}) : \boldsymbol{y} \in [\mathbf{0},\boldsymbol{\infty})\big\} \amalg \big\{(\mathbf{0},\boldsymbol{x_2},\mathbf{0}) : \boldsymbol{x_2} \in \mathbb{R}\big\},$$

where the first component is $C_0(\boldsymbol{X}) \times_{C_0(\boldsymbol{Z})} C_0(\boldsymbol{Y}) \cong \tilde{\boldsymbol{W}}$, and the second is $C_1(\boldsymbol{X}) \times_{C_2(\boldsymbol{Z})} C_1(\boldsymbol{Y})$. Also

$$\begin{aligned}\big\{(x,y) \in X \times Y : g(x) = h(y)\big\} = \big\{&(x_1,x_2,y) \in [0,\infty) \times \mathbb{R} \times [0,\infty) : \\ &\text{either } x_1 = y, x_2 = 0 \text{ or } x_1 = y = 0\big\}.\end{aligned}$$

Considering a neighbourhood of $(0,0,0)$ we see from Theorem 6.27(c) that no fibre product $\boldsymbol{W} = \boldsymbol{X} \times_{\boldsymbol{Z}} \boldsymbol{Y}$ exists in $\mathbf{C^\infty Sch^c_{to,ex}}$. Thus in this case a fibre product $\boldsymbol{X} \times_{\boldsymbol{Z}} \boldsymbol{Y}$ exists in $\mathbf{C^\infty Sch^c_{to}}$, but not in $\mathbf{C^\infty Sch^c_{to,ex}}$.

6.7 B- and c-transverse fibre products in $\mathbf{Man^{gc}_{in}}, \mathbf{C^\infty Sch^c_{to}}$

6.7.1 Results on b- and c-transverse fibre products from [46]

In [46, Def. 4.24] we define transversality for manifolds with g-corners.

Definition 6.31 Let $g : X \to Z$ and $h : Y \to Z$ be morphisms in $\mathbf{Man^{gc}_{in}}$ or $\check{\mathbf{M}}\mathbf{an^{gc}_{in}}$. Then we have the following.

(a) We call g,h *b-transverse* if ${}^bT_xg \oplus {}^bT_yh : {}^bT_xX \oplus {}^bT_yY \to {}^bT_zZ$ is surjective for all $x \in X$ and $y \in Y$ with $g(x) = h(y) = z \in Z$.

(b) We call g, h *c-transverse* if they are b-transverse, and for all $x \in X$ and $y \in Y$ with $g(x) = h(y) = z \in Z$, the linear map ${}^b\tilde{N}_x g \oplus {}^b\tilde{N}_y h : {}^b\tilde{N}_x X \oplus {}^b\tilde{N}_y Y \to {}^b\tilde{N}_z Z$ is surjective, and the submonoid

$$\{(\lambda,\mu)\in \tilde{M}_x X\times \tilde{M}_y Y : \tilde{M}_x g(\lambda)=\tilde{M}_y h(\mu) \text{ in } \tilde{M}_z Z\}\subseteq \tilde{M}_x X\times \tilde{M}_y Y$$

is not contained in any proper face $F \subsetneq \tilde{M}_x X \times \tilde{M}_y Y$ of $\tilde{M}_x X \times \tilde{M}_y Y$.

Here are the main results on b- and c-transversality [46, Ths. 4.26–4.28].

Theorem 6.32 *Let $g : X \to Z$ and $h : Y \to Z$ be b-transverse (or c-transverse) morphisms in $\mathbf{Man^{gc}_{in}}$. Then $C(g) : C(X) \to C(Z)$ and $C(h) : C(Y) \to C(Z)$ are also b-transverse (or c-transverse) morphisms in $\check{\mathbf{M}}\mathbf{an^{gc}_{in}}$.*

Theorem 6.33 *Let $g : X \to Z$ and $h : Y \to Z$ be b-transverse morphisms in $\mathbf{Man^{gc}_{in}}$. Then a fibre product $W = X \times_{g,Z,h} Y$ exists in $\mathbf{Man^{gc}_{in}}$, with* $\dim W = \dim X + \dim Y - \dim Z$. *Explicitly, we may write*

$$W^\circ = \{(x,y) \in X^\circ \times Y^\circ : g(x) = h(y) \text{ in } Z^\circ\},$$

and take W to be the closure $\overline{W^\circ}$ of W° in $X\times Y$, and then W is an embedded submanifold of $X\times Y$, and $e : W \to X$ and $f : W \to Y$ act by $e : (x,y) \mapsto x$ and $f : (x,y) \mapsto y$.

Theorem 6.34 *Suppose $g : X \to Z$ and $h : Y \to Z$ are c-transverse morphisms in $\mathbf{Man^{gc}_{in}}$. Then a fibre product $W = X\times_{g,Z,h}Y$ exists in $\mathbf{Man^{gc}}$, with* $\dim W = \dim X + \dim Y - \dim Z$. *Explicitly, we may write*

$$W = \{(x,y) \in X \times Y : g(x) = h(y) \text{ in } Z\},$$

and then W is an embedded submanifold of $X \times Y$, and $e : W \to X$ and $f : W \to Y$ act by $e : (x,y) \mapsto x$ and $f : (x,y) \mapsto y$. This W is also a fibre product in $\mathbf{Man^{gc}_{in}}$, and agrees with that in Theorem 6.33.

Furthermore, the following is Cartesian in both $\check{\mathbf{M}}\mathbf{an^{gc}}$ and $\check{\mathbf{M}}\mathbf{an^{gc}_{in}}$:

$$\begin{array}{ccc} C(W) & \xrightarrow{\quad C(f) \quad} & C(Y) \\ \big\downarrow{\scriptstyle C(e)} & & {\scriptstyle C(h)}\big\downarrow \\ C(X) & \xrightarrow{\quad C(g) \quad} & C(Z). \end{array} \tag{6.34}$$

Equation (6.34) *has a grading-preserving property, in that if $(w,\beta) \in C_i(W)$ with $C(e)(w,\beta) = (x,\gamma) \in C_j(X)$, and $C(f)(w,\beta) = (y,\delta) \in C_k(Y)$, and $C(g)(x,\gamma) = C(h)(y,\delta) = (z,\epsilon) \in C_l(Z)$, then $i + l = j + k$. Hence*

$$C_i(W) \cong \textstyle\coprod_{j,k,l\geqslant 0: i=j+k-l} C^l_j(X) \times_{C(g)|\dots,C_l(Z),C(h)|\dots} C^l_k(Y),$$

writing $C^l_j(X) = C_j(X) \cap C(g)^{-1}(C_l(Z))$ and $C^l_k(Y) = C_k(Y) \cap C(h)^{-1}$

$(C_l(Z))$, which are open and closed in $C_j(X), C_k(Y)$. When $i = 1$, this gives a formula for ∂W.

6.7.2 $F_{\mathbf{Man^{gc}_{in}}}^{\mathbf{C^\infty Sch^c_{to}}}, F_{\mathbf{Man^{gc}}}^{\mathbf{C^\infty Sch^c_{to,ex}}}$ preserve b-, c-transverse fibre products

We generalize the last part of Theorem 2.54 to the corners case.

Theorem 6.35 **(a)** *The functor $F_{\mathbf{Man^{gc}_{in}}}^{\mathbf{C^\infty Sch^c_{to}}} : \mathbf{Man^{gc}_{in}} \to \mathbf{C^\infty Sch^c_{to}}$ takes b- and c-transverse fibre products in $\mathbf{Man^{gc}_{in}}$ to fibre products in $\mathbf{C^\infty Sch^c_{to}}$.*

(b) *The functor $F_{\mathbf{Man^{gc}}}^{\mathbf{C^\infty Sch^c_{to,ex}}} : \mathbf{Man^{gc}} \to \mathbf{C^\infty Sch^c_{to,ex}}$ takes c-transverse fibre products in $\mathbf{Man^{gc}}$ to fibre products in $\mathbf{C^\infty Sch^c_{to,ex}}$.*

Proof For (a), suppose $g : X \to Z$ and $h : Y \to Z$ are b-transverse in $\mathbf{Man^{gc}_{in}}$, and let $W = X \times_{g,Z,h} Y$ be the fibre product in $\mathbf{Man^{gc}_{in}}$ given by Theorem 6.33, with projections $e : W \to X$ and $f : W \to Y$. Write $\boldsymbol{W}, \ldots, \boldsymbol{h}$ for the images of $W, \ldots, h$ under $F_{\mathbf{Man^{gc}_{in}}}^{\mathbf{C^\infty Sch^c_{to}}}$. Let $\tilde{\boldsymbol{W}} = \boldsymbol{X} \times_{\boldsymbol{g},\boldsymbol{Z},\boldsymbol{h}} \boldsymbol{Y}$ be the fibre product in $\mathbf{C^\infty Sch^c_{to}}$, with projections $\tilde{\boldsymbol{e}} : \tilde{\boldsymbol{W}} \to \boldsymbol{X}$, $\tilde{\boldsymbol{f}} : \tilde{\boldsymbol{W}} \to \boldsymbol{Y}$, which exists by Theorem 5.34. As $\boldsymbol{g} \circ \boldsymbol{e} = \boldsymbol{h} \circ \boldsymbol{f}$, the universal property of $\tilde{\boldsymbol{W}}$ gives a unique morphism $\boldsymbol{b} : \boldsymbol{W} \to \tilde{\boldsymbol{W}}$ with $\boldsymbol{e} = \tilde{\boldsymbol{e}} \circ \boldsymbol{b}$ and $\boldsymbol{f} = \tilde{\boldsymbol{f}} \circ \boldsymbol{b}$. We must prove $\boldsymbol{b}$ is an isomorphism.

First note that as morphisms $\boldsymbol{b} : \boldsymbol{W} \to \tilde{\boldsymbol{W}}$ form a sheaf, this question is local in $\tilde{\boldsymbol{W}}$ and hence in $\boldsymbol{X}, \boldsymbol{Y}, \boldsymbol{Z}$, so it is enough to prove it over small open neighbourhoods of points $x \in \boldsymbol{X}, y \in \boldsymbol{Y}, z \in \boldsymbol{Z}$ with $\boldsymbol{g}(x) = \boldsymbol{h}(y) = z$. Thus we may replace X, Y, Z by small open neighbourhoods of x, y, z in X, Y, Z, or equivalently, by small open neighbourhoods of $(\delta_0, 0)$ in the local models $X_Q \times \mathbb{R}^l$ for manifolds with g-corners, where Q is a toric monoid, as in Remark 3.10. This also replaces W by a small open neighbourhood of $w = (x, y)$ in W. As small open balls about $(\delta_0, 0)$ in $X_Q \times \mathbb{R}^l$ are diffeomorphic to $X_Q \times \mathbb{R}^l$, this means we can suppose that $W, \ldots, Z$ are of the form $X_Q \times \mathbb{R}^l$. So we can take

$$W \cong X_P \times \mathbb{R}^k, \quad X \cong X_Q \times \mathbb{R}^l, \quad Y \cong X_R \times \mathbb{R}^m, \quad Z \cong X_S \times \mathbb{R}^n, \tag{6.35}$$

for toric monoids P, Q, R, S and $k, l, m, n \geqslant 0$, where $e, f, g, h : (\delta_0, 0) \mapsto (\delta_0, 0)$. Then W, X, Y, Z are manifolds with g-faces, so by Theorem 5.22(a) we may take $\boldsymbol{W} = \operatorname{Spec}^{\mathrm{c}}_{\mathrm{in}} \boldsymbol{C}^\infty_{\mathrm{in}}(W)$, and similarly for $\boldsymbol{X}, \boldsymbol{Y}, \boldsymbol{Z}$. Then $\boldsymbol{e} = \operatorname{Spec}^{\mathrm{c}}_{\mathrm{in}}(e^*, e^*_{\mathrm{ex}})$, where $e^* : C^\infty(X) \to C^\infty(W)$, $e^*_{\mathrm{ex}} : \mathrm{In}(X) \amalg \{0\} \to \mathrm{In}(W) \amalg \{0\}$ are the pullbacks, and similarly for $\boldsymbol{f}, \boldsymbol{g}, \boldsymbol{h}$. Form a commutative diagram in $\mathbf{C^\infty Rings^c_{to}}$:

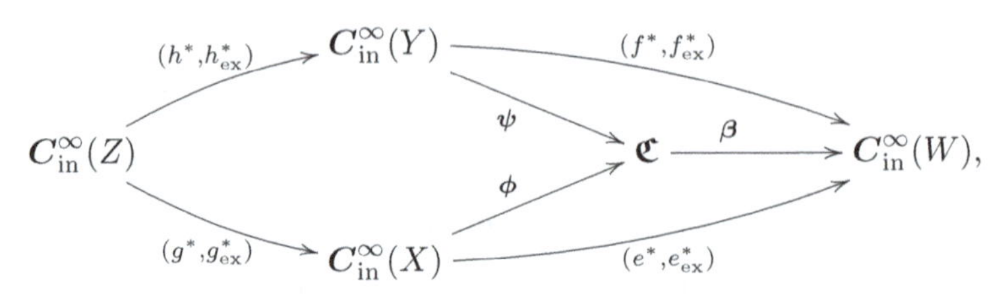

where $\mathfrak{C}$ is the pushout $C^\infty_{\rm in}(X)\amalg_{C^\infty_{\rm in}(Z)} C^\infty_{\rm in}(Y)$ in $\mathbf{C^\infty Rings^c_{to}}$, which exists by Proposition 4.50, and ϕ, ψ are the projections to the pushout, and β exists by the universal property of pushouts. Since $\mathrm{Spec}^{\rm c}_{\rm in}: (\mathbf{C^\infty Rings^c_{to}})^{\rm op} \to \mathbf{C^\infty Sch^c_{to}}$ preserves limits, it follows that $\tilde{\boldsymbol{W}} = \mathrm{Spec}^{\rm c}_{\rm in}\,\mathfrak{C}$ and $\boldsymbol{b} = \mathrm{Spec}^{\rm c}_{\rm in}\,\boldsymbol{\beta}$, and it is enough to prove that $\boldsymbol{\beta}$ is an isomorphism.

The pushout $\mathfrak{C}$ is equivalent to a coequalizer diagram in $\mathbf{C^\infty Rings^c_{to}}$:

$$C^\infty_{\rm in}(Z) \underset{(h^*,h^*_{\rm ex})\circ \boldsymbol{i}_{C^\infty_{\rm in}(Y)}}{\overset{(g^*,g^*_{\rm ex})\circ \boldsymbol{i}_{C^\infty_{\rm in}(X)}}{\rightrightarrows}} C^\infty_{\rm in}(X)\otimes^{\rm to}_\infty C^\infty_{\rm in}(Y) \longrightarrow \mathfrak{C} \tag{6.36}$$

for $C^\infty_{\rm in}(X)\otimes^{\rm to}_\infty C^\infty_{\rm in}(Y)$ the coproduct in $\mathbf{C^\infty Rings^c_{to}}$. Now one can show that the coproduct $C^\infty_{\rm in}(X)\otimes_\infty C^\infty_{\rm in}(Y)$ in $\mathbf{C^\infty Rings^c}$ is $C^\infty_{\rm in}(X\times Y)$. Since this is toric, it coincides with the coproduct in $\mathbf{C^\infty Rings^c_{to}}$. Thus we may replace (6.36) by the coequalizer diagram in $\mathbf{C^\infty Rings^c_{to}}$:

$$C^\infty_{\rm in}(Z) \underset{((h\circ\pi_Y)^*,(h\circ\pi_Y)^*_{\rm ex})}{\overset{((g\circ\pi_X)^*,(g\circ\pi_X)^*_{\rm ex})}{\rightrightarrows}} C^\infty_{\rm in}(X\times Y) \longrightarrow \mathfrak{C}. \tag{6.37}$$

We deduce that

$$\mathfrak{C} \cong \Pi^{\rm sa}_{\rm in}\big(C^\infty_{\rm in}(X\times Y)/\big[(g\circ\pi_X)^*(c')=(h\circ\pi_Y)^*(c'),\ c'\in \mathrm{In}(Z)\big]\big). \tag{6.38}$$

Here the term $C^\infty_{\rm in}(X\times Y)/[\cdots]$ is the coequalizer of (6.37) in $\mathbf{C^\infty Rings^c}$. There is no need to also impose $\mathbb{R}$-type relations $(g\circ\pi_X)^*(c) = (h\circ\pi_Y)^*(c)$ for $x \in C^\infty(Z)$, as these are implied by $(g\circ\pi_X)^*(c') = (h\circ\pi_Y)^*(c')$ for $c' = \Psi_{\rm exp}(c)$ in $\mathrm{In}(X)$. The proof of Theorem 4.20(b) then implies that $\Pi^{\rm sa}_{\rm in}(\cdots)$ is the coequalizer of (6.37) in $\mathbf{C^\infty Rings^c_{to}}$.

In (6.38), we think of $\Pi^{\rm sa}_{\rm in}$ as mapping $\mathbf{C^\infty Rings^c_{fi,in}} \to \mathbf{C^\infty Rings^c_{to}}$. It does not change $C^\infty_{\rm in}(X\times Y)$ as this is already toric, but it can add extra relations. From the construction of $\Pi^{\rm sa}_{\rm in}$ in the proof of Theorem 4.48, we can prove that

$$\begin{aligned}\mathfrak{C} \cong C^\infty_{\rm in}(X\times Y)/\big[&c_1' = c_2' \colon c_1', c_2' \in \mathrm{In}(X\times Y), \text{ for some } c'\in \mathrm{In}(Z)\\ &\text{and } n \geqslant 1 \text{ we have } c_1'(g\circ\pi_X)^*(c')^n = c_2'(h\circ\pi_Y)^*(c')^n\big].\end{aligned} \tag{6.39}$$

Here imposing relations $c'_1 = c'_2$ if $c'_1(g\circ\pi_X)^*(c') = c'_2(h\circ\pi_Y)^*(c')$ for $c'\in \mathrm{In}(Z)$ has the effect of making (6.39) integral, that is, it applies $\Pi^{\mathbb{Z}}_{\mathrm{in}}$. (Note that taking $c'_1 = (h\circ\pi_Y)^*(c')$ and $c'_2 = (g\circ\pi_X)^*(c')$ recovers the relations in (6.38), but $c'_1(g\circ\pi_X)^*(c') = c'_2(h\circ\pi_Y)^*(c')$ also includes relations in the integral completion.) Then forcing $c'_1 = c'_2$ when $c'_1(g\circ\pi_X)^*(c')^n = c'_2(h\circ\pi_Y)^*(c')^n$ for $n\geqslant 1$ also makes (6.39) torsion-free and saturated, that is, it applies $\Pi^{\mathrm{sa}}_{\mathrm{tf}}\circ\Pi^{\mathrm{tf}}_{\mathbb{Z}}$, so overall we apply $\Pi^{\mathrm{sa}}_{\mathrm{in}} = \Pi^{\mathrm{sa}}_{\mathrm{tf}}\circ\Pi^{\mathrm{tf}}_{\mathbb{Z}}\circ\Pi^{\mathbb{Z}}_{\mathrm{in}}$.

Using (6.39), we now show that $\boldsymbol{\beta} = (\beta,\beta_{\mathrm{ex}}) : \mathfrak{C}\to \boldsymbol{C^\infty_{\mathrm{in}}}(W)$ is an isomorphism, where W is an embedded submanifold in $X\times Y$ as in Theorem 6.33. To show that $\beta,\beta_{\mathrm{ex}}$ are surjective, it is enough to show that the restriction maps $C^\infty(X\times Y)\to C^\infty(W)$, $\mathrm{In}(X\times Y)\to \mathrm{In}(W)$ are surjective. That is, all smooth or interior maps $W\to\mathbb{R}$, $[0,\infty)$ should extend to $X\times Y$. Locally near (x,y) in $W, X\times Y$ this follows from extension properties of smooth maps on $X_P\times\mathbb{R}^l$ in [46, Prop. 3.14], and globally it follows from (6.35), which ensures there are no obstructions from multiplicity functions $\mu_{c'}$ as in Remark 3.15.

To show β (or β_{ex}) is injective, we need to show that if $c_1,c_2\in C^\infty(X\times Y)$ (or $c'_1,c'_2\in \mathrm{In}(X\times Y)$), then $c_1|_W = c_2|_W$ (or $c'_1|_W = c'_2|_W$) if and only if $c_1 - c_2$ lies in the ideal I in $C^\infty(X\times Y)$ generated by the relations in (6.39) (or $c'_1\sim c'_2$, where $\sim$ is the equivalence relation on $\mathrm{In}(X\times Y)$ generated by the relations in (6.39)). Since W is an embedded submanifold defined as a subset of $X\times Y$ by the b-transverse relations $(g\circ\pi_X)^*(c') = (h\circ\pi_Y)^*(c')$ for $c'\in\mathrm{In}(Z)$, this is indeed true, though proving it carefully requires a little work on ideals in $C^\infty(X\times Y)$. Therefore $\boldsymbol{\beta}$ is an isomorphism, completing part (a).

For (b), suppose $g: X\to Z$ and $h: Y\to Z$ are c-transverse in $\mathbf{Man^{gc}_{in}}$, which implies g,h b-transverse, and let $W = X\times_{g,Z,h} Y$ be the fibre product in both $\mathbf{Man^{gc}_{in}}$ and $\mathbf{Man^{gc}}$ given by Theorems 6.33–6.34. Then (6.34) is Cartesian in $\mathbf{Man^{gc}_{in}}$ by Theorem 6.34. Write $\boldsymbol{W},\ldots,\boldsymbol{h}$ for the images of $W,\ldots,h$ under $F^{\mathbf{C^\infty Sch^c_{to}}}_{\mathbf{Man^{gc}_{in}}}$. Then part (a), Proposition 6.12, and Theorem 6.16 imply that $\boldsymbol{W} = \boldsymbol{X}\times_{\boldsymbol{Z}}\boldsymbol{Y}$ in $\mathbf{C^\infty Sch^c_{to}}$, and (6.32) is Cartesian in $\mathbf{C^\infty Sch^c_{to}}$. Hence Theorem 6.27(d) says $\boldsymbol{W} = \boldsymbol{X}\times_{\boldsymbol{Z}}\boldsymbol{Y}$ in $\mathbf{C^\infty Sch^c_{to,ex}}$, and the theorem follows. □

Example 6.36 Let $X = Y = Z = [0,\infty)$, and $g: X\to Z$, $h: Y\to Z$ map $g(x) = x^2$, $h(y) = y^2$. Then g,h are b- and c-transverse, and the fibre product $W = X\times_{g,Z,h} Y$ in $\mathbf{Man^{gc}_{in}}$ and $\mathbf{Man^{gc}}$ is $W = [0,\infty)$ with projections $e: W\to X$, $f: W\to Y$ mapping $e(w) = w$, $f(w) = w$.

Let $\boldsymbol{W},\ldots,\boldsymbol{h}$ be the images of $W,\ldots,h$ under $F^{\mathbf{C^\infty Sch^c}}_{\mathbf{Man^c}}$. Then Theorem 6.35 implies that $\boldsymbol{W} = \boldsymbol{X}\times_{\boldsymbol{g},\boldsymbol{Z},\boldsymbol{h}}\boldsymbol{Y}$ in $\mathbf{C^\infty Sch^c_{to}}$ and $\mathbf{C^\infty Sch^c_{to,ex}}$. But what about fibre products $\boldsymbol{X}\times_{\boldsymbol{g},\boldsymbol{Z},\boldsymbol{h}}\boldsymbol{Y}$ in other categories?

As $\boldsymbol{X}, \boldsymbol{Y}, \boldsymbol{Z} = \mathrm{Spec}^{\mathrm{c}}_{\mathrm{in}}\, C^\infty_{\mathrm{in}}([0,\infty))$ are affine, we can show that the fibre product $\hat{\boldsymbol{W}} = \boldsymbol{X} \times_{\boldsymbol{Z}} \boldsymbol{Y}$ in $\mathbf{C^\infty Sch^c_{fi,in}}$ is $\mathrm{Spec}^{\mathrm{c}}_{\mathrm{in}}\, \mathfrak{C}$, where $\mathfrak{C} = \mathbb{R}[x,y]/[x^2 = y^2]$. Then $\mathfrak{C}^\sharp_{\mathrm{in}} \cong \mathbb{Z} \times \mathbb{Z}_2$, identifying $[x] = (1,0)$ and $[y] = (1,1)$ in $\mathbb{Z} \times \mathbb{Z}_2$. We see that $\mathfrak{C}$ is firm and integral, but not torsion-free (as $\mathfrak{C}^\sharp_{\mathrm{in}}$ has torsion $\mathbb{Z}_2$) or saturated (as $x, y \in \mathfrak{C}_{\mathrm{in}}$ with $x \neq y$ but $x^2 = y^2$). Thus $\hat{\boldsymbol{W}}$ is also firm and integral, but not torsion-free or saturated, and $\boldsymbol{W} \not\cong \hat{\boldsymbol{W}}$.

This shows $F^{\mathbf{C^\infty Sch^c_{\mathbb{Z}}}}_{\mathbf{Man^{gc}_{in}}} : \mathbf{Man^{gc}_{in}} \to \mathbf{C^\infty Sch^c_{\mathbb{Z}}}$ and $F^{\mathbf{C^\infty Sch^c_{fi,in}}}_{\mathbf{Man^{gc}_{in}}} : \mathbf{Man^{gc}_{in}} \to \mathbf{C^\infty Sch^c_{fi,in}}$ need not preserve b-transverse fibre products. Example 6.29 above is similar, but $\hat{\boldsymbol{W}}$ is not integral. One conclusion is that in Theorem 6.35 it is essential to work in the categories $\mathbf{C^\infty Sch^c_{to}}, \mathbf{C^\infty Sch^c_{to,ex}}$ of *toric* C^∞-schemes with corners, rather than larger categories such as $\mathbf{C^\infty Sch^c_{fi,tf}}$, $\mathbf{C^\infty Sch^c_{fi,\mathbb{Z}}}$, $\mathbf{C^\infty Sch^c_{fi,in}}$ from §5.6–§5.7. Thus, $\mathbf{C^\infty Sch^c_{to}}$, $\mathbf{C^\infty Sch^c_{to,ex}}$ may be regarded as natural generalizations of $\mathbf{Man^{gc}_{in}}, \mathbf{Man^{gc}}$.

6.7.3 Restriction to manifolds with corners

There is an easy way to restrict Theorem 6.35 to $\mathbf{Man^c_{in}}, \mathbf{Man^c}$.

Definition 6.37 Let $g : h \to Z$ and $h : Y \to Z$ be morphisms in $\mathbf{Man^c_{in}}$. Following the second author [48, §2.5.4], we call g, h *sb-transverse* or *sc-transverse* (short for *strictly b-transverse* or *strictly c-transverse*) if they are b-transverse or c-transverse in $\mathbf{Man^{gc}_{in}}$, respectively, and the fibre product $W = X \times_{g,Z,h} Y$ in $\mathbf{Man^{gc}_{in}}$ is a manifold with corners, not just g-corners. In [48, §2.5.4] we explain the sb- and sc-transverse conditions explicitly. Theorems 6.33–6.34 imply that sb- and sc-transverse fibre products exist in $\mathbf{Man^c_{in}}$ and $\mathbf{Man^c}$, respectively.

Then Theorem 6.35 implies the following.

Corollary 6.38 **(a)** *The functor* $F^{\mathbf{C^\infty Sch^c_{to}}}_{\mathbf{Man^c_{in}}} : \mathbf{Man^c_{in}} \to \mathbf{C^\infty Sch^c_{to}}$ *takes sb- and sc-transverse fibre products in* $\mathbf{Man^c_{in}}$ *to fibre products in* $\mathbf{C^\infty Sch^c_{to}}$.

(b) *The functor* $F^{\mathbf{C^\infty Sch^c_{to,ex}}}_{\mathbf{Man^c}} : \mathbf{Man^c} \to \mathbf{C^\infty Sch^c_{to,ex}}$ *takes sc-transverse fibre products in* $\mathbf{Man^c}$ *to fibre products in* $\mathbf{C^\infty Sch^c_{to,ex}}$.

7

Modules, and sheaves of modules

In this chapter we discuss modules over C^∞-rings with corners, sheaves of $\mathcal{O}_X$-modules on C^∞-schemes with corners $\boldsymbol{X}$, and (b-)cotangent modules and sheaves. This extends the corresponding concepts for C^∞-schemes from §2.3, §2.7 and §2.8.

7.1 Modules over C^∞-rings with corners, (b-)cotangent modules

In §2.3 we discussed modules over C^∞-rings. Here is the corners analogue.

Definition 7.1 Let $\boldsymbol{\mathfrak{C}} = (\mathfrak{C}, \mathfrak{C}_{\rm ex})$ be a C^∞-ring with corners. A *module M over $\boldsymbol{\mathfrak{C}}$*, or *$\boldsymbol{\mathfrak{C}}$-module*, is a module over $\mathfrak{C}$ regarded as a commutative $\mathbb{R}$-algebra as in Definition 2.27, and morphisms of $\boldsymbol{\mathfrak{C}}$-modules are morphisms of $\mathbb{R}$-algebra modules. Then $\boldsymbol{\mathfrak{C}}$-modules form an abelian category, which we write as $\boldsymbol{\mathfrak{C}}$-mod.

The basic theory of §2.3 extends trivially to the corners case. So if $\boldsymbol{\phi} = (\phi, \phi_{\rm ex}) : \boldsymbol{\mathfrak{C}} \to \boldsymbol{\mathfrak{D}}$ is a morphism in $\mathbf{C^\infty Rings^c}$ then using $\phi : \mathfrak{C} \to \mathfrak{D}$ we get functors $\boldsymbol{\phi}_* : \boldsymbol{\mathfrak{C}}\text{-mod} \to \boldsymbol{\mathfrak{D}}\text{-mod}$ mapping $M \mapsto M \otimes_{\mathfrak{C}} \mathfrak{D}$, and $\boldsymbol{\phi}^* : \boldsymbol{\mathfrak{D}}\text{-mod} \to \boldsymbol{\mathfrak{C}}\text{-mod}$ mapping $N \mapsto N$.

One might expect that modules over $\boldsymbol{\mathfrak{C}} = (\mathfrak{C}, \mathfrak{C}_{\rm ex})$ should also include some kind of monoid module over $\mathfrak{C}_{\rm ex}$, but we do not do this.

Example 7.2 Let X be a manifold with corners (or g-corners) and let $E \to X$ be a vector bundle, and write $\Gamma^\infty(E)$ for the vector space of smooth sections e of E. This is a module over $C^\infty(X)$, and hence over the C^∞-rings with corners $\boldsymbol{C}^\infty(X)$ and $\boldsymbol{C}^\infty_{\rm in}(X)$ from Example 4.18.

Section 2.3 discussed cotangent modules of C^∞-rings, the analogues of cotangent bundles of manifolds. As in §3.5 manifolds with corners X have cotangent bundles T^*X which are functorial over smooth maps, and b-cotangent bundles ${}^bT^*X$, which are functorial only over interior maps. In a similar way, for a C^∞-ring with corners $\boldsymbol{\mathfrak{C}}$ we will define the *cotangent module* $\Omega_{\boldsymbol{\mathfrak{C}}}$, and if $\boldsymbol{\mathfrak{C}}$ is interior we will also define the *b-cotangent module* ${}^b\Omega_{\boldsymbol{\mathfrak{C}}}$.

Definition 7.3 Let $\boldsymbol{\mathfrak{C}} = (\mathfrak{C}, \mathfrak{C}_{\rm ex})$ be a C^∞-ring with corners. Define the *cotangent module* $\Omega_{\boldsymbol{\mathfrak{C}}}$ of $\boldsymbol{\mathfrak{C}}$ to be the cotangent module $\Omega_{\mathfrak{C}}$ of §2.3, regarded as a $\boldsymbol{\mathfrak{C}}$-module. If $\boldsymbol{\phi} = (\phi, \phi_{\rm ex}) : \boldsymbol{\mathfrak{C}} \to \boldsymbol{\mathfrak{D}}$ is a morphism in $\mathbf{C^\infty Rings^c}$ then from Ω_ϕ in §2.3 we get functorial morphisms $\Omega_{\boldsymbol{\phi}} : \Omega_{\boldsymbol{\mathfrak{C}}} \to \Omega_{\boldsymbol{\mathfrak{D}}} = \boldsymbol{\phi}^*(\Omega_{\boldsymbol{\mathfrak{D}}})$ in $\boldsymbol{\mathfrak{C}}$-mod and $(\Omega_{\boldsymbol{\phi}})_* : \boldsymbol{\phi}_*(\Omega_{\boldsymbol{\mathfrak{C}}}) = \Omega_{\boldsymbol{\mathfrak{C}}} \otimes_{\mathfrak{C}} \mathfrak{D} \to \Omega_{\boldsymbol{\mathfrak{D}}}$ in $\boldsymbol{\mathfrak{D}}$-mod.

Definition 7.4 Let $\boldsymbol{\mathfrak{C}} = (\mathfrak{C}, \mathfrak{C}_{\rm ex})$ be an interior C^∞-ring with corners, so that $\mathfrak{C}_{\rm ex} = \mathfrak{C}_{\rm in} \amalg \{0_{\mathfrak{C}_{\rm ex}}\}$ with $\mathfrak{C}_{\rm in}$ a monoid. Let M be a $\boldsymbol{\mathfrak{C}}$-module. A *b-derivation* is a map $\mathrm{d_{in}} : \mathfrak{C}_{\rm in} \to M$ satisfying the following corners analogue of (2.5): for any interior $g : [0,\infty)^n \to [0,\infty)$ and elements $c'_1, \ldots, c'_n \in \mathfrak{C}_{\rm in}$, then

$$\mathrm{d_{in}}\bigl(\Psi_g(c'_1, \ldots, c'_n)\bigr) - \sum_{i=1}^{n} \Phi_{g^{-1}x_i \frac{\partial g}{\partial x_i}}(c'_1, \ldots, c'_n) \cdot \mathrm{d_{in}} c'_i = 0. \tag{7.1}$$

Here g interior implies that $g^{-1}x_i \frac{\partial g}{\partial x_i} : [0,\infty)^n \to \mathbb{R}$ is smooth.

It is easy to show that (7.1) implies the following.

(i) $\mathrm{d_{in}} : \mathfrak{C}_{\rm in} \to M$ is a monoid morphism, where M is a monoid over addition.

(ii) $\mathrm{d} = \mathrm{d_{in}} \circ \Psi_{\exp} : \mathfrak{C} \to M$ is a C^∞-derivation in the sense of Definition 2.29.

(iii) $\mathrm{d_{in}} \circ \Psi_{\exp}(\Phi_i(c')) = \Phi_i(c') \cdot \mathrm{d_{in}} c'$ for all $c' \in \mathfrak{C}_{\rm in}$.

As interior $g : [0,\infty)^n \to [0,\infty)$ are of the form $g = x_1^{a_1} \cdots x_n^{a_n} \exp f(x_1, \ldots, x_n)$, with $\Psi_g(c'_1, \ldots, c'_n) = c_1'^{a_1} \cdots c_n'^{a_n} \Psi_{\exp}(\Phi_f(c'_1, \ldots, c'_n))$ for $c'_1, \ldots, c'_n \in \mathfrak{C}_{\rm in}$, we can show that (i)–(iii) are equivalent to (7.1).

We call such a pair $(M, \mathrm{d_{in}})$ a *b-cotangent module* for $\boldsymbol{\mathfrak{C}}$ if it has the universal property that for any b-derivation $\mathrm{d}'_{\rm in} : \mathfrak{C}_{\rm in} \to M'$, there exists a unique morphism of $\boldsymbol{\mathfrak{C}}$-modules ${}^b\lambda : M \to M'$ with $\mathrm{d}'_{\rm in} = {}^b\lambda \circ \mathrm{d_{in}}$.

There is a natural construction for a b-cotangent module: we take M to be the quotient of the free $\boldsymbol{\mathfrak{C}}$-module with basis of symbols $\mathrm{d_{in}}c'$ for $c' \in \mathfrak{C}_{\rm in}$ by the $\mathfrak{C}$-submodule spanned by all expressions of the form (7.1). Thus b-cotangent modules exist, and are unique up to unique isomorphism. When we speak of 'the' b-cotangent module, we mean that constructed earlier, and we write it as $\mathrm{d}_{\boldsymbol{\mathfrak{C}},\rm in} : \mathfrak{C}_{\rm in} \to {}^b\Omega_{\boldsymbol{\mathfrak{C}}}$.

Since $\mathrm{d}_{\boldsymbol{\mathfrak{C}},\rm in} \circ \Psi_{\exp} : \mathfrak{C} \to {}^b\Omega_{\boldsymbol{\mathfrak{C}}}$ is a C^∞-derivation, the universal property of $\Omega_{\mathfrak{C}} = \Omega_{\boldsymbol{\mathfrak{C}}}$ in §2.3 implies that there is a unique $\boldsymbol{\mathfrak{C}}$-module morphism $I_{\boldsymbol{\mathfrak{C}}} : \Omega_{\boldsymbol{\mathfrak{C}}} \to {}^b\Omega_{\boldsymbol{\mathfrak{C}}}$ with $\mathrm{d}_{\boldsymbol{\mathfrak{C}},\mathrm{in}} \circ \Psi_{\exp} = I_{\boldsymbol{\mathfrak{C}}} \circ \mathrm{d}_{\mathfrak{C}} : \mathfrak{C} \to {}^b\Omega_{\boldsymbol{\mathfrak{C}}}$.

Let $\boldsymbol{\phi} : \boldsymbol{\mathfrak{C}} \to \boldsymbol{\mathfrak{D}}$ be a morphism in $\mathbf{C^\infty Rings^c_{in}}$. Then we have a monoid morphism $\phi_{\mathrm{in}} : \mathfrak{C}_{\mathrm{in}} \to \mathfrak{D}_{\mathrm{in}}$. Regarding ${}^b\Omega_{\boldsymbol{\mathfrak{D}}}$ as a $\boldsymbol{\mathfrak{C}}$-module using $\phi : \mathfrak{C} \to \mathfrak{D}$, then $\mathrm{d}_{\boldsymbol{\mathfrak{D}},\mathrm{in}} \circ \phi_{\mathrm{in}} : \mathfrak{C}_{\mathrm{in}} \to {}^b\Omega_{\boldsymbol{\mathfrak{D}}}$ becomes a b-derivation. Thus by the universal property of ${}^b\Omega_{\boldsymbol{\mathfrak{C}}}$, there exists a unique $\boldsymbol{\mathfrak{C}}$-module morphism ${}^b\Omega_{\boldsymbol{\phi}} : {}^b\Omega_{\boldsymbol{\mathfrak{C}}} \to {}^b\Omega_{\boldsymbol{\mathfrak{D}}}$ with $\mathrm{d}_{\boldsymbol{\mathfrak{D}},\mathrm{in}} \circ \phi_{\mathrm{in}} = {}^b\Omega_{\boldsymbol{\phi}} \circ \mathrm{d}_{\boldsymbol{\mathfrak{C}},\mathrm{in}}$. This then induces a morphism of $\boldsymbol{\mathfrak{D}}$-modules $({}^b\Omega_{\boldsymbol{\phi}})_* : \boldsymbol{\phi}_*({}^b\Omega_{\boldsymbol{\mathfrak{C}}}) = {}^b\Omega_{\boldsymbol{\mathfrak{C}}} \otimes_{\mathfrak{C}} \mathfrak{D} \to {}^b\Omega_{\boldsymbol{\mathfrak{D}}}$. If $\boldsymbol{\phi} : \boldsymbol{\mathfrak{C}} \to \boldsymbol{\mathfrak{D}}$, $\boldsymbol{\psi} : \boldsymbol{\mathfrak{D}} \to \boldsymbol{\mathfrak{E}}$ are morphisms in $\mathbf{C^\infty Rings^c_{in}}$ then ${}^b\Omega_{\boldsymbol{\psi}\circ\boldsymbol{\phi}} = {}^b\Omega_{\boldsymbol{\psi}} \circ {}^b\Omega_{\boldsymbol{\phi}} : {}^b\Omega_{\boldsymbol{\mathfrak{C}}} \to {}^b\Omega_{\boldsymbol{\mathfrak{E}}}$.

Remark 7.5 **(a)** We may think of $\mathrm{d}_{\mathrm{in}}c'$ above as $\mathrm{d}(\log c')$. As in Example 7.6(b), if X is a manifold with corners and $c' : X \to [0,\infty)$ is interior, then $\mathrm{d}(\log c')$ is a well-defined section of ${}^bT^*X$. This holds even at ∂X where we allow $c' = 0$ and $\log c'$ is undefined, as ${}^bT^*X$ is spanned by $x_1^{-1}\mathrm{d}x_1, \ldots, x_k^{-1}\mathrm{d}x_k, \mathrm{d}x_{k+1}, \ldots, \mathrm{d}x_m$ in local coordinates $(x_1, \ldots, x_m) \in \mathbb{R}^m_k$, and the factors $x_1^{-1}, \ldots, x_k^{-1}$ control the singularities of $\mathrm{d}\log c'$.

(b) In Definition 7.4 we could have omitted the condition that $\boldsymbol{\mathfrak{C}}$ be interior, and considered monoid morphisms $\mathrm{d}_{\mathrm{ex}} : \mathfrak{C}_{\mathrm{ex}} \to M$ such that $\mathrm{d} = \mathrm{d}_{\mathrm{ex}} \circ \Psi_{\exp} : \mathfrak{C} \to M$ is a C^∞-derivation and $\mathrm{d}_{\mathrm{ex}} \circ \Psi_{\exp}(\Phi_i(c')) = \Phi_i(c')\mathrm{d}_{\mathrm{ex}}c'$ for all $c' \in \mathfrak{C}_{\mathrm{ex}}$. However, since $c' \cdot 0_{\mathfrak{C}_{\mathrm{ex}}} = 0_{\mathfrak{C}_{\mathrm{ex}}}$ for all $c' \in \mathfrak{C}_{\mathrm{ex}}$ we would have $\mathrm{d}_{\mathrm{ex}}c' + \mathrm{d}_{\mathrm{ex}}0_{\mathfrak{C}_{\mathrm{ex}}} = \mathrm{d}_{\mathrm{ex}}0_{\mathfrak{C}_{\mathrm{ex}}}$ in M, so $\mathrm{d}_{\mathrm{ex}}c' = 0$, and this modified definition would give ${}^b\Omega_{\boldsymbol{\mathfrak{C}}} = 0$ for any $\boldsymbol{\mathfrak{C}}$. To get a non-trivial definition we took $\boldsymbol{\mathfrak{C}}$ to be interior, and defined d_{in} only on $\mathfrak{C}_{\mathrm{in}} = \mathfrak{C}_{\mathrm{ex}} \setminus \{0_{\mathfrak{C}_{\mathrm{ex}}}\}$.

If $\boldsymbol{\mathfrak{C}}$ lies in the image of $\Pi^{\mathbf{C^\infty Rings^c_{in}}}_{\mathbf{C^\infty Rings^c}} : \mathbf{C^\infty Rings^c} \hookrightarrow \mathbf{C^\infty Rings^c_{in}}$ from Definition 4.11 then ${}^b\Omega_{\boldsymbol{\mathfrak{C}}} = 0$, since $\mathfrak{C}_{\mathrm{in}}$ then contains a zero element $0_{\mathfrak{C}_{\mathrm{in}}}$ with $c' \cdot 0_{\mathfrak{C}_{\mathrm{in}}} = 0_{\mathfrak{C}_{\mathrm{in}}}$ for all $c' \in \mathfrak{C}_{\mathrm{in}}$.

If $\mathrm{d}_{\mathrm{in}} : \mathfrak{C}_{\mathrm{in}} \to M$ is a b-derivation then it is a morphism from a monoid to an abelian group, and so factors through $\pi^{\mathrm{gp}} : \mathfrak{C}_{\mathrm{in}} \to (\mathfrak{C}_{\mathrm{in}})^{\mathrm{gp}}$. This suggests that b-cotangent modules may be most interesting for *integral* C^∞-rings with corners, as in §4.7, for which $\pi^{\mathrm{gp}} : \mathfrak{C}_{\mathrm{in}} \to (\mathfrak{C}_{\mathrm{in}})^{\mathrm{gp}}$ is injective.

7.2 (B-)cotangent modules for manifolds with (g-)corners

Example 7.6 **(a)** Let X be a manifold with corners. Then $\Gamma^\infty(T^*X)$ is a module over the $\mathbb{R}$-algebra $C^\infty(X)$, and so over the C^∞-rings with corners $\boldsymbol{C^\infty}(X)$ and $\boldsymbol{C^\infty_{\mathrm{in}}}(X)$ from Example 4.18. The exterior derivative $\mathrm{d} : C^\infty(X) \to \Gamma^\infty(T^*X)$ is a C^∞-derivation, and there is a unique morphism $\lambda : \Omega_{C^\infty(X)} \to \Gamma^\infty(T^*X)$ such that $\mathrm{d} = \lambda \circ \mathrm{d}$.

(b) Let X be a manifold with corners (or g-corners), with b-cotangent bundle ${}^bT^*X$ as in §3.5. Example 4.18(b) defines a C^∞-ring with corners $\boldsymbol{C}^\infty_{\rm in}(X) = \mathfrak{C}$ with $\mathfrak{C}_{\rm in} = {\rm In}(X)$, the monoid of interior maps $c' : X \to [0,\infty)$. We have a $C^\infty(X)$-module $\Gamma^\infty({}^bT^*X)$. Define ${\rm d_{in}} : {\rm In}(X) \to \Gamma^\infty({}^bT^*X)$ by

$$\mathrm{d_{in}}(c') = c'^{-1} \cdot {}^b\mathrm{d}c' = {}^b\mathrm{d}(\log c'), \tag{7.2}$$

where ${}^b\mathrm{d} = I_X^* \circ \mathrm{d}$ is the composition of the exterior derivative $\mathrm{d} : C^\infty(X) \to \Gamma^\infty(T^*X)$ with the projection $I_X^* : T^*X \to {}^bT^*X$. Here (7.2) makes sense on the interior X° where $c' > 0$, but has a unique smooth extension over $X \setminus X^\circ$.

We can now show that ${\rm d_{in}} : {\rm In}(X) \to \Gamma^\infty({}^bT^*X)$ is a b-derivation in the sense of Definition 7.4, so there is a unique morphism ${}^b\lambda : {}^b\Omega_{\boldsymbol{C}^\infty_{\rm in}(X)} \to \Gamma^\infty({}^bT^*X)$ such that ${\rm d_{in}} = {}^b\lambda \circ {\rm d_{in}}$.

Theorem 7.7 **(a)** *Let X be a manifold with faces, as in Definition* 3.13, *such that ∂X has finitely many connected components. Then $\Gamma^\infty(T^*X)$ is the cotangent module of $C^\infty(X)$. That is, λ from Example* 7.6(a) *is an isomorphism.*

(b) *Let X be a manifold with faces, or a manifold with g-faces, as in Definition* 4.37, *such that ∂X has finitely many connected components. Then $\Gamma^\infty({}^bT^*X)$ is the b-cotangent module of $\boldsymbol{C}^\infty_{\rm in}(X)$. That is, ${}^b\lambda$ from Example* 7.6(b) *is an isomorphism.*

Proof The first part of the proof is common to both (a) and (b). Let X be a manifold with faces, or with g-faces, and suppose ∂X has finitely many connected components, which we write $\partial_1 X, \ldots, \partial_N X$. We will show that the sharpened monoid $C^\infty_{\rm in}(X)^\sharp$ is finitely generated.

Write $[c'] \in C^\infty_{\rm in}(X)^\sharp$ for the $C^\infty_{\rm in}(X)^\times$-orbit of $c' \in C^\infty_{\rm in}(X)$. Each such c' vanishes to some order $a_i \in \mathbb{N}$ on $\partial_i X$ for $i = 1, \ldots, N$. The map $C^\infty_{\rm in}(X) \to \mathbb{N}^N$, $c' \mapsto (a_1, \ldots, a_N)$ is a monoid morphism, which descends to a morphism $C^\infty_{\rm in}(X)^\sharp \to \mathbb{N}^N$, $[c'] \mapsto (a_1, \ldots, a_N)$, as $\mathbb{N}^N$ is sharp.

It is easy to show that this morphism $C^\infty_{\rm in}(X)^\sharp \to \mathbb{N}^N$ is injective. If X is a manifold with faces it is also surjective, so $C^\infty_{\rm in}(X)^\sharp \cong \mathbb{N}^N$. If X is a manifold with g-faces, the image of $C^\infty_{\rm in}(X)^\sharp$ is the submonoid of $\mathbb{N}^N$ satisfying some linear equations coming from the higher-dimensional g-corner strata. Therefore $C^\infty_{\rm in}(X)^\sharp$ is a toric monoid, and so finitely generated.

Let $x_1, \ldots, x_k \in C^\infty_{\rm in}(X)$ whose images $[x_1], \ldots, [x_k]$ generate $C^\infty_{\rm in}(X)^\sharp$. Following the proof of the Whitney Embedding Theorem, we may choose a smooth map $(x_{k+1}, \ldots, x_l) : X \to \mathbb{R}^{l-k}$ for some l with $l - k \geqslant 2\dim X$ which is a closed embedding (in a weak sense, in which the inclusion $[0,\infty) \hookrightarrow \mathbb{R}$ counts as an embedding). Then $F := (x_1, \ldots, x_k, x_{k+1}, \ldots, x_l) : X \hookrightarrow$

$[0, \infty)^k \times \mathbb{R}^{l-k} = \mathbb{R}^l_k$ is an interior closed embedding in a stronger sense, which embeds the corner strata of X into corner strata of $\mathbb{R}^l_k$.

Here as X has (g-)faces, by Proposition 4.36 and Definition 4.37 the natural morphism $\boldsymbol{\lambda}_x : (\boldsymbol{C}^\infty(X))_{x_*} \to \boldsymbol{C}^\infty_x(X)$ is an isomorphism for all $x \in X$, so as $[x_1], \ldots, [x_k]$ generate $C^\infty_{\rm in}(X)^\sharp$, they also generate $C^\infty_{x,{\rm in}}(X)^\sharp$ for each $x \in X$. Geometrically this means that X near x is nicely embedded in $\mathbb{R}^l_k$ near $F(x)$, in such a way that the corner strata of X near x all come from corner strata of $\mathbb{R}^l_k$ near $F(x)$. Write $\tilde{X} = F(X)$, as a closed, embedded submanifold of $\mathbb{R}^l_k$.

Thus we have a morphism $F^* : \boldsymbol{C}^\infty_{\rm in}(\mathbb{R}^l_k) \to \boldsymbol{C}^\infty_{\rm in}(X)$ in $\mathbf{C^\infty Rings^c_{in}}$. We claim that F^* is surjective. To see this, note that F^* is surjective on C^∞-rings as it is a closed embedding, so every smooth function $F(X) \to \mathbb{R}$ extends to a smooth function $\mathbb{R}^l_k \to \mathbb{R}$ in the usual way. Also $(F^*_{\rm in})^\sharp : C^\infty_{\rm in}(\mathbb{R}^l_k)^\sharp \cong \mathbb{N}^k \to C^\infty_{\rm in}(X)^\sharp$ is surjective as it maps $(a_1, \ldots, a_k) \mapsto [x_1^{a_1} \cdots x_k^{a_k}]$, where $[x_1], \ldots, [x_k]$ generate $C^\infty_{\rm in}(X)^\sharp$. Surjectivity of $F^*_{\rm in} : C^\infty_{\rm in}(\mathbb{R}^l_k) \to C^\infty_{\rm in}(X)$ follows.

Therefore $\boldsymbol{C}^\infty_{\rm in}(X)$ is isomorphic to the quotient of $\boldsymbol{C}^\infty_{\rm in}(\mathbb{R}^l_k)$ by relations generating the kernel of $F^*_{\rm in}$, or equivalently, by equations defining $\tilde{X}$ as a submanifold of $\mathbb{R}^l_k$. If X is a manifold with faces then relations of type $f = 0$ for $f \in C^\infty(\mathbb{R}^l_k)$ are sufficient, so in the notation of Definition 4.24 we have

$$\boldsymbol{C}^\infty_{\rm in}(X) \cong \boldsymbol{C}^\infty_{\rm in}(\mathbb{R}^l_k) / \bigl(f = 0 : f \in C^\infty(\mathbb{R}^l_k),\ f|_{\tilde{X}} = 0\bigr). \tag{7.3}$$

If X is a manifold with g-faces we may also need relations of type $g = h$ for $g, h \in C^\infty_{\rm in}(\mathbb{R}^l_k)$, giving

$$\begin{aligned} \boldsymbol{C}^\infty_{\rm in}(X) \cong \boldsymbol{C}^\infty_{\rm in}(\mathbb{R}^l_k) / \bigl(f = 0 : f \in C^\infty(\mathbb{R}^l_k),\ f|_{\tilde{X}} = 0\bigr) \\ \bigl[g = h : g, h \in C^\infty_{\rm in}(\mathbb{R}^l_k),\ g|_{\tilde{X}} = h|_{\tilde{X}}\bigr]. \end{aligned} \tag{7.4}$$

We now prove part (a), so suppose X is a manifold with faces and ∂X has finitely many boundary components, so that (7.3) holds. Consider the commutative diagram of $C^\infty(X)$-modules with exact rows:

$$\begin{array}{ccccccc} 0 \to K = \mathrm{Ker}(\Omega_{F^*}) & \longrightarrow & \begin{array}{c}\Omega_{C^\infty_{\rm in}(\mathbb{R}^k_l)} \otimes_{C^\infty(\mathbb{R}^k_l)} C^\infty(X) = \\ \langle \mathrm{d}x_1, \ldots, \mathrm{d}x_l \rangle \otimes_{\mathbb{R}} C^\infty(X)\end{array} & \xrightarrow[\Omega_{F^*}]{} & \Omega_{C^\infty_{\rm in}(X)} \to 0 \\ \downarrow{\scriptstyle \mu} & & \downarrow{\scriptstyle \cong} & & \downarrow{\scriptstyle \lambda} \\ 0 \to \Gamma^\infty(\mathrm{Ker}(\mathrm{d}F^*)) & \longrightarrow & \begin{array}{c}\Gamma^\infty(F^*(T^*\mathbb{R}^k_l)) = \\ \langle \mathrm{d}x_1, \ldots, \mathrm{d}x_l \rangle \otimes_{\mathbb{R}} C^\infty(X)\end{array} & \xrightarrow{\Gamma^\infty(\mathrm{d}F^*)} & \Gamma^\infty(T^*X) \to 0. \end{array} \tag{7.5}$$

Here as $F^* : C^\infty(\mathbb{R}^k_l) \to C^\infty(X)$ is surjective, Ω_{F^*} is surjective, and the top row is exact. As the exact sequence $0 \to \mathrm{Ker}(\mathrm{d}F^*) \to F^*(T^*\mathbb{R}^k_l) \to T^*X \to 0$ of vector bundles on X splits, the bottom row is exact. It is easy to check that

$\Omega_{C^\infty_{\rm in}(\mathbb{R}^k_l)} = \Omega_{C^\infty(\mathbb{R}^k_l)} = \langle {\rm d}x_1, \ldots, {\rm d}x_l \rangle \otimes_{\mathbb{R}} C^\infty(\mathbb{R}^k_l)$ satisfies the universal property for cotangent modules. Since $\Gamma^\infty(T^*\mathbb{R}^k_l) = \langle {\rm d}x_1, \ldots, {\rm d}x_l \rangle_{C^\infty(\mathbb{R}^k_l)}$, we see that the two centre terms are $\langle {\rm d}x_1, \ldots, {\rm d}x_l \rangle \otimes_{\mathbb{R}} C^\infty(X)$, and are isomorphic. Clearly the right-hand square commutes. Hence by exactness there is a unique map $\mu : K \to \Gamma^\infty(\mathrm{Ker}({\rm d}F^*))$ making the diagram commute.

As the right-hand square of (7.5) commutes, with the central column an isomorphism, we see that λ is surjective. To prove that λ is injective, and hence an isomorphism, it is enough to show that μ is surjective. Since $\tilde{X}$ is cut out transversely in $\mathbb{R}^k_l$ by equations $f = 0$ for $f \in C^\infty(\mathbb{R}^l_k)$ with $f|_{\tilde{X}} = 0$, as in (7.3), we see that the vector bundle $\mathrm{Ker}({\rm d}F^*) \to X$ is spanned over $C^\infty(X)$ by sections $F^*({\rm d}f)$ for $f \in C^\infty(\mathbb{R}^l_k)$ with $f|_{\tilde{X}} = 0$. But then $({\rm d}f) \otimes 1$ lies in $K = \mathrm{Ker}(\Omega_{F^*})$ with $F^*({\rm d}f) = \mu(({\rm d}f) \otimes 1)$, so the image of μ contains a spanning set, and μ is surjective. This proves (a).

For (b), allow X to have faces or g-faces, so that (7.4) holds, and replace (7.5) by the commutative diagram with exact rows

$$\begin{array}{ccccccc}
0 \to {}^bK = \mathrm{Ker}({}^b\Omega_{F^*}) & \longrightarrow & \begin{array}{c} {}^b\Omega_{C^\infty_{\rm in}(\mathbb{R}^k_l)} \otimes_{C^\infty(\mathbb{R}^k_l)} C^\infty(X) \\ = \langle x_1^{-1}{\rm d}x_1, \ldots, x_k^{-1}{\rm d}x_k, \\ {\rm d}x_{k+1}, \ldots, {\rm d}x_l \rangle \otimes_{\mathbb{R}} C^\infty(X) \end{array} & \xrightarrow{{}^b\Omega_{F^*}} & {}^b\Omega_{C^\infty_{\rm in}(X)} \to 0 \\
\Big\downarrow {}^b\mu & & \Big\downarrow \cong & & \Big\downarrow {}^b\lambda \\
0 \to \Gamma^\infty(\mathrm{Ker}({}^b{\rm d}F^*)) & \longrightarrow & \begin{array}{c} \Gamma^\infty(F^*({}^bT^*\mathbb{R}^k_l)) \\ = \langle x_1^{-1}{\rm d}x_1, \ldots, x_k^{-1}{\rm d}x_k, \\ {\rm d}x_{k+1}, \ldots, {\rm d}x_l \rangle \otimes_{\mathbb{R}} C^\infty(X) \end{array} & \xrightarrow{\Gamma^\infty({}^b{\rm d}F^*)} & \Gamma^\infty({}^bT^*X) \to 0.
\end{array}$$

Here ${}^b\Omega_{C^\infty_{\rm in}(\mathbb{R}^k_l)} = \langle x_1^{-1}{\rm d}x_1, \ldots, x_k^{-1}{\rm d}x_k, {\rm d}x_{k+1}, \ldots, {\rm d}x_l \rangle \otimes_{\mathbb{R}} C^\infty(\mathbb{R}^k_l)$ satisfies the universal property for b-cotangent modules. The argument above gives a unique map ${}^b\mu$ making the diagram commute, and shows ${}^b\lambda$ is surjective. To prove ${}^b\lambda$ is injective, and hence an isomorphism, it is enough to show that ${}^b\mu$ is surjective.

Since $\tilde{X}$ is cut out b-transversely in $\mathbb{R}^k_l$ by equations $f = 0$ for $f \in C^\infty(\mathbb{R}^l_k)$ with $f|_{\tilde{X}} = 0$, and (if X has g-faces) equations $g = h$ for $g, h \in C^\infty_{\rm in}(\mathbb{R}^l_k)$ with $g|_{\tilde{X}} = h|_{\tilde{X}}$, we see that the vector bundle $\mathrm{Ker}({}^b{\rm d}F^*) \to X$ is spanned over $C^\infty(X)$ by sections $F^*({}^b{\rm d}\exp(f))$ for $f \in C^\infty(\mathbb{R}^l_k)$ with $f|_{\tilde{X}} = 0$ and $F^*({}^b{\rm d}g - {}^b{\rm d}h)$ for $g, h \in C^\infty_{\rm in}(\mathbb{R}^l_k)$ with $g|_{\tilde{X}} = h|_{\tilde{X}}$. But then ${}^b{\rm d}\exp(f) \otimes 1$ and $({}^b{\rm d}g - {}^b{\rm d}h) \otimes 1$ lie in ${}^bK = \mathrm{Ker}({}^b\Omega_{F^*})$ with $F^*({}^b{\rm d}\exp(f)) = {}^b\mu({}^b{\rm d}\exp(f) \otimes 1)$ and $F^*({}^b{\rm d}g - {}^b{\rm d}h) = {}^b\mu(({}^b{\rm d}g - {}^b{\rm d}h) \otimes 1)$, so the image of ${}^b\mu$ contains a spanning set, and ${}^b\mu$ is surjective. This proves (b). □

Remark 7.8 The first author [25, Props. 4.7.5 and 4.7.9] proves by a different method that if we allow ∂X to have infinitely many connected components, then λ is still an isomorphism in Theorem 7.7(a), and ${}^b\lambda$ is surjective in (b).

7.3 Some computations of b-cotangent modules

The next proposition generalizes Example 2.31.

Proposition 7.9 **(a)** *Let $A, A_{\rm in}$ be sets and $\mathfrak{F}^{A,A_{\rm in}}$ the free interior C^∞-ring with corners from Definition 4.22, with generators $x_a \in \mathfrak{F}^{A,A_{\rm in}}$, $y_{a'} \in \mathfrak{F}^{A,A_{\rm in}}_{\rm in}$ for $a \in A$, $a' \in A_{\rm in}$. Then there is a natural isomorphism*

$$
{}^b\Omega_{\mathfrak{F}^{A,A_{\rm in}}} \cong \bigl\langle {\rm d_{in}}\Psi_{\rm exp}(x_a), {\rm d_{in}}y_{a'} : a \in A,\ a' \in A_{\rm in}\bigr\rangle_{\mathbb{R}} \otimes_{\mathbb{R}} \mathfrak{F}^{A,A_{\rm in}}. \quad (7.6)
$$

(b) *Suppose $\mathfrak{C}$ is defined by a coequalizer diagram* (4.13) *in* $\mathbf{C^\infty Rings^c_{in}}$. *Then writing $(x_a)_{a\in A},(y_{a'})_{a'\in A_{\rm in}}$ and $(\tilde x_b)_{b\in B},(\tilde y_{b'})_{b'\in B_{\rm in}}$ for the generators of $\mathfrak{F}^{A,A_{\rm in}}$ and $\mathfrak{F}^{B,B_{\rm in}}$, we have an exact sequence*

$$
\begin{matrix}\bigl\langle {\rm d_{in}}\Psi_{\rm exp}(\tilde x_b),\ b\in B, \\ {\rm d_{in}}\tilde y_{b'},\ b'\in B_{\rm in}\bigr\rangle_{\mathbb{R}}\otimes_{\mathbb{R}}\mathfrak{C}\end{matrix} \xrightarrow{\ \gamma\ } \begin{matrix}\bigl\langle {\rm d_{in}}\Psi_{\rm exp}(x_a),\ a\in A, \\ {\rm d_{in}}y_{a'},\ a'\in A_{\rm in}\bigr\rangle_{\mathbb{R}}\otimes_{\mathbb{R}}\mathfrak{C}\end{matrix} \xrightarrow{\ \delta\ } {}^b\Omega_{\mathfrak{C}} \longrightarrow 0 \quad (7.7)
$$

in $\mathfrak{C}$-mod, *where if $\boldsymbol\alpha = (\alpha,\alpha_{\rm ex})$ and $\boldsymbol\beta = (\beta,\beta_{\rm ex})$ in* (4.13) *map*

$$
\begin{aligned}
\alpha : \tilde x_b &\longmapsto e_b\bigl((x_a)_{a\in A},(y_{a'})_{a'\in A_{\rm in}}\bigr),\\
\alpha_{\rm ex} : \tilde y_{b'} &\longmapsto g_{b'}\bigl((x_a)_{a\in A},(y_{a'})_{a'\in A_{\rm in}}\bigr),\\
\beta : \tilde x_b &\longmapsto f_b\bigl((x_a)_{a\in A},(y_{a'})_{a'\in A_{\rm in}}\bigr),\\
\beta_{\rm ex} : \tilde y_{b'} &\longmapsto h_{b'}\bigl((x_a)_{a\in A},(y_{a'})_{a'\in A_{\rm in}}\bigr),
\end{aligned} \quad (7.8)
$$

for $e_b, f_b, g_{b'}, h_{b'}$ depending only on finitely many $x_a, y_{a'}$, then γ,δ are given by

$$
\begin{aligned}
\gamma({\rm d_{in}}\Psi_{\rm exp}(\tilde x_b)) &= \sum_{a\in A}\phi\bigl(\tfrac{\partial e_b}{\partial x_a}\bigl((x_a)_{a\in A},(y_{a'})_{a'\in A_{\rm in}}\bigr) \\
&\qquad - \tfrac{\partial f_b}{\partial x_a}\bigl((x_a)_{a\in A},(y_{a'})_{a'\in A_{\rm in}}\bigr)\bigr){\rm d_{in}}\Psi_{\rm exp}(x_a)\\
&+ \sum_{a'\in A_{\rm in}}\phi\bigl(y_{a'}\tfrac{\partial e_b}{\partial y_{a'}}\bigl((x_a)_{a\in A},(y_{a'})_{a'\in A_{\rm in}}\bigr) \\
&\qquad - y_{a'}\tfrac{\partial f_b}{\partial y_{a'}}\bigl((x_a)_{a\in A},(y_{a'})_{a'\in A_{\rm in}}\bigr)\bigr){\rm d_{in}}y_{a'},\\
\gamma({\rm d_{in}}\tilde y_{b'}) &= \sum_{a\in A}\phi\bigl(g_{b'}^{-1}\tfrac{\partial g_{b'}}{\partial x_a}\bigl((x_a)_{a\in A},(y_{a'})_{a'\in A_{\rm in}}\bigr) \\
&\qquad - h_{b'}^{-1}\tfrac{\partial h_{b'}}{\partial x_a}\bigl((x_a)_{a\in A},(y_{a'})_{a'\in A_{\rm in}}\bigr)\bigr){\rm d_{in}}\Psi_{\rm exp}(x_a)\\
&+ \sum_{a'\in A_{\rm in}}\phi\bigl(y_{a'}g_{b'}^{-1}\tfrac{\partial g_{b'}}{\partial y_{a'}}\bigl((x_a)_{a\in A},(y_{a'})_{a'\in A_{\rm in}}\bigr) \\
&\qquad - y_{a'}h_{b'}^{-1}\tfrac{\partial h_{b'}}{\partial y_{a'}}\bigl((x_a)_{a\in A},(y_{a'})_{a'\in A_{\rm in}}\bigr)\bigr){\rm d_{in}}y_{a'},
\end{aligned} \quad (7.9)
$$

$$
\delta({\rm d_{in}}\Psi_{\rm exp}(x_a)) = {\rm d_{in}}\circ\phi_{\rm in}(\Psi_{\rm exp}(x_a)), \quad \delta({\rm d_{in}}y_{a'}) = {\rm d_{in}}\circ\phi_{\rm in}(y_{a'}).
$$

Hence if $\mathfrak{C}$ is finitely generated (or finitely presented) in the sense of Definition 4.40, *then ${}^b\Omega_{\mathfrak{C}}$ is finitely generated (or finitely presented).*

Proof Part (a) follows from (b) with $B = B_{\rm in} = \emptyset$, so it is enough to prove (b). For (b), first note that by Definition 7.4, we have

$$ {}^b\Omega_{\mathfrak{C}} = \big\langle \mathrm{d}_{\mathrm{in}} c' : c' \in \mathfrak{C}_{\mathrm{in}} \big\rangle_{\mathbb{R}} \otimes_{\mathbb{R}} \mathfrak{C} / \langle \text{all relations (7.1)} \rangle. \tag{7.10} $$

Now let $S \subseteq \mathfrak{C}_{\mathrm{in}}$ be a generating subset, that is, every $c' \in \mathfrak{C}_{\mathrm{in}}$ may be written as $c' = \Psi_g(s_1, \ldots, s_n)$ for some interior $g : [0,\infty)^n \to [0,\infty)$ and $s_1, \ldots, s_n \in S$. For each $c' \in \mathfrak{C}_{\mathrm{in}} \setminus S$, choose interior $g^{c'} : [0,\infty)^{n^{c'}} \to [0,\infty)$ and $s_1^{c'}, \ldots, s_{n^{c'}}^{c'} \in S$ with $c' = \Psi_{g^{c'}}(s_1^{c'}, \ldots, s_{n^{c'}}^{c'})$, using the Axiom of Choice. Then (7.1) gives

$$ \mathrm{d}_{\mathrm{in}} c' = \sum_{i=1}^{n^{c'}} \Phi_{(g^{c'})^{-1} x_i \frac{\partial g^{c'}}{\partial x_i}}(s_1^{c'}, \ldots, s_{n^{c'}}^{c'}) \cdot \mathrm{d}_{\mathrm{in}} s_i^{c'}. \tag{7.11} $$

Hence ${}^b\Omega_{\mathfrak{C}}$ is spanned over $\mathfrak{C}$ by $\mathrm{d}_{\mathrm{in}} s$ for $s \in S$, and we may write ${}^b\Omega_{\mathfrak{C}}$ as $\langle \mathrm{d}_{\mathrm{in}} s : s \in S \rangle_{\mathbb{R}} \otimes_{\mathbb{R}} \mathfrak{C}/(\text{relations})$. The relations required are the result of taking all relations (7.1) in (7.10), and using (7.11) to substitute for all terms $\mathrm{d}_{\mathrm{in}} c'$ for $c' \in \mathfrak{C}_{\mathrm{in}} \setminus S$, so that the resulting relation is a finite $\mathfrak{C}$-linear combination of $\mathrm{d}_{\mathrm{in}} s$ for $s \in S$. A calculation using Definition 4.4(ii) shows that this relation is of the form (7.1) with $c_1', \ldots, c_n' \in S$. Hence

$$ {}^b\Omega_{\mathfrak{C}} \cong \big\langle \mathrm{d}_{\mathrm{in}} c' : c' \in S \big\rangle_{\mathbb{R}} \otimes_{\mathbb{R}} \mathfrak{C} / \big\langle \text{relations (7.1) with } c_1', \ldots, c_n' \in S \big\rangle. \tag{7.12} $$

Now let $\mathfrak{C}$ be as in (4.13). Then $\mathfrak{C}$ has $\mathbb{R}$ type generators $\phi(x_a)$ for $a \in A$ and $[0,\infty)$ type generators $\phi_{\mathrm{in}}(y_{a'})$ for $a' \in A_{\mathrm{in}}$. But (7.12) only allows for $[0,\infty)$ type generators. To convert $\mathbb{R}$ type generators to $[0,\infty)$ type generators observe that, in a similar way to the proof of Proposition 4.7, any smooth function $\mathbb{R} \to \mathbb{R}$ or $\mathbb{R} \to [0,\infty)$ may be extended to a smooth function $[0,\infty)^2 \to \mathbb{R}$ or $[0,\infty)^2 \to [0,\infty)$ by mapping $t \in \mathbb{R}$ to $(x,y) = (e^t, e^{-t}) \in [0,\infty)^2$, which then satisfies $xy = 1$. Thus, we may replace $\mathbb{R}$ type generators $\phi(x_a) \in \mathfrak{C}$ by pairs of $[0,\infty)$ type generators $\phi_{\mathrm{in}} \circ \Psi_{\exp}(x_a), \phi_{\mathrm{in}} \circ \Psi_{\exp}(-x_a)$ in $\mathfrak{C}_{\mathrm{in}}$, satisfying the relation $(\phi_{\mathrm{in}} \circ \Psi_{\exp}(x_a)) \cdot (\phi_{\mathrm{in}} \circ \Psi_{\exp}(-x_a)) = 1$. But applying d_{in} to this relation yields $\mathrm{d}_{\mathrm{in}}(\phi_{\mathrm{in}} \circ \Psi_{\exp}(x_a)) + \mathrm{d}_{\mathrm{in}}(\phi_{\mathrm{in}} \circ \Psi_{\exp}(-x_a)) = 0$. Hence by taking $S = \{\phi_{\mathrm{in}} \circ \Psi_{\exp}(\pm x_a) : a \in A\} \cup \{\phi_{\mathrm{in}}(y_{a'}) : a' \in A_{\mathrm{in}}\}$ in (7.12), we see that

$$ {}^b\Omega_{\mathfrak{C}} \cong \frac{\big\langle \mathrm{d}_{\mathrm{in}}(\phi_{\mathrm{in}} \circ \Psi_{\exp}(\pm x_a)) : a \in A,\ \ \mathrm{d}_{\mathrm{in}}(\phi_{\mathrm{in}}(y_{a'})) : a' \in A_{\mathrm{in}} \big\rangle_{\mathbb{R}} \otimes_{\mathbb{R}} \mathfrak{C}}{\begin{array}{c}\big\langle \mathrm{d}_{\mathrm{in}}(\phi_{\mathrm{in}} \circ \Psi_{\exp}(x_a)) + \mathrm{d}_{\mathrm{in}}(\phi_{\mathrm{in}} \circ \Psi_{\exp}(-x_a)) = 0,\ \ a \in A, \\ \text{other relations (7.1) with } c_i' = \phi_{\mathrm{in}} \circ \Psi_{\exp}(\pm x_a), \phi_{\mathrm{in}}(y_{a'}) \big\rangle\end{array}}. \tag{7.13} $$

Clearly we can omit the generator $\mathrm{d}_{\mathrm{in}} \phi_{\mathrm{in}} \circ \Psi_{\exp}(-x_a)$ and the first relation on the bottom of (7.13) for $a \in A$. We can also omit the 'ϕ_{in}' in

$\mathrm{d_{in}}(\phi_{\mathrm{in}} \circ \Psi_{\mathrm{exp}}(\pm x_a))$, $\mathrm{d_{in}}(\phi_{\mathrm{in}}(y_{a'}))$, as these are only formal symbols anyway. Furthermore, in $\mathfrak{F}_{\mathrm{in}}^{A,A_{\mathrm{in}}}$ in (4.13), the only relations between the generators $\Psi_{\mathrm{exp}}(\pm x_a)$, $a \in A$, $y_{a'}$, $a' \in A_{\mathrm{in}}$, are generated by $\Psi_{\mathrm{exp}}(x_a) \cdot \Psi_{\mathrm{exp}}(-x_a) = 1$ for $a \in A$. Hence in (7.13), the 'other relations' must all come from $\mathfrak{F}^{B,B_{\mathrm{in}}}$. But the relations in $\mathfrak{F}_{\mathrm{in}}^{A,A_{\mathrm{in}}}$ from $\mathfrak{F}^{B,B_{\mathrm{in}}}$ are generated by $\alpha_{\mathrm{in}}(\Psi_{\mathrm{exp}}(\pm\tilde{x}_b)) = \beta_{\mathrm{in}}(\Psi_{\mathrm{exp}}(\pm\tilde{x}_b))$, $b \in B$, and $\alpha_{\mathrm{in}}(\tilde{y}_{b'}) = \beta_{\mathrm{in}}(\tilde{y}_{b'})$, $b' \in B_{\mathrm{in}}$. Putting all this together and using (7.8) yields

$$
{}^b\Omega_{\mathfrak{C}} \cong \frac{\langle \mathrm{d_{in}}\Psi_{\mathrm{exp}}(x_a) : a \in A,\ \ \mathrm{d_{in}}y_{a'} : a' \in A_{\mathrm{in}} \rangle_{\mathbb{R}} \otimes_{\mathbb{R}} \mathfrak{C}}{\begin{array}{l}\big\langle \mathrm{d_{in}} \circ \Psi_{\mathrm{exp}}\big(e_b\big((x_a)_{a\in A}, (y_{a'})_{a'\in A_{\mathrm{in}}}\big)\big) \\ \quad -\mathrm{d_{in}} \circ \Psi_{\mathrm{exp}}\big(f_b\big((x_a)_{a\in A}, (y_{a'})_{a'\in A_{\mathrm{in}}}\big)\big) = 0,\ b \in B, \\ \mathrm{d_{in}}\big(g_{b'}\big((x_a)_{a\in A}, (y_{a'})_{a'\in A_{\mathrm{in}}}\big)\big) \\ \quad -\mathrm{d_{in}}\big(h_{b'}\big((x_a)_{a\in A}, (y_{a'})_{a'\in A_{\mathrm{in}}}\big)\big) = 0,\ b' \in B_{\mathrm{in}}\big\rangle\end{array}}. \tag{7.14}
$$

By (7.1) and (7.9), the relations in (7.14) are $\gamma(\mathrm{d_{in}}\Psi_{\mathrm{exp}}(\tilde{x}_b)) = 0$, $b \in B$, and $\gamma(\mathrm{d_{in}}\tilde{y}_{b'}) = 0$, $b' \in B_{\mathrm{in}}$. Hence (7.14) shows that ${}^b\Omega_{\mathfrak{C}} \cong \operatorname{Coker}\gamma$. Since this isomorphism is realized by mapping $\mathrm{d_{in}}\Psi_{\mathrm{exp}}(x_a) \mapsto \mathrm{d_{in}} \circ \phi_{\mathrm{in}}(\Psi_{\mathrm{exp}}(x_a)$, $\mathrm{d_{in}}y_{a'} \mapsto \mathrm{d_{in}} \circ \phi_{\mathrm{in}}(y_{a'})$, which is δ in (7.9), we see that (7.7) is exact, proving (b). □

We can compute the b-cotangent module of an interior C^∞-ring with corners $\mathfrak{C}$ modified by generators and relations, as in §4.5.

Proposition 7.10 *Suppose $\mathfrak{C} \in \mathbf{C^\infty Rings^c_{in}}$, and we are given sets A, A_{in}, B, B_{in}, so that $\mathfrak{C}(x_a : a \in A)[y_{a'} : a' \in A_{\mathrm{in}}] \in \mathbf{C^\infty Rings^c_{in}}$, and elements $f_b \in \mathfrak{C}(x_a : a \in A)[y_{a'} : a' \in A_{\mathrm{in}}]$ and $g_{b'}, h_{b'} \in \mathfrak{C}(x_a : a \in A)[y_{a'} : a' \in A_{\mathrm{in}}]_{\mathrm{in}}$ for $b \in B$ and $b' \in B_{\mathrm{in}}$. Define $\mathfrak{D} \in \mathbf{C^\infty Rings^c_{in}}$ by*

$$
\mathfrak{D} = \big(\mathfrak{C}(x_a : a\in A)[y_{a'} : a' \in A_{\mathrm{in}}]\big)/\big(f_b = 0,\ b \in B\big)\big[g_{b'} = h_{b'},\ b' \in B_{\mathrm{in}}\big], \tag{7.15}
$$

in the notation of Definition 4.24, with natural morphism $\pi : \mathfrak{C} \to \mathfrak{D}$. Then we have an exact sequence in $\mathfrak{D}$-mod:

$$
\begin{array}{c}\langle \mathrm{d_{in}}\Psi_{\mathrm{exp}}(\tilde{x}_b),\ b\in B\rangle_{\mathbb{R}} \otimes_{\mathbb{R}} \mathfrak{D} \\ \oplus \langle \mathrm{d_{in}}(\tilde{y}_{b'}),\ b' \in B_{\mathrm{in}}\rangle_{\mathbb{R}} \otimes_{\mathbb{R}} \mathfrak{D}\end{array} \xrightarrow{\ \alpha\ } \begin{array}{c}{}^b\Omega_{\mathfrak{C}} \otimes_{\mathfrak{C}} \mathfrak{D} \oplus \\ \langle \mathrm{d_{in}}\Psi_{\mathrm{exp}}(x_a),\ a \in A\rangle_{\mathbb{R}} \otimes_{\mathbb{R}} \mathfrak{D} \\ \oplus \langle \mathrm{d_{in}}(y_{a'}),\ a' \in A_{\mathrm{in}}\rangle_{\mathbb{R}} \otimes_{\mathbb{R}} \mathfrak{D}\end{array} \xrightarrow{\ \beta\ } {}^b\Omega_{\mathfrak{D}} \twoheadrightarrow 0. \tag{7.16}
$$

Here we regard α *as mapping to*

$$
\begin{aligned}
&{}^b\Omega_{\mathfrak{C}(x_a:a\in A)[y_{a'}:a'\in A_{\rm in}]}\otimes_{\mathfrak{C}(x_a:a\in A)[y_{a'}:a'\in A_{\rm in}]}\mathfrak{D}\\
&={}^b\Omega_{\mathfrak{C}\otimes_\infty\mathfrak{F}^{A,A_{\rm in}}}\otimes_{\mathfrak{C}\otimes_\infty\mathfrak{F}^{A,A_{\rm in}}}\mathfrak{D}\\
&\cong\big({}^b\Omega_{\mathfrak{C}}\otimes_{\mathfrak{C}}(\mathfrak{C}\otimes_\infty\mathfrak{F}^{A,A_{\rm in}})\oplus{}^b\Omega_{\mathfrak{F}^{A,A_{\rm in}}}\otimes_{\mathfrak{F}^{A,A_{\rm in}}}(\mathfrak{C}\otimes_\infty\mathfrak{F}^{A,A_{\rm in}})\big)\\
&\qquad\qquad\otimes_{\mathfrak{C}\otimes_\infty\mathfrak{F}^{A,A_{\rm in}}}\mathfrak{D}\\
&\cong{}^b\Omega_{\mathfrak{C}}\otimes_{\mathfrak{C}}\mathfrak{D}\oplus{}^b\Omega_{\mathfrak{F}^{A,A_{\rm in}}}\otimes_{\mathfrak{F}^{A,A_{\rm in}}}\mathfrak{D}\\
&\cong{}^b\Omega_{\mathfrak{C}}\otimes_{\mathfrak{C}}\mathfrak{D}\oplus\langle p_a,\ a\in A\rangle_{\mathbb{R}}\otimes_{\mathbb{R}}\mathfrak{D}\oplus\langle q_{a'},\ a'\in A_{\rm in}\rangle_{\mathbb{R}}\otimes_{\mathbb{R}}\mathfrak{D},
\end{aligned}
$$

using Proposition 7.9(a) *in the last step, and then* α,β *act by*

$$
\alpha=\begin{pmatrix}\mathrm{d}_{\rm in}\Psi_{\rm exp}(\tilde{x}_b)\longmapsto \mathrm{d}_{\rm in}\circ\Psi_{\rm exp}(f_b)\\ \mathrm{d}_{\rm in}(\tilde{y}_{b'})\longmapsto \mathrm{d}_{\rm in}(g_{b'})-\mathrm{d}_{\rm in}(h_{b'})\end{pmatrix},\quad \beta=\begin{pmatrix}({}^b\Omega_\pi)_*\\ \mathrm{id}\\ \mathrm{id}\end{pmatrix}. \tag{7.17}
$$

Proof Proposition 4.23(b) gives a coequalizer diagram for $\mathfrak{C}$ in $\mathbf{C^\infty Rings^c_{in}}$:

$$
\mathfrak{F}^{D,D_{\rm in}}\underset{\delta}{\overset{\gamma}{\rightrightarrows}}\mathfrak{F}^{C,C_{\rm in}}\xrightarrow{\ \ \phi\ \ }\mathfrak{C}. \tag{7.18}
$$

We can extend this to a coequalizer diagram for $\mathfrak{D}$:

$$
\mathfrak{F}^{B\amalg D,B_{\rm in}\amalg D_{\rm in}}\underset{\zeta}{\overset{\epsilon}{\rightrightarrows}}\mathfrak{F}^{A\amalg C,A_{\rm in}\amalg C_{\rm in}}\xrightarrow{\ \ \psi\ \ }\mathfrak{D}. \tag{7.19}
$$

Here ϵ,ζ act as γ,δ do on the generators from $D,D_{\rm in}$, and as $\epsilon(\tilde{x}_b)=\tilde{f}_b$, $\epsilon_{\rm in}(\tilde{y}_{b'})=\tilde{g}_{b'}$, $\zeta(\tilde{x}_b)=0$, $\zeta_{\rm in}(\tilde{y}_{b'})=\tilde{h}_{b'}$ on the generators from $B,B_{\rm in}$, where $\tilde{f}_b,\tilde{g}_{b'},\tilde{h}_{b'}$ are lifts of $f_b,g_{b'},h_{b'}$ along the surjective morphism $\mathfrak{F}^{A\amalg C,A_{\rm in}\amalg C_{\rm in}}\to\mathfrak{C}(x_a:a\in A)[y_{a'}:a'\in A_{\rm in}]$ induced by ϕ. Also ψ acts as ϕ does on the generators from $C,C_{\rm in}$, and as the identity on generators $x_a,y_{a'}$ from $A,A_{\rm in}$.

Thus the effect of enlarging $\mathfrak{F}^{C,C_{\rm in}}$ to $\mathfrak{F}^{A\amalg C,A_{\rm in}\amalg C_{\rm in}}$ is to enlarge $\mathfrak{C}$ to $\mathfrak{C}(x_a:a\in A)[y_{a'}:a'\in A_{\rm in}]$, and the effect of enlarging $\mathfrak{F}^{D,D_{\rm in}}$ to $\mathfrak{F}^{B\amalg D,B_{\rm in}\amalg D_{\rm in}}$ is to quotient $\mathfrak{C}(x_a:a\in A)[y_{a'}:a'\in A_{\rm in}]$ by extra relations $\epsilon(\tilde{x}_b)=\zeta(\tilde{x}_b)$ for $b\in B$, which means $f_b=0$, and $\epsilon_{\rm in}(\tilde{y}_{b'})=\zeta_{\rm in}(\tilde{y}_{b'})$ for $b'\in B_{\rm in}$, which means $g_{b'}=h_{b'}$. So the result is $\mathfrak{D}$ in (7.15).

Now consider the commutative diagram:

$$
\begin{array}{ccccc}
 & \begin{array}{l}\langle \mathrm{d_{in}}\Psi_{\exp}(\tilde v_d),\ d\in D,\\ \mathrm{d_{in}}\tilde w_{d'},\ d'\in D_{\mathrm{in}}\rangle_{\mathbb{R}}\otimes_{\mathbb{R}}\mathfrak{D}\end{array} & \xrightarrow{\binom{0}{\mathrm{id}}} & \begin{array}{l}\langle \mathrm{d_{in}}\Psi_{\exp}(\tilde x_b),\ b\in B,\\ \mathrm{d_{in}}\tilde y_{b'},\ b'\in B_{\mathrm{in}}\rangle_{\mathbb{R}}\otimes_{\mathbb{R}}\mathfrak{D}\oplus\\ \langle \mathrm{d_{in}}\Psi_{\exp}(\tilde v_d),\ d\in D,\\ \mathrm{d_{in}}\tilde w_{d'},\ d'\in D_{\mathrm{in}}\rangle_{\mathbb{R}}\otimes_{\mathbb{R}}\mathfrak{D}\end{array} & \\
 & \Big\downarrow \binom{0}{*} & & \Big\downarrow \begin{pmatrix} * & 0 \\ * & * \end{pmatrix} & \\
 & \begin{array}{l}\langle \mathrm{d_{in}}\Psi_{\exp}(x_a),\ a\in A,\\ \mathrm{d_{in}}y_{a'},\ a'\in A_{\mathrm{in}}\rangle_{\mathbb{R}}\otimes_{\mathbb{R}}\mathfrak{D}\\ \langle \mathrm{d_{in}}\Psi_{\exp}(v_c),\ c\in C,\\ \mathrm{d_{in}}w_{c'},\ c'\in C_{\mathrm{in}}\rangle_{\mathbb{R}}\otimes_{\mathbb{R}}\mathfrak{D}\end{array} & = & \begin{array}{l}\langle \mathrm{d_{in}}\Psi_{\exp}(x_a),\ a\in A,\\ \mathrm{d_{in}}y_{a'},\ a'\in A_{\mathrm{in}}\rangle_{\mathbb{R}}\otimes_{\mathbb{R}}\mathfrak{D}\\ \langle \mathrm{d_{in}}\Psi_{\exp}(v_c),\ c\in C,\\ \mathrm{d_{in}}w_{c'},\ c'\in C_{\mathrm{in}}\rangle_{\mathbb{R}}\otimes_{\mathbb{R}}\mathfrak{D}\end{array} & \\
 & \Big\downarrow \begin{pmatrix} 0 & * \\ \mathrm{id} & 0 \end{pmatrix} & & \Big\downarrow (*\ *) & \\
\begin{array}{l}\langle \mathrm{d_{in}}\Psi_{\exp}(\tilde x_b),\\ b\in B\rangle_{\mathbb{R}}\otimes_{\mathbb{R}}\mathfrak{D}\\ \oplus\langle \mathrm{d_{in}}(\tilde y_{b'}),\\ b'\in B_{\mathrm{in}}\rangle_{\mathbb{R}}\otimes_{\mathbb{R}}\mathfrak{D}\end{array} & \xrightarrow{\alpha} \begin{array}{l}{}^b\Omega_{\mathfrak{C}}\otimes_{\mathfrak{C}}\mathfrak{D}\oplus\\ \langle \mathrm{d_{in}}\Psi_{\exp}(x_a),\ a\in A\rangle_{\mathbb{R}}\otimes_{\mathbb{R}}\mathfrak{D}\\ \oplus\langle \mathrm{d_{in}}(y_{a'}),\ a'\in A_{\mathrm{in}}\rangle_{\mathbb{R}}\otimes_{\mathbb{R}}\mathfrak{D}\end{array} & \xrightarrow{\beta} & {}^b\Omega_{\mathfrak{D}} & \longrightarrow 0 \\
 & \Big\downarrow & & \Big\downarrow & \\
 & 0 & & 0. &
\end{array}
$$

The middle column is the direct sum of $-\otimes_{\mathfrak{C}}\mathfrak{D}$ applied to (7.7) from Proposition 7.9(b) for (7.18) and the identity map on $\langle \mathrm{d_{in}}\Psi_{\exp}(x_a), \mathrm{d_{in}}y_{a'}\rangle_{\mathbb{R}}\otimes_{\mathbb{R}}\mathfrak{D}$, and the right-hand column is (7.7) for (7.19). Thus the columns are exact. Some diagram-chasing in exact sequences and the snake lemma then shows the third row is exact. But this is (7.16), so the proposition follows. □

Here is an analogue of Proposition 2.32.

Proposition 7.11 *Let $\mathfrak{C}$ be an interior C^∞-ring with corners and $A\subseteq\mathfrak{C}$, $A_{\mathrm{in}}\subseteq\mathfrak{C}_{\mathrm{in}}$ be subsets, and write $\mathfrak{D}=\mathfrak{C}(a^{-1}: a\in A)[a'^{-1}: a'\in A_{\mathrm{in}}]$ for the localization in Definition 4.26, with projection $\pi:\mathfrak{C}\to\mathfrak{D}$. Then $({}^b\Omega_\pi)_*: {}^b\Omega_{\mathfrak{C}}\otimes_{\mathfrak{C}}\mathfrak{D}\to{}^b\Omega_{\mathfrak{D}}$ is an isomorphism of $\mathfrak{D}$-modules.*

Proof We may write $\mathfrak{D}$ by adding generators and relations to $\mathfrak{C}$ as in (4.16), where we add an $\mathbb{R}$-generator x_a and $\mathbb{R}$-relation $a\cdot x_a=1$ for each $a\in A$, and a $[0,\infty)$-generator $y_{a'}$ and $[0,\infty)$-relation $a'\cdot y_{a'}=1$ for each $a'\in A_{\mathrm{in}}$. Then Proposition 7.10 gives an exact sequence (7.16) computing ${}^b\Omega_{\mathfrak{D}}$. But in this case the contributions from the generators $x_a, y_{a'}$ in the middle of (7.16) are exactly cancelled by the contributions from the relations $a\cdot x_a=1$, $a'\cdot y_{a'}=1$ on the left of (7.16). That is, α in (7.16) induces an isomorphism from the $\tilde x_b, \tilde y_{b'}$ terms to the $x_a, y_{a'}$ terms. So by exactness we see that the top component of β in (7.16), which is $({}^b\Omega_\pi)_*: {}^b\Omega_{\mathfrak{C}}\otimes_{\mathfrak{C}}\mathfrak{D}\to{}^b\Omega_{\mathfrak{D}}$, is an isomorphism. □

We show that the reflection functors $\Pi_{\mathrm{in}}^{\mathbb{Z}}, \Pi_{\mathbb{Z}}^{\mathrm{tf}}, \Pi_{\mathrm{tf}}^{\mathrm{sa}}, \Pi_{\mathrm{in}}^{\mathrm{sa}}$ in Theorem 4.48 do not change b-cotangent modules, in a certain sense.

Proposition 7.12 *Theorem* 4.48 *gives* $\Pi_{\mathrm{in}}^{\mathbb{Z}}:\mathbf{C^\infty Rings^c_{in}}\to\mathbf{C^\infty Rings^c_{\mathbb{Z}}}$ *left adjoint to* $\mathrm{inc}:\mathbf{C^\infty Rings^c_{\mathbb{Z}}}\hookrightarrow\mathbf{C^\infty Rings^c_{in}}$. *For* $\mathfrak{C}\in\mathbf{C^\infty Rings^c_{in}}$, *there is a natural surjective* $\pi:\mathfrak{C}\to\Pi_{\mathrm{in}}^{\mathbb{Z}}(\mathfrak{C})=:\mathfrak{D}$ *in* $\mathbf{C^\infty Rings^c_{in}}$, *the unit of the adjunction. Thus Definition* 7.4 *gives a morphism* $({}^b\Omega_\pi)_*:\pi_*({}^b\Omega_{\mathfrak{C}})={}^b\Omega_{\mathfrak{C}}\otimes_{\mathfrak{C}}\mathfrak{D}\to{}^b\Omega_{\mathfrak{D}}$ *in* $\mathfrak{D}$-mod. *Then* $({}^b\Omega_\pi)_*$ *is an isomorphism.*

The analogue holds for the other functors $\Pi_{\mathbb{Z}}^{\mathrm{tf}},\Pi_{\mathrm{tf}}^{\mathrm{sa}},\Pi_{\mathrm{in}}^{\mathrm{sa}}$ *in Theorem* 4.48.

Proof The definition of $\mathfrak{D}=\Pi_{\mathrm{in}}^{\mathbb{Z}}(\mathfrak{C})$ in the proof of Theorem 4.48 involves inductively constructing a sequence $\mathfrak{C}=\mathfrak{C}^0\twoheadrightarrow\mathfrak{C}^1\twoheadrightarrow\mathfrak{C}^2\twoheadrightarrow\cdots$ and setting $\mathfrak{D}=\varinjlim_{n=0}^\infty\mathfrak{C}^n$. Here $\mathfrak{C}^{n+1}=\mathfrak{C}^n/[g_{b'}=h_{b'},\ b'\in B^n_{\mathrm{in}}]$, where the relations are $g_{b'}=h_{b'}$ if $g_{b'},h_{b'}\in\mathfrak{C}^n_{\mathrm{in}}$ with $\pi^{\mathrm{gp}}(g_{b'})=\pi^{\mathrm{gp}}(h_{b'})$. This is equivalent to the existence of $i_{b'}\in\mathfrak{C}^n_{\mathrm{in}}$ with $g_{b'}i_{b'}=h_{b'}i_{b'}$. But then in ${}^b\Omega_{\mathfrak{C}^n}$ we have

$$\mathrm{d_{in}}g_{b'}+\mathrm{d_{in}}i_{b'}=\mathrm{d_{in}}(g_{b'}i_{b'})=\mathrm{d_{in}}(h_{b'}i_{b'})=\mathrm{d_{in}}h_{b'}+\mathrm{d_{in}}i_{b'},$$

so $\mathrm{d_{in}}g_{b'}=\mathrm{d_{in}}h_{b'}$. Applying Proposition 7.10 with $\mathfrak{C}^n$ in place of $\mathfrak{C}$ and $A=A_{\mathrm{in}}=B=\emptyset$ gives an exact sequence in $\mathfrak{C}^{n+1}$-mod:

$$\langle\mathrm{d_{in}}(\tilde y_{b'}),\ b'\in B^n_{\mathrm{in}}\rangle_{\mathbb{R}}\otimes_{\mathbb{R}}\mathfrak{D}\xrightarrow{\ \alpha\ }{}^b\Omega_{\mathfrak{C}^n}\otimes_{\mathfrak{C}^n}\mathfrak{C}^{n+1}\xrightarrow{\ \beta\ }{}^b\Omega_{\mathfrak{C}^{n+1}}\to 0.$$

Here α maps $\mathrm{d_{in}}(\tilde y_{b'})\mapsto\mathrm{d_{in}}g_{b'}-\mathrm{d_{in}}h_{b'}=0$ by (7.17), so $\alpha=0$, and β is an isomorphism. Applying $-\otimes_{\mathfrak{C}^{n+1}}\mathfrak{D}$ shows that the natural map ${}^b\Omega_{\mathfrak{C}^n}\otimes_{\mathfrak{C}^n}\mathfrak{D}\to{}^b\Omega_{\mathfrak{C}^{n+1}}\otimes_{\mathfrak{C}^{n+1}}\mathfrak{D}$ is an isomorphism. Thus by induction on n and $\mathfrak{C}^0=\mathfrak{C}$ we see that ${}^b\Omega_{\mathfrak{C}}\otimes_{\mathfrak{C}}\mathfrak{D}\cong{}^b\Omega_{\mathfrak{C}^n}\otimes_{\mathfrak{C}^n}\mathfrak{D}$ for all $n\geqslant 0$.

Now $\mathfrak{D}\cong\mathfrak{C}/[\tilde g_{b'}=\tilde h_{b'},\ b'\in\coprod_{n\geqslant 0}B^n_{\mathrm{in}}]$, where the relations $\tilde g_{b'}=\tilde h_{b'}$ are lifts to $\mathfrak{C}$ of all relations $g_{b'}=h_{b'}$ in $\mathfrak{C}^{n+1}=\mathfrak{C}^n/[g_{b'}=h_{b'},\ b'\in B^n_{\mathrm{in}}]$ for all $n\geqslant 0$. Since $\mathrm{d_{in}}g_{b'}=\mathrm{d_{in}}h_{b'}$ in ${}^b\Omega_{\mathfrak{C}^n}$, and hence in ${}^b\Omega_{\mathfrak{C}^n}\otimes_{\mathfrak{C}^n}\mathfrak{D}$, and ${}^b\Omega_{\mathfrak{C}}\otimes_{\mathfrak{C}}\mathfrak{D}\cong{}^b\Omega_{\mathfrak{C}^n}\otimes_{\mathfrak{C}^n}\mathfrak{D}$, we see that $\mathrm{d_{in}}\tilde g_{b'}=\mathrm{d_{in}}\tilde h_{b'}$ in ${}^b\Omega_{\mathfrak{C}}\otimes_{\mathfrak{C}}\mathfrak{D}$ for all such $\tilde g_{b'},\tilde h_{b'}$. Thus in (7.16) for $\mathfrak{C},\mathfrak{D}$ with $A,A_{\mathrm{in}},B=\emptyset$ and $B_{\mathrm{in}}=\coprod_{n\geqslant 0}B^n_{\mathrm{in}}$ we again have $\alpha=0$, so $\beta=({}^b\Omega_\pi)_*:{}^b\Omega_{\mathfrak{C}}\otimes_{\mathfrak{C}}\mathfrak{D}\to{}^b\Omega_{\mathfrak{D}}$ is an isomorphism.

For $\Pi_{\mathbb{Z}}^{\mathrm{tf}}$ a similar proof works. In this case $\mathfrak{C}^{n+1}=\mathfrak{C}^n/[g_{b'}=h_{b'},\ b'\in B^n_{\mathrm{in}}]$, with relations $g_{b'}=h_{b'}$ if $g_{b'},h_{b'}\in\mathfrak{C}^n_{\mathrm{in}}$ with $\pi^{\mathrm{tf}}(g_{b'})=\pi^{\mathrm{tf}}(h_{b'})$, which (as $\mathfrak{C}^n$ is integral) is equivalent to $g^n_{b'}=h^n_{b'}$ for some $n\geqslant 1$. But then $n\,\mathrm{d_{in}}g_{b'}=\mathrm{d_{in}}(g^n_{b'})=\mathrm{d_{in}}(h^n_{b'})=n\,\mathrm{d_{in}}h_{b'}$, so $\mathrm{d_{in}}g_{b'}=\mathrm{d_{in}}h_{b'}$, and the rest of the proof is unchanged.

For $\Pi_{\mathrm{tf}}^{\mathrm{sa}}$, now $\mathfrak{C}^{n+1}=\mathfrak{C}^n[y_{b'}:b'\in B^n_{\mathrm{in}}]/[y_{b'}^{n_{b'}}=h_{b'},\ b'\in B^n_{\mathrm{in}}]$, for $h_{b'}$ in $\mathfrak{C}^n_{\mathrm{in}}\subseteq(\mathfrak{C}^n_{\mathrm{in}})^{\mathrm{gp}}$ and $n_{b'}\geqslant 2$ for which there exists (unique) $c''\in(\mathfrak{C}^n_{\mathrm{in}})^{\mathrm{gp}}\setminus\mathfrak{C}^n_{\mathrm{in}}$ with $n_{b'}\cdot c''=h_{b'}$. That is, we introduce generators $y_{b'}$ and relations $y_{b'}^{n_{b'}}=h_{b'}$ in pairs. Since $y_{b'}^{n_{b'}}=h_{b'}$ implies that $n_{b'}\cdot\mathrm{d_{in}}y_{b'}=\mathrm{d_{in}}h_{b'}$, again (7.16) implies that ${}^b\Omega_{\mathfrak{C}^n}\otimes_{\mathfrak{C}^n}\mathfrak{C}^{n+1}\to{}^b\Omega_{\mathfrak{C}^{n+1}}$ is an isomorphism, and the rest of the argument is similar. As $\Pi_{\mathrm{in}}^{\mathrm{sa}}=\Pi_{\mathrm{tf}}^{\mathrm{sa}}\circ\Pi_{\mathbb{Z}}^{\mathrm{tf}}\circ\Pi_{\mathrm{in}}^{\mathbb{Z}}$, the result for $\Pi_{\mathrm{in}}^{\mathrm{sa}}$ follows. □

7.4 B-cotangent modules of pushouts

The next theorem generalizes Theorem 2.33.

Theorem 7.13 *Suppose we are given a pushout diagram in the categories* $\mathbf{C^\infty Rings^c_{in}}$, $\mathbf{C^\infty Rings^c_{\mathbb{Z}}}$, $\mathbf{C^\infty Rings^c_{tf}}$, *or* $\mathbf{C^\infty Rings^c_{sa}}$*:*

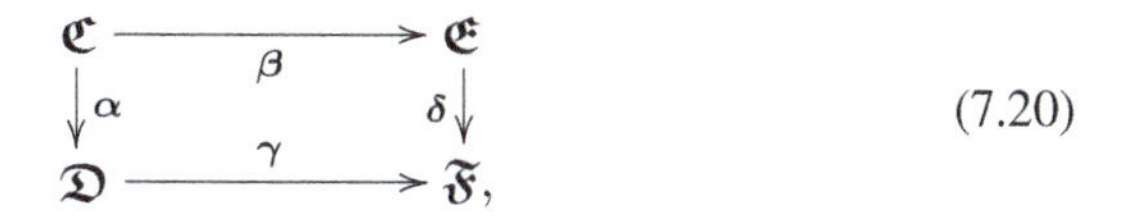

$$\begin{array}{ccc} \mathfrak{C} & \xrightarrow{\beta} & \mathfrak{E} \\ \downarrow \alpha & & \delta \downarrow \\ \mathfrak{D} & \xrightarrow{\gamma} & \mathfrak{F}, \end{array} \qquad (7.20)$$

so that $\mathfrak{F} = \mathfrak{D} \amalg_{\mathfrak{C}} \mathfrak{E}$. *Then the following sequence of* $\mathfrak{F}$*-modules is exact:*

$$ {}^b\Omega_{\mathfrak{C}} \otimes_{\mathfrak{C},\gamma\circ\alpha} \mathfrak{F} \xrightarrow{({}^b\Omega_\alpha)_* \oplus -({}^b\Omega_\beta)_*} \begin{array}{c} {}^b\Omega_{\mathfrak{D}} \otimes_{\mathfrak{D},\gamma} \mathfrak{F} \\ \oplus {}^b\Omega_{\mathfrak{E}} \otimes_{\mathfrak{E},\delta} \mathfrak{F} \end{array} \xrightarrow{({}^b\Omega_\gamma)_* \oplus ({}^b\Omega_\delta)_*} {}^b\Omega_{\mathfrak{F}} \longrightarrow 0. \quad (7.21)$$

Here $({}^b\Omega_\alpha)_* : {}^b\Omega_{\mathfrak{C}} \otimes_{\mathfrak{C},\gamma\circ\alpha} \mathfrak{F} \to {}^b\Omega_{\mathfrak{D}} \otimes_{\mathfrak{D},\gamma} \mathfrak{F}$ *is induced by* ${}^b\Omega_\alpha : {}^b\Omega_{\mathfrak{C}} \to {}^b\Omega_{\mathfrak{D}}$, *and so on. Note the sign of* $-({}^b\Omega_\beta)_*$ *in* (7.21).

Proof First suppose (7.20) is a pushout in $\mathbf{C^\infty Rings^c_{in}}$. By ${}^b\Omega_{\psi\circ\phi} = {}^b\Omega_\psi \circ {}^b\Omega_\phi$ in Definition 7.4 and commutativity of (7.20) we have ${}^b\Omega_\gamma \circ {}^b\Omega_\alpha = {}^b\Omega_{\gamma\circ\alpha} = {}^b\Omega_{\delta\circ\beta} = {}^b\Omega_\delta \circ {}^b\Omega_\beta : {}^b\Omega_{\mathfrak{C}} \to {}^b\Omega_{\mathfrak{F}}$. Tensoring with $\mathfrak{F}$ then gives $({}^b\Omega_\gamma)_* \circ ({}^b\Omega_\alpha)_* = ({}^b\Omega_\delta)_* \circ ({}^b\Omega_\beta)_* : {}^b\Omega_{\mathfrak{C}} \otimes_{\mathfrak{C}} \mathfrak{F} \to {}^b\Omega_{\mathfrak{F}}$. As the composition of morphisms in (7.21) is $({}^b\Omega_\gamma)_* \circ ({}^b\Omega_\alpha)_* - ({}^b\Omega_\delta)_* \circ ({}^b\Omega_\beta)_*$, this implies (7.21) is a complex.

Apply Proposition 4.41 to (7.20), using Proposition 4.23(b) to choose coequalizer presentations (4.22)–(4.24) for $\mathfrak{C}, \mathfrak{D}, \mathfrak{E}$ in $\mathbf{C^\infty Rings^c_{in}}$, so we have a presentation (4.25) for $\mathfrak{F}$. Thus Proposition 7.9(b) gives exact sequences

$$\begin{array}{c}\big\langle \mathrm{d_{in}}\Psi_{\exp}(\tilde s_b),\ b\in B, \\ \mathrm{d_{in}}\tilde t_{b'},\ b'\in B_{\rm in}\big\rangle_{\mathbb{R}} \otimes_{\mathbb{R}} \mathfrak{F}\end{array} \xrightarrow{\epsilon_1} \begin{array}{c}\big\langle \mathrm{d_{in}}\Psi_{\exp}(s_a),\ a\in A, \\ \mathrm{d_{in}}t_{a'},\ a'\in A_{\rm in}\big\rangle_{\mathbb{R}} \otimes_{\mathbb{R}} \mathfrak{F}\end{array} \xrightarrow{\zeta_1} {}^b\Omega_{\mathfrak{C}} \otimes_{\mathfrak{C}} \mathfrak{F} \to 0, \quad (7.22)$$

$$\begin{array}{c}\big\langle \mathrm{d_{in}}\Psi_{\exp}(\tilde u_d),\ d\in D, \\ \mathrm{d_{in}}\tilde v_{b'},\ d'\in D_{\rm in}\big\rangle_{\mathbb{R}} \otimes_{\mathbb{R}} \mathfrak{F}\end{array} \xrightarrow{\epsilon_2} \begin{array}{c}\big\langle \mathrm{d_{in}}\Psi_{\exp}(u_c),\ c\in C, \\ \mathrm{d_{in}}v_{c'},\ c'\in C_{\rm in}\big\rangle_{\mathbb{R}} \otimes_{\mathbb{R}} \mathfrak{F}\end{array} \xrightarrow{\zeta_2} {}^b\Omega_{\mathfrak{D}} \otimes_{\mathfrak{D}} \mathfrak{F} \to 0, \quad (7.23)$$

$$\begin{array}{c}\big\langle \mathrm{d_{in}}\Psi_{\exp}(\tilde w_f),\ f\in F, \\ \mathrm{d_{in}}\tilde x_{f'},\ f'\in F_{\rm in}\big\rangle_{\mathbb{R}} \otimes_{\mathbb{R}} \mathfrak{F}\end{array} \xrightarrow{\epsilon_3} \begin{array}{c}\big\langle \mathrm{d_{in}}\Psi_{\exp}(w_e),\ e\in E, \\ \mathrm{d_{in}}x_{e'},\ e'\in E_{\rm in}\big\rangle_{\mathbb{R}} \otimes_{\mathbb{R}} \mathfrak{F}\end{array} \xrightarrow{\zeta_3} {}^b\Omega_{\mathfrak{E}} \otimes_{\mathfrak{E}} \mathfrak{F} \to 0, \quad (7.24)$$

$$\begin{array}{c}\big\langle \mathrm{d_{in}}\Psi_{\exp}(\tilde y_h),\ h\in H, \\ \mathrm{d_{in}}\tilde z_{h'},\ h'\in H_{\rm in}\big\rangle_{\mathbb{R}} \otimes_{\mathbb{R}} \mathfrak{F}\end{array} \xrightarrow{\epsilon_4} \begin{array}{c}\big\langle \mathrm{d_{in}}\Psi_{\exp}(y_g),\ g\in G, \\ \mathrm{d_{in}}z_{g'},\ g'\in G_{\rm in}\big\rangle_{\mathbb{R}} \otimes_{\mathbb{R}} \mathfrak{F}\end{array} \xrightarrow{\zeta_4} {}^b\Omega_{\mathfrak{F}} \to 0, \quad (7.25)$$

where for (7.22)–(7.24) we have tensored (7.7) for $\mathfrak{C}, \mathfrak{D}, \mathfrak{E}$ with $\mathfrak{F}$, and we write $s_1, \ldots, \tilde z_{h'}$ for the generators in $\mathfrak{F}^{A,A_{\rm in}}, \ldots, \mathfrak{F}^{H,H_{\rm in}}$ as shown.

Now consider the diagram

$$
\begin{array}{ccccccc}
\begin{array}{l}\langle \mathrm{d}_{\mathrm{in}}\Psi_{\exp}(s_a),\ a\in A,\\ \mathrm{d}_{\mathrm{in}}t_{a'},\ a'\in A_{\mathrm{in}}\rangle_{\mathbb{R}}\otimes_{\mathbb{R}}\mathfrak{F}\oplus\\ \langle \mathrm{d}_{\mathrm{in}}\Psi_{\exp}(\tilde u_d),\ d\in D,\\ \mathrm{d}_{\mathrm{in}}\tilde v_{b'},\ d'\in D_{\mathrm{in}}\rangle_{\mathbb{R}}\otimes_{\mathbb{R}}\mathfrak{F}\oplus\\ \langle \mathrm{d}_{\mathrm{in}}\Psi_{\exp}(\tilde w_f),\ f\in F,\\ \mathrm{d}_{\mathrm{in}}\tilde x_{f'},\ f'\in F_{\mathrm{in}}\rangle_{\mathbb{R}}\otimes_{\mathbb{R}}\mathfrak{F}\end{array} & \xrightarrow[\epsilon_4=\begin{pmatrix}\theta_1 & \epsilon_2 & 0\\ -\theta_2 & 0 & \epsilon_3\end{pmatrix}]{} & \begin{array}{l}\langle \mathrm{d}_{\mathrm{in}}\Psi_{\exp}(u_c),\ c\in C,\\ \mathrm{d}_{\mathrm{in}}v_{c'},\ c'\in C_{\mathrm{in}}\rangle_{\mathbb{R}}\otimes_{\mathbb{R}}\mathfrak{F}\oplus\\ \langle \mathrm{d}_{\mathrm{in}}\Psi_{\exp}(w_e),\ e\in E,\\ \mathrm{d}_{\mathrm{in}}x_{e'},\ e'\in E_{\mathrm{in}}\rangle_{\mathbb{R}}\otimes_{\mathbb{R}}\mathfrak{F}\end{array} & \xrightarrow[\zeta_4]{} & {}^b\Omega_{\mathfrak{F}} & \longrightarrow & 0x \\
\Big\downarrow{\scriptstyle(\zeta_1\ 0\ 0)} & & \Big\downarrow{\scriptstyle\begin{pmatrix}\zeta_2 & 0\\ 0 & \zeta_3\end{pmatrix}} & & \Big\|{\scriptstyle\mathrm{id}_{{}^b\Omega_{\mathfrak{F}}}} & & \\
{}^b\Omega_{\mathfrak{C}}\otimes_{\mathfrak{C}}\mathfrak{F} & \xrightarrow{\begin{pmatrix}({}^b\Omega_{\alpha})_*\\ -({}^b\Omega_{\beta})_*\end{pmatrix}} & \begin{array}{l}{}^b\Omega_{\mathfrak{D}}\otimes_{\mathfrak{D}}\mathfrak{F}\oplus\\ {}^b\Omega_{\mathfrak{E}}\otimes_{\mathfrak{E}}\mathfrak{F}\end{array} & \xrightarrow{(({}^b\Omega_{\gamma})_*\ ({}^b\Omega_{\delta})_*)} & {}^b\Omega_{\mathfrak{F}} & \longrightarrow & 0.
\end{array}
\tag{7.26}
$$

Here the top row is (7.25), except that we have substituted for $G, G_{\mathrm{in}}, H, H_{\mathrm{in}}$ in terms of $A, \ldots, F_{\mathrm{in}}$ as in (4.25), and labelled the corresponding generators as in (7.22)–(7.24). Then ϵ_4 in (7.25) is written in matrix form in (7.26) as shown, where θ_1, θ_2 are induced by the morphism $\mathfrak{F}^{A,A_{\mathrm{in}}} \to \mathfrak{F}^{C,C_{\mathrm{in}}} \otimes_\infty \mathfrak{F}^{E,E_{\mathrm{in}}} \cong \mathfrak{F}^{C\amalg E,C_{\mathrm{in}}\amalg E_{\mathrm{in}}} = \mathfrak{F}^{G,G_{\mathrm{in}}}$ lifting $\mathfrak{F}^{A,A_{\mathrm{in}}} \to \mathfrak{D} \otimes_\infty \mathfrak{E}$ chosen during the proof of Proposition 4.41. The bottom line is the complex (7.21).

The left-hand square of (7.26) commutes as $\zeta_2 \circ \epsilon_2 = \zeta_3 \circ \epsilon_3 = 0$ by exactness of (7.23)–(7.24) and $\zeta_2 \circ \theta_1 = ({}^b\Omega_\alpha)_* \circ \zeta_1$, $\zeta_3 \circ \theta_2 = ({}^b\Omega_\beta)_* \circ \zeta_1$ follow from the definition of θ_1, θ_2. The right-hand square commutes as ζ_4 and $({}^b\Omega_\gamma)_* \circ \zeta_2$ map $\mathrm{d}_{\mathrm{in}}\Psi_{\exp}(u_c) \mapsto \mathrm{d}_{\mathrm{in}}\Psi_{\exp} \circ \gamma \circ \chi(u_c)$ and $\mathrm{d}_{\mathrm{in}}v_{c'} \mapsto \mathrm{d}_{\mathrm{in}}\gamma_{\mathrm{in}} \circ \chi_{\mathrm{in}}(v_{c'})$, and similarly for $({}^b\Omega_\delta)_* \circ \zeta_3$. Hence (7.26) commutes. The columns are surjective since $\zeta_1, \zeta_2, \zeta_3$ are surjective as (7.22)–(7.24) are exact and identities are surjective.

The bottom right morphism $\big(({}^b\Omega_\gamma)_*\ ({}^b\Omega_\delta)_*\big)$ in (7.26) is surjective as ζ_4 is and the right-hand square commutes. Also surjectivity of the middle column implies that it maps $\operatorname{Ker}\zeta_4$ surjectively onto $\operatorname{Ker}\big(({}^b\Omega_\gamma)_*\ ({}^b\Omega_\delta)_*\big)$. But $\operatorname{Ker}\zeta_4 = \operatorname{Im}\epsilon_4$ as the top row is exact, so as the left-hand square commutes we see that $\big(({}^b\Omega_\alpha)_* - ({}^b\Omega_\beta)_*\big)^T$ surjects onto $\operatorname{Ker}\big(({}^b\Omega_\gamma)_*\ ({}^b\Omega_\delta)_*\big)$, and the bottom row of (7.26), which is (7.21), is exact. This proves the theorem for $\mathbf{C^\infty Rings^c_{in}}$.

Next suppose (7.20) is a pushout diagram in $\mathbf{C^\infty Rings^c_{\mathbb{Z}}}$. Let $\tilde{\mathfrak{F}} = \mathfrak{D} \amalg_{\mathfrak{C}} \mathfrak{E}$ be the pushout in $\mathbf{C^\infty Rings^c_{in}}$, which exists by Theorem 4.20(b), with projections $\tilde\gamma : \mathfrak{D} \to \tilde{\mathfrak{F}}$, $\tilde\delta : \mathfrak{E} \to \tilde{\mathfrak{F}}$. Since $\Pi^{\mathbb{Z}}_{\mathrm{in}} : \mathbf{C^\infty Rings^c_{in}} \to \mathbf{C^\infty Rings^c_{\mathbb{Z}}}$ in Theorem 4.48 preserves colimits as it is a left adjoint, and acts as the identity on $\mathfrak{C}, \mathfrak{D}, \mathfrak{E}$ as they are integral, we see that $\mathfrak{F} \cong \Pi^{\mathbb{Z}}_{\mathrm{in}}(\tilde{\mathfrak{F}})$. The natural projection $\pi : \tilde{\mathfrak{F}} \to \Pi^{\mathbb{Z}}_{\mathrm{in}}(\tilde{\mathfrak{F}}) \cong \mathfrak{F}$ is the morphism induced by (7.20) and the pushout property of $\tilde{\mathfrak{F}}$, so that $\gamma = \pi \circ \tilde\gamma$ and $\delta = \pi \circ \tilde\delta$.

The first part implies that (7.21) with $\tilde{\mathfrak{F}}, \tilde\gamma, \tilde\delta$ in place of $\mathfrak{F}, \gamma, \delta$ is exact.

Applying $-\otimes_{\tilde{\mathfrak{F}}} \mathfrak{F}$ to this shows the following is exact:

$$
{}^b\Omega_{\mathfrak{C}} \otimes_{\mathfrak{C},\gamma\circ\alpha} \mathfrak{F} \xrightarrow{({}^b\Omega_\alpha)_* \oplus -({}^b\Omega_\beta)_*} {}^b\Omega_{\mathfrak{D}} \otimes_{\mathfrak{D},\gamma} \mathfrak{F} \oplus {}^b\Omega_{\mathfrak{E}} \otimes_{\mathfrak{E},\delta} \mathfrak{F} \xrightarrow{({}^b\Omega_{\tilde{\gamma}})_* \oplus ({}^b\Omega_{\tilde{\delta}})_*} {}^b\Omega_{\tilde{\mathfrak{F}}} \otimes_{\tilde{\mathfrak{F}}} \mathfrak{F} \to 0. \quad (7.27)
$$

Now Proposition 7.12 shows that $({}^b\Omega_\pi)_* : {}^b\Omega_{\tilde{\mathfrak{F}}} \otimes_{\tilde{\mathfrak{F}}} \mathfrak{F} \to {}^b\Omega_{\mathfrak{F}}$ is an isomorphism. Combining this with (7.27) and $({}^b\Omega_\gamma)_* = ({}^b\Omega_\pi)_* \circ ({}^b\Omega_{\tilde{\gamma}})_*$, $({}^b\Omega_\gamma)_* = ({}^b\Omega_\pi)_* \circ ({}^b\Omega_{\tilde{\gamma}})_*$ as $\gamma = \pi \circ \tilde{\gamma}$, $\delta = \pi \circ \tilde{\delta}$ shows that (7.21) is exact. The proofs for $\mathbf{C^\infty Rings^c_{tf}}$, $\mathbf{C^\infty Rings^c_{sa}}$ are the same, using $\Pi^{\rm tf}_{\mathbb{Z}}$, $\Pi^{\rm sa}_{\rm tf}$ in Theorem 4.48. □

7.5 Sheaves of $\boldsymbol{O}_X$-modules, and (b-)cotangent sheaves

We define sheaves of $\boldsymbol{O}_X$-modules on a C^∞-ringed space with corners, as in §2.7.

Definition 7.14 Let $(X, \boldsymbol{O}_X)$ be a C^∞-ringed space with corners. A *sheaf of $\boldsymbol{O}_X$-modules*, or simply an *$\boldsymbol{O}_X$-module*, $\mathcal{E}$ on X, is a sheaf of $\mathcal{O}_X$-modules on X, as in Definition 2.55, where $\mathcal{O}_X$ is the sheaf of C^∞-rings in $\boldsymbol{O}_X$. A *morphism of sheaves of $\boldsymbol{O}_X$-modules* is a morphism of sheaves of $\mathcal{O}_X$-modules. Then $\boldsymbol{O}_X$-modules form an abelian category, which we write as $\boldsymbol{O}_X$-mod. An $\boldsymbol{O}_X$-module $\mathcal{E}$ is called a *vector bundle of rank n* if we may cover X by open $U \subseteq X$ with $\mathcal{E}|_U \cong \mathcal{O}_X|_U \otimes_{\mathbb{R}} \mathbb{R}^n$.

Pullback sheaves are defined analogously to Definition 2.56.

Definition 7.15 Let $\boldsymbol{f} = (f, f^\sharp, f^\sharp_{\rm ex}) : (X, \mathcal{O}_X, \mathcal{O}^{\rm ex}_X) \to (Y, \mathcal{O}_Y, \mathcal{O}^{\rm ex}_Y)$ be a morphism in $\mathbf{C^\infty RS^c}$, and let $\mathcal{E}$ be an $\boldsymbol{O}_Y$-module. Define the *pullback* $\boldsymbol{f}^*(\mathcal{E})$ by $\boldsymbol{f}^*(\mathcal{E}) = \underline{f}^*(\mathcal{E})$, the pullback in Definition 2.56 for $\underline{f} = (f, f^\sharp) : (X, \mathcal{O}_X) \to (Y, \mathcal{O}_Y)$. If $\phi : \mathcal{E} \to \mathcal{F}$ is a morphism of $\boldsymbol{O}_Y$-modules we have a morphism of $\boldsymbol{O}_X$-modules $\boldsymbol{f}^*(\phi) = \underline{f}^*(\phi) = f^{-1}(\phi) \otimes {\rm id}_{\mathcal{O}_X} : \boldsymbol{f}^*(\mathcal{E}) \to \boldsymbol{f}^*(\mathcal{F})$.

Example 7.16 Let X be a manifold with corners, or with g-corners, and $E \to X$ a vector bundle. Write $\boldsymbol{X} = F^{\mathbf{C^\infty Sch^c}}_{\mathbf{Man^{gc}}}(X)$ for the associated C^∞-scheme with corners. For each open set $U \subseteq X$, write $\mathcal{E}(U) = \Gamma^\infty(E|_U)$, as a module over $\mathcal{O}_X(U) = C^\infty(U)$, and for open $V \subseteq U \subseteq X$ define $\rho_{UV} : \mathcal{E}(U) \to \mathcal{E}(V)$ by $\rho_{UV}(e) = e|_V$. Then $\mathcal{E}$ is the $\boldsymbol{O}_X$-module of smooth sections of E.

If $F \to X$ is another vector bundle with $\boldsymbol{O}_X$-module $\mathcal{F}$, then vector bundle morphisms $E \to F$ are in natural 1-1 correspondence with $\mathcal{O}_X$-module morphisms $\mathcal{E} \to \mathcal{F}$. If $f : W \to X$ is a smooth map of manifolds with (g-)corners and $\boldsymbol{f} = F_{\mathbf{Man^{gc}}}^{\mathbf{C^\infty Sch^c}} : \boldsymbol{W} \to \boldsymbol{X}$ then the vector bundle $f^*(E) \to W$ corresponds naturally to the $\mathcal{O}_W$-module $\boldsymbol{f}^*(\mathcal{E})$.

Definition 7.17 Let $\boldsymbol{X} = (X, \mathcal{O}_X)$ be a C^∞-ringed space with corners, and $\underline{X} = (X, \mathcal{O}_X)$ the underlying C^∞-ringed space. Define the *cotangent sheaf* $T^*\boldsymbol{X}$ *of* $\boldsymbol{X}$ to be the cotangent sheaf $T^*\underline{X}$ of $\underline{X}$ from Definition 2.57.

If $U \subseteq X$ is open then we have an equality of $\mathcal{O}_X|_U$-modules

$$T^*\boldsymbol{U} = T^*(U, \mathcal{O}_X|_U) = T^*\boldsymbol{X}|_U.$$

Let $\boldsymbol{f} = (f, f^\sharp, f^\sharp_{\rm ex}) : \boldsymbol{X} \to \boldsymbol{Y}$ be a morphism of C^∞-ringed spaces. Define $\Omega_{\boldsymbol{f}} : \boldsymbol{f}^*(T^*\boldsymbol{Y}) \to T^*\boldsymbol{X}$ to be a morphism of the cotangent sheaves by $\Omega_{\boldsymbol{f}} = \Omega_{\underline{f}}$, where $\underline{f} = (f, f^\sharp) : (X, \mathcal{O}_X) \to (Y, \mathcal{O}_Y)$.

Definition 7.18 Let $\boldsymbol{X} = (X, \mathcal{O}_X)$ be an interior local C^∞-ringed space with corners. For each open $U \subset X$, let $\mathrm{d}_{U,\rm in} : \mathcal{O}_X^{\rm in}(U) \to {}^b\Omega_{\mathfrak{C}_U}$ be the b-cotangent module associated to the interior C^∞-ring with corners $\mathfrak{C}_U = (\mathcal{O}_X(U), \mathcal{O}_X^{\rm in}(U) \amalg \{0\})$, where $\amalg$ is the disjoint union. Here $\mathcal{O}_X^{\rm in}(U)$ is the set of all elements of $\mathcal{O}_X^{\rm ex}(U)$ that are non-zero in each stalk $\mathcal{O}_{X,x}^{\rm ex}$ for all $x \in U$. Note that $\mathcal{O}_X^{\rm in}$ is a sheaf of monoids on X.

For each open $U \subseteq X$, the b-cotangent modules ${}^b\Omega_{\mathfrak{C}_U}$ define a presheaf $\mathcal{P}^bT^*\boldsymbol{X}$ of $\mathcal{O}_X$-modules, with restriction map ${}^b\Omega_{\rho_{UV}} : {}^b\Omega_{\mathfrak{C}_U} \to {}^b\Omega_{\mathfrak{C}_V}$ defined in Definition 7.4. Denote the sheafification of this presheaf the *b-cotangent sheaf* ${}^bT^*\boldsymbol{X}$ of $\boldsymbol{X}$. Properties of sheafification imply that, for each open set $U \subseteq X$, there is a canonical morphism ${}^b\Omega_{\mathfrak{C}_U} \to {}^bT^*\boldsymbol{X}(U)$, and we have an equality of $\mathcal{O}_X|_U$-modules ${}^bT^*(U, \mathcal{O}_X|_U) = {}^bT^*\boldsymbol{X}|_U$. Also, for each $x \in X$, the stalk ${}^bT^*\boldsymbol{X}|_x \cong {}^b\Omega_{\mathcal{O}_{X,x}}$, where ${}^b\Omega_{\mathcal{O}_{X,x}}$ is the b-cotangent module of the interior local C^∞-ring with corners $\mathcal{O}_{X,x}$.

For a morphism $\boldsymbol{f} = (f, f^\sharp, f^\sharp_{\rm ex}) : \boldsymbol{X} \to \boldsymbol{Y}$ of interior local C^∞-ringed spaces, we define the morphism of b-cotangent sheaves ${}^b\Omega_{\boldsymbol{f}} : \boldsymbol{f}^*({}^bT^*\boldsymbol{Y}) \to {}^bT^*\boldsymbol{X}$ by firstly noting that $\boldsymbol{f}^*({}^bT^*\boldsymbol{Y})$ is the sheafification of the presheaf $\mathcal{P}(\boldsymbol{f}^*({}^bT^*\boldsymbol{Y}))$ acting by

$$U \longmapsto \mathcal{P}(\boldsymbol{f}^*({}^bT^*\boldsymbol{Y}))(U) = \lim_{V \supseteq f(U)} {}^b\Omega_{\mathcal{O}_Y(V)} \otimes_{\mathcal{O}_Y(V)} \mathcal{O}_X(U),$$

as in Definition 2.57. Then, following Definition 2.57, define a morphism of presheaves $\mathcal{P}^b\Omega_{\boldsymbol{f}} : \mathcal{P}(\boldsymbol{f}^*({}^bT^*\boldsymbol{Y})) \to \mathcal{P}T^*\boldsymbol{X}$ on X by

$$(\mathcal{P}^b\Omega_{\boldsymbol{f}})(U) = \lim_{V \supseteq f(U)} \big({}^b\Omega_{\boldsymbol{\rho}_{f^{-1}(V)\,U} \circ \boldsymbol{f}_\sharp(V)}\big)_*,$$

where

$$\big({}^b\Omega_{\boldsymbol{\rho}_{f^{-1}(V)\,U} \circ \boldsymbol{f}_\sharp(V)}\big)_* : {}^b\Omega_{\mathcal{O}_Y(V)} \otimes_{\mathcal{O}_Y(V)} \mathcal{O}_X(U) \to {}^b\Omega_{\mathcal{O}_X(U)} = (\mathcal{P}^bT^*\boldsymbol{X})(U)$$

is constructed as in Definition 2.29 from the C^∞-ring with corners morphisms $\boldsymbol{f}_\sharp(V) : \mathcal{O}_Y(V) \to \mathcal{O}_X(f^{-1}(V))$ from $\boldsymbol{f}_\sharp : \mathcal{O}_Y \to f_*(\mathcal{O}_X)$ corresponding to $\boldsymbol{f}^\sharp$ in $\boldsymbol{f}$ as in (2.9), and $\boldsymbol{\rho}_{f^{-1}(V)U} : \mathcal{O}_X(f^{-1}(V)) \to \mathcal{O}_X(U)$ in $\mathcal{O}_X$. Define ${}^b\Omega_{\boldsymbol{f}} : \boldsymbol{f}^*({}^bT^*\boldsymbol{Y}) \to {}^bT^*\boldsymbol{X}$ to be the induced morphism of the associated sheaves.

We prove an analogue of Theorem 2.59(e).

Proposition 7.19 *Let $\mathfrak{C} \in \mathbf{C^\infty Rings^c_{in}}$, and set $\boldsymbol{X} = \mathrm{Spec}^c_{\mathrm{in}}\,\mathfrak{C}$. Then there is a canonical isomorphism ${}^bT^*\boldsymbol{X} \cong \mathrm{MSpec}\,{}^b\Omega_{\mathfrak{C}}$ in $\mathcal{O}_X$-*mod.

Proof By Definitions 2.58 and 7.18, $\mathrm{MSpec}\,{}^b\Omega_{\mathfrak{C}}$ and ${}^bT^*\boldsymbol{X}$ are sheafifications of presheaves $\mathcal{P}\,\mathrm{MSpec}\,{}^b\Omega_{\mathfrak{C}}, \mathcal{P}{}^bT^*\boldsymbol{X}$, where for open $U \subseteq X$ we have

$$\mathcal{P}\,\mathrm{MSpec}\,{}^b\Omega_{\mathfrak{C}}(U) = {}^b\Omega_{\mathfrak{C}} \otimes_{\mathfrak{C}} \mathcal{O}_X(U) \quad \text{and} \quad \mathcal{P}{}^bT^*\boldsymbol{X}(U) = {}^b\Omega_{\mathcal{O}_X^{\mathrm{in}}(U)},$$

writing $\mathcal{O}_X^{\mathrm{in}}(U) = (\mathcal{O}_X(U), \mathcal{O}_X^{\mathrm{in}}(U) \amalg \{0\})$. We have morphisms $\Xi_{\mathfrak{C}} : \mathfrak{C} \to \mathcal{O}_X^{\mathrm{in}}(X)$ in $\mathbf{C^\infty Rings^c_{in}}$ from Definition 5.14, and restriction $\boldsymbol{\rho}_{XU} : \mathcal{O}_X^{\mathrm{in}}(X) \to \mathcal{O}_X^{\mathrm{in}}(U)$ from $\mathcal{O}_X$, and so as in Definition 7.3 a morphism of $\mathcal{O}_X(U)$-modules $\mathcal{P}\rho(U) := (\boldsymbol{\rho}_{XU} \circ \Xi_{\mathfrak{C}})_* : {}^b\Omega_{\mathfrak{C}} \otimes_{\mathfrak{C}} \mathcal{O}_X(U) \to {}^b\Omega_{\mathcal{O}_X^{\mathrm{in}}(U)}$. This defines a morphism of presheaves $\mathcal{P}\rho : \mathcal{P}\,\mathrm{MSpec}\,{}^b\Omega_{\mathfrak{C}} \to \mathcal{P}{}^bT^*\boldsymbol{X}$, and so sheafifying induces a morphism $\rho : \mathrm{MSpec}\,{}^b\Omega_{\mathfrak{C}} \to {}^bT^*\boldsymbol{X}$.

The induced morphism on stalks at $x \in X$ is $\rho_x = (\boldsymbol{\pi}_x)_* : {}^b\Omega_{\mathfrak{C}} \otimes_{\mathfrak{C}} \mathfrak{C}_x \to {}^b\Omega_{\mathfrak{C}_x}$, where $\boldsymbol{\pi}_x : \mathfrak{C} \to \mathfrak{C}_x \cong \mathcal{O}_{X,x}$ is projection to the local C^∞-ring with corners $\mathfrak{C}_x$. But $\mathfrak{C}_x$ is the localization (4.17), so Proposition 7.11 says $(\boldsymbol{\pi}_x)_*$ is an isomorphism. Hence $\rho : \mathrm{MSpec}\,\Omega_{\mathfrak{C}} \to T^*\underline{X}$ is a sheaf morphism which induces isomorphisms on stalks at all $x \in X$, so ρ is an isomorphism. □

The next theorem shows that $T^*\boldsymbol{X}, {}^bT^*\boldsymbol{X}$ are good analogues of the (b-)cotangent bundles of manifolds with (g-)corners.

Theorem 7.20 **(a)** *Let X be a manifold with corners, and $\boldsymbol{X} = F_{\mathbf{Man^c}}^{\mathbf{C^\infty Sch^c}}(X)$ the associated C^∞-scheme with corners. Then there is a natural, functorial isomorphism between the cotangent sheaf $T^*\boldsymbol{X}$ of $\boldsymbol{X}$, and the $\mathcal{O}_X$-module $\boldsymbol{T^*X}$ associated to the vector bundle $T^*X \to X$ in Example* 7.16.

(b) *Let X be a manifold with corners, or g-corners, and $\boldsymbol{X} = F_{\mathbf{Man^{gc}}}^{\mathbf{C^\infty Sch^c}}(X)$ the associated C^∞-scheme with corners. Then there is a natural, functorial isomorphism between the b-cotangent sheaf ${}^bT^*\boldsymbol{X}$ of $\boldsymbol{X}$, and the $\mathcal{O}_X$-module ${}^b\boldsymbol{T^*X}$ associated to the vector bundle ${}^bT^*X \to X$ in Example* 7.16.

Proof For (a), every $x \in X$ has arbitrarily small open neighbourhoods $x \in$

$U \subseteq X$ such that U is a manifold with faces, and ∂U has finitely many boundary components (e.g., we can take $U \cong \mathbb{R}^m_k$). Then we have

$$\mathcal{P}T^*\underline{X}(U) = \Omega_{\mathcal{O}_X(U)} = \Omega_{C^\infty(U)} \cong \Gamma^\infty(T^*X|_U) = \boldsymbol{T}^*\boldsymbol{X}(U).$$

Here in the first step $\mathcal{P}T^*\underline{X}$ is the presheaf in Definition 2.57, whose sheafification is $T^*\boldsymbol{X} = T^*\underline{X}$ by Definition 7.17. In the second step $\mathcal{O}_X(U) = C^\infty(U)$ by Definition 5.21. The third step holds by Theorem 7.7(a), and the fourth by Example 7.16. Hence we have natural isomorphisms $\mathcal{P}T^*\underline{X}(U) \cong \boldsymbol{T}^*\boldsymbol{X}(U)$ for arbitrarily small open neighbourhoods U of each $x \in X$. These isomorphisms are compatible with the restriction morphisms in $\mathcal{P}T^*\underline{X}$ and $\boldsymbol{T}^*\boldsymbol{X}$. Thus properties of sheafification give a canonical isomorphism $T^*\boldsymbol{X} \cong \boldsymbol{T}^*\boldsymbol{X}$. Part (b) is proved in the same way, using Definition 7.18, Example 4.38, and Theorem 7.7(b). □

Here is a corners analogue of Theorem 2.61. It is proved following the proof of Theorem 2.61 in [49, §5.6], but using as input the identity ${}^b\Omega_{\psi\circ\phi} = {}^b\Omega_\psi \circ {}^b\Omega_\phi$ in Definition 7.4 for (a), and Theorem 7.13 for (b). We restrict to *firm* C^∞-schemes with corners so that Proposition 5.32 applies.

Theorem 7.21 **(a)** *Let $\boldsymbol{f} : \boldsymbol{X} \to \boldsymbol{Y}$ and $\boldsymbol{g} : \boldsymbol{Y} \to \boldsymbol{Z}$ be morphisms in $\mathbf{C^\infty Sch^c_{fi,in}}$. Then in $\mathcal{O}_X$*-mod *we have*

$${}^b\Omega_{\boldsymbol{g}\circ\boldsymbol{f}} = {}^b\Omega_{\boldsymbol{f}} \circ \boldsymbol{f}^*({}^b\Omega_{\boldsymbol{g}}) : (\boldsymbol{g}\circ\boldsymbol{f})^*(T^*\boldsymbol{Z}) \longrightarrow T^*\boldsymbol{X}.$$

(b) *Suppose we are given a Cartesian square in $\mathbf{C^\infty Sch^c_{fi,in}}$, $\mathbf{C^\infty Sch^c_{fi,\mathbb{Z}}}$, $\mathbf{C^\infty Sch^c_{fi,tf}}$, or $\mathbf{C^\infty Sch^c_{to}}$:*

$$\begin{array}{ccc} \boldsymbol{W} & \xrightarrow{\quad \boldsymbol{f} \quad} & \boldsymbol{Y} \\ \downarrow{\scriptstyle \boldsymbol{e}} & & \downarrow{\scriptstyle \boldsymbol{h}} \\ \boldsymbol{X} & \xrightarrow{\quad \boldsymbol{g} \quad} & \boldsymbol{Z}, \end{array}$$

so that $\boldsymbol{W} = \boldsymbol{X} \times_{\boldsymbol{Z}} \boldsymbol{Y}$. Then the following is exact in $\mathcal{O}_W$-mod*:*

$$(\boldsymbol{g}\circ\boldsymbol{e})^*({}^bT^*\boldsymbol{Z}) \xrightarrow{\boldsymbol{e}^*({}^b\Omega_{\boldsymbol{g}})\oplus-\boldsymbol{f}^*({}^b\Omega_{\boldsymbol{h}})} \begin{array}{c}\boldsymbol{e}^*({}^bT^*\boldsymbol{X}) \\ \oplus\, \boldsymbol{f}^*({}^bT^*\boldsymbol{Y})\end{array} \xrightarrow{{}^b\Omega_{\boldsymbol{e}}\oplus{}^b\Omega_{\boldsymbol{f}}} {}^bT^*\boldsymbol{W} \to 0. \quad (7.28)$$

Example 7.22 Define morphisms $g : X \to Z$, $h : Y \to Z$ in $\mathbf{Man^c_{in}}$ by $X = [0,\infty)$, $Y = *$, $Z = \mathbb{R}$, $g(x) = x^n$ for some $n > 0$, and $h(*) = 0$. Write $\boldsymbol{X}, \boldsymbol{Y}, \boldsymbol{Z}, \boldsymbol{g}, \boldsymbol{h}$ for the images of $X, \ldots, h$ under $F^{\mathbf{C^\infty Sch^c_{to}}}_{\mathbf{Man^c_{in}}}$. Write $\boldsymbol{W} = \boldsymbol{X} \times_{\boldsymbol{g},\boldsymbol{Z},\boldsymbol{h}} \boldsymbol{Y}$ for the fibre product in $\mathbf{C^\infty Sch^c_{to}}$, which exists by Theorem 5.34 (it is also the fibre product in $\mathbf{C^\infty Sch^c_{fi,tf}}$, $\mathbf{C^\infty Sch^c_{fi,\mathbb{Z}}}$ and $\mathbf{C^\infty Sch^c_{fi,in}}$).

Now ${}^bT^*X = \langle x^{-1}\mathrm{d}x\rangle_{\mathbb{R}}$, ${}^bT^*Y = 0$ and ${}^bT^*Z = \langle \mathrm{d}z\rangle_{\mathbb{R}}$, where ${}^b\Omega_g$ maps $\mathrm{d}z \mapsto nx^n \cdot x^{-1}\mathrm{d}x$. So by Theorem 7.20(b), ${}^bT^*\boldsymbol{X}, {}^bT^*\boldsymbol{Z}$ are trivial line bundles, such that ${}^b\Omega_{\boldsymbol{g}}$ is multiplication by nx^n on $\boldsymbol{X}$, and ${}^bT^*\boldsymbol{Y} = 0$. Since $\boldsymbol{W}$ is defined by $x^n = 0$ in $\boldsymbol{X}$, we see that $\boldsymbol{e}^*({}^b\Omega_{\boldsymbol{g}}) = 0$ on $\boldsymbol{W}$. Thus (7.28) implies that ${}^bT^*\boldsymbol{W} \cong \boldsymbol{e}^*({}^bT^*\boldsymbol{X})$, so ${}^bT^*\boldsymbol{W}$ is a rank 1 vector bundle on $\boldsymbol{W}$.

We might have hoped that if $\boldsymbol{W}$ is an interior C^∞-scheme with corners such that ${}^bT^*\boldsymbol{W}$ is a vector bundle, and $\boldsymbol{W}$ satisfies a few other obvious necessary conditions (e.g. $\boldsymbol{W}$ should be Hausdorff, second countable, toric, locally finitely presented) then $\boldsymbol{W}$ should be a manifold with g-corners. However, in this example ${}^bT^*\boldsymbol{W}$ is a vector bundle, but $\boldsymbol{W}$ is not a manifold with g-corners, and there do not seem to be obvious conditions to impose which exclude $\boldsymbol{W}$.

7.6 (B-)cotangent sheaves and the corner functor

If X is a manifold with (g-)corners, then as in [46, §3.5] and Remark 3.18(d) we have an exact sequence of vector bundles on $C(X)$:

$$0 \longrightarrow {}^bT^*(C(X)) \xrightarrow{{}^b\pi_T^*} \Pi_X^*({}^bT^*X) \xrightarrow{{}^bi_T^*} {}^bN_{C(X)}^* \longrightarrow 0. \quad (7.29)$$

Note that ${}^b\pi_T^*$ is *not* the morphism ${}^b\mathrm{d}\Pi_X^*$ associated to $\Pi_X : C(X) \to X$; this is not defined as Π_X is not interior, and if it were it would map in the opposite direction. Both ${}^b\pi_T^*, {}^bi_T^*$ are specially constructed using the geometry of corners.

We will construct an analogue of (7.29) for (firm, interior) C^∞-schemes with corners. First we define an analogue for (interior) C^∞-rings with corners.

Definition 7.23 Let $\mathfrak{C}$ be an interior C^∞-ring with corners, and $P \in \mathrm{Pr}_{\mathfrak{C}}$ be a prime ideal in $\mathfrak{C}_{\mathrm{ex}}$. Then Example 4.25(b) defines a C^∞-ring with corners $\mathfrak{D} = \mathfrak{C}/{\sim_P}$, which is also interior, with (non-interior) projection $\boldsymbol{\pi}_P : \mathfrak{C} \to \mathfrak{D}$. Writing $P_{\mathrm{in}} = P \setminus \{0\} \subset \mathfrak{C}_{\mathrm{in}}$, define a monoid

$$\mathfrak{C}_{\mathrm{in}}^P = \mathfrak{C}_{\mathrm{in}}/[c' = 1 : c' \in \mathfrak{C}_{\mathrm{in}} \setminus P_{\mathrm{in}}] \cong \mathfrak{C}_{\mathrm{in}}^\sharp/[c'' = 1 : c'' \in \mathfrak{C}_{\mathrm{in}}^\sharp \setminus P_{\mathrm{in}}^\sharp], \quad (7.30)$$

with projection $\lambda^P : \mathfrak{C}_{\mathrm{in}} \to \mathfrak{C}_{\mathrm{in}}^P$. We will define an exact sequence in $\mathfrak{D}$-mod:

$${}^b\Omega_{\mathfrak{D}} \xrightarrow{{}^b\pi_{\mathfrak{C},P}} {}^b\Omega_{\mathfrak{C}} \otimes_{\mathfrak{C}} \mathfrak{D} \xrightarrow{{}^bi_{\mathfrak{C},P}} \mathfrak{C}_{\mathrm{in}}^P \otimes_{\mathbb{N}} \mathfrak{D} \longrightarrow 0. \quad (7.31)$$

Beware of a possible confusion here: in (7.30) we regard $\mathfrak{C}_{\mathrm{in}}^P$ as a monoid under multiplication, with identity 1, but in (7.31) it is better to think of $\mathfrak{C}_{\mathrm{in}}^P$ as a monoid under addition, so that the identity $1 \in \mathfrak{C}_{\mathrm{in}}^P$ is mapped to 0 in the $\mathfrak{D}$-module $\mathfrak{C}_{\mathrm{in}}^P \otimes_{\mathbb{N}} \mathfrak{D}$, that is, $1 \otimes 1 = 0$. This arises as for b-cotangent modules we think of $\mathrm{d}_{\mathrm{in}} = \mathrm{d} \circ \log$, where log maps multiplication to addition and $1 \mapsto 0$.

As in Example 4.25(b), $\mathfrak{D}_{\rm ex} = \mathfrak{C}_{\rm ex}/{\sim_P}$, where for $c_1', c_2' \in \mathfrak{C}_{\rm ex}$ we have $c_1' \sim_P c_2'$ if either $c_1', c_2' \in P$ or $c_1' = \Psi_{\rm exp}(c)c_2'$ for c in the ideal $\langle \Phi_i(P) \rangle$ in $\mathfrak{C}$ generated by $\Phi_i(P)$. Thus $\mathfrak{C}_{\rm in} \setminus P_{\rm in}$ is mapped surjectively to $\mathfrak{D}_{\rm in} = \mathfrak{D}_{\rm ex} \setminus \{0\}$.

Define a map $\mathrm{d}_{\rm in}' : \mathfrak{D}_{\rm in} \to {}^b\Omega_{\mathfrak{C}} \otimes_{\mathfrak{C}} \mathfrak{D}$ by $\mathrm{d}_{\rm in}'[c'] = (\mathrm{d}_{\rm in} c') \otimes 1$, where $c' \in \mathfrak{C}_{\rm in} \setminus P_{\rm in}$, and $[c'] \in \mathfrak{D}_{\rm in}$ is its $\sim_P$-equivalence class, and $\mathrm{d}_{\rm in} c' \in {}^b\Omega_{\mathfrak{C}}$, so that $(\mathrm{d}_{\rm in} c') \otimes 1$ lies in ${}^b\Omega_{\mathfrak{C}} \otimes_{\mathfrak{C}} \mathfrak{D}$. To show this is well defined, note that if $c_1', c_2' \in \mathfrak{C}_{\rm in} \setminus P_{\rm in}$ with $[c_1'] = [c_2']$ then $c_1' = \Psi_{\rm exp}(c)c_2'$ for $c \in \langle \Phi_i(P) \rangle$. We may write $c = a_1\Phi_i(b_1') + \cdots + a_n\Phi_i(b_n')$ for $a_1, \ldots, a_n \in \mathfrak{C}$ and $b_1', \ldots, b_n' \in P$. Hence

$$\mathrm{d}_{\rm in} c_1' = \Psi_{\rm exp}(c) \cdot \mathrm{d}_{\rm in} c_2' + \textstyle\sum_{j=1}^n c_2'\Phi_i(b_j')\big(\mathrm{d}_{\rm in}\Psi_{\rm exp}(a_j) + a_j \cdot \mathrm{d}_{\rm in} b_j'\big)$$

in ${}^b\Omega_{\mathfrak{C}}$, using Definition 7.4. Applying $-\otimes 1 : {}^b\Omega_{\mathfrak{C}} \to {}^b\Omega_{\mathfrak{C}} \otimes_{\mathfrak{C}} \mathfrak{D}$, we have $\pi_P(\Psi_{\rm exp}(c)) = 1$ as $\mathfrak{D} = \mathfrak{C}/I$ with $c \in I$, and $\pi_P(c_2'\Phi_i(b_j')) = 0$ as $b_j' \in P$ so $c_2'\Phi_i(b_j') \in \langle \Phi_i(P) \rangle$. Hence $\mathrm{d}_{\rm in} c_1' \otimes 1 = 1 \cdot (\mathrm{d}_{\rm in} c_2' \otimes 1) + 0 = \mathrm{d}_{\rm in} c_2' \otimes 1$, so $\mathrm{d}_{\rm in}'$ is well defined. As $\mathrm{d}_{\rm in}$ is a b-derivation, $\mathrm{d}_{\rm in}'$ is too. Thus by the universal property of ${}^b\Omega_{\mathfrak{D}}$ there is a unique morphism ${}^b\pi_{\mathfrak{C},P} : {}^b\Omega_{\mathfrak{D}} \to {}^b\Omega_{\mathfrak{C}} \otimes_{\mathfrak{C}} \mathfrak{D}$ in $\mathfrak{D}$-mod with $\mathrm{d}_{\rm in}' = {}^b\pi_{\mathfrak{C},P} \circ \mathrm{d}_{\rm in}$, as required in (7.31).

Next, define a map $\mathrm{d}_{\rm in}'' : \mathfrak{C}_{\rm in} \to \mathfrak{C}_{\rm in}^P \otimes_{\mathbb{N}} \mathfrak{D}$ by $\mathrm{d}_{\rm in}''(c') = \lambda^P(c') \otimes 1$. Regard $\mathfrak{C}_{\rm in}^P \otimes_{\mathbb{N}} \mathfrak{D}$ as a $\mathfrak{C}$-module via $\pi_P : \mathfrak{C} \to \mathfrak{D}$, with trivial $\mathfrak{C}$-action on the $\mathfrak{C}_{\rm in}^P$ factor, so that $c \cdot (\bar c \otimes d) = \bar c \otimes (\pi_P(c)d)$ for $c \in \mathfrak{C}$, $d \in \mathfrak{D}$ and $\bar c \in \mathfrak{C}_{\rm in}^P$. We claim that $\mathrm{d}_{\rm in}''$ is a b-derivation. To prove this, let $g : [0,\infty)^n \to [0,\infty)$ be interior and $c_1', \ldots, c_n' \in \mathfrak{C}_{\rm in}$. Write $g(x_1, \ldots, x_n) = x_1^{a_1} \cdots x_n^{a_n} \exp(f(x_1, \ldots, x_n))$, for $a_1, \ldots, a_n \geqslant 0$ and $f : [0,\infty)^n \to \mathbb{R}$ smooth. Then

$$\begin{aligned}
&\mathrm{d}_{\rm in}''\big(\Psi_g(c_1', \ldots, c_n')\big) = \mathrm{d}_{\rm in}''\big(c_1'^{a_1} \cdots c_n'^{a_n} \Psi_{\rm exp}(f(x_1, \ldots, x_n))\big) \\
&= \lambda^P\big(c_1'^{a_1} \cdots c_n'^{a_n} \Psi_{\rm exp}(f(x_1, \ldots, x_n))\big) \otimes 1 \\
&= \textstyle\sum_{j=1}^n a_j \lambda^P(c_j') \otimes 1 + \lambda^P(\Psi_{\rm exp}(f(x_1, \ldots, x_n))) \otimes 1 \\
&= \textstyle\sum_{j=1}^n a_j \mathrm{d}_{\rm in}''(c_j') + 0,
\end{aligned} \tag{7.32}$$

using in the third step that λ^P is a monoid morphism, and in the fourth that $\lambda^P(\Psi_{\rm exp}(f(x_1, \ldots, x_n))) = 1$ by (7.30) as $\Psi_{\rm exp}(f(x_1, \ldots, x_n)) \in \mathfrak{C}_{\rm in} \setminus P_{\rm in}$, so $\lambda^P(\Psi_{\rm exp}(f(x_1, \ldots, x_n))) \otimes 1 = 0$ as $-\otimes 1$ maps $1 \mapsto 0$. Now

$$g^{-1} x_j \frac{\partial g}{\partial x_j} = a_j + x_j \frac{\partial f}{\partial x_j}(x_1, \ldots, x_n).$$

Hence

$$\Phi_{g^{-1}x_i \frac{\partial g}{\partial x_i}}(c_1', \ldots, c_n') = a_j + \Phi_i(c_j') \Phi_{\frac{\partial f}{\partial x_j}}(c_1', \ldots, c_n'). \tag{7.33}$$

Combining (7.32) with π_P applied to (7.33) we see that

$$\mathrm{d}''_{\mathrm{in}}\bigl(\Psi_g(c'_1,\ldots,c'_n)\bigr) = \textstyle\sum_{j=1}^n \pi_P\bigl(\Phi_{g^{-1}x_j\frac{\partial g}{\partial x_j}}(c'_1,\ldots,c'_n)\bigr)\cdot \mathrm{d}''_{\mathrm{in}}c'_j \qquad (7.34)$$
$$- \textstyle\sum_{j=1}^n \pi_P\bigl(\Phi_i(c'_j)\bigr)\pi_P\bigl(\Phi_{\frac{\partial f}{\partial x_j}}(c'_1,\ldots,c'_n)\bigr)\cdot \mathrm{d}''_{\mathrm{in}}c'_j.$$

For each $j = 1,\ldots,n$, if $c'_j \in P$ then $\Phi_i(c'_j) \in \langle\Phi_i(P)\rangle$ so $\pi_P(\Phi_i(c'_j)) = 0$, and if $c'_j \notin P$ then $\lambda^P(c'_j) = 1$ so $\mathrm{d}''_{\mathrm{in}}c'_j = 0$. Hence the second line of (7.34) is zero. So $\mathrm{d}''_{\mathrm{in}}$ satisfies (7.1), and is a b-derivation.

Thus there exists a unique morphism ${}^b i'_{\mathfrak{C},P} : {}^b\Omega_{\mathfrak{C}} \to \mathfrak{C}^P_{\mathrm{in}} \otimes_{\mathbb{N}} \mathfrak{D}$ in $\mathfrak{C}$-mod with $\mathrm{d}''_{\mathrm{in}} = {}^b i'_{\mathfrak{C},P} \circ \mathrm{d}_{\mathrm{in}}$, by the universal property of ${}^b\Omega_{\mathfrak{C}}$. Applying $- \otimes_{\mathfrak{C}} \mathfrak{D}$ to ${}^b i'_{\mathfrak{C},P}$ and noting that $\mathfrak{D} \otimes_{\mathfrak{C}} \mathfrak{D} \cong \mathfrak{D}$ as $\pi_P : \mathfrak{C} \to \mathfrak{D}$ is surjective gives a morphism ${}^b i_{\mathfrak{C},P} : {}^b\Omega_{\mathfrak{C}} \otimes_{\mathfrak{C}} \mathfrak{D} \to \mathfrak{C}^P_{\mathrm{in}} \otimes_{\mathbb{N}} \mathfrak{D}$, as required in (7.31). Proposition 7.24 shows (7.31) is an exact sequence.

Now suppose also that $\phi : \mathfrak{C} \to \mathfrak{E}$ is a morphism in $\mathbf{C^\infty Rings^c_{in}}$, and $Q \subset \mathfrak{E}_{\mathrm{ex}}$ is a prime ideal with $P = \phi_{\mathrm{ex}}^{-1}(Q)$, and write $\mathfrak{F} = \mathfrak{E}/\!\sim_Q$. As $\phi_{\mathrm{ex}}(P) \subseteq Q$ there is a unique morphism $\psi : \mathfrak{D} \to \mathfrak{F}$ with $\psi\circ\pi_P = \pi_Q\circ\phi : \mathfrak{C} \to \mathfrak{F}$. From the definitions, it is not difficult to show that the following diagram commutes:

$$\begin{array}{ccccccc}
{}^b\Omega_{\mathfrak{D}} \otimes_{\mathfrak{D}} \mathfrak{F} & \xrightarrow[{}^b\pi_{\mathfrak{C},P}\otimes\mathrm{id}]{} & {}^b\Omega_{\mathfrak{C}} \otimes_{\mathfrak{C}} \mathfrak{F} & \xrightarrow[{}^b i_{\mathfrak{C},P}\otimes\mathrm{id}]{} & \mathfrak{C}^P_{\mathrm{in}} \otimes_{\mathbb{N}} \mathfrak{F} & \longrightarrow & 0 \\
\downarrow ({}^b\Omega_\psi)_* & & \downarrow {}^b\Omega_\phi\otimes\mathrm{id} & & \downarrow (\phi_{\mathrm{in}})_*\otimes\mathrm{id} & & \\
{}^b\Omega_{\mathfrak{F}} & \xrightarrow{{}^b\pi_{\mathfrak{E},Q}} & {}^b\Omega_{\mathfrak{E}} \otimes_{\mathfrak{E}} \mathfrak{F} & \xrightarrow{{}^b i_{\mathfrak{E},Q}} & \mathfrak{E}^Q_{\mathrm{in}} \otimes_{\mathbb{N}} \mathfrak{F} & \longrightarrow & 0.
\end{array} \qquad (7.35)$$

Proposition 7.24 *Equation* (7.31) *is an exact sequence.*

Proof Work in the situation of Definition 7.23. Then ${}^b\Omega_{\mathfrak{D}}$ is generated over $\mathfrak{D}$ by $\mathrm{d}_{\mathrm{in}}[c']$ for $c' \in \mathfrak{C}_{\mathrm{in}} \setminus P_{\mathrm{in}}$. But

$${}^b i_{\mathfrak{C},P} \circ {}^b\pi_{\mathfrak{C},P}(\mathrm{d}_{\mathrm{in}}[c']) = {}^b i_{\mathfrak{C},P}\circ \mathrm{d}'_{\mathrm{in}}[c'] = {}^b i_{\mathfrak{C},P}((\mathrm{d}_{\mathrm{in}}c')\otimes 1) = {}^b i'_{\mathfrak{C},P}(\mathrm{d}_{\mathrm{in}}c')$$
$$= \mathrm{d}''_{\mathrm{in}}(c') = \lambda^P(c') \otimes 1 = 1 \otimes 1 = 0.$$

Hence ${}^b i_{\mathfrak{C},P} \circ {}^b\pi_{\mathfrak{C},P} = 0$, so (7.31) is a complex.

By Proposition 4.23(b), $\mathfrak{C}$ fits into a coequalizer diagram in $\mathbf{C^\infty Rings^c_{in}}$:

$$\mathfrak{F}^{B,B_{\mathrm{in}}} \underset{\beta}{\overset{\alpha}{\rightrightarrows}} \mathfrak{F}^{A,A_{\mathrm{in}}} \xrightarrow{\ \phi\ } \mathfrak{C}. \qquad (7.36)$$

Define prime ideals $Q = \phi_{\mathrm{ex}}^{-1}(P) \subset \mathfrak{F}^{A,A_{\mathrm{in}}}_{\mathrm{ex}}$ and

$$R = \alpha_{\mathrm{ex}}^{-1}(Q) = (\phi_{\mathrm{ex}}\circ\alpha_{\mathrm{ex}})^{-1}(P) = (\phi_{\mathrm{ex}}\circ\beta_{\mathrm{ex}})^{-1}(P) = \beta_{\mathrm{ex}}^{-1}(Q) \subset \mathfrak{F}^{B,B_{\mathrm{in}}}_{\mathrm{ex}}.$$

As $\mathfrak{F}^{A,A_{\mathrm{in}}}, \mathfrak{F}^{B,B_{\mathrm{in}}}$ are free, prime ideals in $\mathfrak{F}^{A,A_{\mathrm{in}}}_{\mathrm{ex}}, \mathfrak{F}^{B,B_{\mathrm{in}}}_{\mathrm{ex}}$ are parametrized by subsets of $A_{\mathrm{in}}, B_{\mathrm{in}}$. Thus there are unique subsets $\dot A_{\mathrm{in}} \subseteq A_{\mathrm{in}}$, $\dot B_{\mathrm{in}} \subseteq B_{\mathrm{in}}$ with

$Q = \langle y_{a'} : a' \in \dot{A}_{\rm in} \rangle$ and $R = \langle \tilde{y}_{b'} : b' \in \dot{B}_{\rm in} \rangle$. Set $\ddot{A}_{\rm in} = A_{\rm in} \setminus \dot{A}_{\rm in}$ and $\ddot{B}_{\rm in} = B_{\rm in} \setminus \dot{B}_{\rm in}$. Then

$$\begin{aligned} \mathfrak{F}^{A,A_{\rm in}}/\!\sim_Q &= \mathfrak{F}^{A,A_{\rm in}}/[y_{a'} = 0 : a' \in \dot{A}_{\rm in}] \cong \mathfrak{F}^{A,\ddot{A}_{\rm in}}, \\ \mathfrak{F}^{B,B_{\rm in}}/\!\sim_R &= \mathfrak{F}^{B,B_{\rm in}}/[\tilde{y}_{b'} = 0 : b' \in \dot{B}_{\rm in}] \cong \mathfrak{F}^{B,\ddot{B}_{\rm in}}. \end{aligned}$$

Thus equation (7.36) descends to a coequalizer diagram for $\mathfrak{D}$:

$$\mathfrak{F}^{B,\ddot{B}_{\rm in}} \underset{\boldsymbol{\delta}}{\overset{\boldsymbol{\gamma}}{\rightrightarrows}} \mathfrak{F}^{A,\ddot{A}_{\rm in}} \xrightarrow{\ \boldsymbol{\psi}\ } \mathfrak{D}. \tag{7.37}$$

Consider the diagram in $\mathfrak{D}$-mod:

$$\begin{array}{ccccccc} \begin{array}{l}\langle {\rm d}_{\rm in}\Psi_{\rm exp}(\tilde{x}_b),\, b\in B, \\ {\rm d}_{\rm in}\tilde{y}_{b'},\, b'\in \ddot{B}_{\rm in}\rangle_{\mathbb{R}} \otimes_{\mathbb{R}} \mathfrak{D}\end{array} & \xrightarrow{\rm inc} & \begin{array}{l}\langle {\rm d}_{\rm in}\Psi_{\rm exp}(\tilde{x}_b),\, b\in B, \\ {\rm d}_{\rm in}\tilde{y}_{b'},\, b'\in B_{\rm in}\rangle_{\mathbb{R}} \otimes_{\mathbb{R}} \mathfrak{D}\end{array} & \xrightarrow{\rm project} & \begin{array}{c}\langle {\rm d}_{\rm in}\tilde{y}_{b'},\, b'\in \dot{B}_{\rm in}\rangle_{\mathbb{R}} \\ \otimes_{\mathbb{R}} \mathfrak{D}\end{array} & \rightarrow & 0 \\ \downarrow {\scriptstyle \boldsymbol{\gamma}_* - \boldsymbol{\delta}_*} & & \downarrow {\scriptstyle \boldsymbol{\alpha}_* - \boldsymbol{\beta}_*} & & \downarrow {\scriptstyle \boldsymbol{\alpha}_* - \boldsymbol{\beta}_*} & & \\ \begin{array}{l}\langle {\rm d}_{\rm in}\Psi_{\rm exp}(x_a),\, a\in A, \\ {\rm d}_{\rm in}y_{a'},\, a'\in \ddot{A}_{\rm in}\rangle_{\mathbb{R}} \otimes_{\mathbb{R}} \mathfrak{D}\end{array} & \xrightarrow{\rm inc} & \begin{array}{l}\langle {\rm d}_{\rm in}\Psi_{\rm exp}(x_a),\, a\in A, \\ {\rm d}_{\rm in}y_{a'},\, a'\in A_{\rm in}\rangle_{\mathbb{R}} \otimes_{\mathbb{R}} \mathfrak{D}\end{array} & \xrightarrow{\rm project} & \begin{array}{c}\langle {\rm d}_{\rm in}y_{a'},\, a'\in \dot{A}_{\rm in}\rangle_{\mathbb{R}} \\ \otimes_{\mathbb{R}} \mathfrak{D}\end{array} & \rightarrow & 0 \\ \downarrow {\scriptstyle \boldsymbol{\psi}_*} & & \downarrow {\scriptstyle \boldsymbol{\phi}_*} & & \downarrow {\scriptstyle {\rm d}_{\rm in}y_{a'} \mapsto [y_{a'}]} & & \\ {}^b\Omega_{\mathfrak{D}} & \xrightarrow{{}^b\pi_{\mathfrak{C},P}} & {}^b\Omega_{\mathfrak{C}} \otimes_{\mathfrak{C}} \mathfrak{D} & \xrightarrow{{}^b i_{\mathfrak{C},P}} & \mathfrak{C}^P_{\rm in} \otimes_{\mathbb{N}} \mathfrak{D} & \longrightarrow & 0 \\ \downarrow & & \downarrow & & \downarrow & & \\ 0 & & 0 & & 0. & & \end{array} \tag{7.38}$$

Here the first column is (7.7) for (7.37), and so is exact. The maps $\boldsymbol{\gamma}_*, \boldsymbol{\delta}_*, \boldsymbol{\psi}_*$ are defined as in (7.8)–(7.9) using $\boldsymbol{\gamma}, \boldsymbol{\delta}, \boldsymbol{\psi}$. The second column is $- \otimes_{\mathfrak{C}} \mathfrak{D}$ applied to (7.7) for (7.36) in the same way, and so is exact.

For the third column of (7.38), equation (7.36) and the definition of Q, R induce a coequalizer diagram in $\mathbf{Mon}$:

$$\begin{array}{l}\mathfrak{F}_{\rm in}^{B,B_{\rm in}}/[\tilde{y}' = 1 : \\ \tilde{y}' \in \mathfrak{F}_{\rm in}^{B,B_{\rm in}} \setminus R_{\rm in}]\end{array} \underset{(\beta_{\rm in})_*}{\overset{(\alpha_{\rm in})_*}{\rightrightarrows}} \begin{array}{l}\mathfrak{F}_{\rm in}^{A,A_{\rm in}}/[y' = 1 : \\ y' \in \mathfrak{F}_{\rm in}^{A,A_{\rm in}} \setminus Q_{\rm in}]\end{array} \xrightarrow{(\phi_{\rm in})_*} \begin{array}{l}\mathfrak{C}_{\rm in}^P = \mathfrak{C}_{\rm in}/[c' = 1 : \\ c' \in \mathfrak{C}_{\rm in} \setminus P_{\rm in}].\end{array}$$

Applying $- \otimes_{\mathbb{N}} \mathfrak{D}$, which preserves coequalizers, gives an exact sequence

$$\begin{array}{l}\mathfrak{F}_{\rm in}^{B,B_{\rm in}}/[\tilde{y}' = 1 : \\ \tilde{y}' \in \mathfrak{F}_{\rm in}^{B,B_{\rm in}} \setminus R_{\rm in}] \otimes_{\mathbb{N}} \mathfrak{D}\end{array} \xrightarrow{(\alpha_{\rm in})_* - (\beta_{\rm in})_*} \begin{array}{l}\mathfrak{F}_{\rm in}^{A,A_{\rm in}}/[y' = 1 : \\ y' \in \mathfrak{F}_{\rm in}^{A,A_{\rm in}} \setminus Q_{\rm in}] \otimes_{\mathbb{N}} \mathfrak{D}\end{array} \xrightarrow{(\phi_{\rm in})_*} \mathfrak{C}_{\rm in}^P \otimes_{\mathbb{N}} \mathfrak{D} \rightarrow 0.$$

But this is isomorphic to the third column of (7.38), which is therefore exact.

The top two squares of (7.38) commute by the relation between (7.36) and (7.37). The bottom two squares commute as they are (7.35) for $\boldsymbol{\phi} : \mathfrak{F}^{A,A_{\rm in}} \rightarrow \mathfrak{C}$, using (7.6). Thus (7.38) commutes. The columns are all exact, and the top two rows are obviously exact as $A_{\rm in}=\dot A_{\rm in}\amalg\ddot A_{\rm in}$, $B_{\rm in}=\dot B_{\rm in}\amalg\ddot B_{\rm in}$. The third row is a complex as above. Therefore by some standard diagram-chasing in exact sequences, the third row, which is (7.31), is exact. □

Example 7.25 Set $\mathfrak{C}=\mathbb{R}(x)[y]/[y=y^2,\ e^x y=y]$, and let $P=\langle y\rangle\subset\mathfrak{C}_{\rm ex}$. Then $\mathfrak{D}=\mathfrak{C}/\!\sim_P\,\cong\mathbb{R}(x)$. Proposition 7.10 implies that

$$
\begin{aligned}
{}^b\Omega_{\mathfrak{C}}&\cong\frac{\langle {\rm d_{in}}\Psi_{\exp}(x),{\rm d_{in}}y\rangle_{\mathbb{R}}\otimes_{\mathbb{R}}\mathfrak{C}}{{\rm d_{in}}y=2{\rm d_{in}}y,\ {\rm d_{in}}\Psi_{\exp}(x)+{\rm d_{in}}y={\rm d_{in}}y}=0,\\
{}^b\Omega_{\mathfrak{D}}&\cong\langle {\rm d_{in}}\Psi_{\exp}(x)\rangle_{\mathbb{R}}\otimes_{\mathbb{R}}\mathfrak{D}.
\end{aligned}
$$

Also (7.30) implies that $\mathfrak{C}^P_{\rm in}=\{[1],[y]\}$ with $[y]^2=[y]$, so $\mathfrak{C}^P_{\rm in}\otimes_{\mathbb{N}}\mathfrak{D}=0$. Thus in this case (7.31) becomes the exact sequence $\mathfrak{D}\to 0\to 0\to 0$.

Note in particular that in Example 7.25, in contrast to (7.29), if we add '$0\longrightarrow$' to the left of (7.31), it may no longer be exact. However, this is exact if we also assume $\mathfrak{C}$ is toric.

Proposition 7.26 *In Definition* 7.23, *suppose* $\mathfrak{C}$ *is toric. Then we may extend* (7.31) *to an exact sequence in* $\mathfrak{D}$-mod*:*

$$
0\longrightarrow {}^b\Omega_{\mathfrak{D}}\xrightarrow{{}^b\pi_{\mathfrak{C},P}}{}^b\Omega_{\mathfrak{C}}\otimes_{\mathfrak{C}}\mathfrak{D}\xrightarrow{{}^b i_{\mathfrak{C},P}}\mathfrak{C}^P_{\rm in}\otimes_{\mathbb{N}}\mathfrak{D}\longrightarrow 0. \tag{7.39}
$$

Proof We will explain how to modify the proof of Proposition 7.24. Firstly, as $\mathfrak{C}$ is toric we may take (7.36) to be a coequalizer diagram in $\mathbf{C^\infty Rings^c_{to}}$ rather than in $\mathbf{C^\infty Rings^c_{in}}$, which implies $A_{\rm in},B_{\rm in}$ are finite. We choose it so that $|A_{\rm in}|$ is as small as possible. This implies that we have a 1-1 correspondence

$$
\begin{aligned}
\phi^\sharp_{\rm in}:A_{\rm in}\cong&\bigl\{[y_{a'}]:a'\in A_{\rm in}\bigr\}\xrightarrow{\cong}\\
&\bigl\{[c']\in\mathfrak{C}^\sharp_{\rm in}\setminus\{1\}:[c']\neq[c_1'][c_2']\text{ for }[c_1'],[c_2']\in\mathfrak{C}^\sharp_{\rm in}\setminus\{1\}\bigr\},
\end{aligned}\tag{7.40}
$$

where the second line is the unique minimal set of generators of $\mathfrak{C}^\sharp_{\rm in}$.

Having fixed $A_{\rm in}$, we choose (7.36) so that $|B_{\rm in}|$ is also as small as possible, that is, we define $\mathfrak{C}$ using the fewest possible $[0,\infty)$-type relations in $\mathbf{C^\infty Rings^c_{to}}$. From (7.36) we may form diagrams

$$
\mathbb{N}^{B_{\rm in}}=(\mathfrak{F}^{B,B_{\rm in}}_{\rm in})^\sharp\ \underset{\beta^\sharp_{\rm in}}{\overset{\alpha^\sharp_{\rm in}}{\rightrightarrows}}\ \mathbb{N}^{A_{\rm in}}=(\mathfrak{F}^{A,A_{\rm in}}_{\rm in})^\sharp\xrightarrow{\phi^\sharp_{\rm in}}\mathfrak{C}^\sharp_{\rm in}, \tag{7.41}
$$

$$
0\to\begin{matrix}\mathbb{Q}^{B_{\rm in}}=\\(\mathfrak{F}^{B,B_{\rm in}}_{\rm in})^\sharp\otimes_{\mathbb{N}}\mathbb{Q}\end{matrix}\xrightarrow{\alpha^\sharp_{\rm in}-\beta^\sharp_{\rm in}}\begin{matrix}\mathbb{Q}^{A_{\rm in}}=\\(\mathfrak{F}^{A,A_{\rm in}}_{\rm in})^\sharp\otimes_{\mathbb{N}}\mathbb{Q}\end{matrix}\xrightarrow{\phi^\sharp_{\rm in}}\mathfrak{C}^\sharp_{\rm in}\otimes_{\mathbb{N}}\mathbb{Q}\to 0. \tag{7.42}
$$

Here (7.41) is a coequalizer in the category of toric monoids $\mathbf{Mon_{to}}$, as (7.36) is in $\mathbf{C^\infty Rings^c_{to}}$. This implies (7.42) is exact at the third and fourth terms.

Choosing $|B_{\rm in}|$ as small as possible also forces (7.42) to be exact at the second term. Suppose for a contradiction that $0 \neq (y_{b'} : b' \in B_{\rm in}) \in \mathrm{Ker}(\alpha^\sharp_{\rm in} - \beta^\sharp_{\rm in})$ in (7.42). By rescaling we can suppose all $y_{b'}$ lie in $\mathbb{Z} \subset \mathbb{Q}$. Write $\alpha_{\rm in}(y_{b'}) = g_{b'} \in \mathfrak{F}^{A,A_{\rm in}}_{\rm in}$ and $\beta_{\rm in}(y_{b'}) = h_{b'} \in \mathfrak{F}^{A,A_{\rm in}}_{\rm in}$ for $b' \in B_{\rm in}$. Then $(y_{b'} : b' \in B_{\rm in}) \in \mathrm{Ker}(\alpha^\sharp_{\rm in} - \beta^\sharp_{\rm in})$ implies that in $(\mathfrak{F}^{A,A_{\rm in}}_{\rm in})^\sharp$ we have

$$\prod_{b' \in B_{\rm in}: y_{b'}>0} [g_{b'}]^{y_{b'}} \prod_{b' \in B_{\rm in}: y_{b'}<0} [h_{b'}]^{-y_{b'}} = \prod_{b' \in B_{\rm in}: y_{b'}>0} [h_{b'}]^{y_{b'}} \prod_{b' \in B_{\rm in}: y_{b'}<0} [g_{b'}]^{-y_{b'}}.$$

Therefore there is a unique $f \in \mathfrak{F}^{A,A_{\rm in}}$ such that in $\mathfrak{F}^{A,A_{\rm in}}_{\rm in}$ we have

$$\begin{aligned} &\prod_{b' \in B_{\rm in}: y_{b'}>0} g_{b'}^{y_{b'}} \prod_{b' \in B_{\rm in}: y_{b'}<0} h_{b'}^{-y_{b'}} \\ &= \Psi_{\exp}(f) \prod_{b' \in B_{\rm in}: y_{b'}>0} h_{b'}^{y_{b'}} \prod_{b' \in B_{\rm in}: y_{b'}<0} g_{b'}^{-y_{b'}}. \end{aligned} \tag{7.43}$$

As $g_{b'} = h_{b'}$ in $\mathfrak{C}_{\rm in}$, and $\mathfrak{C}$ is toric, (7.43) implies that $\phi_{\rm in} \circ \Psi_{\exp}(f) = 1$ in $\mathfrak{C}_{\rm in}$, so $\phi(f) = 0$ in $\mathfrak{C}$. Conversely, if we impose the relation $f = 0$ then the other relations involved in (7.43) become dependent. Pick $b' \in B_{\rm in}$ with $y_{b'} \neq 0$. Then we may modify $\mathfrak{F}^{B,B_{\rm in}}_{\rm in}, \boldsymbol{\alpha}, \boldsymbol{\beta}$ in (7.36), replacing $B, B_{\rm in}$ by $B \amalg \{\tilde b\}, B_{\rm in} \setminus \{b\}$, where $\boldsymbol{\alpha}, \boldsymbol{\beta}$ act by $\alpha(x_{\tilde b}) = f$ and $\beta(x_{\tilde b}) = 0$. That is, we replace the $[0,\infty)$-type relation $g_{b'} = h_{b'}$ by the $\mathbb{R}$-type relation $f = 0$. The modified equation (7.36) is still a coequalizer diagram in $\mathbf{C^\infty Rings^c_{to}}$, but we have decreased $|B_{\rm in}|$, a contradiction. Thus (7.42) is exact, and applying $- \otimes_{\mathbb{Q}} \mathbb{R}$ to it is also exact.

Observe that the 1-1 correspondence (7.40) for $A_{\rm in}$ also determines $\dot A_{\rm in}, \ddot A_{\rm in}$ uniquely, as $\dot A_{\rm in}$ is identified with the intersection of the second line of (7.40) with $P^\sharp_{\rm in}$. With $|A_{\rm in}|, |B_{\rm in}|$ fixed and minimal, we choose (7.36) so that $|\dot B_{\rm in}|$ is as small as possible, or equivalently $|\ddot B_{\rm in}|$ is as large as possible. In (7.42) we have $\mathbb{Q}^{A_{\rm in}} = \mathbb{Q}^{\dot A_{\rm in}} \oplus \mathbb{Q}^{\ddot A_{\rm in}}$ and $\mathbb{Q}^{B_{\rm in}} = \mathbb{Q}^{\dot B_{\rm in}} \oplus \mathbb{Q}^{\ddot B_{\rm in}}$, with $\mathrm{Ker}\,\phi^\sharp_{\rm in} \subset \mathbb{Q}^{\rm in}$. The definitions imply that $(\alpha^\sharp_{\rm in} - \beta^\sharp_{\rm in})(\mathbb{Q}^{\ddot B_{\rm in}}) \subseteq \mathbb{Q}^{\ddot A_{\rm in}} \cap \mathrm{Ker}\,\phi^\sharp_{\rm in}$. Choosing $|\ddot B_{\rm in}|$ as large as possible forces $(\alpha^\sharp_{\rm in} - \beta^\sharp_{\rm in})(\mathbb{Q}^{\ddot B_{\rm in}}) = \mathbb{Q}^{\ddot A_{\rm in}} \cap \mathrm{Ker}\,\phi^\sharp_{\rm in}$. Taking the quotient of (7.42) by the subsequence generated by $\mathbb{Q}^{\ddot A_{\rm in}}, \mathbb{Q}^{\ddot B_{\rm in}}$ gives an exact sequence

$$0 \longrightarrow \mathbb{Q}^{\dot B_{\rm in}} \xrightarrow{\dot\alpha^\sharp_{\rm in} - \dot\beta^\sharp_{\rm in}} \mathbb{Q}^{\dot A_{\rm in}} \xrightarrow{\dot\phi^\sharp_{\rm in}} \mathfrak{C}^P_{\rm in} \otimes_{\mathbb{N}} \mathbb{Q} \longrightarrow 0, \tag{7.44}$$

where exactness at $\mathbb{Q}^{\dot B_{\rm in}}$ holds as $(\alpha^\sharp_{\rm in} - \beta^\sharp_{\rm in})(\mathbb{Q}^{\ddot B_{\rm in}}) = \mathbb{Q}^{\ddot A_{\rm in}} \cap \mathrm{Ker}\,\phi^\sharp_{\rm in}$.

Now consider the following commutative diagram extending (7.38):

$$
\begin{array}{ccccccccc}
 & & & & & & 0 & & \\
 & & & & & & \downarrow & & \\
0 & \longrightarrow & \begin{array}{l}\langle \mathrm{d_{in}}\Psi_{\exp}(\tilde{x}_b),\\ b\in B,\ \mathrm{d_{in}}\tilde{y}_{b'},\\ b'\in \ddot{B}_{\mathrm{in}}\rangle_{\mathbb{R}}\otimes_{\mathbb{R}}\mathfrak{D}\end{array} & \longrightarrow & \begin{array}{l}\langle \mathrm{d_{in}}\Psi_{\exp}(\tilde{x}_b),\\ b\in B,\ \mathrm{d_{in}}\tilde{y}_{b'},\\ b'\in B_{\mathrm{in}}\rangle_{\mathbb{R}}\otimes_{\mathbb{R}}\mathfrak{D}\end{array} & \longrightarrow & \begin{array}{c}\langle \mathrm{d_{in}}\tilde{y}_{b'},\ b'\in \dot{B}_{\mathrm{in}}\rangle_{\mathbb{R}}\\ \otimes_{\mathbb{R}}\mathfrak{D}\end{array} & \longrightarrow & 0 \\
 & & \downarrow{\scriptstyle \gamma_*-\delta_*} & & \downarrow{\scriptstyle \alpha_*-\beta_*} & & \downarrow{\scriptstyle \alpha_*-\beta_*} & & \\
0 & \longrightarrow & \begin{array}{l}\langle \mathrm{d_{in}}\Psi_{\exp}(x_a),\\ a\in A,\ \mathrm{d_{in}}y_{a'},\\ a'\in \ddot{A}_{\mathrm{in}}\rangle_{\mathbb{R}}\otimes_{\mathbb{R}}\mathfrak{D}\end{array} & \longrightarrow & \begin{array}{l}\langle \mathrm{d_{in}}\Psi_{\exp}(x_a),\\ a\in A,\ \mathrm{d_{in}}y_{a'},\\ a'\in A_{\mathrm{in}}\rangle_{\mathbb{R}}\otimes_{\mathbb{R}}\mathfrak{D}\end{array} & \longrightarrow & \begin{array}{c}\langle \mathrm{d_{in}}y_{a'},\ a'\in \dot{A}_{\mathrm{in}}\rangle_{\mathbb{R}}\\ \otimes_{\mathbb{R}}\mathfrak{D}\end{array} & \longrightarrow & 0 \\
 & & \downarrow{\scriptstyle \psi_*} & & \downarrow{\scriptstyle \phi_*} & & \downarrow{\scriptstyle \mathrm{d_{in}}y_{a'}\mapsto[y_{a'}]} & & \\
0 & \longrightarrow & {}^b\Omega_{\mathfrak{D}} & \xrightarrow{{}^b\pi_{\mathfrak{C},P}} & {}^b\Omega_{\mathfrak{C}}\otimes_{\mathfrak{C}}\mathfrak{D} & \xrightarrow{{}^b i_{\mathfrak{C},P}} & \mathfrak{C}_{\mathrm{in}}^P\otimes_{\mathbb{N}}\mathfrak{D} & \longrightarrow & 0 \\
 & & \downarrow & & \downarrow & & \downarrow & & \\
 & & 0 & & 0 & & 0, & &
\end{array}
\tag{7.45}
$$

where we have added '$0 \to$' to the rows and third column of (7.38). The first and second rows are clearly exact. As for (7.38), the first and second columns of (7.45) come from (7.7) for (7.37) and (7.36). In this case (7.36)–(7.37) are coequalizers in $\mathbf{C^\infty Rings^c_{to}}$, not in $\mathbf{C^\infty Rings^c_{in}}$. But applying Proposition 7.12 as in the last part of the proof of Theorem 7.13, we can show that Proposition 7.9 also holds for coequalizers in $\mathbf{C^\infty Rings^c_{to}}$. Thus the first and second columns of (7.45) are exact. The third column is exact as it is $-\otimes_{\mathbb{Q}}\mathfrak{D}$ applied to (7.44). Therefore by some standard diagram-chasing in exact sequences, the third row, which is (7.39), is exact. □

Next we generalize (7.29) to C^∞-schemes with corners. We restrict to *firm* interior C^∞-schemes with corners $\mathbf{C^\infty Sch^c_{fi,in}}$ so that we can use Theorem 6.17 and Proposition 6.20, though one can also prove slightly weaker results for $\mathbf{LC^\infty RS^c_{in}}$ and $\mathbf{C^\infty Sch^c_{in}}$.

Definition 7.27 Let $\mathfrak{C}$ be a firm, interior C^∞-ring with corners, and write $\boldsymbol{X} = \mathrm{Spec}^{\mathrm{c}}_{\mathrm{in}}\,\mathfrak{C}$. Let $P \in \mathrm{Pr}_{\mathfrak{C}}$ be a prime ideal in $\mathfrak{C}_{\mathrm{ex}}$. Then Example 4.25(b) defines a C^∞-ring with corners $\mathfrak{D} = \mathfrak{C}/{\sim_P}$, which is also firm and interior, with projection $\boldsymbol{\pi}_P : \mathfrak{C} \to \mathfrak{D}$. Theorem 6.17 shows there is an open and closed C^∞-subscheme $C(\boldsymbol{X})^P$ in $C(\boldsymbol{X})$ with an isomorphism $\boldsymbol{Y} = \mathrm{Spec}^{\mathrm{c}}_{\mathrm{in}}\,\mathfrak{D} \to C(\boldsymbol{X})^P$.

As $\boldsymbol{X}$ is interior, Definition 6.19 defines a sheaf of monoids $\check{M}^{\mathrm{in}}_{C(X)}$ on $C(X)$. The proof of Proposition 6.20 shows that $\check{M}^{\mathrm{in}}_{C(X)}|_{C(X)^P}$ is a constant sheaf on $C(X)^P$ with fibre $\mathfrak{C}^P_{\mathrm{in}}$ in (7.30).

Consider the diagram in $\mathcal{O}_{C(X)^P}$-mod $\cong \mathcal{O}_Y$-mod:

$$\begin{array}{ccccccc}
\mathrm{MSpec}\,{}^b\Omega_{\mathfrak{D}} & \xrightarrow{\mathrm{MSpec}\,{}^b\pi_{\mathfrak{C},P}} & \mathrm{MSpec}({}^b\Omega_{\mathfrak{C}}\otimes_{\mathfrak{C}}\mathfrak{D}) & \xrightarrow{\mathrm{MSpec}\,{}^b i_{\mathfrak{C},P}} & \mathrm{MSpec}(\mathfrak{C}_{\mathrm{in}}^P\otimes_{\mathbb{N}}\mathfrak{D}) & \longrightarrow & 0 \\
\downarrow\cong & & \downarrow\cong & & \downarrow\cong & & \\
{}^bT^*C(\boldsymbol{X})^P & \xrightarrow{{}^b\pi_{\boldsymbol{X}}|_{C(X)^P}} & \boldsymbol{\Pi}_{\boldsymbol{X}}^*({}^bT^*\boldsymbol{X})|_{C(X)^P} & \xrightarrow{{}^b i_{\boldsymbol{X}}|_{C(X)^P}} & \check{M}_{C(X)}^{\mathrm{in}}\otimes_{\mathbb{N}}\mathcal{O}_{C(X)^P} & \longrightarrow & 0.
\end{array} \tag{7.46}$$

Here the top row is the exact functor MSpec : $\mathfrak{D}$-mod $\to \mathcal{O}_Y$-mod from Definition 2.58 applied to (7.31), and so is exact. The isomorphism in the left-hand column comes from Proposition 7.19 for $\mathfrak{D}$, as $\mathrm{Spec}^{\mathrm{c}}_{\mathrm{in}}\,\mathfrak{D}\cong C(\boldsymbol{X})^P$. The isomorphism in the middle column is $\boldsymbol{\Pi}_{\boldsymbol{X}}^*$ applied to the isomorphism $\mathrm{MSpec}\,{}^b\Omega_{\mathfrak{C}}\cong{}^bT^*\boldsymbol{X}$ from Proposition 7.19. The isomorphism in the right-hand column holds as $\check{M}_{C(X)}^{\mathrm{in}}$ is the constant sheaf on $C(X)^P$ with fibre $\mathfrak{C}_{\mathrm{in}}^P$, and $\mathrm{MSpec}\,\mathfrak{D}\cong\mathcal{O}_{C(X)^P}$.

Thus there exist unique morphisms ${}^b\pi_{\boldsymbol{X}}|_{C(X)^P}, {}^b i_{\boldsymbol{X}}|_{C(X)^P}$ making (7.46) commute, and the bottom line of (7.46) is exact. As $C(\boldsymbol{X})=\coprod_{P\in\mathrm{Pr}_{\mathfrak{C}}}C(\boldsymbol{X})^P$ by (6.20), there is a unique exact sequence in $\mathcal{O}_{C(X)}$-mod:

$${}^bT^*C(\boldsymbol{X})\xrightarrow{{}^b\pi_{\boldsymbol{X}}}\boldsymbol{\Pi}_{\boldsymbol{X}}^*({}^bT^*\boldsymbol{X})\xrightarrow{{}^b i_{\boldsymbol{X}}}{}^b\check{N}_{C(X)}=\check{M}_{C(X)}^{\mathrm{in}}\otimes_{\mathbb{N}}\mathcal{O}_{C(X)}\to 0, \tag{7.47}$$

which restricts to the bottom line of (7.46) on $C(X)^P$ for each P. Here we define ${}^b\check{N}_{C(X)}=\check{M}_{C(X)}^{\mathrm{in}}\otimes_{\mathbb{N}}\mathcal{O}_{C(X)}$. As $\check{M}_{C(X)}^{\mathrm{in}}$ is a locally constant sheaf of finitely generated monoids, ${}^b\check{N}_{C(X)}$ is a finite-rank locally free sheaf, that is, a vector bundle on $C(\boldsymbol{X})$, which we call the *b-conormal bundle* of $C(\boldsymbol{X})$.

We have constructed an exact sequence (7.47) in $\mathcal{O}_{C(X)}$-mod whenever $\mathfrak{C}$ is a firm, interior C^∞-ring with corners and $\boldsymbol{X}=\mathrm{Spec}^{\mathrm{c}}_{\mathrm{in}}\,\mathfrak{C}$. We now claim that if $\boldsymbol{X}\in\mathbf{C^\infty Sch^c_{fi,in}}$ then there is a unique exact sequence (7.47) in $\mathcal{O}_{C(X)}$-mod, which restricts to the sequence constructed above on $C(\boldsymbol{U})$ for any open $\boldsymbol{U}\subseteq\boldsymbol{X}$ with $\boldsymbol{U}\cong\mathrm{Spec}^{\mathrm{c}}_{\mathrm{in}}\,\mathfrak{C}$ for some $\mathfrak{C}\in\mathbf{C^\infty Rings^c_{fi,in}}$.

To see this, it is enough to show that the morphisms $({}^b\pi_{\boldsymbol{X}})_1,({}^b i_{\boldsymbol{X}})_1$ and $({}^b\pi_{\boldsymbol{X}})_2,({}^b i_{\boldsymbol{X}})_2$ constructed earlier on affine open subsets $C(\boldsymbol{U}_1),C(\boldsymbol{U}_2)\subset C(\boldsymbol{X})$ agree on overlaps $C(\boldsymbol{U}_1)\cap C(\boldsymbol{U}_2)$ for any two such $\boldsymbol{U}_1,\boldsymbol{U}_2\subset\boldsymbol{X}$ with $\boldsymbol{U}_i\cong\mathrm{Spec}^{\mathrm{c}}_{\mathrm{in}}\,\mathfrak{C}_i$ for $i=1,2$. Further, it is enough to show these morphisms agree on stalks at each $(x,P_x)\in C(\boldsymbol{U}_1)\cap C(\boldsymbol{U}_2)$. But the restriction of (7.47) to the stalk at (x,P_x) is equation (7.31) for the local C^∞-rings with corners $\mathfrak{C}_{1,x}\cong\mathcal{O}_{X,x}\cong\mathfrak{C}_{2,x}$ and $\mathfrak{D}_{1,x}\cong\mathcal{O}_{C(X),(x,P_x)}\cong\mathfrak{D}_{2,x}$. As this depends only on $\mathcal{O}_{X,x},\mathcal{O}_{C(X),(x,P_x)}$, it is independent of the choice of $\mathfrak{C}_1,\mathfrak{C}_2$, as we want.

If we suppose that $\boldsymbol{X}\in\mathbf{C^\infty Sch^c_{to}}$ then we can use (7.39) instead of (7.31), and so get an extended exact sequence in $\mathcal{O}_{C(X)}$-mod:

$$0 \longrightarrow {}^bT^*C(\boldsymbol{X}) \xrightarrow{{}^b\pi_{\boldsymbol{X}}} \boldsymbol{\Pi}_{\boldsymbol{X}}^*({}^bT^*\boldsymbol{X}) \xrightarrow{{}^bi_{\boldsymbol{X}}} {}^b\check{N}_{C(X)} \longrightarrow 0. \quad (7.48)$$

Applying MSpec to Example 7.25 shows that if $\boldsymbol{X}$ is not toric, then (7.48) may not be exact at the second term.

Now let $\boldsymbol{f} : \boldsymbol{X} \to \boldsymbol{Y}$ be a morphism in $\mathbf{C^\infty Sch^c_{fi,in}}$. Consider the diagram

$$\begin{array}{ccccc} C(\boldsymbol{f})^*({}^bT^*C(\boldsymbol{Y})) & \xrightarrow{C(\boldsymbol{f})^*({}^b\pi_{\boldsymbol{Y}})} & \begin{array}{c} C(\boldsymbol{f})^* \circ \boldsymbol{\Pi}_{\boldsymbol{Y}}^*({}^bT^*\boldsymbol{Y}) = \\ \boldsymbol{\Pi}_{\boldsymbol{Y}}^* \circ \boldsymbol{f}^*({}^bT^*\boldsymbol{Y}) \end{array} & \xrightarrow{C(\boldsymbol{f})^*({}^bi_{\boldsymbol{Y}})} & C(\boldsymbol{f})^*({}^b\check{N}_{C(Y)}) \to 0 \\ \downarrow {}^b\Omega_{C(\boldsymbol{f})} & & \downarrow \boldsymbol{\Pi}_{\boldsymbol{X}}^*({}^b\Omega_{\boldsymbol{f}}) & & \downarrow {}^b\check{N}_{C(f)} \\ {}^bT^*C(\boldsymbol{X}) & \xrightarrow{{}^b\pi_{\boldsymbol{X}}} & \boldsymbol{\Pi}_{\boldsymbol{X}}^*({}^bT^*\boldsymbol{X}) & \xrightarrow{{}^bi_{\boldsymbol{X}}} & {}^b\check{N}_{C(X)} \longrightarrow 0, \end{array} \quad (7.49)$$

where we define

$$\begin{aligned} {}^b\check{N}_{C(f)} = \check{M}^{\mathrm{ex}}_{C(f)} \otimes_{\mathbb{N}} \mathrm{id} : C(\boldsymbol{f})^*({}^b\check{N}_{C(Y)}) &\cong C(f)^{-1}(\check{M}^{\mathrm{ex}}_{C(Y)}) \otimes_{\mathbb{N}} \mathcal{O}_{C(X)} \\ &\longrightarrow {}^b\check{N}_{C(X)} = \check{M}^{\mathrm{in}}_{C(X)} \otimes_{\mathbb{N}} \mathcal{O}_{C(X)}, \end{aligned}$$

for $\check{M}^{\mathrm{ex}}_{C(f)}$ as in (6.26). Over affine subsets of $\boldsymbol{X}, \boldsymbol{Y}$ and $C(\boldsymbol{X}), C(\boldsymbol{Y})$, equation (7.49) may be identified with MSpec applied to (7.35), using (7.46). Thus (7.49) commutes on affine open subsets covering $C(\boldsymbol{X})$, and hence it commutes on $C(\boldsymbol{X})$. So the exact sequence (7.47) is functorial over $\mathbf{C^\infty Sch^c_{fi,in}}$.

Let X be a manifold with corners, and set $\boldsymbol{X} = F^{\mathbf{C^\infty Sch^c_{to}}}_{\mathbf{Man^c_{in}}}(X)$ as in Definition 5.21. Proposition 6.12 says $C(\boldsymbol{X}) \cong F^{\mathbf{C^\infty Sch^c_{to}}}_{\mathbf{\check{M}an^c}}(C(X))$. Theorem 7.20(b) identifies ${}^bT^*\boldsymbol{X}, {}^bT^*C(\boldsymbol{X})$ with the sheaves of sections of ${}^bT^*X, {}^bT^*C(X)$. Example 6.22 identifies the sheaf of sections of $M^\vee_{C(X)}$ with $\check{M}^{\mathrm{in}}_{C(X)}$. As ${}^bN^*_{C(X)} = M^\vee_{C(X)} \otimes_{\mathbb{N}} \mathbb{R}$ on $C(X)$ and ${}^b\check{N}_{C(X)} = \check{M}^{\mathrm{in}}_{C(X)} \otimes_{\mathbb{N}} \mathcal{O}_{C(X)}$ on $C(\boldsymbol{X})$, the sheaf of sections of ${}^bN^*_{C(X)}$ is identified with ${}^b\check{N}_{C(X)}$. By comparing definitions, one can show that all these isomorphisms identify the exact sequences (7.29) on $C(X)$ and (7.48) on $C(\boldsymbol{X})$. The analogue holds for manifolds with g-corners.

8

Further generalizations and applications

In this chapter, we discuss four directions in which Chapters 4–7 can be generalized and applied. The first is to develop a theory of Synthetic Differential Geometry with corners. The second is to extend to C^∞-stacks with corners, in a similar way schemes are extended to stacks in classical Algebraic Geometry.

The third is to extend C^∞-rings via other definitions of manifolds with corners. Of particular interest are *manifolds with a-corners* [47], which have a different smooth structure at their boundary and corners. The fourth is to extend to *derived* C^∞-schemes with corners and C^∞-stacks in the sense of Derived Algebraic Geometry, and hence to define *derived manifolds with corners* and *derived orbifolds with corners*, a key motivation for this book.

8.1 Synthetic Differential Geometry with corners

In §2.9.1 we discussed Synthetic Differential Geometry, a subject in which one proves theorems about manifolds in Differential Geometry by reasoning using 'infinitesimals', as in Kock [52, 53]. We explained that C^∞-schemes are used to provide a 'model' for Synthetic Differential Geometry, and so prove that the axioms of Synthetic Differential Geometry are consistent.

It should be possible to develop a theory of *Synthetic Differential Geometry with corners*, for proving theorems about manifolds with corners, and use C^∞-schemes with corners to prove its consistency. As well as the 'number line' $R \supset \mathbb{R}$, one should consider the 'half line' $H \supset [0,\infty)$. As well as the 'double point' D in (2.10), one should consider 'infinitesimals with corners'. For example, we can define the 'half point' HP by

$$HP = \{x \in H \,|\, x = 0 : H \longrightarrow R, \quad x^2 = x : H \longrightarrow H\}.$$

This corresponds to the C^∞-scheme with corners $\mathbf{[0,\infty)}/(x = 0)[x^2 = x]$.

In §2.9.1 we explained that the tangent bundle TX of a manifold X may be written as a mapping space $TX = X^D = \mathrm{Map}_{C^\infty}(D, X)$. In a similar way, we expect that the corners $C(X)$ of a manifold with (g-)corners may be written using the 'half point' HP as $C(X) = X^{HP} = \mathrm{Map}_{C^\infty}(HP, X)$.

8.2 C^∞-stacks with corners

In classical Algebraic Geometry, schemes are generalized to (Deligne–Mumford or Artin) stacks, as in Olsson [79], Laumon and Moret-Bailly [57], and de Jong [40]. This is useful as many moduli spaces naturally have the structure of stacks, but not schemes. Stacks over a field $\mathbb{K}$ form a 2-category $\mathbf{Sta}_{\mathbb{K}}$.

The second author [49, §6–§A] extended the theory of C^∞-schemes in [49, §2–§5] to *C^∞-stacks*, including *Deligne–Mumford C^∞-stacks*, which form 2-categories $\mathbf{DMC^\infty Sta} \subset \mathbf{C^\infty Sta}$. One reason this is interesting is that the category of manifolds $\mathbf{Man}$ generalizes to the 2-category of *orbifolds* $\mathbf{Orb}$, which are roughly Deligne–Mumford stacks in manifolds. The full embedding $\mathbf{Man} \hookrightarrow \mathbf{C^\infty Sch}$ generalizes to a full embedding $\mathbf{Orb} \hookrightarrow \mathbf{DMC^\infty Sta}$.

Thus, it seems an obvious project to generalize our theory of C^∞-schemes with corners to 2-categories $\mathbf{DMC^\infty Sta^c} \subset \mathbf{C^\infty Sta^c}$ of (Deligne–Mumford) C^∞-stacks with corners, with a full embedding $\mathbf{Orb^c} \hookrightarrow \mathbf{DMC^\infty Sta^c}$ of the 2-category of orbifolds with corners.

Now most of the generalization from $\mathbf{C^\infty Sch}$ to $\mathbf{DMC^\infty Sta} \subset \mathbf{C^\infty Sta}$ in [49, §6–§A] is an exercise in the theory of stacks, which depends only on a few properties of C^∞-schemes.

(i) We define a *Grothendieck topology* $\mathcal{J}$ on $\mathbf{C^\infty Sch}$, making $(\mathbf{C^\infty Sch}, \mathcal{J})$ into a *site*, by the usual open sets and open covers for topological spaces.

This is a reasonable definition as affine C^∞-schemes are Hausdorff, and every open subset of a C^∞-scheme is a C^∞-scheme. So there are 'enough' ordinary open sets in C^∞-schemes, and we do not need an exotic notion of 'open set' to define stacks, as for the étale or smooth topologies in ordinary Algebraic Geometry.

(ii) Objects and morphisms in $\mathbf{C^\infty Sch}$ can be glued on covers, and form a sheaf. Thus $(\mathbf{C^\infty Sch}, \mathcal{J})$ is a *subcanonical site*.

(iii) It is convenient, but not essential, that all fibre products exist in $\mathbf{C^\infty Sch}$.

Parts (i) and (ii) work for $\mathbf{C^\infty Sch^c}$ and $\mathbf{C^\infty Sch^c_{in}}$. For (iii), in parts of the theory requiring fibre products, we should restrict to subcategories such as $\mathbf{C^\infty Sch^c_{fi}}$ or $\mathbf{C^\infty Sch^c_{fi,in}}$, in which fibre products exist by Theorem 5.34.

Thus we may define 2-categories $\mathbf{DMC^\infty Sta^c} \subset \mathbf{C^\infty Sta^c}$ of (Deligne–Mumford) C^∞-stacks with corners by following [49, §6–§7] with only cosmetic changes. The material of §5.6 gives interesting 2-subcategories such as toric and interior $\mathbf{DMC^\infty Sta^c_{to}} \subset \mathbf{DMC^\infty Sta^c_{in}} \subset \mathbf{DMC^\infty Sta^c}$.

Theorem 5.34 implies that fibre products exist in 2-categories of C^∞-stacks with corners corresponding to categories in (5.14), such as $\mathbf{DMC^\infty Sta^c_{to}}$, $\mathbf{DMC^\infty Sta^c_{fi,in}}$, $\mathbf{DMC^\infty Sta^c_{fi}}$. Most of the material of §7.1–§7.5 also generalizes immediately, following [49, §8]. Extending Chapter 6 and §7.6 to stacks will require some work.

8.3 C^∞-rings and C^∞-schemes with a-corners

Our notion of C^∞-ring with corners started with the categories $\mathbf{Man^c_{in}} \subset \mathbf{Man^c}$ of manifolds with corners defined in §3.1, and the full subcategories $\mathbf{Euc^c_{in}} \subset \mathbf{Man^c_{in}}$, $\mathbf{Euc^c} \subset \mathbf{Man^c}$ with objects $\mathbb{R}^m_k = [0,\infty)^k \times \mathbb{R}^{m-k}$ for $0 \leqslant k \leqslant m$.

As we explained in Remark 3.3, there are several non-equivalent definitions of categories of manifolds with corners in the literature. So we can consider alternative theories of C^∞-rings and C^∞-schemes with corners starting with one of these categories instead of $\mathbf{Man^c}$. In particular:

(a) If we use the categories $\mathbf{Man^f_{in}} \subset \mathbf{Man^f}$ of *manifolds with faces* in Definition 3.13, following Melrose [69, 70, 71, 72], we do not change $\mathbf{Euc^c_{in}} \subset \mathbf{Euc^c}$, so the theory is unchanged.

(b) The category $\mathbf{Man^c_{st}}$ of manifolds with corners and *strongly smooth maps*, from Definition 3.2 and [41], gives a theory which can be embedded as subcategories $\mathbf{C^\infty Rings^c_{st}} \subset \mathbf{C^\infty Rings^c}$, $\mathbf{C^\infty Sch^c_{st}} \subset \mathbf{C^\infty Sch^c}$, as for the embeddings $\mathbf{C^\infty Rings^c_{in}} \subset \mathbf{C^\infty Rings^c}$, $\mathbf{C^\infty Sch^c_{in}} \subset \mathbf{C^\infty Sch^c}$, since $\mathbf{Man^c_{st}} \subset \mathbf{Man^c}$. The theory also seems less interesting (roughly, $\mathfrak{C}_{\text{st-ex}}$ is a set, compared with $\mathfrak{C}_{\text{ex}}$ being a monoid).

(c) The category $\mathbf{Man^c_{we}}$ of manifolds with corners and *weakly smooth maps*, from Definition 3.2, Cerf [14, §I.1.2], and other authors, yields a theory which is essentially equivalent to ordinary C^∞-schemes. As weakly smooth maps have no compatibility with boundaries, and no notion of 'interior', corner functors in Chapter 6 and b-cotangent modules in Chapter 7 no longer work.

(d) The second author [47] defined the category $\mathbf{Man^{ac}}$ of *manifolds with analytic corners*, or *manifolds with a-corners*, which we explain next.

We recall the definition of manifolds with a-corners from [47, §3].

Definition 8.1 We write $[\![0,\infty)$ to mean $[0,\infty)$, but the notation emphasizes that $[\![0,\infty)$ has a different kind of boundary (an *a-boundary*) at 0, giving a different notion of smooth function on $[\![0,\infty)$. We write other intervals such as $[\![0,1]\!]=[0,1]$ in the same way. For $0\leqslant k\leqslant m$ we write $\mathbb{R}^{k,m}=[\![0,\infty)^k\times\mathbb{R}^{m-k}$.

Let $U\subseteq\mathbb{R}^{k,m}$ be open and $f:U\to\mathbb{R}$ be continuous. Write points of U as $(x_1,\ldots,x_m)$ with $x_1,\ldots,x_k\in[\![0,\infty)$ and $x_{k+1},\ldots,x_m\in\mathbb{R}$. The *b-derivative* of f (if it exists) is a map ${}^b\partial f:U\to\mathbb{R}^m$, written ${}^b\partial f=({}^b\partial_1 f,\ldots,{}^b\partial_m f)$ for ${}^b\partial_i f:U\to\mathbb{R}$, where by definition

$$ {}^b\partial_i f(x_1,\ldots,x_m)=\begin{cases}0, & x_i=0,\ i=1,\ldots,k,\\ x_i\frac{\partial f}{\partial x_i}(x_1,\ldots,x_m), & x_i>0,\ i=1,\ldots,k,\\ \frac{\partial f}{\partial x_i}(x_1,\ldots,x_m), & i=k+1,\ldots,m.\end{cases} \tag{8.1} $$

We say that ${}^b\partial f$ *exists* if (8.1) is well defined, that is, if $\frac{\partial f}{\partial x_i}$ exists on $U\cap\{x_i>0\}$ if $i=1,\ldots,k$, and $\frac{\partial f}{\partial x_i}$ exists on U if $i=k+1,\ldots,m$.

We can iterate b-derivatives (if they exist), to get maps ${}^b\partial^l f:U\to\bigotimes^l\mathbb{R}^m$ for $l=0,1,\ldots$, by taking b-derivatives of components of ${}^b\partial^j f$ for $j=0,\ldots,l-1$.

(i) We say that f is *roughly differentiable*, or *r-differentiable*, if ${}^b\partial f$ exists and is a continuous map ${}^b\partial f:U\to\mathbb{R}^m$.
(ii) We say that f is *roughly smooth*, or *r-smooth*, if ${}^b\partial^l f:U\to\bigotimes^l\mathbb{R}^m$ is r-differentiable for all $l=0,1,\ldots$.
(iii) We say that f is *analytically differentiable*, or *a-differentiable*, if it is r-differentiable and for any compact subset $S\subseteq U$ and $i=1,\ldots,k$, there exist positive constants C,α such that

$$ \left|{}^b\partial_i f(x_1,\ldots,x_m)\right|\leqslant Cx_i^\alpha \qquad \text{for all } (x_1,\ldots,x_m)\in S. $$

(iv) We say that f is *analytically smooth*, or *a-smooth*, if ${}^b\partial^l f:U\to\bigotimes^l\mathbb{R}^m$ is a-differentiable for all $l=0,1,\ldots$.

One can show that f is a-smooth if for all $a_1,\ldots,a_m\in\mathbb{N}$ and for any compact subset $S\subseteq U$, there exist positive constants C,α such that

$$ \left|\frac{\partial^{a_1+\cdots+a_m}}{\partial x_1^{a_1}\cdots\partial x_m^{a_m}}f(x_1,\ldots,x_m)\right|\leqslant C\prod_{i=1,\ldots,k:\,a_i>0}x_i^{\alpha-a_i} $$

$$ \text{for all } (x_1,\ldots,x_m)\in S \text{ with } x_i>0 \text{ if } i=1,\ldots,k \text{ with } a_i>0, $$

where continuous partial derivatives must exist at the required points.

If $f,g:U\to\mathbb{R}$ are a-smooth (or r-smooth) and $\lambda,\mu\in\mathbb{R}$ then $\lambda f+\mu g$ and

$fg : U \to \mathbb{R}$ are a-smooth (or r-smooth). Thus, the set $C^\infty(U)$ of a-smooth functions $f : U \to \mathbb{R}$ is an $\mathbb{R}$-algebra, and in fact a C^∞-ring.

If $I \subseteq \mathbb{R}$ is an open interval, such as $I = (0, \infty)$, we say that a map $f : U \to I$ is *a-smooth*, or just *smooth*, if it is a-smooth as a map $f : U \to \mathbb{R}$.

Definition 8.2 Let $U \subseteq \mathbb{R}^{k,m}$ and $V \subseteq \mathbb{R}^{l,n}$ be open, and $f = (f_1, \ldots, f_n) : U \to V$ be a continuous map, so that $f_j = f_j(x_1, \ldots, x_m)$ maps $U \to [\![0, \infty)$ for $j = 1, \ldots, l$ and $U \to \mathbb{R}$ for $j = l+1, \ldots, n$. Then we say:

(a) f is *r-smooth* if $f_j : U \to \mathbb{R}$ is r-smooth in the sense of Definition 8.1 for $j = l+1, \ldots, n$, and every $u = (x_1, \ldots, x_m) \in U$ has an open neighbourhood $\tilde{U}$ in U such that for each $j = 1, \ldots, l$, either:
  (i) we may write $f_j(\tilde{x}_1, \ldots, \tilde{x}_m) = F_j(\tilde{x}_1, \ldots, \tilde{x}_m) \cdot \tilde{x}_1^{a_{1,j}} \cdots \tilde{x}_k^{a_{k,j}}$ for all $(\tilde{x}_1, \ldots, \tilde{x}_m) \in \tilde{U}$, where $F_j : \tilde{U} \to (0, \infty)$ is r-smooth as in Definition 8.1, and $a_{1,j}, \ldots, a_{k,j} \in [0, \infty)$, with $a_{i,j} = 0$ if $x_i \neq 0$; or
  (ii) $f_j|_{\tilde{U}} = 0$.

(b) f is *a-smooth* if $f_j : U \to \mathbb{R}$ is a-smooth in the sense of Definition 8.1 for $j = l+1, \ldots, n$, and every $u = (x_1, \ldots, x_m) \in U$ has an open neighbourhood $\tilde{U}$ in U such that for each $j = 1, \ldots, l$, either:
  (i) we may write $f_j(\tilde{x}_1, \ldots, \tilde{x}_m) = F_j(\tilde{x}_1, \ldots, \tilde{x}_m) \cdot \tilde{x}_1^{a_{1,j}} \cdots \tilde{x}_k^{a_{k,j}}$ for all $(\tilde{x}_1, \ldots, \tilde{x}_m) \in \tilde{U}$, where $F_j : \tilde{U} \to (0, \infty)$ is a-smooth as in Definition 8.1, and $a_{1,j}, \ldots, a_{k,j} \in [0, \infty)$, with $a_{i,j} = 0$ if $x_i \neq 0$; or
  (ii) $f_j|_{\tilde{U}} = 0$.

(c) f is *interior* if it is a-smooth, and case (b)(ii) does not occur.

(d) f is an *a-diffeomorphism* if it is an a-smooth bijection with a-smooth inverse.

Definition 8.3 We define the category $\mathbf{Man^{ac}}$ of *manifolds with a-corners* X and *a-smooth maps* $f : X \to Y$ following the definition of $\mathbf{Man^c}$ in Definition 3.2, but using *a-charts* (U, ϕ) with $U \subseteq \mathbb{R}^{k,m}$ open, where a-charts $(U, \phi), (V, \psi)$ are *compatible* if $\psi^{-1} \circ \phi$ is an a-diffeomorphism between open subsets of $\mathbb{R}^{k,m}, \mathbb{R}^{l,m}$, in the sense of Definition 8.2(d). We also define an a-smooth map $f : X \to Y$ to be *interior* if it is locally modelled on interior maps between open subsets of $\mathbb{R}^{k,m}, \mathbb{R}^{l,n}$, in the sense of Definition 8.2(c), and we write $\mathbf{Man^{ac}_{in}} \subset \mathbf{Man^{ac}}$ for the subcategory with interior morphisms.

In [47, §3.5] the second author also defines the category $\mathbf{Man^{c,ac}}$ of *manifolds with corners and a-corners*, which are locally modelled on $\mathbb{R}^{l,m}_k = [0, \infty)^k \times [\![0, \infty)^l \times \mathbb{R}^{m-k-l}$, with smooth functions defined as in Definitions 8.1–8.2 for the $[\![0, \infty)$ factors, and as in Definition 3.1 for the $[0, \infty)$ factors. There are full embeddings $\mathbf{Man^c} \hookrightarrow \mathbf{Man^{c,ac}}$ and $\mathbf{Man^{ac}} \hookrightarrow \mathbf{Man^{c,ac}}$.

For $X \in \mathbf{Man^{c,ac}}$, the boundary $\partial X = \partial^c X \amalg \partial^{ac} X$ decomposes into the ordinary boundary $\partial^c X$ and the a-boundary $\partial^{ac} X$.

Remark 8.4 **(a)** Manifolds with a-corners are significantly different to manifolds with corners. Even the simplest examples $[\![0,\infty)$ in $\mathbf{Man^{ac}}$ and $[0,\infty)$ in $\mathbf{Man^c}$ have different smooth structures. For example, $x^\alpha : [\![0,\infty) \to \mathbb{R}$ is a-smooth for all $\alpha \in [0,\infty)$, but $x^\alpha : [0,\infty) \to \mathbb{R}$ is only smooth for $\alpha \in \mathbb{N}$.

(b) Here are two of the reasons for introducing manifolds with a-corners in [47].

Firstly, boundary conditions for partial differential equations are usually of two types.

(i) Boundary 'at finite distance'. For example, solve $\frac{\partial u}{\partial x^2} + \frac{\partial u}{\partial y^2} = f$ on the closed unit disc $D^2 \subset \mathbb{R}^2$ with boundary condition $u|_{\partial D^2} = g$.
(ii) Boundary 'at infinite distance', or 'of asymptotic type'. For example, solve $\frac{\partial u}{\partial x^2} + \frac{\partial u}{\partial y^2} = f$ on $\mathbb{R}^2$ with asymptotic condition $u = O((x^2 + y^2)^\alpha)$, $\alpha < 0$.

We argue in [47] that it is natural to write type (i) using ordinary manifolds with corners, but type (ii) using manifolds with a-corners. In the example, we would compactify $\mathbb{R}^2$ to a manifold with a-corners $\bar{\mathbb{R}}^2$ by adding a boundary circle $\mathcal{S}^1$ at infinity, and then u extends to a-smooth $\bar{u} : \bar{\mathbb{R}}^2 \to \mathbb{R}$.

Secondly, and related, several important areas of geometry involve forming moduli spaces $\overline{\mathcal{M}}$ of some geometric objects, such that $\overline{\mathcal{M}}$ is a manifold with corners, or a more general space such as a Kuranishi space with corners in Fukaya et al. [28, 29, 30, 31], where the 'boundary' $\partial\overline{\mathcal{M}}$ parametrizes singular objects included to make $\overline{\mathcal{M}}$ compact. Examples include moduli spaces of Morse flow lines [4, 82], of J-holomorphic curves with boundary in a Lagrangian [28, 29], and of J-holomorphic curves with cylindrical ends in Symplectic Field Theory [24].

We argue in [47] that the natural smooth structure to put on $\overline{\mathcal{M}}$ in such moduli problems is that of a manifold with a-corners (or Kuranishi space with a-corners, etc.). This resolves a number of technical problems in these theories.

We can now write down a new version of our theory, starting instead with the full subcategories $\mathbf{Euc^{ac}_{in}} \subset \mathbf{Man^{ac}_{in}}$, $\mathbf{Euc^{ac}} \subset \mathbf{Man^{ac}}$ with objects $\mathbb{R}^{l,m} = [\![0,\infty)^l \times \mathbb{R}^{m-l}$ for $0 \leqslant l \leqslant m$. This yields categories of *(pre) C^∞-rings with a-corners* $\mathbf{PC^\infty Rings^{ac}}$, $\mathbf{C^\infty Rings^{ac}}$ and *C^∞-schemes with a-corners* $\mathbf{C^\infty Sch^{ac}}$. We write $\mathfrak{C} \in \mathbf{PC^\infty Rings^{ac}}$ as $(\mathfrak{C}, \mathfrak{C}_{\text{a-ex}})$ satisfying the analogue of Definition 4.4, but for the operations Φ_f, Ψ_g we take $f : \mathbb{R}^m \times [\![0,\infty)^n \to \mathbb{R}$ and $g : \mathbb{R}^m \times [\![0,\infty)^n \to [\![0,\infty)$ to be a-smooth.

Similarly, we could start instead with the full subcategories $\mathbf{Euc^{c,ac}_{in}} \subset$

$\mathbf{Man}^{\mathbf{c,ac}}_{\mathbf{in}}$, $\mathbf{Euc}^{\mathbf{c,ac}} \subset \mathbf{Man}^{\mathbf{c,ac}}$ with objects $\mathbb{R}^{l,m}_k = [0,\infty)^k \times [\![0,\infty)^l \times \mathbb{R}^{m-k-l}$ for $m \geqslant k+l$. This yields categories of *(pre)* C^∞*-rings with corners and a-corners* $\mathbf{PC^\infty Rings^{c,ac}}$, $\mathbf{C^\infty Rings^{c,ac}}$ and C^∞*-schemes with corners and a-corners* $\mathbf{C^\infty Sch^{c,ac}}$. We write $\boldsymbol{\mathfrak{C}} \in \mathbf{PC^\infty Rings^{c,ac}}$ as $(\mathfrak{C}, \mathfrak{C}_{\text{ex}}, \mathfrak{C}_{\text{a-ex}})$, where a functor $F : \mathbf{Euc^{c,ac}} \to \mathbf{Sets}$ corresponds to $\boldsymbol{\mathfrak{C}} = (\mathfrak{C}, \mathfrak{C}_{\text{ex}}, \mathfrak{C}_{\text{a-ex}})$ with $\mathfrak{C} = F(\mathbb{R})$, $\mathfrak{C}_{\text{ex}} = F([0,\infty))$ and $\mathfrak{C}_{\text{a-ex}} = F([\![0,\infty))$.

Much of Chapters 4–7 generalizes immediately to the a-corners case, with only obvious changes. Here are some important differences, though.

(a) There should be a category $\mathbf{Man^{gac}}$ of 'manifolds with generalized a-corners', which relates to $\mathbf{Man^{ac}}$ as $\mathbf{Man^{gc}}$ relates to $\mathbf{Man^{c}}$. But no such category has been written down yet.

(b) The special classes of C^∞-rings with corners $\boldsymbol{\mathfrak{C}} = (\mathfrak{C}, \mathfrak{C}_{\text{ex}})$ in §4.7, and hence of C^∞-schemes with corners in §5.6, relied on treating $\mathfrak{C}_{\text{ex}}$ or $\mathfrak{C}_{\text{in}}$ as monoids, and imposing conditions such as $\mathfrak{C}^\sharp_{\text{ex}}$ finitely generated.

For (interior) C^∞-rings with a-corners, $\mathfrak{C}_{\text{a-ex}}$ (and $\mathfrak{C}_{\text{a-in}}$) are still monoids. However, requiring $\mathfrak{C}^\sharp_{\text{a-ex}}$ to be finitely generated is not a useful condition. For example, if $\boldsymbol{\mathfrak{C}} = C^\infty([\![0,\infty))$ then by mapping $[x^\alpha] \mapsto e^{-\alpha}$, $[0] \mapsto 0$ we may identify $\mathfrak{C}^\sharp_{\text{a-ex}} \cong [0,1]$, $\mathfrak{C}^\sharp_{\text{a-in}} \cong (0,1]$ as monoids under multiplication, and $[0,1]$, $(0,1]$ are not finitely generated monoids.

Instead, we should regard $\mathfrak{C}_{\text{a-ex}}$ and $\mathfrak{C}_{\text{a-in}}$ as $[0,\infty)$*-modules*. That is, $\mathfrak{C}_{\text{a-ex}}$ and $\mathfrak{C}_{\text{a-in}}$ are monoids under multiplication, with operations $c \mapsto c^\alpha$ for c in $\mathfrak{C}_{\text{a-ex}}$ or $\mathfrak{C}_{\text{a-in}}$ and $\alpha \in [0,\infty)$ satisfying $(cd)^\alpha = c^\alpha d^\alpha$, $(c^\alpha)^\beta = c^{\alpha\beta}$, and $c^0 = 1$. The operation $c \mapsto c^\alpha$ is $\Psi_{x^\alpha} : \mathfrak{C}_{\text{a-ex}} \to \mathfrak{C}_{\text{a-ex}}$. Monoids are to abelian groups as $[0,\infty)$-modules are to real vector spaces.

(c) Continuing (b), we should for example define $\boldsymbol{\mathfrak{C}} \in \mathbf{C^\infty Rings^{ac}}$ to be *firm* if $\mathfrak{C}^\sharp_{\text{a-ex}}$ is a *finitely generated* $[0,\infty)$*-module*, that is, there is a surjective $[0,\infty)$-module morphism $[0,\infty)^n \to \mathfrak{C}^\sharp_{\text{a-ex}}$. Key properties of firm C^∞-schemes with corners such as Proposition 5.32, and hence Theorem 5.33, then extend to the a-corners case.

(d) Continuing (b), to define the corners $C(\boldsymbol{X})$ in Definition 6.1 in the a-corners case, for points $(x,P) \in C(X)$ we should take P to be a prime ideal in $\mathcal{O}^{\text{a-ex}}_{X,x}$ considered as a $[0,\infty)$-module, not as a monoid.

We leave it to the interested reader to work out the details.

8.4 Derived manifolds and orbifolds with corners

In §2.9.2 we discussed the subject of Derived Differential Geometry, the study of 'derived manifolds' and 'derived orbifolds', where 'derived' is in the sense of Derived Algebraic Geometry. There are two main approaches to defining ∞-categories or 2-categories of derived manifolds and derived orbifolds $\boldsymbol{X}$, one in which $\boldsymbol{X}$ is a special example of a derived C^∞-scheme or derived C^∞-stack, as in [6, 7, 8, 10, 11, 12, 13, 43, 44, 62, 87, 88, 89], and one in which $\boldsymbol{X}$ is a 'Kuranishi space' defined by an atlas of charts, as in [45, 48, 50]. (See also Fukaya–Oh–Ohta–Ono [28, 29, 30] for an earlier definition of Kuranishi spaces not using derived geometry, in which morphisms between Kuranishi spaces are not defined.)

It is obviously desirable to extend Derived Differential Geometry to theories of *derived manifolds with corners*, and *derived orbifolds with corners*. One reason to want to do this is that many moduli spaces used in Morse theory, as in §3.6.4, and in Floer theories, as in §3.6.6, should naturally be derived manifolds or orbifolds with corners (or possibly with a-corners, see §8.3), leading to important applications of Derived Differential Geometry with corners.

For example, in the 'Folklore Theorem' in §3.6.4 we noted that if $f : X \to \mathbb{R}$ is a Morse function on a manifold X and g is a generic Riemannian metric on X then the moduli spaces $\bar{\mathcal{M}}(x, y)$ of (broken) flow lines of $-\nabla f$ are expected to be manifolds with corners. If we drop the assumption that g is generic, so that the deformation theory of flow lines may be obstructed, we should expect $\bar{\mathcal{M}}(x, y)$ to be a *derived* manifold with corners.

The same applies to moduli spaces used in Floer theories. For Floer theories in Symplectic Geometry based on moduli spaces $\bar{\mathcal{M}}$ of J-holomorphic curves, including Lagrangian Floer cohomology [28, 29], Fukaya categories [3, 83], and Symplectic Field Theory [24], having a good theory of derived orbifolds with corners is particularly important, as in the most general case one *cannot* make $\bar{\mathcal{M}}$ into a manifold or orbifold with corners by a genericity assumption, as $\bar{\mathcal{M}}$ must be singular (and hence 'derived') at J-holomorphic curves which are non-trivial branched covers of other J-holomorphic curves. This is why Kuranishi spaces were developed first in Symplectic Geometry [28, 29, 30].

Some topics in Symplectic Geometry require a theory of *differential geometry* of derived orbifolds with corners. For example, the universal curve over a moduli space $\bar{\mathcal{M}}_{g,h,m,n}(X, J)$ of J-holomorphic curves with marked points in (X, ω) is a forgetful morphism $\bar{\mathcal{M}}_{g,h,m+1,n}(X, J) \to \bar{\mathcal{M}}_{g,h,m,n}(X, J)$ which forgets an interior marked point, and this is a (1-)morphism of derived orbifolds with corners. This is difficult to model in the non-derived approach of [28, 29, 30].

The second author will use the theory of C^∞-algebraic geometry with corners of this book in the final version of [44] to study 2-categories $\mathbf{dMan^c}$ of derived manifolds with corners and $\mathbf{dOrb^c}$ of derived orbifolds with corners, defined as special examples of certain types of derived C^∞-schemes and derived C^∞-stacks with corners. Pelle Steffens [88, §4.1], [89, §5] also uses this book to define an ∞-category of derived manifolds with corners, using Lurie's Derived Algebraic Geometry framework [61, 62]. See the second author [45, 48] for a parallel treatment using Kuranishi spaces with corners.

Although the details of any kind of derived geometry are complex, including corners in some theory of derived manifolds is in principle straightforward. All the discrete aspects of corner structure – for example, the monoids P in the local models X_P for manifolds with g-corners in §3.3, and the monoid bundles $M_{C(X)}, \check M^{\rm ex}_{C(X)}, \check M^{\rm in}_{C(X)}$ in §3.5 and §6.4 – remain classical and do not need to be derived. So the processes of making the geometry derived, and adding corners, are fairly independent. For example, to pass from Kuranishi spaces to Kuranishi spaces with corners in [45, 48], one just replaces manifolds by manifolds with corners, and uses appropriate classes of morphisms in $\mathbf{Man^c}$.

References

[1] A. Adámek, J. Rosický, and E.M. Vitale, *Algebraic theories*, Cambridge University Press, 2010.

[2] M. Atiyah, *Topological Quantum Field Theories*, Publ. Math. IHÉS **68** (1988), 175–186.

[3] D. Auroux, *A beginner's introduction to Fukaya categories*, pages 85–136 in F. Bourgeois et al., editors, *Contact and symplectic topology*, Bolyai Society Mathematical Studies, 26, 2014. arXiv:1301.7056.

[4] D.M. Austin and P.M. Braam, *Morse–Bott theory and equivariant cohomology*, pages 123–183 in *The Floer memorial volume*, Progress in Math. 133, Birkhäuser, 1995.

[5] J.C. Baez and J. Dolan, *Higher-dimensional algebra and Topological Quantum Field Theory*, J. Math. Phys. **36** (1995), 6073–6105. q-alg/9503002.

[6] K. Behrend, H.-Y. Liao, and P. Xu, *Derived Differentiable Manifolds*, arXiv:2006.01376, 2020.

[7] D. Borisov, *Derived manifolds and Kuranishi models*, arXiv:1212.1153, 2012.

[8] D. Borisov and J. Noel, *Simplicial approach to derived differential geometry*, arXiv:1112.0033, 2011.

[9] K. Buchner, M. Heller, P. Multarzyński, and W. Sasin, *Literature on Differentiable Spaces*, Acta Cosmologica **19**, 1993.

[10] D. Carchedi, *Derived manifolds as differential graded manifolds*, arXiv:2303.11140, 2023.

[11] D. Carchedi and D. Roytenberg, *On theories of superalgebras of differentiable functions*, Theory and App. of Cats. **28** (2013), 1022–1098. arXiv:1211.6134.

[12] D. Carchedi and D. Roytenberg, *Homological algebra for superalgebras of differentiable functions*, arXiv:1212.3745, 2012.

[13] D. Carchedi and P. Steffens, *On the universal property of derived manifolds*, arXiv:1905.06195, 2019.

[14] J. Cerf, *Topologie de certains espaces de plongements*, Bull. Soc. Math. France **89** (1961), 227–380.

[15] K. Chen, *Iterated integrals of differential forms and loop space homology*, Ann. Math. **97** (1973), 217–246.

[16] K. Chen, *Iterated integrals, fundamental groups and covering spaces*, Trans. A.M.S. **206** (1975), 83–98.

[17] K. Chen, *Iterated path integrals*, Bull. A.M.S. **83** (1977), 831–879.

[18] K. Chen, *On differentiable spaces*, pages 38–42 in *Categories in Continuum Physics (Buffalo, N.Y., 1982)*, Lecture Notes in Math. 1174, Springer, 1986.

[19] S.K. Donaldson, *Floer homology groups in Yang–Mills theory*, Cambridge Tracts in Math. 147, Cambridge University Press, 2002.

[20] S.K. Donaldson and P.B. Kronheimer, *The geometry of four-manifolds*, Oxford Mathematical Monographs, Oxford University Press, 1990.

[21] A. Douady, *Variétés à bord anguleux et voisinages tubulaires*, Séminaire Henri Cartan **14** (1961–2), exp. 1, 1–11.

[22] E.J. Dubuc, *Schémas* C^∞, Prepub. de la Université de Montreal **80** (1980), Exposé 3.

[23] E.J. Dubuc, C^∞*-schemes*, Amer. J. Math. **103** (1981), 683–690.

[24] Y. Eliashberg, A. Givental, and H. Hofer, *Introduction to Symplectic Field Theory*, Geom. Funct. Anal. 2000, Special Volume, Part II, 560–673. math.SG/0010059.

[25] K.L. Francis-Staite, C^∞*-algebraic geometry with corners*, DPhil thesis, Oxford University, 2019. Available online at https://ora.ox.ac.uk.

[26] D.S. Freed, *The cobordism hypothesis*, Bull. A.M.S. **50** (2013), 57–92. arXiv:1210.5100.

[27] D.S. Freed, *Anomalies and Invertible Field Theories*, pages 25–46 in *String–Math 2013*, Proc. Symp. Pure Math. 88, 2014. arXiv:1404.7224.

[28] K. Fukaya, *Floer homology of Lagrangian submanifolds*, Sugaku Expositions **26** (2013), 99–127. arXiv:1106.4882.

[29] K. Fukaya, Y.-G. Oh, H. Ohta, and K. Ono, *Lagrangian intersection Floer theory – anomaly and obstruction*, Parts I and II. AMS/IP Studies in Advanced Mathematics, 46.1 and 46.2, A.M.S./International Press, 2009.

[30] K. Fukaya, Y.-G. Oh, H. Ohta, and K. Ono, *Kuranishi structures and virtual fundamental chains*, Springer, 2020.

[31] K. Fukaya and K. Ono, *Arnold Conjecture and Gromov–Witten invariant*, Topology **38** (1999), 933–1048.

[32] W.D. Gillam and S. Molcho, *Log differentiable spaces and manifolds with corners*, arXiv:1507.06752, 2015.

[33] R. Godement, *Topologie algébrique et theórie des faisceaux*, Hermann, 1958.

[34] D. Grieser, *Basics of the b-calculus*, pages 30–84 in J.B.Gil et al., editors, *Approaches to Singular Analysis (Berlin, 1999)*, Birkhäuser, 2001. math.AP/0010314.

[35] A. Grothendieck, *Elements de géométrie algébrique I*, Publ. Math. IHES **4**, 1960.

[36] R. Hartshorne, *Algebraic Geometry*, Graduate Texts in Math. 52, Springer, 1977.

[37] H. Hofer, K. Wysocki, and E. Zehnder, *Applications of polyfold theory I: the polyfolds of Gromov–Witten theory*, Mem. A.M.S. **248** (2017), no. 1179. arXiv:1107.2097.

[38] H. Hofer, K. Wysocki, and E. Zehnder, *Polyfold and Fredholm theory*, Ergeb. der Math. und ihrer Grenzgebiete 72, Springer, 2021. arXiv:1707.08941.

[39] P. Iglesias-Zemmour, *Diffeology*, Mathematical Surveys and Monographs, Volume 185, American Mathematical Society, 2013.

[40] A.J. de Jong et al., *The Stacks Project*, online reference available at `http://stacks.math.columbia.edu`.

[41] D. Joyce, *On manifolds with corners*, pages 225–258 in S. Janeczko et al., editors, *Advances in Geometric Analysis*, Advanced Lectures in Mathematics 21, International Press, 2012. arXiv:0910.3518.

[42] D. Joyce, *An introduction to C^∞-schemes and C^∞-algebraic geometry*, Surveys in Differential Geometry **17** (2012), 299–325. arXiv:1104.4951.

[43] D. Joyce, *An introduction to d-manifolds and derived differential geometry*, pages 230–281 in L. Brambila-Paz et al., editors, *Moduli spaces*, L.M.S. Lecture Notes 411, Cambridge University Press, 2014. arXiv:1206.4207.

[44] D. Joyce, *D-manifolds and d-orbifolds: a theory of derived differential geometry*, to be published by Oxford University Press. Preliminary version available at `http://people.maths.ox.ac.uk/~joyce/dmanifolds.html`.

[45] D. Joyce, *A new definition of Kuranishi space*, arXiv:1409.6908, 2014.

[46] D. Joyce, *A generalization of manifolds with corners*, Adv. Math. **299** (2016), 760–862. arXiv:1501.00401.

[47] D. Joyce, *Manifolds with analytic corners*, arXiv:1605.05913, 2016.

[48] D. Joyce, *Kuranishi spaces and Symplectic Geometry*, multiple volume book in progress. Preliminary versions of volumes I, II available at `http://people.maths.ox.ac.uk/~joyce/Kuranishi.html`.

[49] D. Joyce, *Algebraic Geometry over C^∞-rings*, Mem. A.M.S. **260** (2019), no. 1256. arXiv:1001.0023.

[50] D. Joyce, *Kuranishi spaces as a 2-category*, pages 253–298 in J. Morgan, editor, *Virtual Fundamental Cycles in Symplectic Topology*, Mathematical Surveys and Monographs 237, A.M.S., 2019. arXiv:1510.07444.

[51] E. Kalashnikov, *C^∞-Algebraic Geometry*, MSc thesis, Oxford, 2014. Available at `http://people.maths.ox.ac.uk/~joyce/theses/theses.html`.

[52] A. Kock, *Synthetic Differential Geometry*, second edition, L.M.S. Lecture Notes 333, Cambridge University Press, 2006.

[53] A. Kock, *Synthetic geometry of manifolds*, Cambridge Tracts in Math. 180, Cambridge University Press, 2010.

[54] M. Kontsevich, *Enumeration of rational curves via torus actions*, pages 335–368 in R. Dijkgraaf, C. Faber, and G. van der Geer, editors, *The moduli space of curves*, Progr. Math. 129, Birkhäuser, 1995. hep-th/9405035.

[55] C. Kottke and R.B. Melrose, *Generalized blow-up of corners and fibre products*, Trans. A.M.S. **367** (2015), 651–705. arXiv:1107.3320.

[56] P. Kronheimer and T. Mrowka, *Monopoles and three-manifolds*, Cambridge University Press, 2007.

[57] G. Laumon and L. Moret-Bailly, *Champs algébriques*, Ergeb. der Math. und ihrer Grenzgebiete 39, Springer, 2000.

[58] W. Lawvere, *Categorical dynamics*, pages 1–28 in A. Kock, editor, *Topos Theoretic Methods in Geometry*, Chicago Lectures 30, 1967.

[59] P. Loya, *Index theory of Dirac operators on manifolds with corners up to codimension two*, pages 131–166 in J. Gil et al., editors, *Aspects of boundary problems in analysis and geometry*, Birkhäuser, 2004.

[60] P. Loya, *The index of b-pseudodifferential operators on manifolds with corners*, Ann. Global Anal. Geom. **27** (2005), 101–133.

[61] J. Lurie, *Higher Topos Theory*, Annals of Math. Studies 170, Princeton University Press, 2009. math.CT/0608040.

[62] J. Lurie, *Derived Algebraic Geometry V: Structured spaces*, arXiv:0905.0459, 2009.

[63] J. Lurie, *On the classification of Topological Field Theories*, arXiv:0905.0465, 2009.

[64] S. MacLane, *Categories for the Working Mathematician*, second edition, Graduate Texts in Math. 5, Springer, 1998.

[65] S. MacLane and I. Moerdijk, *Sheaves in Geometry and Logic: A first introduction to topos theory*, Springer, 1992.

[66] J. Margalef-Roig and E. Outerelo Dominguez, *Differential Topology*, North-Holland Math. Studies 173, North-Holland, 1992.

[67] S. Ma'u, *Gluing pseudoholomorphic quilted disks*, arXiv:0909.3339, 2009.

[68] S. Ma'u and C. Woodward, *Geometric realizations of the multiplihedra*, Compos. Math. 146 (2010), 1002–1028. arXiv:0802.2120.

[69] R.B. Melrose, *Pseudodifferential operators, corners and singular limits*, pages 217–234 in Proc. Int. Cong. Math. Kyoto, 1990.

[70] R.B. Melrose, *Calculus of conormal distributions on manifolds with corners*, IMRN (1992), 51–61.

[71] R.B. Melrose, *The Atiyah–Patodi–Singer Index Theorem*, A.K. Peters, 1993.

[72] R.B. Melrose, *Differential analysis on manifolds with corners*, unfinished book available at `http://math.mit.edu/~rbm`, 1996.

[73] I. Moerdijk, N. van Quê, and G.E. Reyes, *Rings of smooth functions and their localizations II*, pages 277–300 in D.W. Kueker et al., editors, *Mathematical logic and theoretical computer science*, Lecture Notes in Pure and Applied Math. 106, Marcel Dekker, 1987.

[74] I. Moerdijk and G.E. Reyes, *Rings of smooth functions and their localizations I*, J. Algebra **99** (1986), 324–336.

[75] I. Moerdijk and G.E. Reyes, *Models for smooth infinitesimal analysis*, Springer, 1991.

[76] B. Monthubert, *Groupoids and pseudodifferential calculus on manifolds with corners*, J. Funct. Anal. **199** (2003), 243–286.

[77] J.A. Navarro González and J.B. Sancho de Salas, *C^∞-Differentiable Spaces*, Lecture Notes in Math. 1824, Springer, 2003.

[78] A. Ogus, *Lectures on Logarithmic Algebraic Geometry*, Cambridge Studies in Advanced Mathematics 178, Cambridge University Press, 2018.

[79] M. Olsson, *Algebraic Spaces and Stacks*, A.M.S. Colloquium Publications 62, A.M.S., 2016.

[80] J. Pardon, *Contact homology and virtual fundamental cycles*, J. A.M.S. **32** (2019), 825–919. arXiv:1508.03873.

[81] D. Rees, *On semi-groups*, Math. Proc. Camb. Philos. Soc. **36** (1940), 387–400.

[82] M. Schwarz, *Morse homology*, Progr. Math. 111, Birkhäuser, 1993.

[83] P. Seidel, *Fukaya categories and Picard–Lefschetz theory*, E.M.S., 2008.

[84] R. Sikorski, *Abstract covariant derivative*, Colloquium Mathematicum, XVIII, 1967.

[85] J.M. Souriau, *Groupes différentiels*, pages 91–128 in *Differential geometrical methods in mathematical physics (Proc. Conf., Aix-en-Provence/Salamanca, 1979)*, Lecture Notes in Math. 836, Springer, 1980.

[86] K. Spallek, *Differenzierbare und homomorphe auf analytischen Mengen*, Math. Ann. **161** (1965), 143–162.

[87] D.I. Spivak, *Derived smooth manifolds*, Duke Math. J. **153** (2010), 55–128. arXiv:0810.5174.

[88] P. Steffens, *Derived Differential Geometry*, PhD thesis, Université de Montpellier, 2022. Available at `https://www.imag.umontpellier.fr/~calaque/students-Pelle_Steffens_PhD_Thesis.pdf`.

[89] P. Steffens, *Derived C^∞-Geometry I: Foundations*, arXiv:2304.08671, 2022.

[90] B. Toën, *Higher and derived stacks: a global overview*, pages 435–487 in Proc. Symp. Pure Math. 80, part 1, A.M.S., 2009. math.AG/0604504.

[91] B. Toën, *Derived Algebraic Geometry*, E.M.S. Surv. Math. Sci. **1** (2014), 153–240. arXiv:1401.1044.

[92] B. Toën and G. Vezzosi, *Homotopical Algebraic Geometry II: Geometric Stacks and Applications*, Mem. Amer. Math. Soc. **193** (2008), no. 902. math.AG/0404373.

[93] K. Wehrheim, *Smooth structures on Morse trajectory spaces, featuring finite ends and associative gluing*, pages 369–450 in *Proceedings of the Freedman Fest*, Geom. Topol. Monogr. 18, 2012. arXiv:1205.0713.

[94] K. Wehrheim and C. Woodward, *Quilted Floer cohomology*, Geom. Top. **14** (2010), 833–902. arXiv:0905.1370.

[95] K. Wehrheim and C. Woodward, *Functoriality for Lagrangian correspondences in Floer theory*, Quantum Topol. **1** (2010), 129–170. arXiv:0708.2851.

[96] K. Wehrheim and C. Woodward, *Pseudoholomorphic quilts*, J. Symplectic Geom. **13** (2015), 849–904. arxiv:0905.1369.

Glossary of Notation

$\mathbf{AC^\infty Sch}$ category of affine C^∞-schemes, 39
$\mathbf{AC^\infty Sch^c}$ category of affine C^∞-schemes with corners, 118
$\mathbf{AC^\infty Sch^c_{in}}$ category of interior affine C^∞-schemes with corners, 118
${}^bN_{C(X)}$ b-normal bundle of $C(X)$ in a manifold with (g-)corners X, 64
${}^bN^*_{C(X)}$ b-conormal bundle of $C(X)$ in a manifold with (g-)corners X, 64
${}^b\check N_{C(\boldsymbol{X})}$ b-conormal bundle of $C(\boldsymbol{X})$ in a C^∞-scheme with corners $\boldsymbol{X}$, 192
${}^b\Omega_{\mathfrak{C}}$ b-cotangent module of C^∞-ring with corners $\boldsymbol{\mathfrak{C}}$, 168
${}^bTX$ b-tangent bundle of a manifold with (g-)corners X, 62
${}^bT^*X$ b-cotangent bundle of a manifold with (g-)corners X, 62
${}^bT^*\boldsymbol{X}$ b-cotangent sheaf of C^∞-ringed space with corners $\boldsymbol{X}$, 182
$C:\mathbf{C^\infty Sch^c}\to\mathbf{C^\infty Sch^c_{in}}$ corner functor on C^∞-schemes with corners, 148
$C:\mathbf{LC^\infty RS^c}\to\mathbf{LC^\infty RS^c_{in}}$ corner functor on local C^∞-ringed spaces with corners, 139
$C:\mathbf{Man^c}\to\mathbf{\check{M}an^c_{in}}$ corner functor on manifolds with corners, 61
$C:\mathbf{Man^{gc}}\to\mathbf{\check{M}an^{gc}_{in}}$ corner functor on manifolds with g-corners, 61
$\boldsymbol{\mathfrak{C}}(a^{-1}:a\in A)[a'^{-1}:a'\in A_{\rm ex}]$ localization of C^∞-ring with corners $\boldsymbol{\mathfrak{C}}$, 92
$C^\infty(X)$ C^∞-ring of smooth maps $f:X\to\mathbb{R}$, 53
$C(X)$ corners of a manifold with (g-)corners X, 61
$C(\boldsymbol{X})$ corners of a C^∞-ringed space with corners $\boldsymbol{X}$, 137
$\mathbf{CC^\infty Rings}$ category of categorical C^∞-rings, 23
$\mathfrak{C},\mathfrak{D},\mathfrak{E},\ldots$ C^∞-rings, 23
$\boldsymbol{\mathfrak{C}},\boldsymbol{\mathfrak{D}},\boldsymbol{\mathfrak{E}},\ldots$ (pre) C^∞-rings with corners, 75
$C_k(X)$ k-corners of a manifold with corners X, 59
$C_k(\boldsymbol{X})$ k-corners of a firm C^∞-scheme with corners $\boldsymbol{X}$, 156
$\mathfrak{C}$-mod category of modules over a C^∞-ring $\mathfrak{C}$, 30
$\mathfrak{C}$-mod$_{\rm co}$ category of complete $\mathfrak{C}$-modules, 43
$\boldsymbol{\mathfrak{C}}$-mod category of modules over a C^∞-ring with corners $\boldsymbol{\mathfrak{C}}$, 167

$\mathbf{CPC^\infty Rings^c}$ category of categorical pre C^∞-rings with corners, 74
$\mathbf{CPC^\infty Rings^c_{in}}$ category of categorical interior pre C^∞-rings with corners, 74
$\mathbf{C^\infty Rings}$ category of C^∞-rings, 24
$\mathbf{C^\infty Rings_{co}}$ category of complete C^∞-rings, 40
$\mathbf{C^\infty Rings_{lo}}$ category of local C^∞-rings, 28
$\mathbf{C^\infty Rings^c}$ category of C^∞-rings with corners, 84
$\mathbf{C^\infty Rings^c_{fg}}$ category of finitely generated C^∞-rings with corners, 100
$\mathbf{C^\infty Rings^c_{fi}}$ category of firm C^∞-ring with corners, 102
$\mathbf{C^\infty Rings^c_{fp}}$ category of finitely presented C^∞-rings with corners, 100
$\mathbf{C^\infty Rings^c_{in}}$ category of interior C^∞-rings with corners, 84
$\mathbf{C^\infty Rings^c_{in,lo}}$ category of interior local C^∞-rings with corners, 93
$\mathbf{C^\infty Rings^c_{lo}}$ category of local C^∞-rings with corners, 93
$\mathbf{C^\infty Rings^c_{sa}}$ category of saturated C^∞-rings with corners, 103
$\mathbf{C^\infty Rings^c_{sc}}$ category of semi-complete C^∞-rings with corners, 126
$\mathbf{C^\infty Rings^c_{sc,in}}$ category of interior semi-complete C^∞-rings with corners, 126
$\mathbf{C^\infty Rings^c_{si}}$ category of simple C^∞-rings with corners, 103
$\mathbf{C^\infty Rings^c_{tf}}$ category of torsion-free C^∞-rings with corners, 103
$\mathbf{C^\infty Rings^c_{to}}$ category of toric C^∞-rings with corners, 103
$\mathbf{C^\infty Rings^c_{\mathbb{Z}}}$ category of integral C^∞-rings with corners, 103
$\mathbf{C^\infty RS}$ category of C^∞-ringed spaces, 36
$\mathbf{C^\infty RS^c}$ category of C^∞-ringed spaces with corners, 108
$\mathbf{C^\infty RS^c_{in}}$ category of interior C^∞-ringed spaces with corners, 109
$\mathbf{C^\infty Sch}$ category of C^∞-schemes, 39
$\mathbf{C^\infty Sch^c}$ category of C^∞-schemes with corners, 118
$\mathbf{C^\infty Sch^c_{fg}}$ category of finitely generated C^∞-schemes with corners, 129
$\mathbf{C^\infty Sch^c_{fi}}$ category of firm C^∞-schemes with corners, 129
$\mathbf{C^\infty Sch^c_{fi,in}}$ category of firm interior C^∞-schemes with corners, 129
$\mathbf{C^\infty Sch^c_{fi,in,ex}}$ category of firm interior C^∞-schemes with corners with exterior morphisms, 129
$\mathbf{C^\infty Sch^c_{fi,tf}}$ category of firm torsion-free C^∞-schemes with corners, 129
$\mathbf{C^\infty Sch^c_{fi,tf,ex}}$ category of firm torsion-free C^∞-schemes with corners and exterior morphisms, 129
$\mathbf{C^\infty Sch^c_{fi,\mathbb{Z}}}$ category of firm integral C^∞-schemes with corners, 129
$\mathbf{C^\infty Sch^c_{fi,\mathbb{Z},ex}}$ category of firm integral C^∞-schemes with corners and exterior morphisms, 129
$\mathbf{C^\infty Sch^c_{fp}}$ category of finitely presented C^∞-schemes with corners, 129
$\mathbf{C^\infty Sch^c_{in}}$ category of interior C^∞-schemes with corners, 118

$\mathbf{C^\infty Sch^c_{in,ex}}$ category of interior C^∞-schemes with corners and exterior morphisms, 129
$\mathbf{C^\infty Sch^c_{sa}}$ category of saturated C^∞-schemes with corners, 129
$\mathbf{C^\infty Sch^c_{sa,ex}}$ category of saturated C^∞-schemes with corners and exterior morphisms, 129
$\mathbf{C^\infty Sch^c_{si}}$ category of simple C^∞-schemes with corners, 129
$\mathbf{C^\infty Sch^c_{si,ex}}$ category of simple C^∞-schemes with corners and exterior morphisms, 129
$\mathbf{C^\infty Sch^c_{tf}}$ category of torsion-free C^∞-schemes with corners, 129
$\mathbf{C^\infty Sch^c_{tf,ex}}$ category of torsion-free C^∞-schemes with corners and exterior morphisms, 129
$\mathbf{C^\infty Sch^c_{to}}$ category of toric C^∞-schemes with corners, 129
$\mathbf{C^\infty Sch^c_{to,ex}}$ category of toric C^∞-schemes with corners and exterior morphisms, 129
$\mathbf{C^\infty Sch^c_{\mathbb{Z}}}$ category of integral C^∞-schemes with corners, 129
$\mathbf{C^\infty Sch^c_{\mathbb{Z},ex}}$ category of integral C^∞-schemes with corners and exterior morphisms, 129
depth_X codimension of boundary stratum in a manifold with corners X, 59
∂X boundary of a manifold with (g-)corners X, 59
$\partial \boldsymbol{X}$ boundary of a firm C^∞-scheme with corners $\boldsymbol{X}$, 156
$\mathbf{Euc}$ category of Euclidean spaces $\mathbb{R}^n$, 23
$\mathbf{Euc^c}$ category of Euclidean spaces with corners $\mathbb{R}^m_k$, 74
$\mathbf{Euc^c_{in}}$ category of Euclidean spaces with corners and interior maps, 74
$\mathrm{Ex}(X)$ monoid of exterior maps $g : X \to [0, \infty)$, 53
$\mathfrak{F}^A$ free C^∞-ring generated by a set A, 26
$\boldsymbol{\mathfrak{F}}^{A,A_{\mathrm{ex}}}$ free C^∞-ring with corners generated by (A, A_{ex}), 88
$\boldsymbol{\mathfrak{F}}^{A,A_{\mathrm{in}}}$ free interior C^∞-ring with corners generated by (A, A_{in}), 88
$\Gamma : \mathbf{LC^\infty RS} \to \mathbf{C^\infty Rings}^{\mathrm{op}}$ global sections functor, 38
$\Gamma^{\mathrm{c}} : \mathbf{LC^\infty RS^c} \to (\mathbf{C^\infty Rings^c})^{\mathrm{op}}$ global sections functor, 115
$\Gamma^{\mathrm{c}}_{\mathrm{in}} : \mathbf{LC^\infty RS^c_{in}} \to (\mathbf{C^\infty Rings^c_{in}})^{\mathrm{op}}$ global sections functor, 117
$\mathrm{In}(X)$ monoid of interior maps $g : X \to [0, \infty)$, 53
$\mathbf{LC^\infty RS}$ category of local C^∞-ringed spaces, 36
$\mathbf{LC^\infty RS^c}$ category of local C^∞-ringed spaces with corners, 108
$\mathbf{LC^\infty RS^c_{in}}$ category of interior local C^∞-ringed spaces with corners, 109
$\mathbf{LC^\infty RS^c_{sa}}$ category of saturated local C^∞-ringed spaces with corners, 111
$\mathbf{LC^\infty RS^c_{sa,ex}}$ category of saturated local C^∞-ringed spaces with corners with exterior morphisms, 111
$\mathbf{LC^\infty RS^c_{tf}}$ category of torsion-free local C^∞-ringed spaces with corners, 111

$\mathbf{LC^\infty RS^c_{tf,ex}}$ category of torsion-free local C^∞-ringed spaces with corners with exterior morphisms, 111
$\mathbf{LC^\infty RS^c_{\mathbb{Z}}}$ category of integral local C^∞-ringed spaces with corners, 111
$\mathbf{LC^\infty RS^c_{\mathbb{Z},ex}}$ category of integral local C^∞-ringed spaces with corners with exterior morphisms, 111
$\mathbf{Man}$ category of manifolds, 23
$\mathbf{Man^{ac}}$ category of manifolds with a-corners, 198
$\mathbf{Man^{ac}_{in}}$ category of manifolds with a-corners and interior maps, 198
$\mathbf{Man^c}$ category of manifolds with corners and smooth maps, 52
$\check{\mathbf{M}}\mathbf{an^c}$ category of manifolds with corners of mixed dimension, 52
$\mathbf{Man^{c,ac}}$ category of manifolds with corners and a-corners, 198
$\mathbf{Man^c_{in}}$ category of manifolds with corners and interior maps, 52
$\check{\mathbf{M}}\mathbf{an^c_{in}}$ category of manifolds with corners of mixed dimension and interior maps, 52
$\mathbf{Man^c_{st}}$ category of manifolds with corners and strongly smooth maps, 52
$\mathbf{Man^c_{we}}$ category of manifolds with corners and weakly smooth maps, 52
$\mathbf{Man^f}$ category of manifolds with faces, 60
$\mathbf{Man^f_{in}}$ category of manifolds with faces and interior maps, 60
$\mathbf{Man^{gc}}$ category of manifolds with g-corners, 57
$\check{\mathbf{M}}\mathbf{an^{gc}}$ category of manifolds with g-corners of mixed dimension, 57
$\mathbf{Man^{gc}_{in}}$ category of manifolds with g-corners and interior maps, 57
$\check{\mathbf{M}}\mathbf{an^{gc}_{in}}$ category of manifolds with g-corners of mixed dimension and interior maps, 57
$M_{C(X)}$ monoid bundle of $C(X)$ in a manifold with (g-)corners X, 64
$M^\vee_{C(X)}$ comonoid bundle of $C(X)$ in a manifold with (g-)corners X, 64
$\check{M}^{ex}_{C(X)}, \check{M}^{in}_{C(X)}$ sheaves of monoids on $C(\boldsymbol{X})$, 153
$\mathbf{Mon}$ category of (commutative) monoids, 53
$\mathbf{Mon_{sa,tf,\mathbb{Z}}}$ category of saturated, torsion-free, integral monoids, 104
$\mathbf{Mon_{tf,\mathbb{Z}}}$ category of torsion-free, integral monoids, 104
$\mathbf{Mon_{\mathbb{Z}}}$ category of integral monoids, 104
$\mathrm{MSpec} : \mathfrak{C}\text{-mod} \to \mathcal{O}_X\text{-mod}$ spectrum functor on $\mathfrak{C}$-modules, 43
$\Omega_{\mathfrak{C}}$ cotangent module of C^∞-ring $\mathfrak{C}$, 31
$\Omega_{\boldsymbol{\mathfrak{C}}}$ cotangent module of C^∞-ring with corners $\boldsymbol{\mathfrak{C}}$, 168
$\mathcal{O}_X$ structure sheaf of C^∞-scheme $\underline{X}$, 36
$\boldsymbol{\mathcal{O}}_X = (\mathcal{O}_X, \mathcal{O}^{ex}_X)$ structure sheaf of C^∞-scheme with corners $\boldsymbol{X}$, 107
$P^\sharp$ sharpening of a monoid P, 55
$P^\times$ abelian group of units in a monoid P, 54
P^{gp} groupification of a monoid P, 54
$\mathbf{PC^\infty Rings^c}$ category of pre C^∞-rings with corners, 77

$\mathbf{PC^\infty Rings^c_{in}}$ category of interior pre C^∞-rings with corners, 79
$\mathrm{Pr}_{\mathfrak{C}}$ set of prime ideals $P \subset \mathfrak{C}_{\mathrm{ex}}$ for a C^∞-ring with corners $\mathfrak{C}$, 143
$\mathbf{Sets}$ category of sets, 23
$\mathrm{Spec} : \mathbf{C^\infty Rings}^{\mathrm{op}} \to \mathbf{LC^\infty RS}$ spectrum functor on C^∞-rings, 38
$\mathrm{Spec^c} : (\mathbf{C^\infty Rings^c})^{\mathrm{op}} \to \mathbf{LC^\infty RS^c}$ spectrum functor on C^∞-rings with corners, 113
$\mathrm{Spec^c_{in}} : (\mathbf{C^\infty Rings^c_{in}})^{\mathrm{op}} \to \mathbf{LC^\infty RS^c_{in}}$ spectrum functor on interior C^∞-rings with corners, 115
TX tangent bundle of a manifold with corners X, 62
T^*X cotangent bundle of a manifold with corners X, 62
$T^*\boldsymbol{X}$ cotangent sheaf of C^∞-ringed space with corners $\boldsymbol{X}$, 182
X° interior of a manifold with corners X, 59
$\underline{X}, \underline{Y}, \underline{Z}, \ldots$ C^∞-ringed spaces or C^∞-schemes, 36
$\boldsymbol{X}, \boldsymbol{Y}, \boldsymbol{Z}, \ldots$ C^∞-ringed spaces or C^∞-schemes with corners, 108

Index